From Stars to Life

How did life originate? Is there life beyond Earth? What is the future of life on our planet? The rapidly growing multidisciplinary field of astrobiology deals with life's big questions. This text harnesses the authors' two decades' experience of teaching acclaimed courses in astrobiology, and adopts a novel quantitative approach towards this emergent discipline. It details the physical principles and chemical processes that have shaped the origins and distribution of molecules, stars, planets, and hence habitable environments, life, and intelligence in the Universe. By synthesising insights from domains as diverse as astronomy and physics to microbiology, biochemistry, and geology, the authors provide a cutting-edge summary of astrobiology, and show how answers to many fundamental questions are drawing closer than ever. Geared towards advanced undergraduates and graduate students in the physical sciences, the text contains more than 150 innovative problems designed to enhance students' knowledge and understanding.

Manasvi Lingam is Assistant Professor of Aerospace, Physics and Space Sciences & Chemistry and Chemical Engineering at the Florida Institute of Technology (USA), with more than 100 papers spanning various physical sciences. He is the lead author of *Life in the Cosmos* (2021) and he is a Fellow of the Royal Astronomical Society.

Amedeo Balbi is Associate Professor of Astronomy and Astrophysics at the University of Rome Tor Vergata, Italy, where he has been teaching astrobiology for the past two decades. With an extensive publication record of more than 100 scientific papers, his research encompasses a wide array of topics in theoretical astrophysics.

From Stars to Life
A Quantitative Approach to Astrobiology

Manasvi Lingam
Florida Institute of Technology

Amedeo Balbi
Università degli Studi di Roma 'Tor Vergata'

CAMBRIDGE
UNIVERSITY PRESS

Shaftesbury Road, Cambridge CB2 8EA, United Kingdom

One Liberty Plaza, 20th Floor, New York, NY 10006, USA

477 Williamstown Road, Port Melbourne, VIC 3207, Australia

314–321, 3rd Floor, Plot 3, Splendor Forum, Jasola District Centre, New Delhi – 110025, India

103 Penang Road, #05–06/07, Visioncrest Commercial, Singapore 238467

Cambridge University Press is part of Cambridge University Press & Assessment, a department of the University of Cambridge.

We share the University's mission to contribute to society through the pursuit of education, learning and research at the highest international levels of excellence.

www.cambridge.org
Information on this title: www.cambridge.org/9781009411219

DOI: 10.1017/9781009411257

First published 2024

A catalogue record for this publication is available from the British Library.

A Cataloging-in-Publication data record for this book is available from the Library of Congress.

ISBN 978-1-009-41121-9 Hardback

To our families, with love and gratitude

Contents

Preface

A myriad of multifaceted spectres are haunting humankind: anthropogenic climate change, biodiversity collapse(s), wars and armed conflicts, and artificial intelligences (AIs), to name a few. In these volatile times, the refrain of '*Are we alone?*' – which has echoed through millennia of human history – reverberates and resonates stronger than ever, in the wake of growing alienation and chaos. Against this sober backdrop, seeking answers to this question and others of its ilk might seem a trivial pursuit. And so one may well ask: why write this book, and why undertake it now? There are two responses that spring to our minds: a pragmatic one that we elucidate below and a deeper, philosophical one we defer to the end.

Astrobiology strives to explore three questions of fundamental import: *Where did we come from? Are we alone? Where are we going?* We have witnessed significant progress in the twenty-first century on these fronts, especially within the last decade. Synthetic chemistry and systems chemistry have accomplished a host of breakthroughs in producing and assembling the building blocks of cells. Missions to worlds in our Solar System (e.g., Mars and Enceladus) have revealed several ingredients vital for life, such as liquid water, (free) energy, and bioessential elements. The search for extraterrestrial life and intelligence has also received a major impetus from the exoplanet revolution, which is unearthing a spate of promising targets for future investigations by next-generation telescopes. Lastly, with renewed interest from governmental and private agencies, space exploration is swiftly picking up steam.

Thus, the rapid growth and innate significance of astrobiology constitute a compelling and timely case for authoring a textbook on this subject. There are, however, a number of excellent works in this area, so the question of determining the rationale behind our book remains. Based on an extensive literature survey, the available textbooks are characterised by one or more of the following criteria: (1) they are introductory-level and targeted towards first- or second-year undergraduates; (2) they are qualitative in nature, with an emphasis on the biological aspects; and (3) they are not sufficiently abreast of the latest developments in the field.

In contrast, driven by the remarkable blossoming of astrobiology, universities all over the world are increasingly offering tailored courses at the advanced undergraduate and beginning (post)graduate levels, often in physics and/or astronomy departments, since astrobiology draws heavily on these areas. Despite this fact, a major lacuna in the domain of astrobiology textbooks is discernible: there is no book that evinces a quantitative approach, is pitched towards advanced undergraduates and (post)graduates in the physical sciences, and chronicles the current state of the field. It is our hope and expectation that this textbook will address the above lacuna.

Our book is explicitly intended to be employed in a one-semester course at the forenamed levels, comprising about 30 lectures of 75 minutes each; this is roughly the format adopted in our institutions. We anticipate that 80–90% of the total content in the textbook could be covered within this time frame, with chapters of around 12, 24, and 35 pages warranting one, two, and three lectures, respectively. The prerequisites for comprehending this book are high school biology, two to three years of undergraduate coursework in physics or a closely related discipline (e.g., astronomy), and general chemistry at the freshman undergraduate level. The structure and motifs of the book, divided into six parts, are sketched next.

In Part I, we dive into the astrophysical origins of astrobiology, because it is crucial to understand the phenomena and steps that created the celestial worlds (e.g., planets) for the emergence of lifeforms. We commence with a brief historical introduction in Chapter 1. By starting from the Big Bang, we trace the formation of the Universe, of stars, elements, and molecules in Chapter 2, after which we embark on an in-depth exposition of the multiple stages and processes of planet formation in Chapter 3.

Currently, only a single world is confirmed to host life in the Cosmos: our planet Earth. Taking cue from this simple datum, we focus on Earth in Part II. We describe the potential physicochemical conditions on early Earth in Chapter 4, as they shape the milieu in which living systems originated. This account is followed by a detailed analysis of *how* and *where* the origin(s) of life occurred on Earth in Chapter 5. Building on this narrative, we summarise the co-evolution of life and its environment(s) in Chapter 6.

Habitability is a cardinal notion in astrobiology, since it enables us to determine which settings are conducive for harbouring life, thereby aiding us in target selection for seeking putative signatures of life. Hence, in Part III, we elucidate the set of physicochemical constraints that must be fulfilled at any given moment in time to maintain habitable conditions for life-as-we-know-it (Chapter 7), as well as the physical mechanisms that modulate and/or sustain habitability on geological timescales (Chapter 8).

In Part IV, we centre our attention on the noteworthy matter of astrobiological targets inside and outside our solar system. We report the prospects for life-as-we-know-it on Mars in Chapter 9, and on icy worlds with subsurface oceans of liquid water (e.g., Europa and Enceladus) in Chapter 10. Next, we delve into the possibility of 'exotic' life on worlds like Venus and Titan in Chapter 11. Moving past the solar system, we recount methods of detecting and characterising (rocky) exoplanets in Chapter 12.

Given that one of the key objectives of astrobiology is to look for evidence of extraterrestrial life, we tackle this topic in Part V. We delineate the various techniques and associated concepts that may help us discover markers of non-technological life (i.e., biosignatures) in Chapter 13, and traces of extraterrestrial intelligence (i.e., technosignatures) in Chapter 14. The search for extraterrestrial life is intertwined with our own future on this planet and beyond (e.g., space exploration), owing to which we briefly discuss this aspect in Chapter 15 of Part VI at the end of the book.

Each chapter is accompanied by $\sim$10 problems on average. Instead of regurgitating facts, calculations, or principles in that chapter, we sought to design the problems such that they complement, supplement, or enhance the comprehension of the corresponding material(s). We are quite aware that many, perhaps even most (but *not* all), of the problems could be solved by ChatGPT and its cognate descendants. However, even in this scenario, we trust that the students may still learn something substantive from the AI-generated answers, and that they can duly boost their understanding.

This textbook contains extensive references, albeit without any pretensions of being exhaustive. This unusual decision was made consciously on account of the following reasons: (1) astrobiology synthesises knowledge from a bevy of disparate fields, and it seemed inappropriate to quote data without proper citations; (2) because astrobiology is so diverse, our readership (chiefly advanced students) can peruse these publications to further educate themselves in the domains that interest them; and (3) the references underscore the fast-moving and multidisciplinary nature of astrobiology.

At this juncture, we wish to highlight a couple of important caveats concerning this textbook. In expounding a subject of this scope and magnitude, countless topics had to be either excluded or severely truncated to keep the book within a strict limit ($\sim$400 pages). Needless to say, this winnowing was subjective, and thus coloured to an extent by our limitations and biases. In a similar vein, both of us hail from backgrounds in the physical sciences (NB: the book is also oriented towards such an audience), implying that our coverage of pivotal areas in geology, biology, and chemistry may be insufficiently nuanced and/or inaccurate. We accordingly take this opportunity to gladly solicit corrections and feedback from readers.

We will round off our exposition by circling back to the central question posed at the outset: why compose this book in this disruptive epoch? As a multitude of people experience feelings of estrangement (*Entfremdung*) and loss, humanity is progressively beset by three profound existential questions that have engrossed our species since the dawn of history: *Where did we come from? Who and what are we, humans, and what is our place in the Cosmos? Where are we going?* On comparing this trio with the core goals of astrobiology outlined in the second paragraph, the *fil rouge* or *roter Faden* (i.e., common thread) weaving them together is manifest.

Hence, though our era may seem uniquely ridden with overwhelming difficulties and riven with acute schisms *prima facie*, its anxieties and troubles could be considered universal and ever-present, to a certain degree. Astrobiology, by virtue of grappling with the aforementioned fundamental questions, might have the potential to situate our zeitgeist in a cosmic context, and/or forge bonds and bridges across myriad groups, especially if we unearth evidence that we are not alone in this vast Universe. The eminent biologist Carl Woese (1928–2012) wrote of biology that (Woese, 2004, pg. 185):

[S]ociety will come to see that biology is here to understand the world, not primarily to change it. Biology's primary job is to teach us. In that realisation lies our hope of learning to live in harmony with our planet.

It is our opinion that astrobiology can assume and fulfil an analogous role for humankind in due time. Therefore, books in this subject are rendered valuable and opportune, arguably all the more so in this turbulent age.

Last but not least, gazing beyond concrete advantages or intangible benefits ensuing from astrobiology in general and our book specifically, we believe that the quest for knowledge and the holistic growth thence kindled is a worthy endeavour in its own right – the human propensity for learning is a wondrous thing, and could ultimately engender its own rewards. This theme was eloquently articulated over a millennium ago by the renowned Japanese writer Lady Murasaki Shikibu (973?–1014?) in her abiding and poignant masterpiece *Genji monogatari* (Shikibu, (c. 1010) 2003, pg. 329), perceived by many as the first 'modern' novel in recorded history:

No art or learning is to be pursued halfheartedly …and any art worth learning will certainly reward more or less generously the effort made to study it.

We are grateful to the diverse throng of individuals – hailing from multifarious walks of life and sundry locales sprinkled across the world – who have contributed to this textbook, to our ways of thinking, and ultimately to our very psyches. Above all, we are indebted to our families for their profound love, support, and patience: this book is as much the fruits of their labour as it is ours.

We also thank colleagues (faculty, staff, and students) in our respective workplaces for their helpful assistance vis-à-vis our textbook.

We appreciate the permissions and technical content (e.g., figures) generously provided by multiple publishers and individuals for inclusion in the textbook. We thank Germano D'Abramo, Kate Adamala, Robin Canup, Marjorie Chan, Charles Cockell, Bruce Damer, David Deamer, Charles Diamond, Chuanfei Dong, Jason Dworkin, Yuka Fujii, Vishal Gajjar, Alexander Halliday, Trinity Hamilton, Frode Hansen, Paul Hartogh, Lisa Kaltenegger, Sebastiaan Krijt, Joshua Krissansen-Totton, Helmut Lammer, Andrew Lincowski, Timothy Lyons, Christoph Mordasini, Marc Neveu, Martina Preiner, Edward Schwieterman, Hector Socas-Navarro, Jorge Vago, Kevin Walsh, and Joana Xavier in this regard. Finally, we express our gratitude to the staff at Cambridge University Press (especially our editor, Vince Higgs) for commissioning our book and shepherding it to a successful conclusion.

Although the process of writing this textbook was occasionally arduous, it was a deeply rewarding experience. Likewise, we hope that our readers will enjoy their intellectual forays into the fascinating world of astrobiology.

Part I

Astronomical Origins

1   The Foundations of Astrobiology

Before embarking on our voyage into the vast and multifaceted domain of astrobiology, every budding practitioner and student of this field should be equipped with a few basic tools to venture forth on this grand journey. First, it is vital to grasp some foundational terms that crop up often, such as 'life', 'habitability', and 'astrobiology' itself. Although these concepts may seem self-evident at first glimpse, all of them are inherently complex in actuality, and have accordingly attracted intense debate since at least the twentieth century, sometimes even commencing centuries and millennia earlier. In fact, a universal definition of life remains elusive to this day, and philosophers and scientists continue to debate this matter.

Second, it is important (arguably even essential) to understand the historical development and growth of astrobiology. Pursuing this historical path is valuable for a minimum of two reasons, though it may appear to deviate from the scientific goal(s) of this textbook. For starters, it will help us comprehend and appreciate how, on the one hand, astrobiology has ancient roots and, on the other hand, it is a remarkably young and dynamic discipline. Next, from a broader perspective, the history of science can expand our horizons, and consequently enable us to gain a better picture of where this field might be headed towards in the turbulent twenty-first century.

Thus, the purpose of this chapter is to fulfil the aforementioned objectives. In the first part, we provide working definitions for some fundamental terms encountered in astrobiology, after which we explore an abbreviated history of astrobiology in the second part.

1.1   Key concepts and definitions

In this section, we will carefully examine some of the central concepts that are often encountered in astrobiology.

1.1.1   Astrobiology and exobiology

Since this book deals with astrobiology, it is natural to start with posing the question: *what is astrobiology?* This sweeping question can be broken up into additional segments: what answers are astrobiologists seeking? What targets do they strive to survey? What are some of the disciplines that astrobiology draws on? To answer this plethora of questions, the working definition of astrobiology delineated by the NASA Astrobiology Institute (NAI) is reproduced in its entirety,[1] because it is fairly succinct, yet thorough.

---

[1]   https://astrobiology.nasa.gov/nai/about/index.html.

Figure 1.1 The major fields employed in astrobiology (small dark circles) to analyse its key areas of research (large light circles).

Astrobiology is the study of the origins, evolution, distribution, and future of life in the Universe. This interdisciplinary field requires a comprehensive, integrated understanding of biological, geological, planetary, and cosmic phenomena. Astrobiology encompasses the search for habitable environments in our solar system and on planets around other stars; the search for evidence of prebiotic chemistry or life on solar system bodies such as Mars, Jupiter's moon Europa, and Saturn's moon Titan; and research into the origin, early evolution, and diversity of life on Earth. Astrobiologists address three fundamental questions: How does life begin and evolve? Is there life elsewhere in the Universe? What is the future of life on Earth and beyond?

The themes expressed in this quote are also depicted in Figure 1.1, which illustrates the topics overlapping with astrobiology.

To reiterate the points conveyed in the prior paragraph, astrobiology grapples with three overarching and truly fundamental questions, which may be colloquially expressed in the following fashion.

1. *Where did we come from?* [How does life begin and evolve?]
2. *Are we alone?* [Is there life elsewhere in the Universe?]
3. *Where are we going?* [What is the future of life on Earth and beyond?]

This trio of questions governs the organisation of the book. In Parts I and II, we address the astronomical, physical, chemical, and geological processes heralding the origin(s) and evolution of life on Earth. Next, in Parts III, IV, and V, we describe astrobiological targets in our solar system and beyond, their potential for supporting extraterrestrial life, and how to detect the latter. Finally, in Part VI, we touch on the future of humanity, especially in relation to space exploration. We do not tackle the third question in as much depth as the preceding duo because uncovering and knowing the past and the present is difficult, but feasible, whereas forecasting and knowing the future is deeply challenging.

As the previous definition of astrobiology implies, the Earth serves as a bedrock for this discipline. The rationale is simple: Earth is currently the only world unequivocally confirmed to host life, although this status might change in the future. Hence, to varying degrees, most attempts to assess the possibility of extraterrestrial life extrapolate from life on Earth. It may be argued that such an approach runs the risk of geocentrism (i.e., overly relying on data from Earth). While this objection is valid to an extent, in the absence of any other samples, the Earth remains the only (and therefore the best available) guide to resolving profound questions such as: how did the origin(s) of life occur? how would life and its environment coevolve together? what are the physical and chemical extremes tolerable by organisms?

Thus, it is apparent that the Earth is an essential ingredient of astrobiology. In this respect, astrobiology is distinguishable from the earlier term *exobiology*, which encapsulates the discipline that aims to infer the '*cosmic distribution of life*' (Lederberg, 1960, pg. 393). Given that exobiology emphasises the study of extraterrestrial life, it can be said to primarily focus on the second question (are we alone?), and might be regarded as a major subset of astrobiology. An early prominent advocate of pursuing exobiology was Joshua Lederberg (1925–2008), who received the 1958 Nobel Prize in Physiology or Medicine for his work on genetic transfer in bacteria.

Lederberg is widely credited with having coined the term exobiology, although this word might predate his seminal 1960 paper (Lederberg, 1960). As an interesting aside, the word 'astrobiology' was coined several decades earlier, towards the end of the nineteenth century; however, this field has much older roots, as outlined in Section 1.2. A short review of the etymology and twentieth-century history of astrobiology is furnished in Lingam and Loeb (2020b).

1.1.2 Habitability

If we inspect the NAI definition of astrobiology in Section 1.1.1, we encounter the word 'habitable', which brings up the question: how do we define 'habitability' or a 'habitable environment'? As per the 2015 NASA Astrobiology Strategy,[2] these two terms may be understood as follows:

Habitability has been defined as the potential of an environment (past or present) to support life of any kind. . . . A habitable environment is one with the ability to generate life endogenously – solely using available resources – or support the survival of life that may arrive from elsewhere.

A closely related definition of habitability is furnished in a comprehensive review of this subject by Cockell et al. (2016, pg. 89):

In this review on habitability, we define it as the ability of an environment to support the activity of at least one known organism.

---

[2] https://astrobiology.nasa.gov/nai/media/medialibrary/2015/10/NASA_Astrobiology_Strategy_2015_151008.pdf.

These definitions appear straightforward at first glimpse, but in reality, a number of ambiguities persist. For example, is habitability discrete (e.g., binary) or continuous? On the one hand, it may be argued that an environment can either support one or more organisms or that it cannot do so, thereby conferring a binary basis (Cockell et al., 2019). Yet, on the other hand, the complexity of organisms and the associated environments, as well as their coevolution, could effectively impart continuity to the notion of habitability (Space Studies Board, 2019; Heller, 2020). Another subtlety pertaining to habitability is spelt out next because of its significance.

Broadly speaking, we can envision habitability as some set of characteristics that are imperative at any given instant in time to render an environment suitable for hosting life (Cockell et al., 2016; Domagal-Goldman et al., 2016). Such factors would contribute to what might be dubbed *instantaneous* habitability. The variables that enter the picture insofar as life-as-we-know-it is concerned would therefore include the likes of a solvent (specifically water), energy sources for metabolism, essential elements (in nutrient form), and appropriate physicochemical conditions. We shall delve into the multiple aspects that constitute instantaneous habitability in Chapter 7.

Alternatively, since life should necessitate a certain amount of time to originate, evolve, and create a biosphere, it is evident that environments must retain clement properties for life over a sufficiently long timescale. Hence, we must also engage with the notion of *continuous* habitability, which encapsulates the potential of a particular world (typically a planet) to sustain conditions amenable to life over an extended period of time. The variables that modulate continuous habitability are many and variegated, ranging from planetary characteristics such as size, axial tilt, and plate tectonics to stellar factors (e.g., winds and flares) and even galactic processes like gamma ray bursts and supermassive black hole activity. We will touch on these components, depicted in Figure 1.2 in Chapter 8.

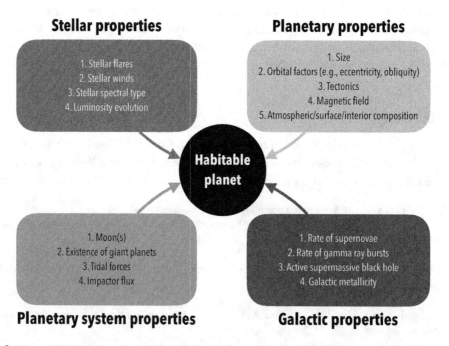

Figure 1.2 The myriad variables that influence the habitability of a planet; they are either properties intrinsic to the planet, host star(s), and planetary system, or regions of the Milky Way.

1.1.3   Life

Hitherto, we have often employed the term *life* without explicitly clarifying what we mean by it. *What is the definition of life?* As the reader may imagine, this question has engaged intellectuals for millennia, and still remains unresolved. This subject has become so extensive that it has been thoroughly examined in entire books, which can be consulted for more information (e.g., Schrödinger, 1944; Cleland, 2019; Smith and Mariscal, 2020).

If we wish to discover extraterrestrial life or ascertain how the origin(s) of life may transpire, which are two prime goals of astrobiology, it is natural to contend that we must define life and demarcate it from non-life. Even though this stance seems straightforward, it faces several subtle drawbacks, as summarised in Cleland (2019). For each proposed definition of life, it is feasible to come up with counterexamples and/or exceptions; this issue is illustrated a few paragraphs hereafter. Hence, in place of 'universal' formulations of life, some authors have advocated for operational definitions and heuristics, while others have suggested that the category of 'life' itself is problematic (Smith and Mariscal, 2020).

One of the earliest recorded definitions of life was provided by Aristotle (384–322 BCE) in the fourth century BCE. In the famous *De Anima* (On the Soul), Aristotle (1907, pg. 49) postulated:

Of natural bodies some possess life and some do not: where by life we mean the power of self-nourishment and of independent growth and decay.

On inspecting this definition, it is apparent that Aristotle highlighted the metabolic facets of life – to wit, its capacity to perform self-sustaining activities such as maintenance and growth by employing energy obtained from its environment. As we shall witness in Section 5.3.1, a prominent set of hypotheses for the origin(s) of life has, likewise, attempted to trace the steps comprising this transition from a metabolic perspective.

The above definition illustrates the pitfalls of subscribing to a single formulation of life without including any accompanying caveats. Fire and crystals, both of which are not living systems, are capable of maintenance and growth when they are granted access to suitable energy sources and/or raw materials. Now, let us turn our attention to the so-called NASA definition of life (Joyce, 1994, pg. xi), because it is perhaps the closest that we have to a consensus, although its acceptance is by no means universal.

Life is a self-sustaining chemical system capable of Darwinian evolution.

If we scrutinise this definition carefully, we see that most of the terms therein are relatively straightforward, except for *Darwinian evolution*. Singling out this phrase is problematic due to the following three reasons, and additional objections could be raised (refer to Question 1.3).

• In referring to Darwinian evolution, multiple theories are clubbed together under the same umbrella (Mayr, 2004, Chapter 6), which makes it challenging to define this term. Darwinian evolution is often equated with the theory of natural selection, although they are not synonymous; the latter may be summarised as follows (Mayr, 2004, pg. 31):

It is rather a shock for some biologists to learn that natural selection, taken strictly, is not a selection process at all, but rather a process of elimination and differential reproduction. It is the least adapted individuals that in every generation are eliminated first, while those that are better adapted have a greater chance to survive and reproduce.

- While Darwinian evolution is conventionally held to be preeminent and predominant in the evolutionary history of Earth, there is growing recognition that alternative (i.e., non-Darwinian) modes of evolution might have played vital roles on our planet. For instance, Lamarckian evolution involves '*non-randomly acquired, beneficial phenotypic changes*' (Koonin and Wolf, 2009).[3] If other modes of evolution can operate in isolation, then the emphasis on Darwinian evolution may be misplaced.
- As per the NASA definition, it is conceivable that viruses and artificial intelligences (AIs) might need to be excluded: the former because they may not count as self-sustaining in the strict sense, and the latter because they are not necessarily subject to Darwinian evolution. Yet, persuasive arguments could be (and have been) made as to why both these entities are classifiable as 'living' systems, in which case the NASA definition would be rendered incomplete, if not incorrect.

As already mentioned, in place of concrete definitions, some authors have opted for identifying general characteristics of living systems. Such lists have proliferated over the years, owing to which we shall concentrate on just one of them. Koshland Jr. (2002) hypothesised that life may consist of seven pillars, collectively called PICERAS: (1) **P**rogram (a blueprint encoding the information pertaining to the organism); (2) **I**mprovisation (achieved via mutation and selection); (3) **C**ompartmentalisation (to mitigate dilution); (4) **E**nergy (to sustain the system); (5) **R**egeneration (transport of chemicals to replenish losses); (6) **A**daptability (behavioural responses to stimuli); and (7) **S**eclusion (allowing many reactions to occur in tandem).

However, even transitioning from rigorous definitions to lists does not altogether solve the riddle of understanding life. In connection with the PICERAS scheme, it is possible that rudimentary (perhaps the first) life forms on Earth or elsewhere lacked well-defined compartments while retaining the looser property of spatial localisation; in such a scenario, they would be labelled as non-living, despite being the opposite. Moreover, the likes of viruses and AIs may not exactly fulfil all the criteria, though some scientists consider them living systems.

This brief foray into the realm of delineating life is obviously not meant to be comprehensive. Instead, it is meant to convey a handful of commonly employed formulations of life, and illustrate how they are limited in scope.

1.2   A brief history of astrobiology

As already remarked, the word 'astrobiology' was coined merely about 125 years ago, but the history of speculations about extraterrestrial life stretches back millennia. The excellent monographs by Dick (1982), Crowe (1986, 2008), and Weintraub (2014) may be consulted to obtain further details. In the subsequent pages, we will furnish a highly abbreviated and selective timeline of some notable developments in this arena.

It has become customary to attribute the concept of *plurality of worlds* and the existence of extraterrestrial life to Anaximander (sixth century BCE) and the Greek atomists, who believed that matter was composed of indivisible units, namely, atoms. In *Letter to Herodotus*, Epicurus (341–270 BCE) conjectured that the Universe is infinite and that other inhabited worlds abound, as evidenced by the following text (Bailey, 1957):

---

[3] Simply put, the phenotype encompasses the set of observable traits of a particular organism.

Furthermore, there are infinite worlds both like and unlike this world of ours. . . . We must believe that in all worlds there are living creatures, and plants and other things that we see here in this world.

In Europe, the Epicurean standpoint was mostly relegated to the margins by the worldview of Aristotle, which held that the Earth is the only inhabited world in the Cosmos. The arguments that underpinned Aristotle's thesis are rather intricate, stemming from the notion that the elements constituting the Universe possess an intrinsic order, culminating in the Earth emerging as the centre of the Cosmos and the only world with life (Dick, 1982).

As the Aristotelian worldview and its variants held sway in Europe, most narratives tend to jump nearly 1,500 years ahead in time, that is a few decades or centuries prior to the Copernican revolution (which dethroned the Earth from its place as the centre of the Universe). Intellectuals such as Albertus Magnus (1200–1280), Hasdai Crescas (1340–1410), and Nicholas of Cusa (1401–1464) wrote persuasively about the possibility of extraterrestrial life. However, undertaking such a skip in time would manifestly paint an inaccurate picture because it ignores parallel musings unfolding in Asia and Africa; we shall concisely explore the former herein.

The Śānti Parva (book 12) of the renowned Indian epic *Mahābhārata* – whose precise date of (oral) composition is unclear, but might be roughly 2,000 years old – contains the following passage (Ray, 1891, pg. 34):

The sky thou seest above is Infinite. It is the abode of persons crowned with ascetic success and of divine beings. It is delightful, and consists of various regions.

Additional examples of similar speculations from ancient India and China are furnished in Selin and Sun (2000) and Nazé (2009). Many of the major religions currently concentrated (or with origins) in South Asia, such as Hinduism, Buddhism, Jainism, and Sikhism, are distinguished by writings supporting the existence of extraterrestrial life (Weintraub, 2014). Moving to west Asia, multiple authors wrote about extraterrestrial life during the Islamic Golden Age. One of them was Muhammad al-Bāqir (676–733), who asserted that (Weintraub, 2014, pg. 165):

Maybe you see that God created only this single world and that God did not create Homo sapiens besides you. Well, I swear by God that God created thousands and thousands of worlds and thousands and thousands of humankind.

The polymath Fakhr al-Dīn al-Rāzī (1149–1209) also presented counterarguments against the Aristotelian worldview, and advocated for the plurality of worlds, and even the potentiality of a multiverse.

Of the early post-Copernican thinkers, the most famous among them is Giordano Bruno (1548–1600). Much of Bruno's fame is attributable to the fact that he was burnt at the stake for religious heresy. It is often claimed that Bruno's gruesome fate was directly caused by his beliefs regarding the plurality of worlds and extraterrestrial life, but other matters of religious doctrine appear to have been partly (if not chiefly) responsible. In his treatise *De l'infinito, universo e mondi* (published in 1584), Bruno strikingly asserted (Boulting, 1914, pg. 144):

There are countless suns and an infinity of planets which circle round their suns as our seven planets circle round our Sun . . . and there must be plants and minerals in the worlds of space like those of our Earth or different. We can attribute life to worlds with better reason than we can to our own Earth.

In the centuries after Bruno, the idea of extraterrestrial life slowly gained acceptance, as chronicled in Crowe (1986, 2008). By the time we reach the commencement of the twentieth century, extraterrestrial life was quite widely embedded in the public consciousness through science fiction books such as H. G. Wells' *The War of the Worlds* (1898) and the writings of Percival Lowell (1855–1916), who erroneously claimed that Mars' surface was crisscrossed by 'canals' constructed by extraterrestrial technological species.

It would be a mistake, however, to presume that astrobiology at this stage was confined only to mere speculations. One of the central concepts in habitability is the circumstellar *habitable zone*, which we shall introduce in Section 8.1, and employ in subsequent chapters. Recent research has established that the habitable zone was already cast into semi-modern (albeit qualitative) form by the start of the twentieth century, thanks to contributions from Sir Isaac Newton (1643–1727), William Whewell (1794–1866), and Alexander Winchell (1824–1891), among others (Lingam, 2021a).

As we have seen thus far, astrobiology was predominantly (yet not exclusively) confined to the realm of philosophical musings until the start of the twentieth century. In some respects, this situation started to change in the nineteenth century, but it was the first half of the twentieth century that birthed many crucial breakthroughs. We will underscore only a handful of them, using the trio of questions posed in Section 1.1.1 as our guide; we will encounter more such milestones in the upcoming chapters.

- In 1913, Edward Maunder (1851–1928) authored a book with the self-explanatory title *Are the planets inhabited?*, wherein he presented systematic calculations to demonstrate that the surface of Mars is unlikely to host long-lived bodies of liquid water. Furthermore, the habitable zone concept was delineated, and a heuristic estimate of the number of inhabited worlds in the Milky Way was provided.
- As described in Section 5.1.1, amino acids are the building blocks of proteins. Hence, producing these molecules from simple, widespread, inorganic compounds is relevant to the origin(s) of life. This synthesis was accomplished in 1913 independently by two chemists: Walther Löb (1872–1916) and Oskar Baudisch (1881–1950) (Lazcano and Bada, 2003).
- Identifying signatures generated by biology is vital since they can pave the way for the discovery of extraterrestrial life. Vladimir Martynovitch Artsikhovski (1876–1931) advocated in 1912 that chlorophylls (utilised in photosynthesis) may constitute biological indicators that could be detected by telescopes. Likewise, Sir James Jeans (1877–1946) suggested in 1930 that molecular oxygen – a product of oxygenic photosynthesis (i.e., of potentially biological origin) – might be discernible by telescopes.
- In 1935, Ary Shtérnfeld (1905–1970) authored an article in the science magazine *La Nature*, which was notable because he outlined one of the first modern definitions of astrobiology. In this publication, he predicted that Saturn's moon Titan hosts an atmosphere, and delved into topics as diverse as the origin(s) of life and organisms in extreme habitats.
- Gavriil Adrianovich Tikhov (1875–1960) ranks among the forgotten pioneers of astrobiology (Briot et al., 2004). He conducted experiments on plant physiology under extreme conditions, developed spectroscopic methods to detect chlorophylls and analogous pigments on the surface of Mars, and analysed light from Earth reflected back by the Moon.

From the second half of the twentieth century (i.e., 1950s and later), a multitude of experimental and theoretical advances propelled astrobiology into the mainstream, despite some pushback from detractors. Many of the subfields that comprise astrobiology – such as origin-of-life studies, microbiology in extreme environments, habitability of worlds in our solar system, and the search for extraterrestrial technological intelligences (ETIs) – were characterised by such rapid progress that even handpicking a select few examples from each domain would significantly expand our narrative; some of the salient milestones are, instead, covered in the appropriate chapters.

It is tempting to suppose that astrobiology entered its 'mature' phase with the discovery of planets beyond our solar system (exoplanets) and the establishment of NAI in the 1990s. In reality, however, astrobiology has been growing in complexity and depth since the twentieth century, as already remarked. With that said, the modern period of astrobiology (which is ongoing) might be credibly dated to the 1990s on account of the aforementioned two reasons. The state of funding, educational programmes, data, modelling, and personnel is more robust and substantial than ever before, owing to which the future seems bright, as summarised next.

- We are combining sophisticated physical and chemical models with state-of-the-art experiments and field analyses to ascertain how, where, and when the origin(s) of life was actualised on Earth, and perhaps other worlds. This combined approach may enable us to tackle the first fundamental question: where did we come from?
- We are performing laboratory experiments and field studies to gauge the physicochemical limits tolerable by organisms, as well as the signatures and markers they generate. Future missions to promising targets in our solar system (e.g., Mars) will inform us about their habitability, and might uncover traces of life. In concurrence, forthcoming space- and ground-based telescopes may help us detect atmospheric and surface signatures of biological activity on exoplanets. These avenues could collectively help us resolve the second major question: are we alone?
- We are developing a deeper global awareness of the direct impact of human activity on the Earth system (including, but not limited to, climate change), and of the existential threats that humanity will face in the immediate and distant future. Considerable attention is now devoted to addressing such challenges, and devising sustainable trajectories for prolonging the lifespan of human civilisation. In parallel, a renewed interest in space exploration, promoted not only by national space agencies but also by private entrepreneurs, is perceived by some as the dawn of human expansion beyond our planet. Exploring such issues might help us gain insights into the final question: where are we going?

1.3   Problems

Question 1.1: The website you will be utilising for this problem is Google Scholar.[4] Enter the word 'astrobiology' and use the *Custom Range* feature to determine how many times this term appears in 10-year intervals over the past 100 years, that is quantify how many instances of this word are recorded in 1920–1930, 1930–1940, and so on until 2010–2020. Report your results in tabular form and/or generate a histogram. What do these results suggest concerning the progress of astrobiology in the last 100 years?

Question 1.2: After consulting the references cited in Section 1.1.2, discuss whether you would regard habitability as binary, discrete but multi-valued, continuous, or none of these categories. Make sure to explain your answer by drawing on peer-reviewed sources.

[4] https://scholar.google.com/.

Question 1.3: In Section 1.1.3, some of the limitations associated with including 'Darwinian evolution' in the NASA definition of life were addressed. Identify at least one more reason as to why the NASA definition is either incomplete or incorrect. Draw judiciously on internet resources, and restrict your sources to peer-reviewed publications.

Question 1.4: To begin with, familiarise yourself with the review of viruses in an astrobiological context by Berliner et al. (2018). Next, equipped with this information, peruse the three formulations of life furnished in Section 1.1.3. Would viruses count as 'life' as per any of this trio? Justify your reasoning and back it up with peer-reviewed sources, if and when necessary.

Question 1.5: Aside from the proponents of the plurality of worlds (also called cosmic pluralism) in Section 1.2, select another individual from ≥500 years ago who subscribed to cosmic pluralism and the existence of extraterrestrial life. Provide a short biography of this person, and mention which of their work(s) explicitly conveyed their belief in extraterrestrial life.

2   From the Big Bang to Molecules

One of the major scientific advancements of the past century is the formulation of an evidence-based description of the origin and evolution of our Universe. This cosmological model, conventionally termed the *hot Big Bang model*, rests on the solid foundations of Albert Einstein's (1879–1955) general relativity, and displays impressive agreement with a large set of observations and empirical evidence. According to the Big Bang model, the Universe started 13.8 billion years ago in a hot, dense state, with all matter in the form of free particles uniformly distributed in space: from such extraordinarily remote and unfamiliar conditions, the forces of nature slowly assembled stars and galaxies, planets, complex molecules, and life itself.

It may be worth pointing out that throughout this book we will always use the expression 'the Universe' to embody the region of spacetime that is accessible to direct observation, that is, the volume contained in the *cosmic horizon*. The latter represents the distance that light could have travelled in the time elapsed from the Big Bang. According to some scenarios postulated in theoretical physics, there might exist a *multiverse* of countless disconnected spacetime regions beyond the one encompassed by our cosmic horizon. These regions could, in principle, possess vastly different physical conditions from our own, and life may only appear in those with the right properties. The implications of such a scenario are tangential to the scope of astrobiology, which seeks to comprehend life in the only Universe that we can directly observe. The connections between life and the multiverse are elaborated in Lewis and Barnes (2016) and Adams (2019).

The affirmation of the Big Bang model has introduced an 'evolutionary' theme in the description of the Universe, albeit with a different slant than its strict biological use (Chaisson, 2001). Time and change play a crucial role in our modern understanding of the Cosmos: not only was the Universe radically different in the past but also it has only existed for a finite amount of time. The history of the Universe can thus be regarded as a sequence of transitions from one physical state to the next, following an overarching trend that proceeds from the simple towards the complex; this is, however, not to say that these transitions are an inexorable march towards some predetermined end. Broadly speaking, we now know that the early Universe was smooth, and in almost perfect thermal equilibrium (i.e., uniform temperature), while the present Universe is clumpy, structured, and capable of producing and sustaining local disequilibrium.

This grand picture has direct implications for our attempts to pursue the goals of astrobiology, and especially to understand the appearance of life in the Universe and its distribution. In the Big Bang universe, none of the basic ingredients of life were available at the beginning. Atoms and molecules had to be built from scratch, and the same goes for stars and planets. Even the flow of free energy that is necessary to fuel the activity of any complex physical system, including living organisms, was unavailable in the undifferentiated primordial conditions from which cosmic evolution started.

In other words, the Universe has not been equally habitable at all epochs in the past, and it will probably not remain so in the far future (Adams and Laughlin, 1997). By linking the average conditions of the Universe to the passage of time, the Big Bang model amplifies the need for a satisfactory explanation of the existence of even a single inhabited planet. This scenario strikingly contrasts the other cosmological models that were entertained in the past, especially the idea of an eternal and stationary Cosmos that possessed uniform average conditions, and had an indeterminate amount of time to produce even the most unlikely outcome somewhere.

Therefore, elucidating the physical processes that gave rise to the situation we observe in the current Universe is essential if we want to frame the problem of life in the right context. This objective is what we set out to accomplish in this chapter. In the first part, we will review the basic features of the Big Bang model, and the overall history of the Universe. Next, we will witness how the evolution of the Cosmos 'sets the stage' for the appearance of life, by forming stars, atoms, and molecules.

2.1   The expanding universe

In the first half of the twentieth century, the discovery of the expansion of the Universe was instrumental in establishing the Big Bang model. This characteristic was inferred by the observation that distant galaxies appear to recede with a velocity v that is directly proportional to their distance from us (D), as mathematically expressed by *Hubble's Law*:

$$v = H_0 D, \tag{2.1}$$

where the proportionality factor H_0 is called the *Hubble constant*. State-of-the-art direct measurements of the distances and recession velocities of galaxies have yielded a best-fit value of $H_0 = 73.0 \pm 1.0$ km s^{-1} per Mpc (1 Mpc, denoting 1 million parsecs, is equivalent to a distance of 3.26 million light-years or 3.1×10^{22} m) (Riess et al., 2022).[1]

When complemented with the commonsensical presupposition that our physical location in the Universe is not peculiar or privileged (an assumption often termed the *Copernican* or *cosmological principle*), a clear explanation for Hubble's law is that the distance between any two unbound points in space grows with time t in the following fashion:

$$D(t) = a(t)D(t_0), \tag{2.2}$$

where t_0 is the present time, and $a(t)$ is a dimensionless scale factor. By definition, $a = 1$ at $t = t_0$, and it is smaller in the past. Taking the derivative of (2.2) with respect to time, we see that any two points in the Universe move away from each other with relative velocity $v(t) = \dot{D}(t) = \dot{a}(t)D(t_0) = \dot{a}(t)D(t)/a(t)$ (where the dot represents the time derivative). On defining

$$H(t) \equiv \frac{\dot{a}(t)}{a(t)}, \tag{2.3}$$

[1] Other estimates, based on indirect inferences from observations of the early Universe, have resulted in lower values of H_0. The apparent tension between the values obtained via these two different approaches is currently unexplained. We refer the interested reader to a recent review on the subject by Kamionkowski and Riess (2023).

we obtain $v(t) = H(t)D(t)$, which, when evaluated at the present time with $H_0 \equiv H(t_0)$, is simply (2.1). Note, however, that the expansion of the Universe introduced in (2.2) implies that the exact same law is valid at any moment in time, and at any location in space.

The commonly adopted convention is to define the beginning of time ($t = 0$) as the moment when $a = 0$, generally termed the *Big Bang*. However, it should be borne in mind that this extremely early epoch requires an extrapolation beyond the regime of well-tested physical theories. The detailed manner in which the scale factor $a(t)$ evolves with time depends on the matter and energy content of the Universe, which are measurable by astrophysical observations. The present age of the Universe is determinable by integrating the scale factor over the entire history of the Universe, $t_0 = \int_0^1 da/\dot{a}$. A current best estimate (Planck Collaboration, 2020) is:

$$t_0 = 13.787 \pm 0.020 \text{ Gyr.} \tag{2.4}$$

A straightforward consequence of the expansion is that the content of the Universe becomes progressively diluted with time or, equivalently, that its density was higher in the past. Specifically, if we assume that mass is conserved, and recognise that the volume of any given region of space grows in time as the third power of the scale factor, we can deduce that the average density of matter, ρ_{mat}, should scale as

$$\rho_{\text{mat}} \propto a^{-3} \tag{2.5}$$

and was nominally infinite at the Big Bang (whether this condition, termed the *singularity*, actually occurred, still remains an open question). The average energy density of photons (denoted by γ here), ρ_γ, was also larger in the past, albeit with a different dependence,[2]

$$\rho_\gamma \propto a^{-4}. \tag{2.6}$$

Therefore, at early times, the energy density of photons was substantially larger than the density of matter, and sufficiently high to ionise atoms, so that all matter existed in a plasma state. The high density and the frequent interactions between photons and charged particles (electrons and ions) kept the primordial plasma in thermal equilibrium with the electromagnetic radiation, very much akin to the interior of a star: in such conditions, radiation exhibited a blackbody energy spectrum. Given that the temperature of a blackbody is related to its energy density via the *Stefan–Boltzmann law*, $\rho_\gamma \propto T^4$, from (2.6) we can deduce that the average temperature of the early Universe obeyed the relation

$$T \propto a^{-1}. \tag{2.7}$$

In other words, as the Universe expanded over time, it would have simultaneously cooled down. This behaviour has important consequences for determining the average physical conditions existing in the Cosmos, and the processes that could take place at different epochs. A summary of some major milestones in the history of the Universe is given in Table 2.1.

A notable repercussion of the decreasing temperature after the Big Bang was the formation of neutral hydrogen atoms. The bond between protons and electrons could only stabilise when the average

---

[2] The extra factor of a in (2.6) arises from the fact that, in addition to increasing the volume, the expansion also stretches the wavelength of photons by a, thereby reducing their energy (photon energy is inversely proportional to wavelength), a phenomenon known as *redshift*.

Table 2.1 Rough timeline of key events occurring in the Universe

t	T	ρ_{mat} (g/cm^3)	Description
0	∞ (?)	∞ (?)	Big Bang
$\sim 10^3$ s	$\sim 10^7$ K	$\sim 10^{-8}$	Primordial nucleosynthesis ends
~ 380 kyr	$\sim 3{,}000$ K	$\sim 10^{-19}$	Hydrogen atoms form (recombination)
~ 150 Myr	~ 100 K	$\sim 10^{-25}$	First stars are born
~ 13.8 Gyr	~ 3 K	$\sim 10^{-29}$	Present

Note: The age of the Universe (first column) and its temperature (second column) and matter density (third column) at that moment in time are approximate.

temperature T declined well below the ionisation threshold, or $k_B T \ll 13.6$ eV, where k_B is the Boltzmann constant and 1 eV = 1.6×10^{-19} J; this may have transpired when T was a few thousands of Kelvin. This transition, called *recombination*, occurred $\sim 380{,}000$ years after the Big Bang. After recombination, radiation decoupled from matter, and the Universe became transparent to the former.

Photons kept filling space with an almost-uniform density, maintaining a precise blackbody spectral distribution, with a decreasing temperature still given by (2.7). In the present day, this nearly perfect blackbody radiation, the so-called *cosmic microwave background* (CMB), possesses an average temperature of 2.72548 ± 0.00057 K (Fixsen, 2009). The precise observation of the CMB constitutes strong evidence in support of the Big Bang model, and of the hot, dense conditions prevalent in the early Universe.

2.1.1 The emergence of complexity

Before proceeding further, it is worth pausing to describe some general features of the expanding Universe that have a profound connection with the emergence of structure and complexity in the Universe, including life. We just learnt that the early Universe was hot, dense and undifferentiated, while the current Cosmos contains a panoply of complex structures, on scales ranging from the molecular to the astrophysical, with enormous variations in density and temperature, spanning the entire interval between the huge values found in compact objects (such as neutron stars or black holes), and the coldness and emptiness of interstellar and intergalactic space.

The transition of the Universe from a simple state into a structured condition may seem counterintuitive at first because we are accustomed to witnessing the opposite behaviour in physical systems; we touch on this notion in Section 5.6. In fact, we generally observe that the passage of time washes out density and temperature differences, and creates more equilibrium. The classic example is that of a gas of free particles contained in a closed volume: if the particles are initially clumped in one corner of the box, with completely random velocities, they will end up uniformly occupying the entire volume. The system spontaneously evolves into a progressively simpler, smoother and diluted state, with the same temperature everywhere.

The reason why the evolution of the Universe seems to contradict the aforementioned behaviour is attributable to the role of gravity. This force affects the picture in two different but related ways. First of all, gravity governs the expansion of the Universe: this feature, as we saw, leads to the emergence of markedly different environmental conditions at different epochs. To be more specific, the average temperature of the Universe (in the sense of the thermal radiation blackbody temperature), and the density of matter and energy all decrease with time, implying that any reaction involving the frequent interaction of matter with radiation eventually goes out of equilibrium. For example, consider the generic reaction:

$$A + B \leftrightarrow AB \tag{2.8}$$

that could proceed in either direction as long as the prevalent conditions exceed a certain energy threshold. As the density decreases, interactions become more infrequent; furthermore, there is not enough thermal energy to maintain the reaction in balance. Therefore, the process can only occur in the 'forward' direction in due course.[3] We already saw a powerful example of this mechanism in action when discussing the recombination of hydrogen atoms. We shall encounter other instances later.

The second way in which gravity contributes to the emergence of complexity is by amplifying initial inhomogeneities in the distribution of matter. Accurate measurements of tiny variations in the temperature of the CMB show that the density of the early Universe had random fluctuations, of order 10^{-5} about the average value. These minuscule density perturbations increased with time, as gravity aggregated more matter around them, eventually resulting in a complex network of cosmic structures. It is worth underscoring that, in an expanding universe, collapse will prevail over diffusion, given enough time (Lineweaver and Egan, 2008).

This statement can be understood by considering that the average temperature of a gas of matter particles is $T_m \propto PV$, and for an adiabatic expansion of a monoatomic gas, we have $PV^{5/3} = \text{const}$; here, T_m, P, and V are the gas temperature, pressure, and volume, respectively. Therefore, as the Universe expands, the gas temperature obeys $T_m \propto V^{-2/3} \propto a^{-2}$. The kinetic energy K of a cloud of particles (which is proportional to the thermal energy) in an expanding Universe scales as $K \propto k_B T_m \propto a^{-2}$. On the other hand, for fixed mass, the gravitational binding energy U is inversely proportional to the size of the cloud, and scales as $U \propto 1/a$. Thus, eventually $U \gg K$, and the random agitation of particles does not suffice to overcome gravity. In contrast to the 'gas in a box' picture, the Universe consequently becomes clumpier, rather than increasingly uniform, with time.

All this, of course, is crucial for the appearance of one particular kind of clump of matter, to wit, biological systems. Living organisms may be visualised as a type of *dissipative structure* (Russell et al., 2013), a term coined by Nobel Prize winner Ilya Prigogine to identify coherent, ordered systems that are far from thermodynamical equilibrium, but maintain a 'steady state' by exchanging matter and energy with the environment (Prigogine, 1978). Dissipative structures typically emerge in the presence of an energy gradient (e.g., thermal, chemical, gravitational) – or, equivalently, when thermodynamic disequilibrium is existent – and include not just living systems but also convection cells, turbulent flows, cyclones, tornadoes, galaxies, and so on.

In fact, life thrives on the disequilibrium produced by the overall evolution of the Universe. For example, contemplate sunlight striking the surface of Earth. Photons from the Sun are produced by nuclear reactions that only occur at substantial densities and temperatures, in the highly concentrated state of matter produced by gravitational collapse. Such energy can then radiate from the

---

[3] An alternative way to see this is to compare the timescale of the reaction – which depends on the interaction between particles, and therefore scales with density as $t_{\text{int}} \propto \rho^{-1}$ – with the timescale over which the Universe significantly changes its average conditions (which depends on the expansion rate H, and obeys $t_H \propto \rho^{-1/2}$). Eventually, the first grows larger than the second, and the reaction is not fast enough to maintain equilibrium.

high temperature of the Sun's surface ($T \simeq 6,000$ K) into empty space ($T \gtrsim 2.7$ K), exploiting the extreme disuniformity of the present Universe. Earth finds itself in the midst of this flux of energy, which predominantly fuels physical processes taking place on the planet, including life. We will see in Section 7.1 that there are various kinds of free energy (in the thermodynamical sense) that living systems could exploit in order to maintain their 'machinery' in operation, far from equilibrium, much like engines (Russell et al., 2013). Ultimately, however, any source of free energy in the Universe is traceable back to gravity, in its double capacity of driving the expansion of the Universe and the aggregation of structures.

We conclude by highlighting the remarkable fact that the existence of self-organising structures in the present Universe is an ephemeral feature, one that not only deviates significantly from the simpler conditions prevalent in the past but also will not persist indefinitely in the future. If the Universe keeps expanding at an accelerated rate (as current observations suggest), eventually it will become cold and empty, thus approaching perfect thermodynamic equilibrium (Adams and Laughlin, 1997). An eternity where basically nothing complex will happen could subsequently come into being, since the Universe would have converted all its free energy into unusable thermal radiation, a state ominously termed *heat death*. No new structures will be able to manifest, and the existing ones will dissolve into chaos. This picture situates the present era in a peculiar (even unique) position, when measured against the sweeping unfolding of cosmic time.

We are living in a fleeting epoch amidst the overall history of the Universe: a transient interim period wherein stars, planets, molecules, and life are possible, enclosed between two equally feature-less aeons. One potentially profound consequence of this picture is that, while the Copernican principle might be a valid guiding principle when applied to space, it is most certainly wrong when applied to time. Our era is markedly atypical, and we should take this into account when forecasting the frequency and distribution of life in the Universe, or judging its likelihood (Ćirković and Balbi, 2020).

2.1.2   Primordial nucleosynthesis

As a further example of how the expansion of the Universe produces significant changes in the state of matter, and prepares the initial building blocks for additional transformations, let us illustrate the formation of the nuclei of helium in the early Universe, also appropriately termed *primordial nucleosynthesis* (see Steigman, 2007 for details).

In the first moments after the Big Bang, temperatures and densities were so extreme that protons and neutrons were unbound. Furthermore, they could be converted into each other by the following reactions:

$$n + e^+ \leftrightarrow p + \bar{\nu}_e, \tag{2.9}$$

$$n + \nu_e \leftrightarrow p + e^-, \tag{2.10}$$

$$n \leftrightarrow p + e^- + \bar{\nu}_e. \tag{2.11}$$

In (2.9)–(2.11), n and p are neutrons and protons, e^- and e^+ are electrons and positrons (antiparticles of electrons), and ν_e and $\bar{\nu}_e$ are electron neutrinos and antineutrinos. Such reactions, involving the weak nuclear interaction, proceed in both directions as long as $k_B T \sim 0.8$ MeV (where 1 MeV is 10^6 eV), which was the case until $t \sim 1$ s after the Big Bang. Before that time, the relative abundance of neutrons to protons was given by

$$\frac{N_n}{N_p} \sim e^{-\Delta E/k_B T}, \tag{2.12}$$

where N_n and N_p are the number densities of neutrons and protons, respectively, and $\Delta E = (m_n - m_p)c^2 = 1.3$ MeV is the neutron–proton mass difference. After the thermal energy became insufficient to maintain equilibrium, only the third reaction remained effective, in the forward direction, that is, manifesting as free neutron decay with a characteristic timescale of $\sim 10^3$ s.

A crucial preliminary step towards the formation of helium nuclei was the production of deuterium (D) nuclei through the nuclear fusion reaction:

$$p + n \rightarrow D + \gamma. \tag{2.13}$$

Although temperatures and densities in the first few seconds were high enough for deuterium fusion to occur in the primordial plasma, the reaction was not effective until the energy became much smaller than the binding energy of deuterium, $E_D \approx 2.2$ MeV: the reason is that there were $\sim 10^9$ photons per each nucleon (i.e., proton or neutron), so that dissociation of newly formed nuclei was much more efficient than their production. As soon as temperatures were low enough for deuterium to become stable, around $t \sim 200$ s after the Big Bang, further fusion reactions created ^{4}He nuclei (containing two neutrons and two protons) efficiently. When this happened, all the neutrons quickly ended up bound into helium nuclei, with the surviving free protons constituting the nuclei of future hydrogen atoms.[4]

Because there are two neutrons per helium nuclei, the number density of ^{4}He simply translates to $N_n/2$. Finally, since the neutron to proton ratio at the time of helium synthesis had decayed to $N_n/N_p \simeq 1/7$, it is straightforward to calculate the mass of ^{4}He nuclei relative to the total mass of nucleons in the Universe (denoted by Y_P):

$$Y_P = \frac{4N_n/2}{N_n + N_p} = \frac{2N_n/N_p}{[1 + N_n/N_p]} = \frac{2/7}{[1 + 1/7]} = 0.25. \tag{2.14}$$

The factor of 4 in the numerator above appears because the mass of ^{4}He is approximately four times that of a single neutron. The final result displays excellent agreement with the observed 24% mass abundance of ^{4}He in the Universe, which is yet another line of evidence in support of the hot Big Bang model, because nuclear fusion reactions taking place in stars alone cannot account for such copious production.

The absence of stable nuclides composed of five and eight nucleons, combined with the ongoing depletion (and unavailability) of free neutrons as well as the drop in temperature and density due to the cosmological expansion, collectively stymied primordial fusion reactions around $t \sim 1,000$ s after the Big Bang. No heavier nuclei were synthesised in the early Universe, except for trace amounts of ^{6}Li, ^{7}Li, and ^{7}Be. As a final remark, it is worth emphasising that hydrogen remained the most abundant element (about 74% of the total mass) after primordial nucleosynthesis ended.

This outcome has important consequences for astrobiology, as hydrogen is an essential ingredient for life (via water) and diverse chemical reactions. Moreover, if nuclear fusion in the early Universe had converted all protons into helium, there would have been no material left to assemble the most common and long-lived kind of stars in the Universe, that is, those fusing hydrogen in their cores; the fact that stars chiefly consist of hydrogen and helium was advanced by Cecilia Payne-Gaposchkin (1900–1979) in her seminal 1925 dissertation. The existence of stars is of vital

---

[4] Trace amounts of deuterium and ^{3}He also remained in the aftermath of this process.

importance to astrobiology not only because they generate immense amounts of usable energy but also since they facilitate the synthesis of elements, as described shortly.

We now proceed to discuss how these two related crucial steps (namely, the formation of stars and of atomic elements) took place.

2.2   The formation of stars

The first stars in the Universe probably formed crudely $\sim$100–200 Myr after the Big Bang, as matter accumulated in overdense regions through the action of gravity (Barkana, 2016). The precise timing is still unresolved, and so far no direct observations of the first generation of stars exist, although galaxies have been detected as early as a few 100 Myr subsequent to the Big Bang (Castellano et al., 2022; Whitler et al., 2023).

It was originally supposed the first stars were very massive ($\sim$$10^2$–$10^3$ $M_\odot$) and short-lived (a few Myr), but the average mass is now predicted to be lower, that is, comparable to $10\,M_\odot$ (Hosokawa et al., 2016; Ishigaki et al., 2018); note that $M_\odot$ is the Sun's mass. Their appearance marked the end of the *dark ages* following recombination, during which the Universe contained no sources of light, and the beginning of the *stelliferous era*, that is, the epoch of star formation: this period continues to the present day, although at a declining rate. In fact, approximately half of the stars currently shining in the Universe formed between 8 and 11 Gyr ago, and the rate of star formation today is a few per cent of that past peak value. About 95% of all the stars that the Universe will ever create already exist (Sobral ct al., 2013).

2.2.1   The Hertzsprung–Russell diagram

A useful starting point for discussing the properties of stars is the so-called *Hertzsprung–Russell (HR) diagram*, reviewed in Harwit (2006, Chapter 1.5), which displays the relation between stellar luminosities and temperatures (Figure 2.1). In essence, the location of a star in the HR diagram completely defines its key features, and its evolutionary stage. Stars radiate their energy as that of a near-perfect blackbody: therefore, their temperature dictates their spectral energy distribution (i.e., energy emitted per unit wavelength), which, in turn, governs the colour of the light they emit. In general, hotter stars appear bluer, while cooler stars are redder.

An immediately noticeable feature of the HR diagram is the presence of a narrow diagonal region, called the *main sequence*. The majority of stars in the Universe (namely, those that burn hydrogen into helium through nuclear fusion reaction taking place at their core) lie along the main sequence, arranged according to their mass. The most massive main sequence stars are also hotter and more luminous, and are located in the upper-left corner of the HR diagram; conversely, smaller stars arrange themselves along the lower-right positions. An empirical mass-luminosity relation valid for most main sequence stars is (Harwit, 2006, pg. 317):

$$\frac{L_\star}{L_\odot} = \left(\frac{M_\star}{M_\odot}\right)^{3.5}, \tag{2.15}$$

where $M_\star$ and $L_\star$ are stellar mass and luminosity, whereas the symbol '$\odot$' represents the associated solar quantities. The minimal mass for hydrogen fusion to occur is $\sim$0.08 $M_\odot$, about 100 times the mass of Jupiter. The observed upper limit for the mass of main sequence stars is $\gtrsim$200 $M_\odot$.

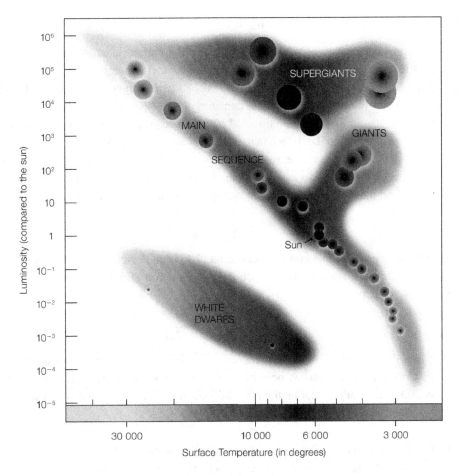

Figure 2.1 Hertzsprung–Russell diagram illustrating the luminosity (y-axis) of various types of stars as a function of their temperature (x-axis). (Credit: Adapted from ESO; www.eso.org/public/images/eso0728c/; CC-BY-4.0 license)

The initial mass of main sequence stars also affects their projected lifetime. More massive stars are hotter, more luminous, and burn faster, consequently lasting for much less time than smaller, dimmer stars. An estimate of the lifetime of a main sequence star is found by dividing $M_\star$ by the rate of mass loss (due to energy generation), which is proportional to $L_\star$, thus yielding

$$t_{ms} \approx 10^{10} \left(\frac{M_\star}{M_\odot} \right)^{-2.5} \text{yr}, \tag{2.16}$$

where t_{ms} is the stellar main-sequence lifetime. While the smallest main sequence stars could exist for as much as 10^{13} years, the most massive stars remain on the main sequence for merely a few million years.

Once a star has exhausted hydrogen in its core, it exits the main sequence and, depending on its mass, temporarily occupies the regions of *giants* or *supergiant* stars in the HR diagram. After nuclear reactions completely shut down, low mass stars end up as *white dwarfs*: small, dense remnants occupying the lower-left region of the HR diagram.

2.2.2   Gravitational collapse and Jeans mass

Star formation occurs in *molecular clouds*. These denser regions in interstellar space are composed of gas (mostly molecular hydrogen, H_2) and dust. They have a diameter of tens to hundreds of light years and a mass of $\sim 10^3$–$10^6 \, M_\odot$. In the densest areas of a typical molecular cloud, there are roughly 10^5 particles per cm^3 (corresponding to mass densities of $\sim 10^{-19} \, g \, cm^{-3}$ for H_2), while temperatures are relatively low at ~ 10 K (e.g., Shimajiri et al., 2019). Such regions are the preferred site for initiating the process of gravitational collapse that eventually leads to star formation.

We can gain some insights on the mechanism of collapse by considering a simplified model in which the molecular cloud is a homogeneous and isothermal (constant temperature) spherical region of radius R and mass M. It can be shown that its gravitational binding energy is

$$U = -\frac{3}{5} \frac{GM^2}{R}, \tag{2.17}$$

where G is the gravitational constant. Equation (2.17) can be verified to possess the correct dimensions (of energy). Because of the random thermal agitation of the gas particles, the cloud will also have a kinetic energy:

$$K = \frac{1}{2} M v^2 = \frac{3}{2} \frac{M}{m} k_B T, \tag{2.18}$$

where m is the average mass of the particles, $v = (3k_B T/m)^{1/2}$ is their root mean square speed, and T is the cloud temperature. It can be shown that when the cloud is in hydrostatic equilibrium, that is, neither collapsing nor expanding, the gravitational and kinetic energies have to satisfy the condition

$$2K + U = 0, \tag{2.19}$$

which is known as the *virial theorem* (Harwit, 2006, Chapter 3.15).

For collapse to happen, gravity has to overwhelm the internal pressure of the cloud, so that $|U| > 2K$. After expressing the mass of the cloud in terms of its mean density (denoted by ρ), the condition for collapse becomes

$$R > R_J \equiv \left(\frac{15 k_B T}{4 \pi G m \rho} \right)^{1/2}, \tag{2.20}$$

where we have introduced the *Jeans radius*, R_J, the minimal size for a uniform, isothermal spherical cloud to start collapsing. Note that the Jeans radius depends only on the ratio between the density and temperature of the cloud. Denser and colder regions thus have a smaller Jeans radius.

Equivalently, this condition can be expressed in terms of a mass cutoff, known as the *Jeans mass* (M_J), in the following manner:

$$M > M_J \equiv \frac{4\pi}{3} \rho R_J^3. \tag{2.21}$$

The densest regions of a large molecular cloud have a Jeans mass of $M_J \sim 10 \, M_\odot$, which is small enough to make these clumps of matter unstable to disturbances in density or pressure, consequently initiating collapse. This process is often triggered by some external event, such as a stellar explosion or a collision with another cloud.

Although the previous simplified treatment gives a general idea about the onset of star formation, many additional complex factors need to be introduced to explain the emergence of multiple stellar-size objects within a cloud, including temperature and density inhomogeneities, turbulence, fragmentation, chemical composition, magnetic fields, radiation and angular momentum transfer, and so on. Some of these aspects are still not fully understood, and constitute the subject of ongoing research. Further details concerning the current understanding of star formation are reviewed by McKee and Ostriker (2007) and Girichidis et al. (2020).

2.2.3 Protostars and entrance in the main sequence

Once the collapse starts, the cloud loses gravitational energy. If this energy were converted into kinetic energy, the cloud would heat up, which would result in an increased Jeans mass, effectively stopping the collapse. However, initially the cloud is thin enough to be transparent to radiation. Therefore, it is able to quickly radiate away its excess heat, and the contraction can proceed at a nearly constant temperature. During this phase, there is essentially no resistance from internal pressure, and the cloud collapses quickly in free-fall, that is, on timescales of $t_{\text{ff}} \sim (G\rho)^{-1/2}$.

Eventually, however, the inner region of a collapsing molecular cloud becomes very dense, and opaque to its own radiation. When this happens, temperature and pressure rise rapidly towards the centre, until they compensate gravity, giving rise to a small core in hydrostatic equilibrium. The core now keeps contracting in a quasi-static fashion: its total energy variation is $\Delta E = \Delta K + \Delta U$, so that, according to the virial theorem, half of the gravitational energy released is converted into kinetic (thermal) energy, $\Delta K = -\Delta U/2$, while half is radiated away, $\Delta E = \Delta U/2$. Therefore, although the core is losing energy, it nevertheless heats up.

Around this stage, the contracting core may be considered a *protostar*, having only about 1% of the final mass of the star. As the infall of material from the outer cloud continues, pressure and temperature in the protostar keep increasing. When $T \sim 2,000$ K, molecular hydrogen gets dissociated: this process absorbs the energy produced by contraction, disrupting the existing equilibrium, so that the core becomes unstable again and collapses into a new equilibrium state. Heating resumes then, until the ionisation of hydrogen (at $T \sim 10^4$ K) absorbs further energy and causes another collapse.

At this point, the surrounding material has either been collected or swept away, so that accretion is essentially over. The star enters the so-called *pre-main-sequence* phase: it is still too cool to fuse hydrogen, and generates its energy by gravitational contraction. As compression progresses in quasi-equilibrium, the internal temperature increases as dictated by the virial theorem, until it reaches the threshold for the ignition of thermonuclear fusion of hydrogen nuclei, $T \sim 10^7$ K. When this happens, the newly formed star finally enters the main sequence region of the HR diagram.

The timescale for the initial free-fall collapse of a spherical cloud and the production of a protostar is rather transient, of the order of a few 10^4 yr. The accretion of the remaining mass into the forming main sequence star, however, takes a longer time, of the order $\gtrsim 10^6$ yr.

2.3 Nucleosynthesis in stars

When the first stars formed, the only significant available element in the Universe, besides hydrogen, was the helium produced during primordial nucleosynthesis. The formation of stars facilitated the production of heavier elements, through processes happening during their life and in the final stages of their existence, as illustrated in Figure 2.2. This figure depicts the various pathways whereby the

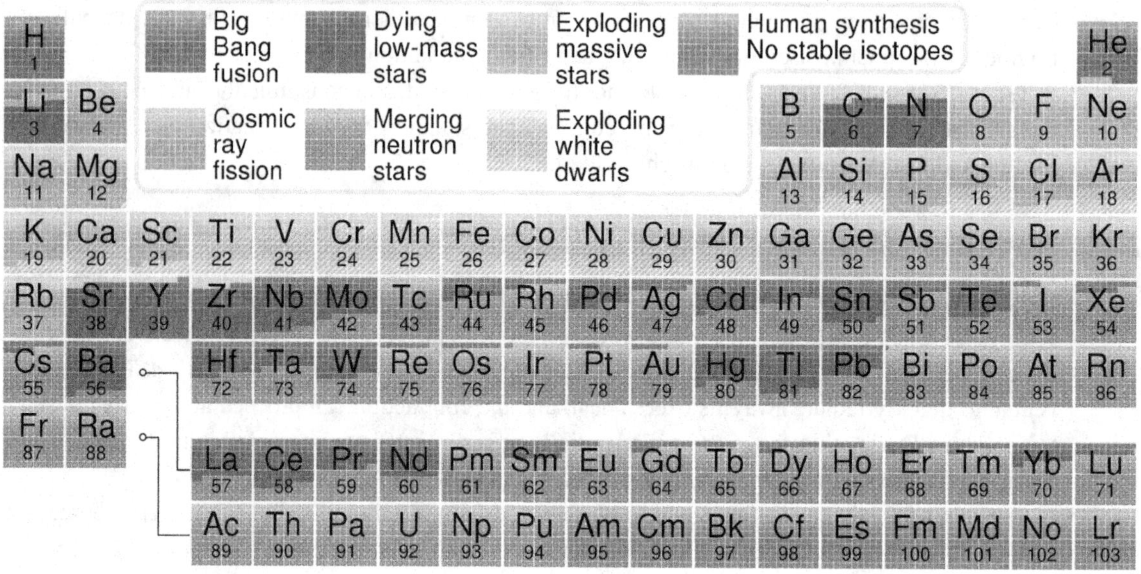

Figure 2.2 The 'nucleosynthesis periodic table' of the elements, which encapsulates the channels through which elements are synthesised. (Credit: https://commons.wikimedia.org/wiki/File:Nucleosynthesis_periodic_table.svg; CC-BY-SA 3.0 license)

synthesis of elements takes place. Eventually, successive generations of stars manufactured the entirety of atomic elements in the Cosmos, including those that constitute planets, complex molecules, and living systems. Reviews of the formation of chemical elements in the Universe are provided by Rauscher and Patkós (2011), Johnson (2019), and Arcones and Thielemann (2023).

The vast majority of all energy produced in stars is on account of the fusion of hydrogen into helium. There are essentially two main routes to this process: the *proton–proton chain*, and the *CNO cycle*. Another important reaction, relevant to the production of carbon, is the *triple–alpha process*. We will now describe these processes in turn, with most of the salient equations and parameters drawn from Lamers and Levesque (2017, Chapter 8).

2.3.1   The proton–proton chain

This process is predominant in the core of main sequence stars whose masses are less than twice that of the Sun. It requires a temperature of $\sim 5 \times 10^6$ K to be initiated. The chief reactions involved in the proton–proton chain, along with their mean reaction times (denoted by τ) in the centre of a star at $\sim 10^7$ K, are expressible as (Harwit, 2006, pg. 342):

$$p + p \rightarrow D + e^+ + \nu_e + 1.44 \, \text{MeV} \quad \tau \sim 1.4 \times 10^{10} \, \text{yr} \tag{2.22}$$

$$p + D \rightarrow {}^3\text{He} + \gamma + 5.49 \, \text{MeV} \quad \tau \sim 6 \, \text{s} \tag{2.23}$$

$$^3\text{He} + {}^3\text{He} \rightarrow {}^4\text{He} + 2p + 12.85 \, \text{MeV} \quad \tau \sim 10^6 \, \text{yr}. \tag{2.24}$$

In (2.22)–(2.24), D is deuterium (^{2}H), while ^{3}He and ^{4}He are isotopes of helium. The last reaction is one of the possible paths to produce ^{4}He via ^{3}He, and it is known as the *pp1* branch. It serves as the dominant process below 1.8×10^7 K. Overall, the proton–proton chain is characterised by a total energy release of 26.7 MeV. The first reaction is obviously very slow compared to the others (i.e., billions of years), and requires the mediation of the weak interaction, to wit, one of the four fundamental forces of nature. For this reason, if C and O nuclei are present, another process becomes competitive: the CNO cycle, which we illustrate next.

2.3.2   The CNO cycle

The carbon–nitrogen–oxygen cycle, or CNO cycle, is an alternative way to convert hydrogen nuclei into helium in the core of main sequence stars. Its efficiency is strongly dependent on temperature, owing to which it is the dominant process in high-mass main sequence stars with their higher temperatures. It starts at $T \sim 1.5 \times 10^7$ K, and it becomes the main energy production mechanism at $T \gtrsim 1.7 \times 10^7$ K (which is achievable in the cores of stars about 1.3 times more massive than the Sun). With a core temperature of $T \sim 1.6 \times 10^7$ K, the Sun only produces about 1% of its total energy by means of this process (The Borexino Collaboration, 2020b). Given that the core temperature increases as the stars age, the contribution from the CNO cycle is rendered more important with time.

As the name suggests, the CNO cycle involves carbon, nitrogen, and oxygen nuclei. However, they are not actually consumed but are continually transformed into each other in a cycle, essentially acting as a catalyst for the burning of hydrogen into helium. The main reactions of the CNO cycle, and the typical reaction times in the centre of a star at 1.5×10^7 K, are broadly given by (Harwit, 2006, pg. 342)

$$^{12}\text{C} + \text{p} \rightarrow {}^{13}\text{N} + \gamma + 1.94\,\text{MeV} \quad \tau \sim 10^6\,\text{yr}, \tag{2.25}$$

$$^{13}\text{N} \rightarrow {}^{13}\text{C} + \text{e}^+ + \nu_e + 2.22\,\text{MeV} \quad \tau \sim 9 \times 10^2\,\text{s}, \tag{2.26}$$

$$^{13}\text{C} + \text{p} \rightarrow {}^{14}\text{N} + \gamma + 7.55\,\text{MeV} \quad \tau \sim 2 \times 10^5\,\text{yr}, \tag{2.27}$$

$$^{14}\text{N} + \text{p} \rightarrow {}^{15}\text{O} + \gamma + 7.29\,\text{MeV} \quad \tau \sim 2 \times 10^8\,\text{yr}, \tag{2.28}$$

$$^{15}\text{O} \rightarrow {}^{15}\text{N} + \text{e}^+ + \nu_e + 2.76\,\text{MeV} \quad \tau \sim 1.8 \times 10^2\,\text{s}, \tag{2.29}$$

$$^{15}\text{N} + \text{p} \rightarrow {}^{12}\text{C} + {}^4\text{He} + 4.97\,\text{MeV} \quad \tau \sim 10^4\,\text{yr}. \tag{2.30}$$

This sequence releases a total energy of 26.7 MeV (Wiescher et al., 2010). Note that the slowest reaction in the cycle (i.e., the fourth in the sequence) requires that the other nuclei from the preceding reactions are converted into ^{14}N. Moreover, we notice that the ^{12}C needed in the first reaction is regenerated in the last reaction, hence the cyclical nature of the process.

2.3.3   The triple–alpha process

Once all the hydrogen in the core of a star has been transformed into helium, fusion reactions come to an interim halt. Hydrostatic equilibrium cannot be maintained, and the star starts contracting and heating up. Eventually, temperatures rise up sufficiently for hydrogen to restart fusing in a shell surrounding the temporarily inert helium core. Pressure then pushes the outer layers of the star outward, causing it to leave the main sequence, and become a *red giant*. For the Sun, this transition will commence approximately five to six billion years from now (Lamers and Levesque, 2017, Figure 16.1).

When the temperature in the core is $T \sim 10^8$ K, two helium nuclei can fuse to ^{8}Be, which is very unstable and decays with a half-life of $\sim 10^{-16}$ s. A small fraction of beryllium nuclei, however, manages to fuse with another helium nucleus, thereby producing carbon and sufficient energy to restore hydrostatic equilibrium. The overall process is

$$^4\text{He} + {}^4\text{He} + 92\,\text{keV} \rightarrow {}^8\text{Be}, \tag{2.31}$$

$$^8\text{Be} + {}^4\text{He} \rightarrow {}^{12}\text{C}^* \rightarrow {}^{12}\text{C} + 7.4\,\text{MeV}. \tag{2.32}$$

Although the first reaction requires energy to proceed, the net energy generated via this process is approximately 7.3 MeV. The chain of reactions, (2.31) and (2.32), is called the triple–alpha process, as it involves the fusion of three helium nuclei, also termed *alpha particles*.

The triple-alpha process is conventionally deemed very unlikely (compare Adams and Grohs, 2017). It only occurs in significant amounts in the Universe because an excited state of carbon (^{12}C*) – whose energy (7.65 MeV) is relatively close to the energy associated with the ^{8}Be+^{4}He reaction – called the *Hoyle state* exists (Freer and Fynbo, 2014); carbon is thus first produced in the excited state, and it subsequently decays into the ground state, releasing energy. As a side product of the triple-alpha process, ^{12}C can fuse with helium to produce a stable, and the most abundant, isotope of oxygen, ^{12}C+^{4}He→^{16}O, with an energy release of 7.16 MeV.

The triple–alpha process is especially important for astrobiology, as it allows the production of two of the most relevant atomic elements for life: C and O. However, further processes have to take place in order to synthesise the remaining elements of the periodic table.

2.3.4   The synthesis of heavier nuclei

After helium also runs out, the interior of a red giant star contains an inert core made of carbon and oxygen. At this stage, low mass stars cannot reach temperatures high enough to ignite further nuclear reactions, and so they end up as white dwarfs. The cores of stars with masses larger than $M_\star \simeq 8M_\odot$, however, can compress enough to reach temperatures of $T \sim 6 \times 10^8$ K, enough for fusion reactions to proceed beyond carbon. A sequence of successive burning stages ensues, starting with the fusion of carbon nuclei into neon, followed by the fusion of neon, magnesium, and silicon.

Each stage occurs faster than the previous one, with the silicon burning phase merely lasting about one day. During these aptly named *advanced burning phases*, heavier and heavier nuclei are built, and the products of previous reactions provide the fuel for the subsequent ones. Additional nuclei are also produced through reactions involving the capture of helium nuclei, known as *alpha processes*. A fully evolved massive star has an internal structure made of concentric shells, with temperatures declining from the innermost to the outermost, in which different burning phases prevail; this structure is evident on inspecting Figure 2.3.

In the central core, iron accumulates through silicon burning. It is the final product of fusion in stars, because iron has the highest binding energy per nuclear particle. In other words, nuclei heavier than iron are liable to break down, while elements lighter than iron are favoured to fuse together and synthesise iron. Once the core of the star is totally composed of iron, a final catastrophic collapse takes place, triggering a supernova explosion and the formation of either a neutron star or a black hole as the end product, depending on the initial mass of the star.

Because fusion in stars cannot proceed beyond iron, other mechanisms must be invoked to explain the synthesis of even heavier elements. The most likely mechanism involves successive capture of

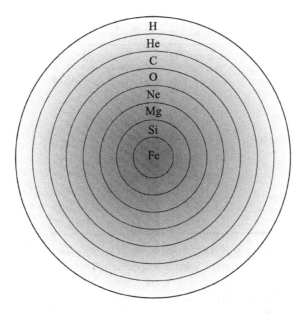

Figure 2.3 Multiple burning shells (i.e., advanced burning phases) of a massive star towards the end of its lifetime, showing only the dominant elements, collectively constituting an onion-like structure.

neutrons by pre-existing nuclei, which can then be followed by the decay of neutrons into protons. One such route, rapid neutron capture (or *r-process*), requires an intense source of free neutrons that could be provided by a supernova explosion or the wind from newborn neutron stars. A similar, but alternative, route is slow neutron capture (or *s-process*), which may occur chiefly inside asymptotic giant branch (AGB) stars.[5] The r-process and s-process might have each produced roughly one-half of the nuclei heavier than iron in the Universe.

2.4   The formation of molecules

As stellar nucleosynthesis enriched the Universe with atomic elements, the possibilities for the formation of the first molecules diversified over time. Although hydrogen and helium are still overwhelmingly more abundant than any other atom, elements like carbon, oxygen, and nitrogen are prevalent in significant amounts; they are, in fact, among the most common in the Universe. The chemical composition of low-mass stars broadly resembling the Sun is summarised in Table 2.2. Many of the elements therein are the building blocks of organic compounds, and their universal availability has profound implications for astrobiology.

Interstellar space is certainly a challenging environment for chemistry to operate, but we have discovered that the synthesis and survival of molecules is not only possible but can also lead to complex products. Molecules in space are detected remotely by astronomical observations of their emission and absorption lines, as elaborated in Section 12.2. These observations are mostly carried out at long wavelengths, ranging from the infrared to radio. Major advancements in collecting data

---

[5] AGB stars are stars of less than eight solar masses at late stages in their evolutionary history comprising a carbon–oxygen core, an intermediate He shell, and an outer layer of H.

Table 2.2 Chemical composition of stars in the Milky Way disc

Element	Atomic number (Z)	Abundance
Hydrogen (H)	1	0.00
Helium (He)	2	−1.01
Oxygen (O)	8	−3.07
Carbon (C)	6	−3.44
Neon (Ne)	10	−3.91
Nitrogen (N)	7	−3.95
Magnesium (Mg)	12	−4.42
Silicon (Si)	14	−4.45
Iron (Fe)	26	−4.46
Sulfur (S)	16	−4.79

Note: The abundance of element X is expressed as $\log_{10}(N_X/N_H)$, where N_X and N_H are the number of atoms of X and H, respectively. These *approximate* values are adopted from Lamers and Levesque (2017, Table 1.1), where it was assumed that stars in the Milky Way disc have a loosely similar composition.

have been made in the last decade through the use of new facilities from the ground (e.g., *ALMA*), in the stratosphere (e.g., *SOFIA*) and in space (e.g., *Spitzer* and *Herschel*).

As of 2021–2022, astronomical surveys detected 241 individual molecular species in the interstellar medium, composed of 19 different elements and ranging in size from 2 to 70 atoms (McGuire, 2022). Many of them are of direct astrobiological significance (see Table 2.3), which immediately begs the question: where and how did they form? This is the subject of *astrochemistry*, a field that has made impressive strides in recent years, both in theoretical understanding and in laboratory experiments.

From a theoretical perspective, elucidating the synthesis of molecules in interstellar regions requires a versatile toolbox of modelling techniques (see Öberg, 2016; Biczysko et al., 2018 for overviews). Time-consuming approaches can involve ab initio calculations of reaction potentials, photodissociation cross sections, excitation coefficients, and so forth, which are subsequently fed into molecular dynamics simulations or kinetic models to calculate how single atoms, ions, or molecules interact. Higher-level and less-demanding techniques deal with average properties of chemical species populations (such as the so-called Monte Carlo methods), or solve rate equations to determine time-dependent abundances based on ambient physical conditions.

Experimental studies are an essential complement to theoretical models, and such approaches have been pursued in earnest for a longer duration. Detailed laboratory measurements of molecular spectra and excitation properties are needed in order to properly identify the fingerprints of diverse chemical species in astronomical observations. Furthermore, many gas-phase and solid-state reactions that are anticipated to occur in interstellar space may be simulated in laboratory conditions, where their rates and expected products are determined in controlled settings.

Astrochemistry is a vast and rapidly evolving topic, and we will only sketch some of its findings in the following sections. More details are provided in the state-of-the-art reviews by Herbst and van Dishoeck (2009), Öberg (2016), Jørgensen et al. (2020), Sandford et al. (2020), and Öberg and Bergin (2021).

Table 2.3 Selective list of molecules of astrobiological significance identified in interstellar space

Molecule	Name	Significance
HCN	Hydrogen cyanide	Building block for prebiotic chemistry
HCHO	Formaldehyde	Building block for sugars
H_2C_2O	Ketene	Building block for prebiotic chemistry
CH_3OH	Methanol	Intermediate in interstellar chemistry
$HCONH_2$	Formamide	Building block for prebiotic chemistry
CH_3COOH	Acetic acid	Member of short-chain fatty acids
CH_2OHCHO	Glycolaldehyde	First interstellar (pre-)sugar
$(CH_3)_2CO$	Acetone	Building block for prebiotic chemistry
CH_3CHCH_2O	Propylene oxide	First interstellar chiral molecule
CH_3CONH_2	Acetamide	Molecule with peptide (–CO–NH–) bond
C_3HN	Cyanoacetylene	RNA precursor/building block
NH_2CONH_2	Urea	RNA precursor/building block
NH_2OH	Hydroxylamine	RNA precursor/building block
NH_2CH_2CN	Aminoacetonitrile	Amino acid precursor (of glycine)
$NH_2CH_2CH_2OH$	Ethanolamine	Precursor of (phospho)lipids

Note: The list of molecules is chiefly constructed from McGuire (2022, Tables 2 & 3). Some of the molecules take part in multiple pathways leading to the synthesis of the building blocks of life, owing to which their description is generic. For other molecules, their significance is easier to express in specific terms (as done above). RNA (ribonucleic acid), sugars, fatty acids, lipids, peptides, amino acids, and (homo)chirality are covered in Section 5.1.

2.4.1   The sites of molecule formation

Molecule formation is a variegated process, and one that does not merely occur in a small, selective, and isolated environment. It is, rather, part of a vast and complex cycle that involves myriad locations, and a broad range of physical conditions (as depicted in Figure 2.4).

1. *Stellar ejectae:* During their life, and especially in their terminal stages, stars lose significant fractions of their mass to the surrounding space, either via gradual outflows or through violent supernova explosions. Some chemistry is operational at these stages, resulting in a wide array of products based on the ambient conditions. When low mass stars exhaust hydrogen in the core and enter the red giant phase, the outer shells expand, and are typically ejected over the course of $\sim 10^4$ years. Chemical reactions occurring in the outflow can form simple compounds such as carbon monoxide (CO) and – depending on the relative abundance of carbon and oxygen in the expelled stellar material – silicates (i.e., compounds comprising Si and O) and other minerals (if C/O < 1), or organic molecules and carbides like SiC or TiC (if C/O > 1). More complex compounds, notably polycyclic aromatic hydrocarbons (PAHs), may also be synthesised in stellar winds during the last stages of carbon-rich stars.

    Massive stars eject their material more violently, that is, as supernovae explosions. As mentioned in Section 2.3.4, this stage is of great importance for the synthesis of heavier nuclei. However, while the shock waves produced by supernovae can increase the density in the gas, and provide

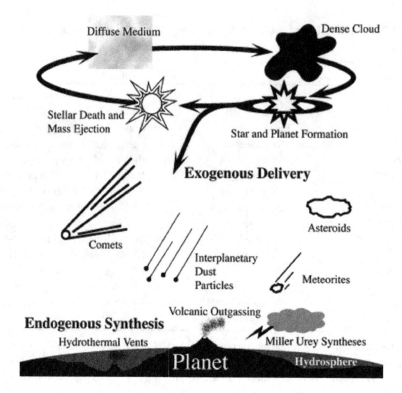

Figure 2.4 The multifaceted cycle of complex molecule formation in varied environments, and the delivery of such molecules to terrestrial planets. (Credit: Deamer et al., 2002, Figure 1; reproduced with permission of Mary Ann Liebert, Inc. publishers)

enough energy for some simple chemical reactions involving neutral molecules to occur, this environment is, in general, not ideal for complex chemistry.

2. *Diffuse interstellar medium:* The diffuse interstellar medium (ISM) has a very low number density ($n \sim 10^2$ cm^{-3}) and temperature ($T \sim 100$ K). It mostly contains atoms, in addition to a few simple molecular species like CO, CH, OH, and CN. As the diffuse ISM is subject to high fluxes of stellar radiation and shocks from supernova explosions, small molecules are easily ionised or destroyed. The energy of typical covalent bonds (with shared electrons) corresponds to radiation in the ultraviolet (UV) region of the electromagnetic spectrum. Because the diffuse medium is mostly transparent to UV photons, the splitting of a molecular bond via the absorption of a photon with the right energy (or photodissociation, see Figure 2.5) is effective in this setting. Therefore, only robust complex molecules such as PAHs are capable of long-term survival in gas phase in the ISM. Most other matter that does not exist in atomic form is found as either silicate or carbonaceous (carbon-rich) grains.

3. *Molecular clouds:* The most interesting environment for astrochemistry and complex molecule formation is the dense cores of molecular clouds. As we saw earlier, this region is also where star formation takes place. Various polyatomic species have been observed in such clouds, where many of them are carbon-chain molecules such as HC$_{2n+1}$N, H$_2$C$_n$, C$_n$O, C$_n$S, HC$_n$, with n usually of order unity (Millar, 2015). From a chemistry standpoint, the low temperature ($T \sim 10$ K) and density ($n \sim 10^4$–10^6 atoms per cm^3) of molecular clouds make gas-phase reactions rather ineffective. On the other hand, these clouds have the advantage of being opaque to visible and

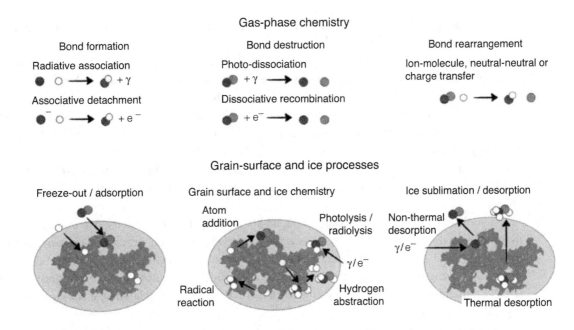

Figure 2.5 Schematic of common gas-phase and grain-surface chemical reactions in molecular clouds. (Credit: Öberg and Bergin, 2021, Figure 3; reproduced with permission from Elsevier)

UV radiation (hence, the frequent designation 'dark clouds'), thereby protecting small molecules from photodissociation. Moreover, most volatile species (i.e., which vaporise at relatively modest temperatures) will condense to form ice mantles around dust grains, whose surfaces function as the sites of interesting chemistry.

4. *Protostars and protoplanetary discs:* The material processed in dense molecular clouds can be subsequently incorporated in newly formed stars and the surrounding protoplanetary discs. Additional chemistry takes place in such environments (Jørgensen et al., 2020), and the resulting material may constitute the bulk of planets. Further delivery on planetary surfaces continues afterwards, through dust and meteorites, as well as larger impacts from asteroids and comets. After the death of the host star, the expelled material is once again dispersed into the ISM and in dense clouds, and the cycle continues, as illustrated in Figure 2.4.

For the purpose of this section, we will restrict our attention to the chemistry of dark molecular clouds, deferring a detailed treatment of the conditions in protoplanetary discs and their associated chemistry to Section 3.1. In the conditions prevalent in molecular clouds, there are two main paths to the formation of complex molecules (see Figure 2.5): (1) reactions occurring in gas-phase, and chiefly driven by cosmic rays that are able to penetrate the clouds; (2) reactions involving the catalysing and concentrating pathways offered by dust grains. We will outline these two paths hereafter.

2.4.2   Gas-phase chemistry

Although dense molecular clouds are effectively screened from external stellar radiation, cosmic rays (viz., protons and ions accelerated close to the speed of light by high-energy mechanisms) can easily penetrate them, and generate secondary UV radiation via the excitation and de-excitation of

H_2 molecules. The primary and secondary ionising radiation from cosmic rays may, in turn, produce ions in the gas phase, such as H^+, H_2^+, He^+, and C^+.

Ionisation is a crucial prerequisite for molecule formation in dense clouds, as it could greatly enhance the likelihood that random collisions result in chemical reactions. Reactions between neutral atoms generally require large activation energies, and are therefore less likely to take place in the cold environment of molecular clouds. Furthermore, although densities are higher than in the diffuse ISM, collisions are still very infrequent, so that radiative association reactions (refer to Figure 2.5) have extremely low rates; the reason is because the typical dissociation timescale ($t_{diss} \sim 10^{-13}$ s) is much smaller than the time it takes for the molecule to radiate away its bond energy ($t_{rad} \sim 10^{-3}$ s) (Öberg and Bergin, 2021), thus needing an extremely large number of collisions for this mechanism to be effective.

Ion–molecule reactions, on the other hand, do not typically necessitate activation energy because they are often exothermic (i.e., they release heat), so that their reaction rates tend to be temperature-independent, and are essentially dictated by the collision rates; the latter are as high as 10^{-9} cm^3 s^{-1}. They can consequently lead to the formation of noteworthy molecules such as hydrogen cyanide (HCN), formaldehyde (HCHO), and methanol (CH_3OH), which are vital building blocks for complex organic compounds.

An example of such processes in action is the assemblage of hydrocarbon molecules (Millar, 2015; Herbst, 2017), which are kickstarted by the initial ionisation of H_2 by cosmic rays (cr):

$$H_2 + cr \rightarrow H_2^+ + e^-, \tag{2.33}$$

which then reacts with more H_2 forming H_3^+:

$$H_2^+ + H_2 \rightarrow H_3^+ + H. \tag{2.34}$$

This ion enables the formation of carbon–hydrogen bonds via

$$C + H_3^+ \rightarrow CH^+ + H_2, \tag{2.35}$$

followed by further reactions of CH^+ with H_2 to form CH_3^+ and then CH_5^+. The latter does not react directly with H_2, but rather produces CH_2, CH_3, and CH_4 via dissociative recombination (see Figure 2.5), as illustrated below:

$$CH_5^+ + e^- \rightarrow CH_4 + H. \tag{2.36}$$

The growth of hydrocarbon molecules may then proceed by adding one C or C^+ at a time, which eventually leads to ions like $C_mH_n^+$; the latter, after undergoing dissociative recombination, can thus give rise to linear carbon chains with m carbon atoms (expressed as C_m).

Interestingly, reactions of O with H_3^+ may proceed akin to what we just witnessed with C, paving the way for the production of water via:

$$H_3^+ + O \rightarrow OH^+ + H_2, \tag{2.37}$$
$$OH_n^+ + H_2 \rightarrow OH_{n+1}^+ + H, \tag{2.38}$$
$$OH_3^+ + e^- \rightarrow H_2O + H. \tag{2.39}$$

A comment regarding the formation of the previously introduced PAHs is warranted. These molecules are observed in a variety of astrophysical environments, ranging from protoplanetary discs around young stars to the interstellar medium of our Milky Way, and even in external galaxies (Li, 2020; Kipfer et al., 2022). Polycyclic aromatic hydrocarbons are estimated to contain 10–25% of all interstellar carbon, and are of crucial importance to astrobiology, since they might have played a role in the origin(s) of life as precursors to relevant biological molecules. Their multiple carbon–carbon double bonds make PAHs more stable against photodissociation in the ISM compared to most other molecules. However, how and where PAHs are synthesised is currently unclear. As stated earlier, one scenario involves carbon-rich star outflows, followed by injection into the ISM. It has also been suggested that PAHs form directly in the ISM via gas-phase ion–molecule reactions.

2.4.3   Grain-surface chemistry

Although gas-phase reactions, especially those with ion–molecule components, can produce many simple molecules observed in interstellar space, they cannot fully explain the observed abundances, and especially the synthesis of certain complex compounds (see Table 2.4). It is now believed that many important reactions take place in the solid phase, to wit, on the surface of small, cold dust grains (at $\sim$10 K in dense clouds), which are made of silicates and carbonaceous material. Grain sizes range from $\sim$1 nm (comparable to the size of a large molecule) to roughly 1 μm. The size distribution is approximately (size)$^{-3.5}$, that is, an inverse power law, so that there are far more small grains, although most of the mass is locked in the bigger ones.

In the environment of cold dark clouds, all volatile species, except for H_2, He, and their ions, freeze and condense on the surface of dust grains, resulting in the formation of a layered ice mantle containing myriad compounds of O, C, and N in these settings. The composition of such mantles has been ascertained via space-based infrared spectroscopy, which revealed water ice to be the dominant component, followed by CO, CO_2, CH_3OH, NH_3, and CH_4 (Öberg et al., 2011). A host of appealing properties of dust grains make them promising sites for astrochemistry. The most notable is their capability to aggregate chemical species on their surface, thereby enhancing concentrations and, therefore, the likelihood and rates of chemical reactions. The chief factor that limits species interactions is their diffusion on the grain surface and their mobility in ice.

Table 2.4 Representative examples of biomolecular building blocks produced from ices in laboratory experiments

Composition	Ice T	Irradiation	Study
Sugars	12 K	UV	Nuevo et al. (2018)
Amino acids	12 K	UV	Muñoz Caro et al. (2002)
Nucleobases	10 K	UV	Oba et al. (2019)
Lipid-like molecules	15 K	UV	Dworkin et al. (2001)
Sugars	5.5 K	Electrons	Maity et al. (2015)
Amino acid(s)	10 K	Electrons	Holtom et al. (2005)
Nucleobase precursors	5 K	Electrons	Marks et al. (2022)
Lipid-like molecules	10 K	Electrons	Zhu et al. (2020)
Amino acid(s)	13–14 K	None	Ioppolo et al. (2021)

Note: Ice T is ice temperature. This list is not definitive because many other experimental studies abound in the literature, as reviewed in Sandford et al. (2020). Nucleobases, sugars, lipids, and amino acids are covered in Section 5.1.

Another crucial aspect of gas-grain chemistry is the catalysing action of dust grains. This feature is exemplified by the formation of H_2 from two hydrogen atoms, which is strongly suppressed in gas phase, at the low densities of interstellar space, because the excess energy of a molecule formed in an excited state is not efficiently radiated away, leading to its rapid dissociation (i.e., returning back to square one). It has long been recognised that, by absorbing this excess energy, the mediation of a dust grain could increase the formation rate, and thus account for the high observed abundance of interstellar H_2 (see, for example, Wakelam et al., 2017).

The general mechanism for gas-grain chemistry involves the accumulation of one or more atoms, molecules, or ions on the surface of a grain (known as *adsorption*), their subsequent reaction, and, finally, the *desorption* (i.e., the inverse of adsorption) of the reaction products (shown in Figure 2.5). For desorption to occur, a molecule has to gain enough energy to overcome the activation barrier that hinders its escape. This process can happen either thermally – with a rate $\propto \exp(-E_d/k_B T)$, where E_d is the minimal energy for desorption – or through non-thermal processes. Thermal desorption is favoured in warmer regions, such as in protostellar envelopes. Hypervolatile species like CH_4, CO, and N_2 readily sublimate at $T \sim 20\text{–}40$ K, depending on the pressure, whereas volatile species such as water, CH_3OH, CO_2, and NH_3 require higher temperatures. Non-thermal desorption mechanisms dominate at the low temperatures of cold cores of molecular clouds: these include photodesorption (i.e., interaction with a photon of the right energy), electron stimulated desorption, and cosmic ray bombardment.

Gas-grain chemistry is, at least to begin with, dominated by H atom addition. Besides H_2, reactions in regions predominantly composed of atomic H yield hydrides like CH_3OH, CH_4, H_2O, and NH_3. The two main types of gas-grain reactions are the *Langmuir–Hinshelwood* (LH) and the *Eley–Rideal* (ER) mechanisms. In LH reactions, two species stick to the surface of the grain and then diffuse across the surface, until they come into contact and react. In ER reactions, a gas-phase species collides with a reactant that was previously adsorbed on the surface. A third type of reaction, the so-called *hot atom reaction*, is also possible: in this case, a gas-phase species lands on the surface and travels along it due to its high energy (hence 'hot atom'), until it encounters and reacts with an adsorbed species.

An in-depth understanding of gas-grain chemistry is difficult because multiple factors are still poorly constrained, including some physical properties of the grains themselves. Theoretical modelling is also challenging, as it canonically calls for stochastic approaches that are computationally intensive. In spite of these caveats, however, significant progress has been accomplished, both in numerical methods (Biczysko et al., 2018) and in laboratory studies (Linnartz et al., 2015). In particular, numerous laboratory studies have shown that the exposure of ices to high-energy radiation and particles may produce a vast assortment of interesting compounds.

The typical setup of such experiments entails very low temperatures (5–20 K) and very low pressures ($<10^{-7}$ bar), wherein simple gas-phase molecules that are common in the ISM (such as H_2O, NH_3, HCN, CO, CO_2, CH_3OH) are deposited onto a solid substrate. This mixture is subsequently irradiated by either UV radiation or charged particles (similar to the action of cosmic rays). A summary of select noteworthy compounds of astrobiological interest (including amino acids) produced in laboratories by such methods is provided in Table 2.4. Needless to say, these studies are exceedingly relevant for advancing astrobiology, since they could help clarify the origin(s) of many prebiotic compounds, and whether they may have formed outside Earth, and were eventually delivered to our planet.

To round off this chapter, let us take a step back and take stock of where we stand. In a nutshell, we have witnessed how the emergence of 'order' (or 'structure') was facilitated in the Cosmos, with an emphasis on the genesis of stars, chemical elements, and molecules (some of which are the building blocks of life itself); intriguingly, the former are essential prerequisites, in one form or another, for the latter duo. However, the potential nurseries of life are planets, as well as moons and small bodies, owing to which we must next comprehend the intricate physical and chemical mechanisms mediating the formation of planets; the forthcoming chapter spans this broad topic.

2.5   Problems

Question 2.1: An important quantity in systems dominated by gravity is the free-fall time, which embodies the characteristic time for a body to collapse under its own gravity in the absence of other forces. An exact calculation of the free-fall time can be performed in the case of a collapsing zero-pressure uniform sphere of gas, with initial density ρ_0 and radius r_0. Suppose that a gas particle of mass m, initially at rest at distance $r = r_0$ from the centre, starts falling inward at time $t = 0$, arriving at $r = 0$ when $t = t_{ff}$. At each instant, the particle experiences a gravitational force as if all mass $M(r)$ internal to r is concentrated at the centre. From the conservation of energy, show that $t_{ff} = (3\pi/32G\rho_0)^{1/2}$.

Question 2.2: The sound speed of an ideal gas is $c_s = (\gamma_m k_B T/m)^{1/2}$, where $\gamma_m \sim 1$ is the adiabatic index (5/3 for monoatomic gases, 7/3 for diatomic gases). When a self-gravitating gas cloud of radius R is perturbed, the time it takes for pressure to react to compression is of order $t_s \sim R/c_s$. Demonstrate that a good estimate of the Jeans radius (2.20) can be obtained by requiring that the condition for the cloud to collapse is that its free-fall time (see the previous question) is smaller than the pressure reaction time, $t_{ff} < t_s$.

Question 2.3: Draw on the definitions of the Jeans radius and the Jeans mass to answer the following questions.

a. The Jeans mass M_J obeys the relationship $M_J \propto T^\alpha \rho^\beta$, where T and ρ are the cloud temperature and density, respectively. Determine the associated power-law exponents α and β, and then verify that $\alpha + \beta = 1$.

b. By perusing the discussion in Section 2.1.1, describe how the gas temperature and density scale with a (i.e., the scale factor). From these results and the expression for M_J, derive the scaling of Jeans mass with a and the cosmological redshift z; the latter is defined as $z = a^{-1} - 1$.

Question 2.4: The Sun has mass $M_\odot = 2 \times 10^{30}$ kg, radius $R_\odot = 7 \times 10^8$ m, and intrinsic luminosity $L_\odot = 3.83 \times 10^{26}$ W.

a. Estimate its average density and the free-fall time (see Question 2.1).

b. Next, compute the time required to radiate away its entire gravitational binding energy, which is encapsulated by the *thermal*, or *Kelvin–Helmholtz*, timescale, $t_{KH} \sim GM_\odot^2/R_\odot L_\odot$.

c. Finally, assume that nuclear fusion reactions convert a fraction 0.007 of the mass of hydrogen into energy via the renowned mass-energy equivalence formula $E = mc^2$, and compute the time needed to convert all the Sun's mass into energy at the current luminosity in this fashion.

d. How do these three timescales compare with each other, and with the expected lifetime of the Sun, which is given by (2.16)? Comment on the significance of your findings.

Question 2.5: Consider the following five elements: (a) carbon; (b) phosphorus; (c) iron; (d) molybdenum; and (e) uranium. For each of these elements, briefly answer these questions: What is the significance of the given element in biology or human technology on Earth (you may consult Section 5.1 if necessary)? What is the predominant channel through which this element is synthesised in the Cosmos? Summarise how this channel works.

If you utilise any internet sources, ensure that they are peer-reviewed scholarly publications, and cite them accordingly.

Question 2.6: Estimate the temperatures at which the following species would be dissociated or ionised (the required energies are given in parenthesis): (a) H (13.6 eV); (b) He (24.59 eV); (c) C (11.26 eV); (d) C-C (3.6 eV/bond); (e) C-O (3.99 eV/bond); and (f) C-H (4.3 eV/bond). Are any of these values higher than the effective temperature ($\sim 5{,}780$ K) of the Sun?

Question 2.7: The frequency of collisions per unit volume between particles in a gas goes as $\propto n^2 T^{1/2}$, where n is the number density and T is the temperature. Estimate the frequency of collisions in the diffuse ISM ($n \sim 10^2$ cm^{-3}, $T \sim 100$ K) and in dense molecular clouds ($n \sim 10^6$ cm^{-3}, $T \sim 10$ K) relative to air at sea level ($n \sim 10^{19}$ cm^{-3}, $T \sim 300$ K).

Question 2.8: The photodissociation rate of the CO molecule by UV radiation in the unattenuated interstellar radiation field is $R_{\text{diss}} \approx 2 \times 10^{-10} n_{\text{CO}}$ s^{-1} (Visser et al., 2009), while its formation rate in gas phase from collisions is $R_{\text{form}} = n_{\text{C}} n_{\text{O}} \sigma_{\text{CO}} v$ (Öberg and Bergin, 2021, pg. 5), where n_{X} represents the number density of species X, $\sigma_{\text{CO}} \sim 10^{-16}$ cm^2 is the cross section for the CO reaction, and $v \sim 10^5$ cm/s (corresponding to a temperature of $T \sim 100$ K). Assuming that $n_{\text{C}} \sim n_{\text{O}} \sim 10^{-2}$ cm^{-3}, calculate the number density of CO at equilibrium, that is, when $R_{\text{diss}} = R_{\text{form}}$. Does the result suggest that the formation of CO in the ISM is an efficient process?

Question 2.9: A dust grain has a mass of $\sim 10^{12}\, m_H$ (where $m_H \approx 1.7 \times 10^{-27}$ kg is the mass of a hydrogen atom). When molecular hydrogen (H$_2$) is formed from two hydrogen atoms on the surface of a grain, the release of the molecule bond energy ($E \approx 4.5$ eV) thrusts the molecule into space. Assuming that momentum is conserved, and that all the bond energy is converted into kinetic energy, estimate (a) the velocity of the molecule relative to the grain; and (b) the velocity change of the grain.

Question 2.10: It was stated that Table 2.3 is not exhaustive. Identify another molecule of prebiotic relevance discovered in interstellar space by perusing peer-reviewed sources. After citing the study, briefly describe the (a) telescope employed; (b) wavelength range surveyed; (c) spectral feature(s) detected; and (d) prebiotic significance of the molecule.

Question 2.11: By browsing peer-reviewed literature, cite two or more other publications (aside from Table 2.4) for amino acid synthesis via UV irradiation of interstellar ices and delineate: (a) ice temperature; (b) ice components; (c) UV flux or fluence (time-integrated flux); and (d) amino acid(s) detected.

3   Planet Formation and Migration

As far as we know, life is an intricate process that needs a planetary environment (broadly interpreted) to begin and to sustain itself. For this reason, this chapter is devoted to a detailed exposition of how planets form and evolve. This is a vast topic, whose understanding has developed considerably in the last few decades, and is still the subject of active and ground-breaking research. We will endeavour to condense a large body of knowledge in a relatively restricted space, thus following a pedagogical approach that will prioritise basic physical arguments rather than complete and accurate numerical calculations, which are arguably the only theoretical avenue to obtain a realistic picture of planetary formation, and to sort out the complicated interplay of phenomena involved. Our treatment will largely draw on the review by Armitage (2007), the lecture notes by Alexander (2022), and the advanced monograph by Armitage (2020). We urge the reader to consult those works for in-depth coverage of the relevant areas.

Historically, myriad hypotheses were proposed to explain the formation of the solar system and, as a corollary, the possible occurrence of planetary systems around other stars. Some of them were based on rare events, like close encounters between stars, and therefore had dire implications for the abundance of planets in the Galaxy. Perhaps the most notable among them was the *tidal theory* advanced by Sir James Jeans (Jeans, 1916), which postulated that the passage of a massive star thence stripped material from the Sun by way of tidal interactions, creating an orbiting filament that would be unstable and subsequently break down to form planets. In Jeans' view, such an event was potentially so uncommon that the formation of the solar system (and thereby Earth) might be possibly unique in the entire Galaxy. Variations of this catastrophic theme, entailing the capture of material via tidal interactions with a low-mass protostar, were developed and examined subsequently, even until as late as the mid- and late-twentieth century (e.g., Woolfson, 1964; Dormand and Woolfson, 1989).

Today, however, planetary formation is considered an almost inescapable consequence of star formation, and is therefore confirmed to be a common occurrence throughout the Galaxy, as corroborated by observations of extrasolar planets (Perryman, 2018). The present view is akin to the hypothesis laid down in the eighteenth century by the French polymath Pierre-Simon Laplace. In this scenario, called the *nebular hypothesis*, the same cloud that formed the Sun also produced a surrounding disc of gas and dust, within which clumps of matter grew and evolved to eventually become the planets. Despite its elegance, the nebular hypothesis initially struggled to explain how the total mass of the solar system was dominated by that of the Sun, while its angular momentum was almost fully contained in the orbital motion of the planets. This delayed the affirmation of the theory until, as we shall shortly witness, modern physical treatments managed to find a satisfactory solution to the conundrum. The discovery of extrasolar planets and observations of protoplanetary discs around young stellar objects have furnished conclusive evidence for the overall scenario, although many details of the picture are under active investigation.

This chapter is organised into three main subjects. First, we describe the structure and chemistry of the discs around forming stars that constitute the environment where planets take shape. Next, we explore the main physical mechanisms presiding planetary formation, and show how they can explain the rich variety of features observed in actual planets. Finally, we delve into the dynamical evolution of planetary systems after their genesis.

3.1   Protoplanetary disc structure and chemistry

Our own solar system is the obvious starting point for acquiring a more general understanding of planetary formation. Some key physical properties of the eight planets orbiting the Sun are summarised in Table 3.1.

A preliminary classification can be made into *terrestrial* (or *rocky*) planets and *giant* planets: the latter category is often further divided into *gas* and *ice* giants. Sizes and compositions are relatively homogeneous within each category, while markedly differing from one another. Rocky planets occupy the innermost region of the solar system: they have small masses and high densities, with thin atmospheres and a metallic core. Giant planets are found at larger orbital separations: they all have thick gaseous atmospheres, and are predominantly made of hydrogen and helium, with the addition of methane, ammonia, and water for the ice giants. The latter also comprise a comparatively larger solid core containing frozen volatiles (hence the name).[1] All planets orbit within a few degrees of the equatorial plane of the Sun and have small eccentricities (<0.1), with the notable exception of Mercury. The coplanarity of the solar system planets was an early line of evidence in favour of their formation from a single disc of material.

Most of the mass in the solar system is contained in the Sun, which holds true even if only elements heavier then helium, generically called *metals* in astronomical parlance, are considered. We know from observations that metals constitute a fraction $Z \simeq 0.02$ of the total solar mass, and it is easy to verify that $ZM_\odot \gg \sum M_p$, where M_p's are the planetary masses. Conversely, the angular momentum

Table 3.1 Key physical parameters of solar system planets

		a (AU)	T_p (yr)	$R_p/R_\oplus$	$M_p/M_\oplus$	ρ_p (g/cm^3)
Rocky planets	Mercury	0.387	0.24	0.38	0.06	5.4
	Venus	0.723	0.61	0.95	0.82	5.2
	Earth	1.000	1.00	1.00	1.00	5.5
	Mars	1.524	1.88	0.53	0.11	3.93
Gas giants	Jupiter	5.203	11.86	11.19	317.89	1.33
	Saturn	9.537	29.45	9.46	95.18	0.71
Ice giants	Uranus	19.191	84.02	4.01	14.54	1.24
	Neptune	30.069	164.79	3.81	17.13	1.67

Note: a – orbital semi major axis; T_p – sidereal orbital period (time taken for one orbit with respect to 'fixed' stars); R_p – mean radius; M_p – total mass; ρ_p – mean density; 1 AU $\approx 1.5 \times 10^8$ km; $R_\oplus = 6371$ km; $M_\oplus = 5.97 \times 10^{24}$ kg.

[1] Put simply, volatiles are chemical substances that can undergo vaporisation readily.

of the Sun's rotation, $\ell_\odot \simeq 3 \times 10^{48}$ g cm^2 s^{-1}, is much smaller than the angular momentum contained in planetary orbits; for Jupiter alone, it is $\mathcal{L}_J \simeq 2 \times 10^{50}$ g cm^2 s^{-1}.

Hence, we can highlight two lessons gleaned from the solar system that indicate general features of the planetary formation process:

1. The aggregation of mass into planets need not be very efficient, as it involves only a small fraction of the initial mass of the nebula.
2. Mass and angular momentum are differentiated during the formation of the planetary system, with most of the material ending up in the star and most of the angular momentum inherited by the planets.

We will expound these two features in the subsequent sections and show how they arise naturally from the physical processes involved in the nebular model of planetary formation.

3.1.1   The Minimum Mass Solar Nebula

We have seen in Section 2.2.2 that the process of star formation starts from the collapse of a nebula made predominantly of hydrogen and helium, enriched by small amounts of heavier elements and dust. We may now ask: how much of the initial mass should end up in a disc surrounding the star, providing the building blocks for planetary formation? Again, we can obtain some clues from the solar system, by computing an approximate lower bound for the gas and dust needed to form the planets. This is known as *Minimum Mass Solar Nebula* (MMSN), which is estimated as follows:

1. Quantify the mass of elements heavier than helium (i.e., confusingly labelled metals) contained in each planet.
2. Scale it appropriately in order to match the (relative) solar abundance of hydrogen and helium, and determine the enhanced planetary mass.
3. Calculate the sum of this boosted mass over all the planets.

The resulting mass is none other than the MMSN.

Rocky planets are almost entirely made of non-volatile materials (e.g., iron and silicates), comprising $\sim$0.4% of the solar composition. Therefore, their mass would have to be multiplied by a factor of $\sim$100%/0.4% = 250, with more precise estimates varying from $\sim$350 (Mercury) to $\sim$235 (Mars). Gas giants have very similar composition to the original nebula, merely requiring a small enhancement factor (5 for Jupiter and 8 for Saturn), while ice giants contain larger amounts of icy volatile compounds (such as water, ammonia, or methane, constituting $\sim$1.4% of the original nebula), and have enhancement factors of 15 (Uranus) and 20 (Neptune) (Lewis, 2004, Table IV.8). With these values, the resulting MMSN is $\sim$3500$M_\oplus$, or $\sim$0.01$M_\odot$.

An estimate of the MMSN surface density (denoted by Σ), that is, the projected mass per unit area at a given radial distance, can be obtained by distributing the enhanced mass computed for each planet over an annulus centred around the current planetary orbit, with the boundaries halfway between the neighbouring planets' orbits. The annulus centred on Earth, for instance, ranges from 0.86 to 1.26 AU (Lewis, 2004, Table IV.8). A widely adopted result for Σ was first derived by Hayashi (1981):

$$\Sigma = 1.7 \times 10^3 \text{ g cm}^{-2} \left(\frac{R}{1 \text{ AU}} \right)^{-3/2}, \tag{3.1}$$

where $R \equiv \sqrt{x^2 + y^2}$ is the radial distance (in the cylindrical coordinate system). More comprehensive theoretical models predict less steep slopes, namely, closer to $\Sigma \propto R^{-1}$. This scaling is supported by observations of actual protoplanetary discs (e.g., Zhang et al., 2017), which have become increasingly detailed over the past few decades. We shall scrutinise these findings further in the forthcoming section.

3.1.2   Protostellar envelopes and discs

Protoplanetary discs around newly formed stars have been widely observed in recent years, and in many cases with high enough resolution to detect small-scale features that are relevant for understanding the planetary formation process, such as rings and gaps (see, for example, Andrews et al., 2018).

Broadly speaking, the observed protoplanetary discs have sizes of tens to hundreds of AU, and masses of $\lesssim 10^{-3}\,M_\odot$ to $\gtrsim 0.1\,M_\odot$; as we have seen in Section 3.1.1, the estimated MMSN sits roughly in between this range. Active discs, powered by the inward flow of gas, have accretion rates ranging from $\lesssim 10^{-10}\,M_\odot\,\mathrm{yr}^{-1}$ to $\gtrsim 10^{-7}\,M_\odot\,\mathrm{yr}^{-1}$. The epoch when the disc stops collecting gas is roughly concurrent with the dispersal (clearance) of dust, which occurs when the protostar phase ends, and the star enters the pre–main sequence stage. When combined with the observed mass range, this feature suggests that disc lifetimes are on the order of ~ 1–10 Myr, but there are significant variations beyond this interval. The lifetime of protoplanetary discs sets a limit on the viable timescale for planet formation.

The formation of a disc from a collapsing (spherical) molecular cloud stems from the fact that the nebula carries some initial angular momentum. As the cloud shrinks, its rotational velocity has to increase to conserve angular momentum, thereby causing flattening wherein a fraction of the gas and dust is concentrated in a flat disc orthogonal to the axis of rotation. The presence of a large disc around a forming star can partly solve the so-called *angular momentum problem*, that is, the strong mismatch, of order $\sim 10^7$, between the angular momentum of a dense cloud core and that of a typical star (Ray, 2012). This issue can be mitigated by transferring much of the initial rotation to the disc: in fact, the characteristic angular momentum per unit mass of a cloud is comparable to that of a circular Keplerian orbit around a Sun-like star, $\sqrt{GM_\odot R}$, for an orbital radius of $R \sim 100$ AU.

The disc of gas and dust surrounding a young stellar object (YSO), namely, a developing star, leaves distinct imprints in its infrared (IR) emission spectrum. The shape of the IR spectral energy distribution (SED) is chiefly controlled by the parameter

$$\alpha_{\mathrm{IR}} = \frac{d\log(\lambda F_\lambda)}{d\log\lambda}, \tag{3.2}$$

where F_λ is connected to the Planck blackbody function B_λ (elaborated in Section 12.2.3), and the measurements are usually made in the near- and mid-IR (wavelengths of $2\,\mu\mathrm{m} \lesssim \lambda \lesssim 20\,\mu\mathrm{m}$). Different values of α_{IR} can be linked to successive stages of disc accretion, and constitute the basis of the empirical classification scheme for YSOs (Adams et al., 1987) in Figure 3.1.

- *Class 0:* No IR emission is detected, and α_{IR} is undefined. This corresponds to the initial phase ($t = 0$) when collapse is happening radially and the cloud is obscuring the YSO.
- *Class I:* $\alpha_{\mathrm{IR}} \gtrsim 0.3$. The SED shape is nearly flat, or slowly rising towards the mid-IR. A circumstellar disc has started forming, but the YSO is still surrounded by a substantial gaseous envelope ($10^4 \lesssim t \lesssim 10^5$ yr).

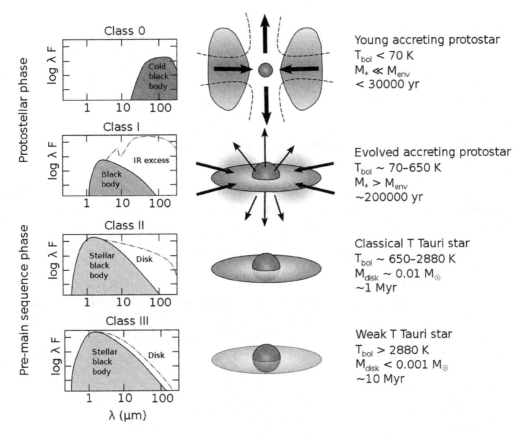

Figure 3.1 The four classes of the infrared spectral energy distribution (SED) of a young stellar object (YSO) shown alongside a simplified depiction of the corresponding evolutionary stages of the surrounding disc. (Credit: Wikipedia, https://commons.wikimedia.org/wiki/File:Evolution_of_young_stellar_objects.svg; CC-BY-SA 4.0 license)

- *Class II:* $-1.5 \lesssim \alpha_{IR} \lesssim 0$. The SED declines towards mid-IR wavelengths. This object is also labelled a 'classical T Tauri' star. The envelope has dissipated, and the YSO is surrounded by a thick and observable gaseous disc, often accompanied by energetic outflows occurring along the rotation axis ($10^5 \lesssim t \lesssim 3 \times 10^6$ yr).
- *Class III:* $\alpha_{IR} \simeq -1.5$. The gaseous disc has dissipated, leaving only a debris disc ($3 \times 10^6 \lesssim t \lesssim 5 \times 10^7$ yr). This object is also called a 'weak-lined T Tauri' star. There is virtually no excess IR emission, and the spectral signature of the YSO is essentially the same as a pre–main sequence star.

3.1.3 Disc density and temperature profile

Some insights concerning the structure of protoplanetary discs can be obtained from simple physical arguments (Armitage, 2020).

We first consider the vertical density profile $\rho(z)$ of the disc, choosing a reference frame where the star occupies the origin, the coordinate R is aligned with the middle plane of the disc, and the

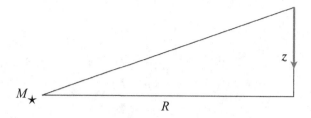

Figure 3.2 Coordinate system for a protoplanetary disc.

vertical coordinate z points towards the centre (as illustrated in Figure 3.2). We impose the condition of hydrostatic equilibrium along the vertical direction:

$$\frac{1}{\rho}\frac{dP}{dz} = -\frac{d\Phi}{dz}, \tag{3.3}$$

where P is the pressure and $\Phi = -GM_\star/\sqrt{R^2 + z^2}$ is the gravitational potential, with $M_\star$ denoting the mass of the central star. In qualitative terms, hydrostatic equilibrium is a statement about force balance, in which the inward-directed force of gravity is balanced by the outward force exerted by the (gas) pressure. Self-gravity of the disc can be ignored, since its mass is much smaller than $M_\star$, as implied by the MMSN estimate presented earlier. After using $z \ll R$, the previous equation simplifies to

$$\frac{1}{\rho}\frac{dP}{dz} = -\Omega_K^2 z, \tag{3.4}$$

where $\Omega_K \equiv \sqrt{GM_\star/R^3}$ is the Keplerian angular velocity.

Next, we can insert the equation of state $P = c_s^2 \rho$, where c_s is the speed of sound in the disc, and solve the differential equation. The final result (refer to Question 3.3) is a Gaussian density profile of the form

$$\rho = \rho_0 \exp\left(-\frac{z^2}{2H^2}\right), \tag{3.5}$$

where $H \equiv c_s/\Omega_K$ is the *vertical scale height*, and the middle plane (midplane) density $\rho_0 = \Sigma/(\sqrt{2\pi}H)$ follows from mass conservation, where Σ is the disc surface density. In general, the speed of sound will be a function of the location in the disc. If $c_s \propto R^{-\beta}$, it can be verified that $H/R \propto R^{-\beta+1/2}$. Thus, if $\beta < 1/2$ holds true, then the aspect ratio of the disc (H/R) increases further away from the centre, which is known as *disc flaring*.

The radial temperature profile of the disc can be calculated by assuming that the material surrounding the star absorbs all of the intercepted stellar radiation. We will consider a thin flat disc, and adopt spherical coordinates centred around a star of radius $R_\star$ and temperature $T_\star$, as shown in Figure 3.3. The flux absorbed by an infinitesimal disc element at radial distance R is given by $dF_{\mathrm{abs}} = I_\star \sin\theta \cos\phi \, d\Omega$, where $I_\star = \sigma_{\mathrm{SB}}T_\star^4/\pi$ is the surface brightness (viz., flux per unit solid angle) of the star, and σ_{SB} is the Stefan–Boltzmann constant. The factor $\sin\theta \cos\phi$ is added because we are specifically interested in the flux intercepted by the disc element; loosely speaking, it may be envisioned as the 'x-component' of the total flux. Note that $d\Omega = \sin\theta \, d\theta \, d\phi$ is the infinitesimal solid angle.

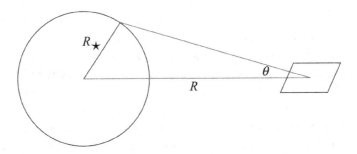

Figure 3.3 Geometry of incident radiation from a star (left) on an infinitesimal part of the protoplanetary disc (right).

The total absorbed flux F_{abs} is found by integrating dF_{abs} as follows:

$$F_{\text{abs}} = I_\star \int_{-\pi/2}^{+\pi/2} \cos\phi \, d\phi \int_0^{\Theta} \sin^2\theta \, d\theta. \tag{3.6}$$

Limits on ϕ are dictated by the setup geometry, as only radiation from the upper hemisphere is intercepted by the upper surface of the disc. The angle Θ is set by Figure 3.3, which implies $\sin\Theta \approx R_\star/R$. Finally, in thermodynamic equilibrium, the absorbed flux has to equal the radiated flux $F_{\text{rad}} = \sigma_{\text{SB}}T^4$, where T is the disc temperature, allowing us to solve the integral (3.6) to determine the radial temperature dependence (consult Question 3.4):

$$\left(\frac{T}{T_\star}\right)^4 = \frac{1}{\pi}\left[\sin^{-1}\left(\frac{R_\star}{R}\right) - \frac{R_\star}{R}\sqrt{1 - \left(\frac{R_\star}{R}\right)^2}\right], \tag{3.7}$$

which, in the physically meaningful limit of $R_\star/R \ll 1$, reduces to

$$T \propto R^{-3/4}. \tag{3.8}$$

If we suppose that $c_s^2 \propto T$, then it can be shown that $H/R \propto R^{1/8}$, which fulfils the flaring criterion derived earlier. A more sophisticated treatment would account for the effect of flaring itself, which enables the outer regions of the disc to intercept a larger fraction of stellar radiation. The outcome is a flatter temperature profile roughly characterised by $T \propto R^{-1/2}$.

3.1.4 Disc evolution

The orbital velocity $v(R)$ of gas in the disc must obey the force balance between centripetal acceleration, gravity, and radial pressure gradient:

$$\frac{v^2}{R} = \frac{GM_\star}{R^2} + \frac{1}{\rho}\frac{dP}{dR}. \tag{3.9}$$

If pressure decreases outward as a power law, $P \propto R^{-n}$, with $n > 0$, it is easy to demonstrate (see Question 3.6) that

$$v = v_K(1 - \eta)^{1/2}, \tag{3.10}$$

where $v_K = \Omega_K R = \sqrt{GM_\star/R}$ is the Keplerian velocity, and we have introduced $\eta \equiv n(c_s/v_K)^2 = n(H/R)^2$. Therefore, in the thin disc approximation corresponding to $H/R \ll 1$, η can be neglected and the gas orbits at a speed nearly equal to the Keplerian velocity.

If the gravitational potential remains constant in time, the surface density Σ cannot change, unless angular momentum is redistributed within the disc; in other words, some portions of the gas can gain angular momentum and move outward only at the expense of other portions that lose angular momentum and move inward: such process is usually described as *viscosity*. It should be recognised, however, that this term is a bit of a misnomer, as inter-molecular forces are not responsible for generating this effective viscosity. Rather, angular momentum transport is generally associated with turbulence in the disc, which creates an effective viscosity of the form

$$\nu = \alpha c_s H, \tag{3.11}$$

where $\alpha \leq 1$ is a dimensionless parameter that encapsulates the magnitude of turbulence in the given astrophysical system (Shakura and Sunyaev, 1973). A typical value for protoplanetary discs would be $\alpha \sim 0.01$.

Analytic solutions for the evolution of the surface density Σ can be found if the disc viscosity ν scales as a power law of the radial distance (Lynden-Bell and Pringle, 1974). When ν is constant, the evolution of Σ is such that, as $t \to \infty$, the mass flows towards the centre, while the angular momentum is transported outward – this behaviour correctly reproduces the mass/angular momentum segregation that we alluded to previously. More generally, for $\nu \propto R^\gamma$, one obtains self-similar solutions, whose qualitative behaviour is shown in Figure 3.4. Initially, the surface density mirrors the steady-state solution $\Sigma \propto R^{-\gamma}$ for $R < R_0$ (a characteristic distance), and decays exponentially beyond that radius. At later times, the disc mass decreases, while its typical size increases to conserve angular momentum. The characteristic timescale of the process is estimated to be

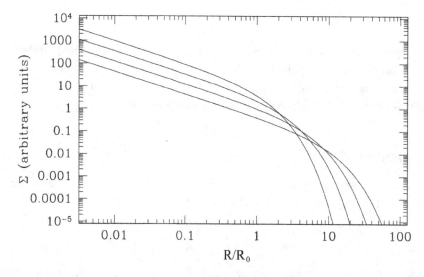

Figure 3.4 Self-similar solutions for $\nu \propto R$: from top to bottom, the respective curves correspond to times of $t = 0$, $t = t_\nu$, $t = 3t_\nu$, and $t = 7t_\nu$. (Credit: Modified from Armitage, 2020, Figure 3.4; reproduced with permission from Cambridge University Press)

$$t_\nu = \frac{R_0^2}{3(2-\gamma)^2\nu_0},$$ (3.12)

where $\nu_0 \equiv \nu(R_0)$. The inward flow of the mass occurs with an accretion rate given by the formula

$$\dot{M} = \frac{M_0}{2(2-\gamma)t_\nu}\left(\frac{t}{t_\nu}+1\right)^{-(2.5-\gamma)/(2-\gamma)},$$ (3.13)

where M_0 represents the initial disc mass.

A final note is warranted with respect to the source of instability in the disc, which is responsible for creating turbulence and the corresponding viscosity. Gravity is generally not relevant for facilitating angular momentum transport, except for very massive discs. Purely hydrodynamic turbulence necessitates the *Rayleigh instability criterion* to be satisfied, namely,

$$\frac{d}{dR}(R^2\Omega) < 0,$$ (3.14)

where Ω is the angular velocity at radius R in the disc. This condition is not fulfilled by Keplerian discs, as the term inside the parentheses (i.e., the angular momentum per unit mass) increases with the radius. However, if the disc is weakly magnetised, it can become unstable to perturbations via the magnetorotational instability, and the corresponding criterion is

$$\frac{d\Omega^2}{dR} < 0,$$ (3.15)

which is satisfied in Keplerian discs. A full-fledged treatment of this magnetohydrodynamic (MHD) instability and turbulence in protoplanetary discs requires complex analytical or numerical calculations, and constitutes an active area of modern research (Balbus, 2003).

3.1.5 Snow lines, volatile budget, and chemistry

The chemical structure of protoplanetary discs is crucial for understanding planetary formation, as it governs the availability of various materials at different locations and times. It has further direct repercussions for astrobiology, since the basic building blocks for life found in the disc can be incorporated into planets during their formation, or delivered at later times by impacts with remaining debris. The subject of disc chemistry has been thoroughly reviewed by the likes of Henning and Semenov (2013), Öberg and Bergin (2021), and Broadley et al. (2022). While there are certain similarities with the chemistry in molecular clouds, covered in Section 2.4, substantial differences also exist, which we will highlight here.

A useful starting point is the radial and vertical structure of the disc, depicted in Figure 3.5. Broadly speaking, temperature increases as one approaches closer to the central star and away from the midplane, and so does the ionising stellar and cosmic radiation. Therefore, photodissociation is prevalent in the central portions and exterior layers of the disc, while gas–dust interactions and richer molecular chemistry are rendered more significant in the outer regions and in the internal layers. Typical reaction timescales also increase while moving downwards and outwards in the disc, such that the outer disc midplane is dominated by chemical species inherited from the previous stages of star formation, rather than by local chemistry. This crude picture, however, ignores the complications arising from

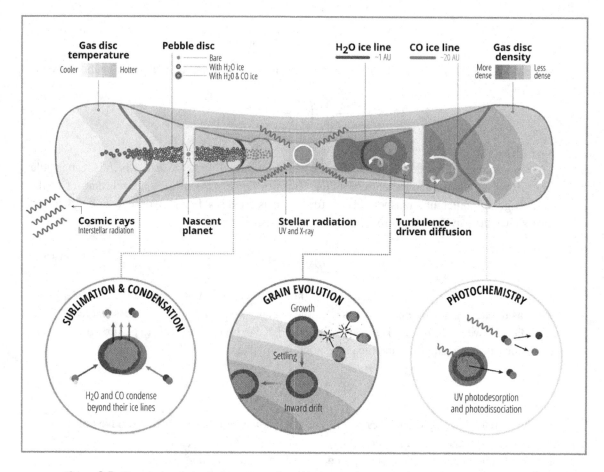

Figure 3.5 Chemical and physical structure of a typical protoplanetary disc. (Credit: Modified from Öberg and Bergin, 2021, Figure 15; reproduced with permission from Elsevier)

the redistribution of material within the disc, driven by grain growth and by turbulent diffusion. In particular, as dust grains become larger, they tend to settle into the midplane, and towards the inner disc regions.

An important factor in disc chemistry is the solidification of volatile compounds in the regions where temperature falls below the sublimation point, namely, beyond the aptly named *snow line*, also called the *ice line*. For water, this temperature occurs around 150–170 K (Penteado et al., 2017). In the early solar nebula, the water snow line was manifested at radial distances between 2 and 3 AU, potentially at ~ 2.7 AU (Armitage, 2020, pg. 15). Similar values are expected in the protoplanetary discs around Sun-like stars (Lecar et al., 2006), as sketched in Section 10.1.

In the case of carbon monoxide (CO), the sublimation temperature is around 20 K, and the corresponding snow line in the early solar nebula occurred at $\sim$20 AU. Condensation temperatures for the major volatiles – especially the C, N, and O carriers – determine at what locations in the disc they could be incorporated by the planets in solid form, rather than only through gas accretion. It should be noted, however, that snow lines are not static throughout the history of a disc. First, the thermal profile of the disc evolves with time, as a result of variations in stellar irradiation and in the efficiency of heating from accretion. Second, angular momentum transfer between solids and gases, as well as diffusive flows, can shift the snow line location over time. In general, however, it is expected that the outer disc regions ought to be depleted of O and C in gas phase.

The prior $\sim$20 AU radial distance may also be interpreted as a rough division between the inner and outer disc. In the inner disc, temperatures and densities are often high enough ($T \sim 100$–5000 K and $n_{disc} \gtrsim 10^{12}$ cm^{-3}) for molecules to be abundant in gas phase, and chemistry is near equilibrium. Theoretical models predict H_2O and CO to be prevalent in vapour form at $\sim$1 AU, as well as ammonia (NH_3), hydrogen cyanide (HCN),[2] and hydrocarbons such as methane (CH_4) and acetylene (C_2H_2).

In the outer disc, the dominant chemical pathways depend on the vertical location. Photochemistry dominates the exterior layers, driven by stellar and interstellar UV radiation and X-rays. Ionisation, dissociation, and ion–molecule interactions are all active, but they impact different molecules in distinct ways, which are also highly dependent on the attributes of the radiation field and, therefore, on the stellar type. Moving to intermediate layers, they have $T \sim 30$–70 K, and are mostly shielded from UV and X-rays. While CO can still be found in gas phase here, O is depleted because water is frozen into grains. This aspect gives rise to high C/O ratios and the possibility of organic chemistry. Finally, closer to the midplane, temperatures are lower than 20 K, and external ionising radiation (other than cosmic rays) cannot penetrate this region. Here, the prevalent mechanism is grain-surface chemistry, which was presented in Section 2.4.3.

Overall, the abundances of molecules in protoplanetary discs is regulated by a complicated interplay between gas-phase and grain-surface chemistry, photodesorption and photodissociation processes, and dynamical transport and mixing in both the radial and vertical directions.

3.1.6 Observation of molecules in protoplanetary discs

Molecules in protoplanetary discs are chiefly detectable at millimetre and sub-millimetre wavelengths, where dust is optically thin, that is, relatively transparent (with the exception of regions closer to the middle plane). Current sensitivities and resolutions have primarily limited such observations to the outer disc (beyond $\sim$30 AU). Infrared observations are disadvantageous because dust is optically thick at those wavelengths, although they can be used to study solids, and discern molecules without a strong permanent electric dipole moment, such as C_2H_2 or polycyclic aromatic hydrocarbons.

A multitude of organic molecules and species are confirmed in protoplanetary discs, including HCN, CN, C_2H_2, C_2H, c-CH_3H_2, HCHO, CH_3OH, HCOOH, HC_3N, H_3CN, CS, and H_2CS. As we have seen in the previous chapter (consult Table 2.3), many other complex organic molecules have been found in interstellar space, but so far they have mostly eluded detection in protoplanetary discs. This limitation could partly be explained by observational constraints, and partly by the fact that heavier compounds are segregated into solids in the outer disc regions.

Infrared observations performed by the *Spitzer* telescope revealed 'forests' of emission lines linked to the presence of hot H_2O and OH in the inner discs around YSOs. Needless to say, the chemistry of water in discs is paramount for astrobiology, as it has obvious implications for the habitability of planets. We will address possible mechanisms for water delivery later in this chapter. In the outer disc regions, water is found mostly in the solid state, frozen on dust grains surfaces, from which it can be photodesorbed by UV radiation. In the inner regions, above a few hundreds of K, water is presumably

[2] The sublimation temperature for HCN in the disc midplane is $\sim$85–100 K, and the snow line for HCN is $\sim$1–2 AU beyond the water snow line (Bergner et al., 2022).

the main oxygen carrier in gas phase. In Section 2.4.2, we learnt that water may be produced through ion–molecule reactions (below $\sim$100 K). In the warm conditions of the inner disc, however, water formation happens rapidly via neutral–neutral reactions:

$$O + H_2 \rightarrow OH + H, \tag{3.16}$$

$$OH + H_2 \rightarrow H_2O + H. \tag{3.17}$$

In general, the observation of water in gas phase is rendered difficult in discs around Sun-like stars, due to the aforementioned spatial resolution limitations. However, the detection of water has been accomplished in warm discs around very luminous protostars, with snow line locations in excess of several tens of AU (see, for example, Tobin et al., 2023).

3.1.7   Isotopic fractionation

An important tool for gaining insight into processes of planetary formation involves studying *isotopic fractionation*, to wit, alterations (e.g., enrichment) of isotopic ratios, which represent the relative abundances of stable isotopes of the same element found in various compounds – most notably hydrogen, carbon, nitrogen, and oxygen. *Isotopologues* are compounds that differ in their isotopic composition. Isotopic fractionation is set at molecule formation, as it depends on environmental conditions such as temperature or radiation flux. Therefore, measurements of isotopic fractionation can furnish some clues as to when and where various molecules originated.

At low temperatures, isotopic fractionation is facilitated by the ground state energy difference between isotopologues comprising lighter and heavier isotopes of an element. For example, chemical species containing deuterium are documented to form slightly stronger bonds than their hydrogenated counterparts, owing to the larger mass of the former; similar energy differences exist between isotopologues hosting ^{12}C, ^{14}N, and ^{16}O and those comprising ^{13}C, ^{15}N, and ^{18}O, respectively. As a consequence, in cold environments, the transfer of heavier isotopes into ions and molecules is favoured because of the ensuing stronger bond interactions.

In the case of deuterium, fractionation is actualised via reactions such as

$$H_3^+ + HD \leftrightarrow H_2D^+ + H_2, \tag{3.18}$$

$$CH_3^+ + HD \leftrightarrow CH_2D^+ + H_2, \tag{3.19}$$

which strongly favour the forward direction at low temperatures. The outcome would therefore be an elevated deuterium-to-hydrogen (D/H) ratio in cold settings with respect to the primordial nucleosynthesis value of $\sim 10^{-5}$. In contrast, the deuterium abundance tends to approach the cosmic value in warmer conditions like the interiors of giant planets and stars. Therefore, as elaborated in Section 3.3.3, careful measurements of the D/H ratio in water, when undertaken at different spatial locations, could help trace its complex trajectory from the interstellar medium to discs to planets.

With regard to C, N, and O, an additional fractionation mechanism based on photodissociation is feasible. Molecules in the external layers of a given setting (e.g., gas cloud) can protect other molecules of the same species further removed from the source of radiation from dissociation, a phenomenon known as *self-shielding*. Since heavier isotopologues are generally less abundant than their common lighter analogues, they tend to be self-shielded in deeper layers relative to the latter. Hence, photodissociation may create a selective isotope stratification, with a deficit of heavier isotopologues in the outer layers, and a corresponding internal excess of certain heavier isotopologues instantiated by means of photodissociation.

As a result, the isotopic ratios of C, N, and O found in ices are modified, depending on the epoch and location of their formation, yielding information about their birth environments. For instance, water formed in UV irradiated regions, such as the outer layers of protoplanetary discs, tends to be enriched in ^{18}O. Isotopic enrichment in some C, N, and O compounds, including amino acids, is also confirmed in cometary and meteoritic materials.

3.2 The formation of planets

Having delineated the physical and chemical conditions prevalent in the protoplanetary disc, we shall now delve into the intricate array of processes that, starting from solid material in dust form, culminate in the formation of a full-fledged planetary system. This growth in size, spanning some 12 orders of magnitude, may be arranged into separate stages, each entailing distinct physical mechanisms and regimes.

1. *From dust to rocks:* The first stage involves the accretion of *dust*, with initial grain sizes ranging from the sub-micron scale up to a few centimetres. In this phase, growth is dominated by direct collisions between individual grains, and their subsequent adhesion through short-range van der Waals interactions. Small grains are strongly coupled to the gas in the disc and experience aerodynamic drag (resistance). As the dust grains get progressively larger, they settle towards the midplane of the disc.

2. *From rocks to planetesimals:* In the second stage, *rocks* of a few metres grow up to the size of *planetesimals*, that is, larger than a kilometre. During this phase, the solid material is still coupled to the gas, and is therefore affected by phenomena such as fluid turbulence and resistance, as well as fragmentation, bouncing, radial drift, and so on. The physical modelling at this stage is more complicated and uncertain, but it may be approximated by a combination of gravitational dynamics and aerodynamic drag.

3. *From planetesimals to (proto)planets:* During the third and final stage, planetesimals grow up to the scale of *protoplanets* (also called *planetary embryos*), that is, hundreds of kilometres in size. These large objects keep colliding and merging at an accelerated pace to produce rocky planets and gas giant cores, having masses comparable to or higher than the Earth. This phase can be analysed by purely modelling gravitational interactions between planetesimals. The coupling with the remaining gas in the disc also occurs via gravity rather than drag, driving the formation of giant planets through gas accretion by solid cores, and enabling the phenomenon of *migration*, which we will tackle in Section 3.3.

We will now proceed to examine the physical processes involved in each of these stages in moderate detail.

3.2.1 Aerodynamic drag

As a spherical object of radius S moves with velocity v relative to a gas with density ρ_g, it is subject to *aerodynamic drag*, that is, a resistance force given by

$$F_D = -\frac{1}{2}C_D(\pi S^2)(\rho_g v^2). \tag{3.20}$$

This force can be understood as the result of a pressure $\rho_g v^2$ (termed the *ram pressure*) acting on the cross-sectional area πS^2. The dimensionless quantity C_D, referred to as the *drag coefficient*, depends on the velocity of the object and its size relative to the mean free path in the gas (λ_{mfp}). In the limit $S < \lambda_{\mathrm{mfp}}$ (the so-called *Epstein regime*), which is applicable to small particles, the drag force simplifies to

$$F_D = -\frac{4}{3}\pi S^2 v_{th}\rho_g v, \tag{3.21}$$

where v_{th} is the mean thermal velocity of the gas molecules, which is close to the speed of sound c_s. We can define a typical timescale for the aerodynamic drag (known as the stopping time) by dividing the total change in the particle momentum by the drag force, or $t_s \equiv mv/F_D$. By expressing the particle mass in terms of its density ρ_d, we obtain

$$t_s = \frac{\rho_d}{\rho_g}\frac{S}{v_{th}}. \tag{3.22}$$

We can estimate t_s for dust grains in the protoplanetary disc at ~ 1 AU by inserting the fiducial values of the corresponding physical quantities:

$$t_s \sim 1\,\mathrm{s}\left(\frac{\rho_d}{1\,\mathrm{g\,cm^{-3}}}\right)\left(\frac{\rho_g}{10^{-9}\,\mathrm{g\,cm^{-3}}}\right)^{-1}\left(\frac{v_{th}}{1\,\mathrm{km\,s^{-1}}}\right)^{-1}\left(\frac{S}{1\,\mu\mathrm{m}}\right), \tag{3.23}$$

which shows that the coupling between dust and gas is very strong for small grains; in other words, the grains are rapidly slowed down, so that they move in conjunction with the gas. As the radius S increases up to the order of metres, the drag timescale approaches Ω_d^{-1}, where Ω_d is the angular velocity of dust grains in the disc. When $t_s \gg \Omega_d^{-1}$, an ordering that is valid for planetesimals (i.e., at sizes larger than ~ 1 km), the drag processes become too slow to play a significant role on such objects.

3.2.2 Dust settling

Earlier, in the RHS of (3.4) we stated that the gravitational force acting on a dust grain of mass m_d in the direction z perpendicular to a thin disc is given by $-m_d\Omega_K^2 z$. By equating this expression with the drag force (3.21) we obtain an expression for the terminal velocity, namely, the constant speed v_f reached by a dust grain in the direction z (after a time of order t_s):

$$v_f = \frac{\Omega_K^2}{v_{th}}\frac{\rho_d}{\rho_g(z)}Sz, \tag{3.24}$$

The typical time it takes for a grain to settle in the disc midplane from a height z can then be estimated as $t_f = z/v_f$, which after substitution yields

$$t_f = \frac{v_{th}}{\Omega_K^2}\frac{\rho_g(z)}{\rho_d}\frac{1}{S}. \tag{3.25}$$

We should recall from (3.5) that the gas density has a Gaussian dependence on z. The settling time is roughly inversely proportional to the size of the grain, and is much lower in the upper disc layers since ρ_g declines considerably (i.e., settling is much faster in those regions and slows down towards the midplane). For grain radii of $S \sim 1$ μm at $z = H$ and radial distances of ~ 1 AU, the settling time is $\sim 10^5$ years, which is short compared to the disc lifetime. When additional effects such as turbulence are taken into account, however, the settling time is generally longer than this estimate.

The net effect is that even small grains can, in principle, eventually settle into the disc midplane before its dispersal.

3.2.3 Radial drift

We learnt from (3.10) in Section 3.1.4 that the velocity of gas in the disc is Keplerian up to a term of order $(H/R)^2$. This small difference becomes vital when we consider the coupling of dust grains to the gas.

The coupling is stronger for smaller grains, indicating that they would orbit at the same velocity of the gas. However, because this velocity is not fully Keplerian, the centrifugal force and gravity do not balance each other, thereby causing these particles to drift inward at a terminal radial velocity. Larger grains, on the other hand, are only weakly coupled to the gas, and exhibit Keplerian velocity. As a result of the relative velocity with the gas (effectively, a headwind) stemming from the non-Keplerian nature of the latter, they experience aerodynamic drag and lose angular momentum, which again causes a net radial drift towards the centre.

If the gas has nearly zero radial drift velocity, then the radial drift velocity of the particles (v_r) can be expressed as

$$\frac{v_r}{v_K} = -\frac{\eta}{T_s + T_s^{-1}}, \qquad (3.26)$$

where $\eta = n(c_s/v_K)^2$ was introduced in Section 3.1.4, and we have defined the dimensionless parameter $T_s \equiv t_s \Omega_K$.

Given that $\eta > 0$, the radial velocity is rendered negative, that is, directed inward. The maximum value of v_r is attained for $T_s = 1$ (i.e., when $t_s = \Omega_K^{-1}$), and has a magnitude of $\eta v_K/2$. As intimated in Section 3.2.1, the condition $t_s = \Omega_K^{-1}$ is typically valid for grains with $S \sim 1$ m. The peak radial velocity at $R \sim 1$ AU is $v_r \gtrsim 10$ m s^{-1}, which translates to an infall time towards the central star of $\lesssim 500$ years. Hence, the growth of dust grains close to the aptly named *metre barrier* must be very rapid for planet formation to take place. In other words, the formation of planetesimals has to be swift, as otherwise the constituent particles will spiral inward and get destroyed. Overall, when considered over the entire lifetime of the disc, radial drift could be significant for particles of all sizes, implying considerable variations in the ratio of solids to gas at different epochs and locations.

It is also worth pointing out that the previous treatment implicitly supposed a monotonic decrease in the gas pressure moving outwards from the centre. If, however, pressure perturbations are present elsewhere in the disc, the same mechanism will act around the local gradient, with the radial drift directed towards the direction of increasing pressure. As a consequence, pressure maxima in the disc, generated by turbulence or other physical mechanisms, behave as locations where dust grains can get concentrated and aggregate to form planetesimals (Izidoro et al., 2022).

3.2.4 Dust coagulation

The initial growth of solid bodies in the disc is thought to occur via *coagulation*, caused by the direct collision of two dust grains and their subsequent sticking due to inter-molecular forces.

For this mechanism to work successfully, the physical properties and velocity distribution of solids have to be of such a kind as to favour aggregation rather than fragmentation or bouncing. Furthermore, the process has to be rapid enough to occur before radial drift causes the infall of most or all solid material towards the star.

Growth via coagulation can be described by a simple mathematical model

$$\frac{dS}{dt} \sim \frac{\rho_g}{4\rho_d} Z\sigma,$$ (3.27)

where Z is the dust-to-gas mass ratio and σ is the velocity dispersion (crudely, the typical relative velocity) of the solid particles. In the case of Brownian motion, energy equipartition leads to

$$\sigma \sim \sqrt{\frac{m_H}{m_d}} c_s,$$ (3.28)

where m_H is the mass of the hydrogen atom (comprising the gas), m_d is the mass of the grain, and c_s is the sound speed.

Experimental studies conducted on materials representative of the disc composition have provided valuable insights into the effectiveness of coagulation (Blum and Wurm, 2008; Birnstiel et al., 2016; Wurm and Teiser, 2021). Such studies, complemented with theoretical models, suggest that growth via coagulation can proceed quite unhindered from sizes of microns up to a few centimetres for a diverse range of collision velocities. At typical values of the physical parameters at radii of AU, (3.27) and (3.28) could support growth rates of $dS/dt \sim 10^{-4}$ cm yr^{-1}, thereby permitting the formation of cm-sized particles on timescales of $\lesssim 10^4$ yr.

However, at larger sizes of $\gtrsim 10$ cm, collisions lead to detrimental effects like bouncing and fragmentation instead of adhesion, so that further growth is essentially stalled (Blum and Wurm, 2008, Figure 12). This hurdle may be less pronounced beyond the snow line, because of the higher resistance of icy material and its capability to stick at higher collision speeds. However, even at such locations, growth via coagulation does not seem capable of overcoming the metre barrier and explaining the formation of bigger objects.

To sum up, when the growing particles approach a size threshold of ~ 1 m, the twin obstacles of radial drift and collisional mass loss hamper further growth. Hence, it is necessary to investigate alternative routes for overcoming this metre barrier and forming planetesimals, as done next.

3.2.5   Growth by instability

One classic avenue proposed for overcoming the metre barrier is based on the Goldreich–Ward mechanism (Goldreich and Ward, 1973). The central idea is that the combined action of settling and radial drift produces a very thin and dense dusty sub-disc, which subsequently fragments into clusters of particles due to *gravitational instability*. Such solid aggregates then proceed to form planetesimals by means of collisions.

The onset of gravitational instability in discs (with dust) requires that the following criterion be satisfied:

$$Q \equiv \frac{\sigma \Omega_d}{\pi G \Sigma_{\text{dust}}} \lesssim 1,$$ (3.29)

where Q is the so-called *Toomre parameter* (Toomre, 1964) and Σ_{dust} is the surface density of the dust layer. Typically, $\Sigma_{\text{dust}} \sim 10^{-2}\Sigma_{\text{gas}} \sim 10$ g cm^{-2} at 1 AU. On substituting the values for a solar mass star, the criterion (3.29) translates into a dust velocity dispersion of $\sigma \sim 10$ cm s^{-1}. Because the appropriate scale height is proportional to the characteristic velocity ($H \propto \sigma$ for dust and $H \propto c_s$ for gas), and the typical velocity of gas is $\sim 10^5$ cm s^{-1}, we find that solid disc should be extremely thin, namely, $\sim 10^4$ times thinner than the gas disc. If such a thin layer were achieved, the mass of unstable fragments (M_{frag}) would be

$$M_{\text{frag}} \sim 4\pi^5 G^2 \frac{\Sigma_{\text{dust}}^3}{\Omega_d^4}, \tag{3.30}$$

and substituting the characteristic values at 1 AU yields M_{frag} of the order 10^{18} g, amounting to km-sized planetesimals. This mechanism may therefore seem an attractive way to produce substantial growth. Unfortunately, though, this process cannot operate in its basic form, as such thin layers are not realisable in practice on account of turbulence triggered by the ensuing velocity shear (i.e., the Kelvin–Helmholtz instability).

However, a different kind of instability that does not require self-gravity may come to the rescue: the *streaming instability* delineated in Youdin and Goodman (2005). This instability necessitates a weaker condition to occur, which is that the density of solid material is comparable to that of gas. Such a condition can be achieved by a combination of coagulation and settling towards the midplane. The mechanism underlying the streaming instability stems from mutual gas–dust interactions mediated by aerodynamic forces.

As solids clump, the net surface area of the system decreases (as the ratio of surface area to volume is $\propto S^{-1}$), and so does the drag. In turn, this outcome permits more clumping to occur (because large clumps decouple from the gas and trap other solid particles) and growth could rapidly escalate. The process is rendered most efficient for $T_s \sim 1$, and numerical simulations have demonstrated that the formation of large (~ 100 km) planetesimals may be feasible over short timescales. One possible drawback, however, is that not many particles (i.e., seeds) exist in the size range of $T_s \sim 1$. In general, much remains unknown at this stage (e.g., MHD turbulence).

3.2.6 Gravitational focusing and growth rates

One way or another, we will now suppose that the metre barrier has been overcome, thus furnishing the planetary system with small planetesimals. After most solid material in the disc has aggregated into planetesimals, gravity starts playing an important role in further growth, as it influences the rate of pairwise collisions. This happens through a mechanism known as *gravitational focusing*, which essentially enhances the collision cross section of an object of radius S from its projected area of πS^2 to $\pi S^2 F_g$, where $F_g > 1$ is the *focusing factor* defined as

$$F_g = 1 + \frac{v_e^2}{\sigma^2}, \tag{3.31}$$

and $v_e = \sqrt{2G\tilde{m}/S}$ is the escape velocity from the surface of the body endowed with mass $\tilde{m}$. The ratio $v_e^2/\sigma^2 = 2G\tilde{m}/S\sigma^2$ also goes by the name of the *Safronov number*. All other factors being equal, a smaller velocity dispersion σ will induce a larger focusing effect. Therefore, in the colder regions of the disc with lower σ, collisions between planetesimals would be more frequent, stimulating faster growth.

The mass growth rate from pairwise collisions can be expressed as

$$\frac{d\tilde{m}}{dt} = \tilde{\rho}_p \sigma \pi S^2 F_g, \tag{3.32}$$

where $\tilde{\rho}_p \sim \Sigma_p / H_p \sim \Sigma_p \Omega / \sigma$ is the mass density of planetesimals, and Σ_p and H_p are the associated surface density and scale height. This equation may be interpreted, in physical terms, as the product of the density ($\tilde{\rho}_p$) times the volume rate (i.e., volume per unit time), where the latter is the product of the cross-sectional area $\pi S^2 F_g$ and the characteristic velocity σ.

It is interesting to explore some limiting cases for the rate equation. First, if gravitational focusing is neglected ($F_g \approx 1$), the equation becomes

$$\frac{d\tilde{m}}{dt} \propto S^2. \tag{3.33}$$

On using the fact that $\tilde{m} \propto S^3$, we can rewrite (3.33) as

$$\frac{1}{\tilde{m}} \frac{d\tilde{m}}{dt} \propto \tilde{m}^{-1/3}, \tag{3.34}$$

which implies that the fractional growth of mass slows down as mass increases. This regime is also known as *orderly growth* for this reason. Rewriting (3.33) in terms of the size, we have

$$\frac{dS}{dt} = \text{constant}. \tag{3.35}$$

Therefore, when the collision rate is not enhanced by gravitational interaction, the size of planetesimals grows only linearly in time, $S \propto t$. It can be shown that, for the typical values of the physical parameters involved, this orderly growth is too slow to reach the size of terrestrial planets or gas giant cores in the available timescale. This discrepancy strongly suggests that gravitational focusing must play an important role.

We can therefore look at the opposite limiting case of $F_g \gg 1$. In this regime, we are free to write

$$F_g = 1 + \frac{v_e^2}{\sigma^2} \simeq \frac{v_e^2}{\sigma^2} \tag{3.36}$$

so that if the velocity dispersion and the surface density of planetesimals are held fixed, then (3.32) is transformed into

$$\frac{1}{\tilde{m}} \frac{d\tilde{m}}{dt} \propto S \propto \tilde{m}^{1/3}. \tag{3.37}$$

In this regime, the fractional growth of mass accelerates as mass increases, thus giving rise to the so-called *runaway growth*, with mass becoming infinite in a finite time (which is termed a finite-time singularity).

In practice, as we will see shortly, the condition $F_g \gg 1$ cannot be sustained indefinitely, and eventually growth comes to a halt. Nonetheless, a period of runaway growth is essential for explaining the rapid accretion of planetesimals into planetary-sized objects.

3.2.7   Isolation mass

In order to have $F_g \gg 1$, the desired condition from earlier, the velocity dispersion σ has to be small with respect to the escape velocity v_e. Only planetesimals occupying orbits close to each other fulfil this criterion, and could thus collide effectively via gravitational focusing over time. As a consequence, only a finite amount of objects are accessible for runaway growth of a principal body at a given location, and therefore its mass can only reach a limiting value, known as the *isolation mass* $M_{\rm iso}$.

We can compute the value of $M_{\rm iso}$ by starting from an estimate of the maximum distance at which an object of mass $\tilde{m}$ could significantly influence other bodies through its gravity, which is called the *Hill radius*:

$$r_H \approx a \left(\frac{\tilde{m}}{3M_\star} \right)^{1/3}, \tag{3.38}$$

where a is the orbital distance of the object around a star of mass $M_\star$. A heuristic means of deriving (3.38) is that the density of matter contributed by the star in proximity to the object (proportional to $M_\star/a^3$) must be approximately equal to the density of matter due to the object at the Hill radius (proportional to $\tilde{m}/r_H^3$).

When applied to a growing planetesimal, this picture establishes a *feeding zone* – an annular region around the star containing the bodies that could be perturbed up to the point of collision. As per theoretical calculations, the extent of the feeding zone is $\Delta a_{\rm max} \approx 4\sqrt{3}r_H$, that is, several times the Hill radius. The total mass contained within an annulus of size $\Delta a_{\rm max}$ at distance a from the star is the isolation mass we are looking for:

$$M_{\rm iso} = (2\pi a)(\Delta a_{\rm max})\Sigma_p \approx (2\pi a) \left[4\sqrt{3}a \left(\frac{M_{\rm iso}}{3M_\star} \right)^{1/3} \right] \Sigma_p. \tag{3.39}$$

This equation is solved for $M_{\rm iso}$, to arrive at

$$M_{\rm iso} \approx 165.8\, M_\star^{-1/2}\, \Sigma_p^{3/2}\, a^3. \tag{3.40}$$

Note that the scalings in the formula suggest that less massive stars and larger orbital radii might result in higher values of $M_{\rm iso}$. On substituting the relevant physical quantities, it can be shown that, at 1 AU from a solar mass star, we obtain $M_{\rm iso} \sim 0.07 M_\oplus$, which is roughly the mass of Mars. Hence, the implication is that runaway growth is crucial in the formation of terrestrial planets. On the other hand, at 5 AU (i.e., the distance of Jupiter from the Sun), it turns out that $M_{\rm iso} \sim 9 M_\oplus$, which is roughly comparable to the estimated mass of Jupiter's core; the isolation mass may, however, not always be reached during the formation of giant planets.

3.2.8 The formation of terrestrial planets

Having reviewed the main physical mechanisms involved in the growth of solids, from grains up to planetary embryos, we are now in a position to summarise the process that leads to the formation of terrestrial planets.

A convenient mathematical tool for organising our description of collision-driven growth is the *coagulation equation* (Smoluchowski, 1916):

$$\frac{dn_k}{dt} = \frac{1}{2} \sum_{i+j=k} A_{ij} n_i n_j - n_k \sum_{i=1}^{\infty} A_{ik} n_i. \tag{3.41}$$

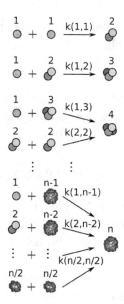

Figure 3.6 Aggregation of discrete particles according to the coagulation equation. (Credit: Wikimedia Commons. https://commons.wikimedia.org/wiki/File:Smoluchowski_Aggregation_Kinetics.svg; CC-BY-SA 3.0 license)

This equation describes the time evolution of the number of objects n_k with mass $m_k = km_0$, where m_0 is the mass of some small basic 'building block'. The growth of larger objects happens through repeated aggregation of these building blocks (see Figure 3.6). The two terms on the RHS of the equation account for: (1) the number of bodies of mass m_k arising from the coagulation of all possible pairs of objects having masses m_i and m_j such that $i + j = k$; and (2) the loss of objects with mass m_k in collisions with other objects. The effect of fragmentation is neglected.

The usefulness of the coagulation equation is that it can trace the evolution of an initial population of objects across different stages of growth in a unified fashion: the details of the physical processes involved enter the formula via the rate coefficients A_{ij}. Two notable choices of A_{ij} that give rise to distinct classes of solutions are:

- A_{ij} = constant: In this case, initially all objects possess the elementary mass m_0, and subsequently the average mass increases, while the shape of the mass distribution maintains the same form. This regime corresponds to the mode of *orderly growth*, delineated in Section 3.2.6.
- $A_{ij} \propto m_i m_j$: In this scenario, the shape of the mass distribution progressively acquires a long tail at large masses. This reflects the stage of runaway growth, with a few objects growing much larger than average.

To recapitulate, the growth of terrestrial planets may be schematically organised into four broad phases as follows.

1. *Orderly growth*: Gravitational focusing is still unimportant, and grain-sized objects initially grow through direct collisions and sticking. Difficulties arise in the transition from the centimetre to the metre scale, which could be bypassed by the onset of instabilities. The relative growth of objects

slows down with time, and virtually all tend to approach similar sizes. This phase is perhaps the most uncertain and challenging to model.

2. *Runaway growth*: Once objects reach planetesimal size, gravitational focusing starts coming into play. The growth rate accelerates for the most massive objects, with one or a few bodies going on to subsequently dominate a particular region of the disc.

3. *Oligarchic growth*: At this stage, the largest objects are massive enough to modulate the relative velocity distribution of the surrounding planetesimals. The growth of the dominant bodies slows down relative to the runaway phase, but still exceeds that of smaller objects. This process generates a bifurcation in the mass distribution, with ~ 100–$1,000$ objects in the disc having sizes comparable to that of the Moon or Mars, and all the others remaining on the scale of smaller planetesimals. This phase ends once the isolation mass limit is attained.

4. *Post-oligarchic growth*: The planetary embryos resulting from the previous phase keep interacting via sporadic collisions initiated by large-scale stochastic mixing of the orbits. This chaotic phase culminates in the final assembly of a limited number of terrestrial planets.

3.2.9 The formation of giant planets

What we have learnt hitherto for the aggregation of terrestrial planets can be generalised, with some notable caveats, to understand the formation of the rocky/icy cores of giant planets. We will now turn our attention to the mechanism(s) by which such planets acquire their thick gaseous envelopes; one popular hypothesis suggests that this occurs via gravitational accretion of gas around the central solid object, or *core accretion*.

To retain a gaseous envelope, the escape speed from the core must exceed the gas sound speed c_s. The former, as we have already seen, is given by

$$v_e = \sqrt{\frac{2G\tilde{m}}{S}},$$ (3.42)

and we will utilise $H = c_s/\Omega_K \approx c_s R/v_K$ to obtain

$$c_s \sim v_K \left(\frac{H}{R}\right).$$ (3.43)

For the limiting case of $v_e \sim c_s$, the minimal mass $M_{\min}$ required to retain the gas is determined to be

$$M_{\min} \sim \sqrt{\frac{3}{32\pi}} \left(\frac{H}{R}\right)^3 \sqrt{\frac{M_\star^3}{\rho_d a^3}},$$ (3.44)

where we have used $\tilde{m} = 4\pi\rho_d S^3/3$ to derive the above equation. For a Sun-like star and a disc aspect ratio of $H/R \sim 0.1$, typical values of the relevant physical parameters at Jupiter's location ($a = 5$ AU) yield $M_{\min} \sim 4 \times 10^{-3} M_\oplus$, which is a few times smaller than the mass of the Moon. However, we emphasise that this estimate is quite crude, and only a thin gaseous envelope could be acquired in this way. A more refined treatment described in Armitage (2020, pg. 224) leads to

$$M_{\min} \approx \left(\frac{3}{4\pi\rho_d}\right)^{1/2} \left(\frac{c_s^2}{G}\right)^{3/2} \left[\ln\left(\frac{\varepsilon\rho_d}{\rho_0}\right)\right]^{3/2},$$ (3.45)

where ρ_0 is the midplane density introduced previously, and ε is the envelope mass divided by the total mass, and is taken to have a fiducial value of ~ 0.1. At a distance of 1 AU, it is found that $M_{\min} \sim M_\oplus$ from (3.45).

A realistic treatment would take the full hydrostatic structure of the envelope into account, and consist of various stages. Initially, after the core has reached the isolation mass, the accretion of gas happens slowly at hydrostatic equilibrium, and is simultaneously accompanied by dissipation of excess energy and contraction by means of the Kelvin–Helmholtz mechanism. The rate of growth is limited by the radiative timescale for energy dissipation. However, as the total mass increases, the forming planet is able to sweep out a progressively larger feeding zone, driving an increase of the core mass and further accretion. Once the envelope dominates the total mass, its growth enters a short runaway phase shaped by the self-gravity of the gas, ending with the dispersal of the gaseous disc component.

Only planetary cores that are massive enough to initiate these stages constitute viable progenitors of gas giants. Therefore, the model has a critical mass M_{crit} above which hydrostatic equilibrium can be disrupted such that the gravitational contraction and runaway accretion phases are operational. This value depends on a number of poorly constrained parameters, such as the magnitude of the opacity of the gas. A rough fit to numerical simulations of the critical mass (Ikoma et al., 2000) yielded

$$\frac{M_{\text{crit}}}{M_\oplus} \approx 12 \left(\frac{\kappa_R}{1\,\text{cm}^2\,\text{g}^{-1}} \right)^{1/4} \left(\frac{\dot{M}_{\text{core}}}{10^{-6}\,M_\oplus\,\text{yr}^{-1}} \right)^{1/4}, \tag{3.46}$$

where κ_R represents the characteristic opacity of the envelope, and $\dot{M}_{\text{core}}$ is the accretion rate associated with the core.

Core accretion is conventionally presumed to explain most of the features observed in giant planets, and is currently the favoured scenario for their formation. However, it also suffers from a number of drawbacks. One major uncertainty concerns the opacity of the accreting gas, since this gas is 'contaminated' by dust, and therefore difficult to estimate. Another obstacle is that it requires a plentiful supply of planetesimals as a starting point, which may not be realistic in all cases. Loosely speaking, the cumulative timescale to enter the runaway phase is $\lesssim 10$ Myr, which is close to the timescale over which gas is available in the protoplanetary disc. Finally, the effects of migration due to gravitational torques, which influence the time taken for core formation, are not addressed in classical core accretion models.

Taking a slight detour, we highlight that the process of formation of giant planet cores and/or terrestrial planets could involve *pebble accretion*, wherein particles with $T_s \sim 1$ (refer to Section 3.2.3) are termed pebbles. In pebble accretion, the aforementioned cores or planets are formed through the direct accretion of pebbles that drift radially inward because of drag forces. The physical principles of the pebble accretion model are reviewed in Ormel (2017). Numerical simulations of pebble accretion have concluded that the formation of giant planet cores and perhaps even terrestrial planets in the solar system is feasible through this paradigm (Johansen et al., 2021), although the latter has been challenged (Burkhardt et al., 2021).

Circling back to our theme, an alternative process has been considered owing to the preceding difficulties: gravitational instability. We touched on this mechanism earlier in connection with dust coagulation (Section 3.2.5) and implied that the condition for instability in the disc demands

$$Q \equiv \frac{c_s \Omega}{\pi G \Sigma} \lesssim 1. \tag{3.47}$$

We will now employ the approximate relations $H \approx c_s/\Omega_K$, $\Omega_K^2 \approx GM_\star/R^3$, and $M_{\mathrm{disc}} \approx \pi \Sigma R^2$, where M_{disc} is the disc mass. Upon substitution in (3.47), the instability condition becomes

$$\frac{M_{\mathrm{disc}}}{M_\star} \gtrsim \frac{H}{R} \sim 0.1. \tag{3.48}$$

This limit is close to the upper threshold of observed disc masses, and is therefore not unattainable. Analogous to the Jeans radius from Section 2.2.2, the length scale at which the instability is triggered in a disc is $\lambda_{\mathrm{frag}} = 2c_s^2/(G\Sigma)$ (Armitage, 2007, Equation 165), allowing us to estimate the mass of the resulting disc fragments as

$$M_{\mathrm{frag}} \sim \pi \Sigma \lambda_{\mathrm{frag}}^2 \sim \frac{4\pi c_s^4}{G^2 \Sigma}, \tag{3.49}$$

which translates to a typical value of $M_{\mathrm{frag}} \sim 5M_J$, where M_J denotes the mass of Jupiter. An even cruder estimate can be obtained by neglecting factors of order unity, thus yielding

$$M_{\mathrm{crit}} \sim \left(\frac{H}{R}\right)^3 M_\star \sim M_J \tag{3.50}$$

after assuming $M_\star \sim M_\odot$ for the star. This fragment could continue to accrete mass from the disc and grow further. Hence, the outcome of the gravitational instability scenario may be very massive planets.

It should also be recognised that the onset of gravitational instability depends on the disc's thermodynamics. Colder discs, with lower values of c_s, are more unstable. However, once fragmentation is triggered, heating may ensue from angular momentum transport (and gravity), and the disc could self-regulate towards $Q \sim 1$, therefore returning to stability. As a rule of thumb, for instability to occur, it is necessary that the characteristic disc cooling timescale (t_{cool}) is shorter than the self-gravity heating timescale, which is of order Ω^{-1}. If cooling takes place quickly, then the heating arising from adiabatic compression due to gravitational collapse can be radiated away, leaving the disc susceptible to fragmentation. Numerical and analytical studies suggest that fragmentation is initiated when the criterion $t_{\mathrm{cool}}\Omega \lesssim 3$ is met. In real-world contexts, this condition is not always valid.

Figure 3.7 constitutes a summary of the main stages and physical mechanisms that we have described so far, and includes rough estimates of the associated timescales. The entire transition from a disc of gas and dust to the final full-fledged planets is completed in a period of $\sim 10^7$–10^8 years. In the upcoming section, we will explore the events unfolding afterwards.

3.3   Evolution and migration

Planet formation is not the final word on the architecture of a planetary system. In fact, the latter may vary substantially over time due to the dynamical evolution of orbits, a phenomenon known as *migration*. The orbit evolution can result from various physical processes:

- *Planet–disc interactions*: Angular momentum exchange between a planet and the disc may lead to the migration of the former; this mechanism is applicable to both terrestrial planets and giant planets.

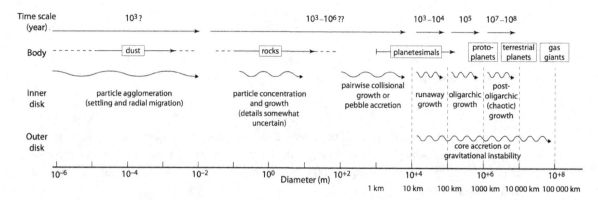

Figure 3.7 Bird's-eye view of the processes comprising planet formation, and their fiducial timescales. (Credit: Perryman, 2018, Figure 10.4; reproduced with permission from Cambridge University Press)

- *Planet–planetesimals interactions*: This mechanism is especially relevant for giant planets that can eject residual planetesimals from the disc and thereby exchange angular momentum. In our solar system, the ice giants and maybe Saturn might have migrated in this fashion.
- *Planet–planet scattering*: According to numerical simulations, orbital instabilities may lead to the ejection of smaller planets from the system, with the survivors persisting on high-eccentricity orbits. Interaction of planets with stellar binary companions could play a similar role.
- *Tidal interactions*: This may occur between a planet and the host star, and is significant for giant planets in close orbits (also called *Hot Jupiters*), as their orbital radii are only a few times larger than the stellar radii.

We will now focus our attention on unpacking the first two mechanisms and discussing their ramifications.

3.3.1 Planet–gaseous disc interaction

The simplest model for planet migration because of the interaction with gas is the so-called *impulse approximation*. Let us assume that a gas 'parcel' in the disc is flowing horizontally with a velocity Δv relative to a planet of mass M_p and impact parameter b; the latter roughly embodies the perpendicular distance between the trajectory of this parcel and the planet. It can be shown (Armitage, 2020, Chapter 7.1.1) that, due to gravity, the gas parcel acquires a velocity perpendicular to the initial direction, given by

$$|\delta v_\perp| = \frac{2GM_p}{b\Delta v}.$$ (3.51)

Being radially directed (i.e., normal to the orbit), this does not correspond to a change of angular momentum. Since we are in the rest frame of the planet, the kinetic energy of the parcel is conserved, so that

$$(\Delta v)^2 = |\delta v_\perp|^2 + \left(\Delta v - \delta v_\parallel\right)^2.$$ (3.52)

Here, $\delta v_\parallel$ is the velocity change in the parallel direction, which produces a variation in angular momentum. It is reasonable to assume that the deflection angle is small, so that $\delta v_\parallel \ll \Delta v$. We can therefore solve the above equation via Taylor expansion and invoking (3.51), to obtain:

$$\delta v_{\parallel} \sim \frac{2G^2 M_p^2}{b^2 (\Delta v)^3}. \tag{3.53}$$

If the gas parcel possesses semimajor axis a (effectively the orbital radius about the star), the change in specific angular momentum (angular momentum per mass) Δj is expressed as

$$\Delta j \approx a \, \delta v_{\parallel} \sim \frac{2G^2 M_p^2 a}{b^2 (\Delta v)^3}. \tag{3.54}$$

Let us now consider a thin annulus of gas (of infinitesimal thickness db) in the vicinity of the planet, where the surface gas density is Σ. The annulus will contain a mass $dm = 2\pi a \, db \, \Sigma$. Next, let us suppose that the angular velocity of the gas and the planet are Ω and Ω_p, respectively. The typical timescale Δt for the gas to encounter the planet is 2π divided by the relative frequency, so that we have

$$\Delta t \approx \frac{2\pi}{|\Omega - \Omega_p|} \approx 2\pi \left(\left| \frac{d\Omega_p}{da} \right| b \right)^{-1} \approx 2\pi \left(\frac{3\Omega_p}{2a} b \right)^{-1}, \tag{3.55}$$

where the second equality follows from the Taylor expansion of $|\Omega - \Omega_p|$ about a, and then imposing the ordering $b \ll a$.

We can now discuss the consequence of the change in angular momentum. Given that the orbital velocity of the gas exterior to the planet is slower than that of the planet itself, a reduction of the relative velocity in the parallel direction implies that the gas has gained angular momentum. Consequently, the planet has to lose angular momentum and migrate inwards. The converse is true for gas interior to the planet's orbit – in this case, the interaction decreases the angular momentum of the gas and increases that of the planet, which will then migrate outward. The direction of migration will be set by the dominance of one of these two effects.

The net torque attributable to the planet is determined from the rate of change of angular momentum as per classical mechanics, thus yielding

$$\frac{dJ}{dt} = - \int \frac{\Delta j}{\Delta t} \, dm. \tag{3.56}$$

The negative sign is introduced because the previous calculation was undertaken for the change of angular momentum associated with the gas, creating an equal and opposite torque for the planet. We may further assume that the orbits are nearly Keplerian, which translates to $\Delta v \approx a \left(|d\Omega_p/da| b \right) \approx 3\Omega_p b/2$. On solving the above integral, we arrive at

$$\frac{dJ}{dt} = - \frac{8G^2 M_p^2 \Sigma a}{27 \Omega_p^2 b_{\mathrm{min}}^3}, \tag{3.57}$$

where b_{min} represents a suitable cutoff impact parameter. This variable is typically comparable to the Hill radius (for small planets) or to the disc scale height (for massive planets). If all other factors are held equal, which is not always true for giant planets, we notice that $dJ/dt \propto M_p^2$. Hence, heavier planets may tend to migrate relatively rapidly.

Going beyond the impulse approximation, a more sophisticated treatment would have to account for the evolution of linear perturbations within the disc. The planet excites waves in the gas, which in turn are distinguished by special resonant locations. The total torque can be computed by summing over the torques generated at these resonant points. One noteworthy resonance is produced at radii

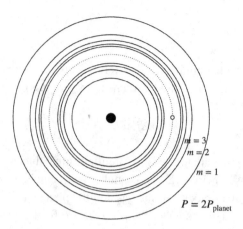

Figure 3.8 Corotation (dotted) and inner/outer Lindblad (continuous) resonances at a given location in the disc.

in the disc wherein $\Omega(R) = \Omega_p$, called the *corotation resonance*. For Keplerian discs, this radius is equal to a. Another notable category of resonances are manifested in Keplerian discs at

$$R_L = \left(1 \pm \frac{1}{m}\right)^{2/3} a, \tag{3.58}$$

where m is an integer. These are known as the *Lindblad resonances* (see Figure 3.8). The angular momentum exchange between the planet and the gas takes place at such locations. Akin to our discussion in the impulse approximation, interactions with the gas situated at exterior resonant locations result in inward motion of the planet, while interactions with the gas at interior resonant locations cause outward motion of the planet.

When the planet is not massive enough to perturb the disc structure significantly, the ensuing angular momentum exchanges engender *Type I migration*. In this case, the total torque $\mathcal{T}$ can be expressed as

$$\mathcal{T} = \sum \mathcal{T}_{\text{ILR}} + \sum \mathcal{T}_{\text{OLR}} + \mathcal{T}_{\text{CR}}, \tag{3.59}$$

where ILR and OLR are the inner and outer Lindblad resonances, respectively; they correspond to negative and positive signs in (3.58). The last term is the contribution from co-orbital gas (i.e., nearly in the same orbit). As stated earlier, the ILR and OLR terms respectively promote outward and inward migration. As they contribute almost equally, simplified models may fail to accurately predict not only the magnitude of the migration but even its direction. Detailed numerical simulations must be used for this purpose, which often find that Type I migration is directed inward.

A simple analytical model developed by Tanaka et al. (2002) estimated the Type I migration timescale to be

$$\tau \approx 2 \frac{M_\star}{M_p} \frac{M_\star}{\Sigma a^2} \left(\frac{c_s}{a\Omega_p}\right)^2 \Omega_p^{-1}. \tag{3.60}$$

For typical physical parameters in the disc, a short migration timescale of merely $\tau \sim 0.5$ Myr for a gas giant core at 5 AU is anticipated. We caution, however, that state-of-the-art simulations have challenged this paradigm, which remains the subject of ongoing research.

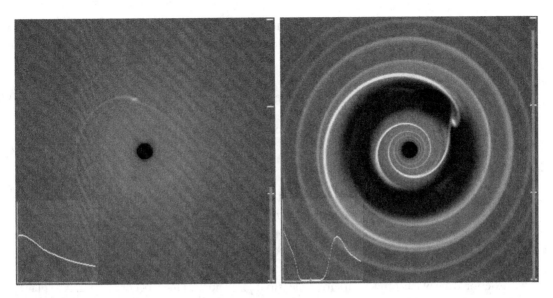

Figure 3.9 Comparison of Type I (left) and Type II (right) migration from numerical simulations, for a low-mass planet and a giant ($10\,M_J$) planet, respectively. The lower left insert in each image depicts the azimuthally averaged surface density profile. (Credit: Armitage and Rice, 2008, Figure 1; reproduced with permission from Cambridge University Press)

When the planetary mass is large enough to result in significant effects on the density structure of the surrounding gas, the outcome is *Type II migration* shown in Figure 3.9. As we have seen in (3.57), the torque exerted by a planet has a M_p^2 dependence, so we would expect that giant planets can cause a substantial repulsion of the neighbouring gas, up to the point of opening a gap in the disc. More formally, two conditions must be fulfilled in order to create such a gap. First, the planet's Hill radius must exceed the disc scale height, to ensure that gas present above or below the planet cannot be accreted to (re)fill the gap. In quantitative terms, this criterion is given by

$$\frac{M_p}{M_\star} \gtrsim 3 \left(\frac{H}{a}\right)^3 . \tag{3.61}$$

The second condition is that the timescale for creating a gap due to the torques must be shorter than the time needed by viscous diffusion to refill this opening. A simplified analysis undertaken by Takeuchi et al. (1996) concluded that this constraint is expressible as

$$\frac{M_p}{M_\star} \gtrsim \left(\frac{c_s}{a\Omega_p}\right)^2 \alpha^{1/2}, \tag{3.62}$$

where α is a dimensionless measure of the viscosity introduced in Section 3.1.4. For typical values of the salient physical quantities, we find that both conditions are satisfied for $M_p/M_\star \sim 10^{-4}$–10^{-3}, that is, a mass intermediate between that of Saturn and Jupiter for $M_\star \sim M_\odot$.

The characteristic timescale attributed to Type II migration is roughly given by (Armitage, 2007, Section IV.5):

$$\tau \approx \frac{2}{3\alpha} \left(\frac{H}{a}\right)^{-2} \Omega_p^{-1}, \tag{3.63}$$

and for characteristic values at 5 AU, the timescale again turns out to be short, $\tau \sim 0.5$ Myr. However, note that this formula was derived under the premise that the local disc mass is greater than that of the planet. This assumption has diminished accuracy when

$$a \lesssim 6 \, \text{AU} \left(\frac{M_p}{M_J} \right), \tag{3.64}$$

and migration is expected to be much slower inside this orbital radius.

3.3.2   Planet–planetesimal interactions

After gas dispersal, the planetary disc would still host numerous solid objects the size of planetesimals or even larger. These will continue to gravitationally interact with newly formed planets, affecting their orbital dynamics.

To see how such interactions may happen and trigger migration, consider the simple case of a planet of mass M_p at distance a from the star, and assume that there exists a disc of planetesimals exterior to the planet orbit, with surface density Σ_p, but none of them occur at $r < a$. As a planetesimal is scattered inward due to gravitational interaction, the planet has to migrate outward to conserve angular momentum. We will suppose that, once a planetesimal is scattered, it will not interact again with the planet. The total mass of planetesimals which are inside the sphere of influence of the planet (i.e., within the Hill radius r_H) is

$$\Delta m = 2\pi a r_H \Sigma_p, \tag{3.65}$$

which is found by multiplying the area of the annulus with the surface density of planetesimals. The total change in the angular momentum of planetesimals in this region is $\Delta J \approx \Delta m \Delta j \sim \Delta m \, |dj/da| \, r_H$, where $j = \sqrt{GM_\star a}$ is the specific angular momentum; this expression is obtained from a Taylor expansion of j. On simplification, we end up with

$$\Delta J \sim \Delta m \left(\frac{1}{2} \sqrt{\frac{GM_\star}{a}} \right) r_H. \tag{3.66}$$

As we mentioned, this change in angular momentum of the planetesimals is equal in magnitude (although opposite in sign) to the change in angular momentum of the planet. The latter is estimated by assuming that the planet remains on a circular orbit, so that

$$\Delta J \sim M_p \left| \frac{dj}{da} \right| \Delta a \sim M_p \left(\frac{1}{2} \sqrt{\frac{GM_\star}{a}} \right) \Delta a, \tag{3.67}$$

where Δa is the extent of the radial migration; we have performed a Taylor expansion of j about a. On combining (3.66) and (3.67), we have

$$\Delta a \sim \frac{2\pi a \Sigma_p r_H^2}{M_p}, \tag{3.68}$$

after solving for the desired quantity Δa.

The outward motion of the planet can be sustained if the planet enters a new zone with existing planetesimals. In quantitative terms, the radial migration must be such that we cross the Hill radius, that is, the condition $\Delta a \gtrsim r_H$ must be satisfied, and therefore we have

$$M_p \lesssim 2\pi a \Sigma_p r_H. \tag{3.69}$$

In other words, efficient migration driven by interaction with planetesimals necessitates that the planetary mass is no larger than the mass of planetesimals occupying its region of influence. The timescale for migration is estimated by means of $\Delta t \sim \Delta\Omega^{-1}$, where the latter,

$$\Delta\Omega \sim \left|\frac{d\Omega_p}{da}\right| r_H \sim \frac{3\Omega_p r_H}{2a}, \tag{3.70}$$

emerges from a Taylor expansion of Ω_p about a, broadly along the lines of (3.55). Finally, the migration rate is calculated as follows:

$$\frac{da}{dt} \sim \frac{\Delta a}{\Delta t} \sim a\Omega_p \frac{\Sigma_p a^2}{M_\star}. \tag{3.71}$$

This rate is independent of the planet mass, and scales linearly with the disc's planetesimal mass $\sim \pi\Sigma_p a^2$, if the regime of sustained migration is valid. For a sufficiently massive disc (viz., two to three orders of magnitude higher than the Kuiper Belt mass), migration may be quite rapid. Planets as massive as Saturn are capable of substantial outward migration. This scenario is compatible with evidence from the solar system, where many Kuiper Belt objects (i.e., small bodies at $\sim$30–50 AU) are documented in 3:2 orbital resonance with Neptune, which might have occurred when Neptune migrated outward, and consequently swept planetesimals into resonance.

Migration driven by interaction with planetesimals plays a crucial role in the *Nice model* (Gomes et al., 2005; Tsiganis et al., 2005), which reproduces many features of the orbital architecture of the giant planets of the solar system. According to this model, Jupiter and Saturn passed through their 2:1 orbital resonance while they migrated by virtue of their interactions with a disc of planetesimals. The migrations and instabilities predicted by the Nice model may also explain the ostensibly enhanced frequency of impacts on the Moon (chronicled by the crater history) dated to $\sim$0.6 billion years after the genesis of the solar system. This so-called *Late Heavy Bombardment* is conjectured to have affected the Earth as well, with vital ramifications for its early environment and habitability (see Section 4.3).

A further development, known as the *Grand Tack model* (Walsh et al., 2011), formulated the initial conditions pertaining to the Nice model by positing an inward Type II migration of Jupiter to $\sim$1.5 AU, followed by the onset of the 2:1 resonance with Saturn, and the subsequent outward migration of both planets, as illustrated in Figure 3.10. This scenario predicts that the inward excursion of Jupiter caused a depletion of planetesimals in the inner solar system, and reportedly constitutes a satisfactory explanation for features like the low mass of Mars, and the population of small bodies in the main asteroid belt (Raymond and Morbidelli, 2022).

This heuristic picture is not definitive, but it suggests that manifestly dynamic – if not outright chaotic – behaviour may be a generic property of the evolution of planetary systems. The observation of a sizeable number of exoplanetary systems comprising giant planets with very short orbital periods (namely, *Hot Jupiters*) lends credence to the (inward) migration scenario. The astrobiological significance is apparent, as the evolution of orbits and the migration history of giant planets could shape the frequency and physical characteristics of terrestrial planets. Models suggest that inward migration and gravitational scattering by massive planets might jointly inhibit the formation of rocky planets in the inner, temperate regions of the disc, often by putting them (or their building blocks) on highly eccentric orbits and into a collision course with the star (e.g., Matsumura et al., 2013; Carrera

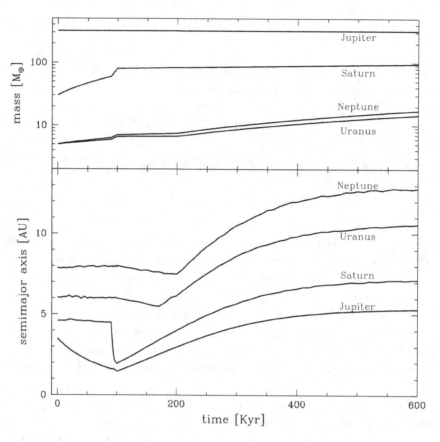

Figure 3.10 Mass growth (top panel) and orbital distance (bottom panel) during the simulated evolution of the four giant planets of the solar system, as per the Grand Tack model. (Credit: Walsh et al., 2011, Figure 1; reproduced with permission from Springer Nature)

et al., 2016). The precise extent to which such effects modulate the habitability of planetary systems remains an open question.

3.3.3   Water delivery

One aspect of planetary formation that has a direct bearing on habitability is the mechanism by which terrestrial planets can acquire their water reservoirs. As we have witnessed, rocky worlds typically assemble inside the snow line of protoplanetary discs, where the availability of icy materials is restricted compared to the outer regions. The final water inventory of such worlds will depend on a complicated interplay of the disc chemistry and its temperature profile, accretion of volatile materials, photoevaporation by high-energy radiation, and the dynamical evolution of the system.

Since Earth is the only confirmed world with life, and is the best understood of all the planets, we can use it as a starting point to frame the problem, keeping in mind that what we observe on our planet may not be emblematic of the norm. Earth's surface water has a mass of $\sim 2 \times 10^{-4} M_{\oplus}$, while the content trapped in its interior is very uncertain, but possibly on the order of 10^{-3} $M_{\oplus}$ (Raymond and Morbidelli, 2022). We are confronted by the key question: what was the primary source of Earth's water?

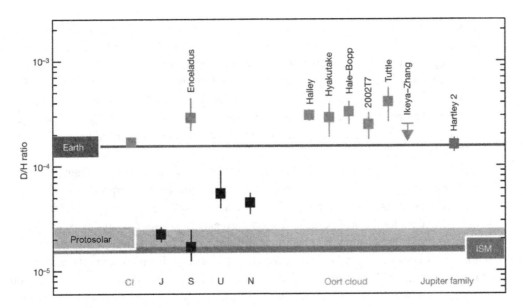

Figure 3.11 The D/H ratios are shown for various objects in the solar system. (Credit: Hartogh et al., 2011, Figure 2; reproduced with permission from Springer Nature)

Isotopic fractionation, tackled in Section 3.1.7, is the main tool for investigating the origin(s) of Earth's water. The canonical D/H ratio in water from Earth's oceans (1.56×10^{-4}) is around an order of magnitude higher than that of the Sun and gas giants, which are probably representative of the gas component of the protoplanetary disc. On the other hand, icy bodies in the solar system exhibit D/H ratios comparable to, or higher than, that of Earth's value (Hartogh et al., 2011; Lis et al., 2019), as revealed in Figure 3.11. As we saw earlier in Section 3.1.7, the D/H ratio is sensitive to temperature, and planetesimals formed within the snow line ought not be much enriched in deuterium. Thus, the Earth may have conceivably assimilated some of its water from cold objects with origin(s) beyond the snow line, and a portion from warmer objects formed inside the snow line.

A plethora of hypotheses have sprung up to elucidate how and when water arrived on Earth and, by extension, how similar mechanisms could operate in exoplanetary systems. The reader can consult Öberg and Bergin (2021) and Izidoro and Piani (2022) for thorough reviews of the subject; we summarise merely a handful of salient candidates here.

- *Wet formation:* In this scenario, the Earth accreted water-rich material during its formation. This process requires the existence of water in dust grains within the snow line, well above its nominal evaporation temperature. In principle, adsorption of hydrogen into silicate grains is a viable pathway leading to the formation of hydrated minerals even at high temperatures (D'Angelo et al., 2019). These grains can end up in rocky planets during accretion, and provide several Earth's oceans worth of water, according to simulations. This scenario, however, does not seem to explain the high D/H ratio of terrestrial water; the delivery of substantial material from beyond the snow line may still be needed towards this end.
- *Pebble 'snow':* As stated in Section 3.1.5, the snow line location is not fixed, but tends to move inward as the disc evolves and cools down. A rocky planetary embryo forming within the snow line

could therefore find itself outside the snow line at subsequent times during the accretion process. Radial drift (discussed in Section 3.2.3) can then transport icy dust and cm-sized pebbles from the outer regions, and facilitate their gradual incorporation within the forming planet (Sato et al., 2016). This water delivery mechanism would be compatible with the D/H ratio observed on Earth, but it is potentially unlikely to have occurred in our solar system, as the formation of Jupiter presumably hindered the drift of icy pebbles. However, it might constitute a viable avenue for water delivery in exoplanetary systems lacking a Jupiter analogue, which are estimated to be a distinct majority (Wittenmyer et al., 2020).

- *Impact delivery:* Terrestrial planets can be bombarded by planetesimals originating from beyond the snow line whose orbit is perturbed by giant planets' growth (especially during the runaway gas accretion phase) and migration. It has been shown, for example, that Jupiter's and Saturn's outward migration in the Grand Tack model triggers the inward scattering of a large number of planetesimals. The ensuing bombardment of Earth could have delivered an amount of water akin to today's oceans, with a D/H ratio matching the observed value (Raymond and Izidoro, 2017; O'Brien et al., 2018). Hence, this mechanism is considered an appealing explanation for the origin of Earth's water. It is, however, unclear whether it would be applicable to exoplanetary systems devoid of giant planets analogous to those of our solar system.

The mechanisms sketched above are not mutually exclusive, and in fact may act in tandem. Their relative weight and efficiency would influence the final water content of rocky exoplanets. Numerical simulations have demonstrated that a wide distribution of planetary water inventories is feasible, ranging from desiccated terrestrial planets to ocean worlds (Raymond et al., 2004; Mulders et al., 2015; Kimura and Ikoma, 2022). How this distribution is linked to planetary formation processes, the architecture of planetary systems, and subsequent dynamical processes (e.g., photoevaporation) is an ongoing area of active research.

3.4   Final remarks: a population synthesis approach

We will round off this chapter with a pictorial summary of the putative distribution of planetary masses and orbital distances in Figure 3.12, predicted by modern numerical models of planetary formation and evolution after inputting a variety of initial conditions. These global models incorporate the various physical mechanisms encountered hitherto in the context of protoplanetary discs, the accretion of solid and gaseous planetary objects, and their dynamical interactions. Thus, a statistical population of planets is synthesised that can be evaluated against actual observations to constrain theoretical scenarios (Mordasini, 2018; Drazkowska et al., 2023; Emsenhuber et al., 2023); for instance, variations in populations emerge depending on whether pebbles or planetesimals are accreted (Brügger et al., 2020).

Many notable features arising from state-of-the-art population synthesis models (Emsenhuber et al., 2023) are apparent upon inspecting Figure 3.12: (1) a vast number of low-mass, rocky planets that accreted interior to the snow line and occupy close-in orbits; (2) a sizeable population of more massive icy planets that formed beyond the snow line, and migrated towards the inner regions in many cases; (3) a transition at $\sim 30\,M_\oplus$, where the shift from solid- to gas-dominated accretion occurs; (4) a population of gas giants at orbital distances below 1 AU, resulting from inward migration; and (5)

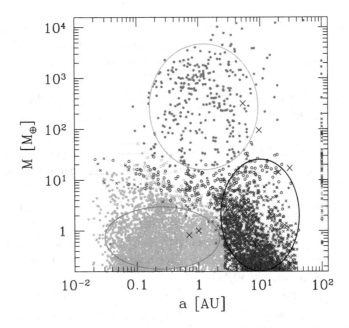

Figure 3.12 Results of planet population synthesis from a global model of planetary formation. Top group is composed of giant planets with a gaseous envelope dominating the total mass; bottom right group comprises volatile-rich planets that accreted icy material from beyond the snow line; bottom left group is composed of rocky planets. The solar system planets (black crosses) are shown for reference. (Credit: Modified from Mordasini, 2018, Figure 8; reproduced with permission from Springer Nature)

a population at 100 AU, which conventionally embodies objects that were scattered away from the planetary system. Most of these characteristics have been observed in actual exoplanetary systems (Zhu and Dong, 2021). The success of this framework testifies to the exciting progress accomplished in the broad arena of theoretical exoplanetary science, although much remains to be done in resolving the numerous open questions.

In closing, it is helpful to take stock of what we have learnt so far. In the previous chapter and the current one, we have delved into how the formation of stars, protoplanetary discs, planets, elements, and molecules has unfolded in the Universe through a panoply of physical and chemical processes. With the stage now set for the birth of our solar system, and the Earth in particular, in the forthcoming chapter we will explore the crucial events in Earth's early history, some of which were fundamentally shaped by the mechanisms covered in this chapter.

3.5   Problems

Question 3.1: Using the data from Table 3.1, compute the angular momentum contained in the orbits of planets. Next, given that the Sun makes a full rotation on its axis in $\sim$27 days, compute its angular momentum (look up the mass and radius of the Sun). Discuss the implications of these results.

Question 3.2: The angular momentum of a cloud with mass $\sim 1 M_\odot$ and diameter ~ 1 pc, endowed with an edge-to-edge velocity difference of ~ 1 km s^{-1}, is $\sim 0.25 M_\odot$ km s^{-1} pc, which is $\sim 10^7$

times higher than the angular momentum of the Sun. Show that the specific angular momentum of the cloud is roughly equal to the specific angular momentum of the Keplerian orbit around the Sun at 100 AU. What does this result suggest with regard to the angular momentum problem discussed in Section 3.1.2?

Question 3.3: Solve the ODE provided in (3.4), with the prescribed equation of state, and obtain (3.5). Integrate the latter equation to derive the middle plane density ρ_0 in terms of the other variables.

Question 3.4: Integrate (3.6) and use the result to derive the temperature profile of the disc given by (3.7). Utilise the Taylor approximation to determine the temperature dependence on R in the limit $R_\star/R \ll 1$.

Question 3.5: Refer to (3.7) and insert the Sun's surface temperature ($T_\odot = 5{,}780$ K) and radius to calculate the location of the snow line in the midplane for water (sublimation temperature of ~ 170 K) and CO (sublimation temperature of ~ 20 K). Are your results a good approximation to the values mentioned in the text? How much hotter should the surface temperature be to shift the water snow line location to 30 AU?

Question 3.6: A good approximation for the pressure of the gaseous component of the protoplanetary disc is $P \propto R^{-n}$. Assuming this scaling law, employ (3.9) to derive the velocity correction in (3.10), and show that it is given by $\eta = nc_s^2/v_K^2$. Next, show that the radial pressure gradient term in (3.9) has the scaling $(GM_\star/R^2)(H/R)^2$.

Question 3.7: In Section 3.1.4, we stated that inter-molecular forces cannot be responsible for the viscosity and angular momentum exchange in protoplanetary discs. Justify this assertion by computing the characteristic viscous timescale from inter-molecular forces at $R = 10$ AU in the disc. This timescale is estimated to be $\sim R^2/\nu_m$, where $\nu_m \sim c_s/(n_H \sigma_H)$ is the molecular viscosity. Utilise $n_H \sim 10^{12}$ cm^{-3} and $\sigma_H \sim 2 \times 10^{-15}$ cm^2 for the number density and cross-sectional area of molecular hydrogen. Is this viscous timescale much larger or smaller than the lifetime of the disc?

Question 3.8: Show that (3.32), the mass growth rate equation for an object of radius S, can be rewritten as

$$\frac{dS}{dt} \approx \frac{\Sigma_p \Omega}{4\rho_d} F_g, \qquad (3.72)$$

where the variables on the RHS are clarified below (3.32). Next, suppose that gravitational focusing is neglected ($F_g \approx 1$) and solve this ODE to determine $S(t)$, after adopting the values $\rho_d = 1$ g cm^{-3}, $\Sigma_p = 10$ g cm^{-2}, and $R = 5$ AU (the distance of Jupiter) from the Sun. At $t = 0$, if the object has $S = 10^3$ cm, how long does it take to reach $S = 10^3$ km? Is this duration compatible with the timescale of a typical protoplanetary disc? How could the same final radius be attained in a shorter interval?

Question 3.9: In the strong gravitational focusing regime, consider two objects with radii S_1 and S_2 (such that $S_1 > S_2$) and masses of m_1 and m_2. How do the fractional increases in mass ($\Delta \tilde{m}/\tilde{m}$) of the two objects compare with each other? What is the physical interpretation of this result?

Question 3.10: If the velocity dispersion and surface density of planetesimals do not vary in time or with the mass, we have shown in (3.37) that the mass growth rate obeys $dm/dt = Cm^{4/3}$, where C is a constant. Prove that this nonlinear ODE admits a solution of the form $m(t) = (m_0^{-1/3} - Ct)^{-3}$. At what time does the mass become infinite? Justify physically why the mass cannot diverge this way in a real-world situation.

Question 3.11: Draw on (3.40) to compute the isolation mass at $a = 1$ AU and $a = 5$ AU for $\Sigma_p = 10$ g cm^{-2} and $M_\star = M_\odot$. Verify that your results match those in the paragraph below (3.40).

Question 3.12: Calculate the isolation mass (M_{iso}) for the following cases. In all instances, you may select $\Sigma_p = 10$ g cm^{-2}, but you will need to consult the internet (e.g., Wikipedia) for the relevant values of $M_\star$ and a.

- Locations of TRAPPIST-1c, TRAPPIST-1f, and TRAPPIST-1h.
- Locations of Proxima Centauri b and Proxima Centauri c.
- Locations of Kepler-186f, Kepler-62e, and Kepler-61b.

Why are these results likely to be inaccurate (see Williams and Cieza, 2011)?

Question 3.13: In Section 3.3.1, we estimated the timescales for Type I and Type II migration. These two equations are both expressible as $\tau \propto M_\star^{n_1} a^{n_2}$. Determine the values of n_1 and n_2 for Type I and Type II migration. For the sake of simplicity, suppose that α and c_s in these two equations are independent of a, although these assumptions are incorrect. Employ the relation $\Sigma \propto a^{-3/2}$ and the definition $H \approx c_s/\Omega_K$ to obtain your answers.

Question 3.14: It can be verified that (3.71) has the form $da/dt \propto M_\star^{n_1} a^{n_2}$. By utilising the information from Question 3.13, find the values of n_1 and n_2 for this migration rate driven by planetesimal interactions.

Question 3.15: Draw on the data provided in Question 3.12, and calculate the migration rate da/dt given by (3.71) for the eight planets specified therein. Obtain estimates for the corresponding migration timescales of these planets by validating the formula $a \times (da/dt)^{-1}$ on dimensional grounds.

Question 3.16: Suppose that we contemplate two protoplanetary discs with the following properties: (a) Disc A has constant angular velocity (Ω is independent of R), (b) Disc B has Keplerian angular velocity. Are Discs A and B stable according to the Rayleigh instability criterion? Are Discs A and B stable if they are weakly magnetised?

Part II

Earth

4 Conditions on Early Earth

In the previous chapter, we investigated the general principles by which planets and planetary systems are assembled together, and dynamically evolve thereafter. We will now pivot to early Earth, and describe some of its key characteristics in the *Hadean eon* – the first geological interval in the history of our planet, spanning $\sim$4,567 Myr ago (Ma) to $\sim$4,000 Ma. This emphasis on Earth is motivated by the simple, yet profound, fact that our planet is the only world confirmed to host life; moreover, it is clearly the most extensively studied world in the solar system or beyond.

In the next chapter, we will delve into the chemical pathways that may have paved the way for the origin of life (i.e., *abiogenesis*) to occur on Earth. However, just as virtually every play is inherently reliant on a theatre for enacting it, understanding abiogenesis requires us to understand the complex settings and circumstances in which this phenomenon played out on Earth (Westall et al., 2023). In other words, the origin of life is deeply rooted in, and intertwined with, the physical, chemical, and geological history of early Earth, and can therefore be envisioned as a planetary-scale process (Smith and Morowitz, 2016; Sasselov et al., 2020).

We tackle this objective in the current chapter, with the caveat that our exposition is brief and selective out of necessity. While our focus is on Earth in the Hadean eon, we generalise our treatment to generic terrestrial planets where possible. For example, we address the evolution of internal heat over time, and the manifold roles played by impacts, on terrestrial planets.

4.1 Channels for internal heating and heat transport

We begin by outlining the various avenues through which planets generate and transport heat, because they have major ramifications for habitability, and they set the stage for delving into the specifics of Hadean Earth.

4.1.1 Sources of internal heat

The first source of internal heat is the *gravitational energy of formation*, to wit, the *accretional heat*. The gravitational potential energy dE_g associated with the protoplanet of mass $M(r)$ and an infinitesimal mass dM at the point of assimilation (i.e., separated by the radius r) is

$$dE_g = -\frac{G M(r)\, dM}{r}, \qquad (4.1)$$

and after integrating this expression between the initial and final radii (see Question 4.1), we end up with the total gravitational energy of formation:

$$E_g = -\frac{3}{5}\frac{GM_p^2}{R_p} = -\frac{16\pi^2 G\rho_p^2 R_p^5}{15}, \qquad (4.2)$$

where M_p, R_p, and ρ_p are the mass, radius, and mean density of the planet, respectively. If we assume that $R_p \propto M_p^{1/3}$, which is true only if mean density is held fixed, we obtain $E_g \propto M_p^{5/3}$ from (4.2), that is, a superlinear scaling of E_g with R_p. Hence, larger worlds possess much higher accretional heat.

This source of heat alone is sufficient to cause melting in some instances. The melting that ensues leads to the formation of a *magma ocean*, wherein a sizeable fraction of the planet ($\gtrsim 10\%$ of the radius) exists in a molten state; by definition, magmas consist of molten rocky materials. This magma ocean is thought to last for an extended period of time, with the estimate for Earth ranging from 50 to 150 Myr (Lichtenberg et al., 2023, Figure 7) based on analysis of isotope ratios. The intricate process of the solidifcation of the magma ocean regulates the composition (especially volatile compounds such as water) and interior structure of the world (Elkins-Tanton, 2012; Krijt et al., 2023), thereby exerting a vital influence on putative habitability.

Let us suppose that all the accretional heat is utilised to increase the planetary temperature by ΔT_p; in actuality, only a fraction of 0.3–0.5 may be available as internal heat (Melosh, 2011, pg. 113). As per thermodynamics, the energy needed to perform this action is $C_p M_p \Delta T_p$, where C_p is the specific heat capacity (i.e., amount of heat needed to raise the temperature of 1 kg of material by 1 K) of the homogeneous planet. On equating this expression with the accretional heat given by (4.2), we arrive at

$$\Delta T_p \approx \frac{3}{5} \frac{GM_p}{C_p R_p} \approx \frac{4\pi}{5} \frac{G\rho_p R_p^2}{C_p}, \tag{4.3}$$

where we have used $M_p = 4\pi \rho_p R_p^3 / 3$ for this homogeneous object. For the Earth, it is estimated that $\Delta T_p \approx 3.8 \times 10^4$ K, whereas $\Delta T_p \approx 1.7 \times 10^3$ K for the Moon (Melosh, 2011, pg. 113); compare with Question 4.2.

A second heat source worthy of mention is the differentiation of the initially (molten) homogeneous planet into distinct regions when high-density material sinks deeper than its low-density counterpart(s); this process corresponds to metal–silicate differentiation and core formation. For the sake of simplicity, consider a planet with only two layers (inner core and outer mantle), denoted by subscripts 'c' and 'm', respectively. When the radius obeys $r < R_c$, the associated mass M_c is given by

$$M_c = \frac{4\pi}{3} \rho_c r^3, \tag{4.4}$$

where ρ_c is the core density. When $r > R_c$, the total mass M_m is

$$M_m = \frac{4\pi}{3} \rho_c R_c^3 + \frac{4\pi}{3} \rho_m \left(r^3 - R_c^3 \right), \tag{4.5}$$

where R_c is the core radius, and ρ_m is the mantle density. The new gravitational energy of formation (refer to Question 4.3), denoted by E_g', is

$$E_g' = -\frac{16\pi^2 G R_p^5}{15} \left[\rho_m^2 + \left(\rho_c^2 - \rho_m^2 \right) \alpha_c^5 + \frac{5}{2} \rho_m \left(\rho_c - \rho_m \right) \alpha_c^3 \left(1 - \alpha_c^2 \right) \right], \tag{4.6}$$

where $\alpha_c \equiv R_c / R_p$ is the core radius fraction. For the Earth, it can be shown that the heat released during differentiation is about one order of magnitude lower than the accretion heat (see Question 4.3). The energy liberated via this process is typically deposited deep within the planetary interior.

The third source that we highlight is the heat derived from radioactive decay of isotopes, known as *radiogenic heat*. This category is further divided into short- and long-lived radioactive isotopes. The most prominent member of the former group is aluminium-26 (^{26}Al), whereas uranium-235 (^{235}U) and potassium-40 (^{40}K) are two examples from the latter class. The power P_r released by radioactive decay is conventionally represented as

$$P_r = \sum_i P_{ri}, \quad P_{ri} \equiv Q_i(t) \times M_p, \tag{4.7}$$

where Q_i (units of W/kg) is the energy liberated per unit time per unit mass (of the object under consideration) for the i-th isotope, which is time-dependent because the abundances of radioactive isotopes exponentially decline with time. In the building blocks of planets around 4.6 Gyr ago, it is estimated that $Q_i \sim 1.7 \times 10^{-11}$ W/kg for potassium-40 (Melosh, 2011, pg. 119), while $Q_i \sim 2.5 \times 10^{-7}$ W/kg might be valid for aluminium-26. On modern Earth, as much as 80% of the total internal heat budget may stem from long-lived radioactive isotopes (The Borexino Collaboration, 2020a).

The corresponding energy produced by the i-th radioactive isotope is roughly $E_{ri} \sim P_{ri}\tau_i$, where τ_i is the half-life of this isotope; this ansatz follows from recognising that Q_i declines substantially beyond timescales of order τ_i. The above relation implies that $E_{ri} \propto M_p$, and we have previously indicated that $E_g \propto M_p^{5/3}$. Hence, smaller objects are more likely to experience initial melting due to (short-lived) radioactive isotopes in comparison to accretion heat, but the converse is plausible for larger worlds. By equating the released energy E_{ri} with $C_p M_p \Delta T_p$ as before, we can gauge the temperature increase ΔT_p of the body caused by radiogenic heating:

$$\Delta T_p \sim \frac{Q_i \tau_i}{C_p}. \tag{4.8}$$

This expression is independent of M_p to leading order, suggesting that the temperature increase arising from radiogenic heating might be comparable for most worlds of a given composition.

4.1.2 Avenues for heat transport and cooling

There are three primary modes of heat transport: conduction, convection, and radiation. In principle, all these processes can operate in tandem, and consequently enable the planet in question to undergo cooling. The reader may consult the detailed reviews of this subject by Schubert et al. (2001) and Douce (2011, Chapter 3) for additional information.

Conduction involves heat transport from hotter to colder regions, typically attributed to solid objects. This mechanism is prominent for small bodies (e.g., planetesimals) at temperatures below melting point. It is possible to determine the temperature T as a function of time by numerically solving the heat conduction equation. We can, however, gain a rough estimate of the conduction timescale t_{cond} by treating it as a purely diffusive process in a uniform medium, in which case dimensional analysis corroborates

$$t_{cond} \sim \frac{L^2}{\kappa}, \tag{4.9}$$

where L is the length scale of the system, and κ is the thermal diffusivity.

Given that $t_{cond} \propto L^2$, it is evident that smaller objects cool faster (and vice versa) via conduction. If we consider $L \sim R_\oplus$ for Earth and $\kappa \sim 10^{-6}$ m^2/s for representative rocks, we end up with $t_{cond} \sim 1.3 \times 10^3$ Gyr, which is much longer than the age of the Universe; conduction would, therefore, be ineffective for worlds of this size. On the other hand, for an asteroid with $L \sim 100$ km, we obtain a reasonable timescale of $t_{cond} \sim 0.3$ Gyr. This length scale is close to the thickness of Earth's *lithosphere*, the rigid and solid outermost layer (comprising the crust and part of the mantle) where conduction is documented to be prevalent.

Convection entails heat transport by means of physical motion(s) of material(s). This process is predicted to be widespread and important in rocky planets (and planetesimals) at various stages of their evolution. In qualitative terms, a 'blob' (technically, parcel) of fluid material subject to heating experiences a decrease in density (i.e., becomes lighter), thereby leading to its upward movement towards the cooler regions. The ensuing cooling therein increases the density of the parcel, and causes it to sink downward into the hotter regions. This cycle is repeated, thus facilitating the transfer of heat.

It is possible to construct a timescale t_{conv} for the overturn caused by convection. The dimensionless Rayleigh number (Ra) is equivalent to t_{cond}/t_{conv} for the same layer thickness. Hence, high values of Ra indicate the existence of strong convection. The onset of convection occurs when Ra exceeds a critical Rayleigh number Ra_c, whose value depends on the particular system, but is typically of order 10^3–10^4. The Rayleigh number is defined as

$$\mathrm{Ra} = \frac{g\alpha\rho_0 \Delta T L^3}{\eta\kappa}, \tag{4.10}$$

where ρ_0 is the fiducial fluid density, α and η are the thermal expansion coefficient and dynamic viscosity of the fluid, ΔT is the temperature difference across the fluid layer, and g is the acceleration due to gravity of the given world. In the Earth's mantle (viz., a layer made up chiefly of silicate rocks), vigorous convection is operational (refer to Question 4.5).

All objects ultimately emit blackbody radiation, which gives rise to radiative cooling. In the case of airless worlds, the radiation is directly emitted to space. The presence of an atmosphere complicates this stage of heat transport somewhat, as the latter can occur via convection or radiation. The timescale associated with radiative cooling t_{rad} is estimated as $t_{rad} \sim E/(dE/dt)_{rad}$, where E is the internal heat content and $(dE/dt)_{rad} \equiv P_{rad}$ is the power lost via radiation. We write P_{rad} as the product of the total surface area $4\pi R_p^2$ and the net radiative flux $\sigma_{SB}\left(T_p^4 - T_{eq}^4\right)$, where T_p is the planet's emission temperature and T_{eq} is the equilibrium temperature (Salyk and Lewis, 2020, Chapter 6). After simplification, we end up with

$$t_{rad} \sim \frac{E}{4\pi\sigma_{SB}R_p^2 \left(T_p^4 - T_{eq}^4\right)}. \tag{4.11}$$

In Section 4.1.1, we noted that heat of accretion and differentiation scale as $E \propto R_p^5$, whereas radiogenic heating obeys $E \propto R_p^3$. On substituting these relations into (4.11), we arrive at $t_{rad} \propto R_p^3$ and $t_{rad} \propto R_p$, depending on the context. In either scenario, however, it is apparent that smaller worlds are susceptible to swifter temperature decrease.

To sum up, as per the heuristic analysis in this section, smaller planets could end up cooling faster (in relative terms), if all other factors are held equal. The enhanced loss of internal heat is presumed to partially explain why these worlds can possess larger crusts, and may often fail to develop *plate tectonics* (see Section 8.3.2), as the latter requires sufficient internal heat to permit the existence of mobile plates in the lithosphere. The cooling might also explain why smaller planets potentially experience early cessation of volcanism, and lack long-term strong magnetic fields, the latter of which arise from the so-called *geodynamo* mechanism involving convection in the outer core (explored further in Section 8.3.3). We will revisit both plate tectonics and magnetic fields in Section 4.2.4.

4.2 A rough timeline of Hadean Earth

The first, and foremost, point to appreciate with regard to the Hadean eon is that Earth's geological record is incredibly scarce. For example, the Acasta Gneiss Complex (Canada) harbours possibly the oldest whole rocks, which are from 4.03 Gyr ago (Ga). Prior to this period, knowledge of Earth's crust is almost exclusively confined to hardy mineral fragments called detrital *zircons*, namely, zirconium(IV) silicate ($ZrSiO_4$). Hence, owing to the paucity of data, the dates and processes delineated hereafter must be viewed with a healthy dose of scepticism and caution.

4.2.1 The steps up to the Moon-forming impact

The solar system clock starts ($t = 0$) at ∼4,567 Ma, because this is when calcium–aluminum-rich inclusions (CAIs), the oldest solid materials in the solar system, formed as per lead–lead dating (Connelly et al., 2012). In this method, the ratios of lead isotopes derived from different radioactive decay pathways are measured and plotted, from which the age is inferred. A succinct review of dating samples via computing the appropriate isotope ratios is furnished in de Pater and Lissauer (2015, Chapter 8.6).

By the time that the dispersal of gas in the protoplanetary disc had occurred ($t \sim 4$ Myr), proto-Earth may have grown to ∼$0.5\,M_\oplus$; isotopic evidence suggests that Earth accreted a mass of ∼$0.8\,M_\oplus$ by $t \sim 10$–30 Myr (Lammer et al., 2021, Section 4). The process of accretion continued nearly until the formation of the Moon, which we tackle below. During this entire period, which may have lasted up to ∼150 Myr, the formation of Earth's core and differentiation of its interior is expected to have taken place; however, the associated timescale may have been as low as $t \sim 30$ Myr (Lichtenberg et al., 2023, Section 3.2.1). The duration of Earth's magma ocean was broadly coeval, as it could have persisted for $t \sim 50$–150 Myr.

Now, we turn our attention to the formation of the Moon. The precise timing of this event has proven hard to pin down. The range of possible times, which are sensitive to the radiometric dating methodologies involved (e.g., uranium–lead; hafnium–tungsten), is $t \approx 30$–150 Myr (i.e., 4,417–4,537 Ma); the recent review by Halliday and Canup (2023) has advocated a slightly narrower interval of $t \approx 70$–120 Myr. In what follows, we shall adopt a fiducial timing of ∼4.5 Ga ($t \sim 70$ Myr) for the formation of the Moon, as well as substantial completion of Earth's accretion.

Many hypotheses have been advanced for the formation of the Moon (consult Question 4.6), of which the *Giant Impact hypothesis* is considered the most plausible (Canup and Asphaug, 2001; Jacobsen, 2005), wherein the Moon formed from the ejecta of a collision between the early Earth and a roughly Mars-sized body (called *Theia*). A number of reasons support this hypothesis, such as: (1) the Earth–Moon system has an unusually high angular momentum, which is compatible with a giant impact; (2) the small size of Moon's core (i.e., its iron-poor nature) is explainable through numerical simulations whereby the impactor's core merges with that of the Earth; and (3) similar relative abundances of certain non-volatile elements and isotopes, which is suggestive of a common origin.[1]

A tentative timeline of major events leading up to the Moon-forming impact and beyond is depicted in Figure 4.1.

---

[1] On the other hand, some mysteries concerning isotopic similarities between the Earth and Moon remain, owing to which several variants of the classical Giant Impact hypothesis have sprung up (Halliday and Canup, 2023, Table 1).

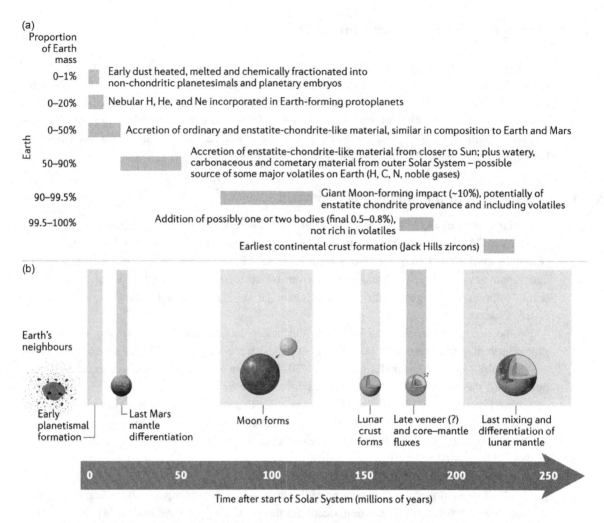

Figure 4.1 (a) Stages in Earth's accretion history (expressed as per cent of Earth's mass), along with the relevant timescales and pathways. (b) Major stages of formation and differentiation of Earth, Moon, and Mars. (Credit: Halliday and Canup 2023, Figure 4; reproduced with permission from Springer Nature)

4.2.2   Evolution of volatiles and continental crust

We have seen that it may have taken $t \approx 70$–120 Myr for the Moon-forming impact, and the majority of Earth's accretion to occur. In the aftermath of the Moon-forming impact, it is thought that $\sim 0.5\%$ of Earth's mass was acquired through collisions with a small number of relatively dry planetesimals in the solar system, known as the *late veneer*. During this entire interval and even earlier, volatiles (substances with low condensation temperatures) were not just assimilated but also subjected to diverse phenomena, some of which caused their depletion. A schematic illustration of these processes at various stages commencing with planetesimal formation is depicted in Figure 4.2. In particular, the so-called *catastrophic outgassing* of volatiles, presumed to have occurred during the course of crystallisation of the magma ocean (Lammer et al., 2018, Section 3.3), is shown in the fourth panel.

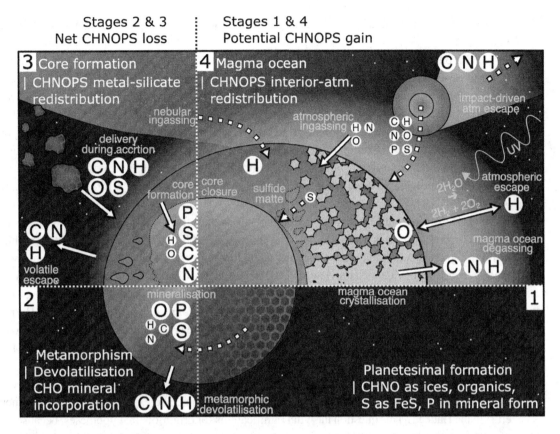

Figure 4.2 Fluxes of biologically essential elements (C, H, N, O, P, and S) during planet formation: (1) formation of planetesimals; (2) planetesimal heating by short-lived radioactive isotopes; (3) planet accretion, core formation and differentiation; and (4) magma ocean cooling and crystallisation on Earth-sized worlds. (Credit: Krijt et al. 2023; figure by O. Shorttle, reproduced with permission from Astronomical Society of the Pacific)

We explored the delivery of water and other volatiles in Section 3.3.3, but we caution that many uncertainties remain: Broadley et al. (2022), Izidoro and Piani (2022), and Krijt et al. (2023) provide state-of-the-art reviews of this subject. It was originally thought that comets (owing to their water-rich inventories) could have delivered the majority of Earth's water, but this hypothesis is considered unlikely on isotopic grounds; for example, the deuterium-to-hydrogen (D/H) ratio in comets is a few times higher than Earth's oceans (see Figure 3.11). It is now thought that comets may have contributed only ∼1% of Earth's water (Krijt et al., 2023, Section 2.2.2). However, it is plausible that comets supplied a significant fraction of certain volatiles such as noble gases (Marty et al., 2016).

The D/H ratio of *carbonaceous chondrites* – a category of meteorites with significant water inventories (up to ≳10% by weight) and organics, derived from parent bodies potentially in the outer asteroid belt – closely matches that of water from the Moon and Earth (Saal et al., 2013), thus bolstering hypotheses that these objects supplied the bulk of Earth's water. Recent work has, however, established that the progenitors of *enstatite chondrites* – meteorites with parent bodies possibly closer to Earth – may have supplied enough hydrogen to form about three times the water in Earth's oceans (Piani et al., 2020). Hence, Earth's water seems to have originated from both inner and outer solar system materials, mediated by snow line migration, dynamics of giant planets, and more (Izidoro and Piani, 2022).

The Moon-forming impact boosted temperatures significantly, thus driving ocean evaporation (refer to Section 4.3) and planetary melting. Models suggest that Earth cooled to a temperature of around 500 K in merely ~10 Myr after the Moon-forming impact, followed by the condensation of water vapour to form oceans (Sleep et al., 2014).[2] However, the temperature was initially too hot for life-as-we-know-it, on account of ~100 bar of atmospheric carbon dioxide (CO_2), a potent greenhouse gas (refer to Section 8.1.2), outgassed in the magma ocean epoch. The sequestration of most CO_2 in the mantle was regarded as difficult, but recent research indicates that a wet and heterogeneous mantle might have completed this process in a timescale of ~160 Myr (Miyazaki and Korenaga, 2022).

The earliest traces of liquid water on Earth's surface are from ~4.3–4.4 Ga, as per analyses of detrital zircons from Jack Hills, Australia (Wilde et al., 2001). The detection of ^{18}O-enrichment in these minerals was interpreted to signify their low-temperature interactions with liquid water; other geochemical evidence is reviewed in Harrison (2020, Chapter 7). Likewise, the existence of continental crust by ~4.3–4.4 Ga was inferred from these zircons. The chemical composition, isotope ratios, and the presence of certain trapped materials (*inclusions*) support this statement (Harrison, 2020, Chapter 7), but the evidence is not sans ambiguities. The crust and oceans perhaps occasionally experienced melting and evaporation, respectively, in the Hadean because of large impacts (Marchi et al., 2014).

However, while (proto)continents may have existed by 4.4 Ga, or even earlier (Santosh et al., 2017), an enduring mystery to this day is: what was the volume and exposed area of the continental crust in the Hadean? This question has vital ramifications for origin-of-life hypotheses, because some of them necessitate land-based environments (as surveyed in Section 5.5). We will not delve into this debate, but it suffices to say that some studies have posited a Hadean *water world* with minimal land coverage (see reviews by Hawkesworth et al. 2020; Russell 2021), while others have contended that this eon featured significant continental crust, namely, $\gtrsim 20\%$ of its current volume (Harrison, 2020; Korenaga, 2021).

4.2.3   Atmosphere and faint young Sun paradox

In Sections 4.2.1 and 4.2.2, we have already encountered a plethora of controls on the early atmosphere: (1) accretion of matter (including gaseous H and He); (2) large impacts, including the Moon-forming impact; and (3) crystallisation of the magma ocean. The latter, for instance, is linked with the catastrophic outgassing of volatiles, as highlighted in Section 4.2.2. Large impacts can not only deplete or replenish atmospheres but may also deliver metals (e.g., iron) and induce chemical reactions that create transiently reducing atmospheres (Zahnle et al., 2020), to wit, rich in electron donors. In addition, mechanisms such as atmospheric escape are tackled in Sections 8.2.3 and 9.3.1, which alter the atmospheric composition.

Hence, in light of these complexities, definitive statements concerning the inventory of gases in the Hadean atmosphere is challenging. With that being said, multiple lines of evidence point to very

---

[2] Ocean properties such as the temperature and pH are poorly constrained in the Hadean; for instance, the proposed oceanic pH at ~4.3 Ga ranges from acidic at ~5 (Ueda and Shibuya, 2021) to seawater-like alkaline at ~8 (Kadoya et al., 2020).

low abundances of molecular oxygen (O_2) throughout this eon (Catling and Kasting, 2017, Part II). Once the lighter elements H and He acquired during accretion were lost via atmospheric escape, the dominant components of the Hadean atmosphere were molecular nitrogen (N_2) and CO_2 (Catling and Zahnle, 2020). The concentration of the former is uncertain, but may have been comparable to the present-day value or higher/lower by a factor of a few. As for CO_2, the initial inventory of $\sim$100 bar (Sleep et al., 2014) might have declined to $\lesssim 10^{-2}$ bar due to weathering reactions with ejecta produced as a result of frequent impacts in the Hadean (Kadoya et al., 2020).

An oxidised magma ocean, expected on Hadean Earth (Pahlevan et al., 2019), facilitates the outgassing (e.g., through volcanoes) of aforementioned oxidised species such as CO_2 and H_2O (Trail et al., 2011). However, noteworthy concentrations of reduced gases like molecular hydrogen (H_2), methane (CH_4), and ammonia (NH_3) were likely available in the Hadean, even if for transient periods. Two avenues identified in this context include: (1) (volcanic) outgassing of these gases from accreted chondritic materials (mentioned in Section 4.2.2); and (2) reduced materials delivered during the late veneer that reacted with Earth's oceans and atmosphere (Zahnle et al., 2020). Figure 4.3 illustrates the major events that unfolded in the Hadean and the subsequent Archean eon (2.5–4.0

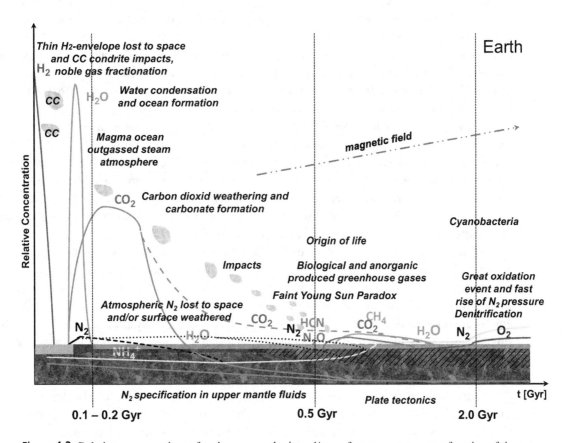

Figure 4.3 Relative concentrations of various atmospheric and/or surface components as a function of time, reflecting Earth's geological evolution in its first 2 Gyr of history. 'CC' represents carbonaceous chondrites, a class of meteorites accreted by Earth. (Credit: Lammer et al. 2018, Figure 24; reproduced with permission from Springer Nature)

Ga), as well as the dynamical evolution of atmospheric constituents; note that the qualitative trends are more reliable than the delineated quantitative values.

Lastly, as elaborated in Section 8.1.4, the Sun's luminosity in the Hadean was merely $\sim$75% of its present-day value. In turn, the lower luminosity would translate to a surface temperature below freezing if a similar albedo and greenhouse effect as modern Earth is assumed (Sagan and Mullen, 1972), yet there is considerable evidence that Hadean Earth hosted some liquid water. This apparent discrepancy is known as the *faint young Sun paradox* (FYSP), although this paradox is canonically invoked with respect to the Archean eon. As indicated, higher abundances of greenhouse gases in the Hadean (and Archean) – primarily CO_2, but possibly CH_4 and nitrous oxide (N_2O) – may suffice to resolve the FYSP, in tandem with other effects such as albedo and additional heat sources (Charnay et al., 2020).

4.2.4 Plate tectonics and magnetic field

In plate tectonics, the lithosphere consists of multiple distinct plates, which move at non-zero velocities relative to each other. These plates undergo recycling into the mantle via the mechanism of *subduction*, which occurs when two plates converge, and the heavier of the two sinks beneath its lighter counterpart. While it is often reported that stagnant-lid tectonics (wherein the lithosphere comprises a single plate) and plate tectonics are the only two tectonic modes, alternative modes such as episodic lid and plutonic-squishy lid tectonics are feasible. Furthermore, a given planet can transition from one mode to another at some stage. We refer the reader to Palin et al. (2020) and Brown et al. (2020) for detailed overviews of plate tectonics.

It is widely, albeit not universally, held that plate tectonics may be essential for long-term habitability because it functions as a thermostat, among other functions, as described later in Section 8.3.2. This raises the crucial question: when did plate tectonics commence on Earth? A broad spectrum of answers have been proposed, ranging from as recently as $\sim$0.85 Ga (Stern, 2018) to shortly after the crystallisation of the magma ocean (Harrison, 2020), while others have posited a timing of $\sim$3 Ga (Hawkesworth et al., 2020); this variation reflects the diversity of processes at play, as well as some subtle differences in defining what constitutes plate tectonics.

Publications that have advocated the early onset of plate tectonics (in/near the Hadean eon) have invoked specialised mechanisms such as: (1) initiation of subduction (and plate tectonics) by features known as *mantle plumes* that are associated with convection at high Rayleigh numbers (Gerya et al., 2015); and (2) initiation of subduction by large impacts that serve to thin and/or weaken the lithosphere (O'Neill et al., 2017). In terms of empirical evidence, analysis of 3.3–4.3 Ga Jack Hills zircons suggests that they may have crystallised from magmas formed near subduction zones (Turner et al., 2020; Chen et al., 2023). We caution, however, that sparse data make it challenging to draw robust conclusions.

We now turn our attention to the planetary magnetic field, which is generated by the geodynamo mechanism, elucidated in Section 8.3.3 – as outlined therein, there is some tentative, albeit debated, evidence that a strong planetary magnetic field is valuable for habitability. Just as with plate tectonics, ascertaining when the onset of the geodynamo happened is rendered vital. The analysis of magnetic inclusions recovered from Jack Hills zircons of ages 3.3–4.2 Ga have led certain authors to conclude that Earth possessed a magnetic field in the Hadean (e.g., Tarduno et al., 2015, 2023), but others have contended that these inclusions are of later age or poor recorders of the magnetic field (e.g., Tang et al., 2019; Borlina et al., 2020). While early Earth may have indeed hosted a magnetic field, its strength and timing are thus still unsettled (Tikoo and Evans, 2022).

If a Hadean geodynamo was indeed operational, appropriate energy sources for powering this phenomenon must be identified. Potential candidates in this ongoing area of research include the precipitation of magnesium-containing minerals from the core, the precipitation of silicon dioxide from the core, thermal cooling of the core, radioactive decay, and tidal forces.

4.3 The timeline and consequences of impacts

In this section, we will furnish a general treatment of the upsides and downsides of impacts, which has relevance for not only Earth but also potentially habitable worlds in the solar system and beyond. Before doing so, we offer a brief account of the *Late Heavy Bombardment* (LHB) that might have been initiated towards the end of the Hadean.

4.3.1 The Late Heavy Bombardment

The frequency of impactors colliding with the Earth loosely declined with time, which is intuitive since the number of leftover bodies likewise decreased over time owing to collisions, scattering and ejection, and so forth. However, this trend is not necessarily monotonic, because it was mediated by multiple factors such as the potential migration(s) of giant planets in the solar system (Nesvorný, 2018). Therefore, a sawtooth profile for the impactor rate – with modest spikes superimposed upon an overall decline – has been proposed in some papers, as reviewed in Bottke and Norman (2017).

The inner solar system harbours worlds such as the Moon, Mercury, and Mars with extensive cratering records (owing to lower geological activity), thus functioning as recorders of the frequency of impacts over time. In the 1960s and 1970s, isotopic signatures (extracted from *Apollo* rock samples) mapped to intense metamorphic events compatible with impacts were interpreted as evidence for a *lunar cataclysm* on account of their apparent clustering at $\sim$3.9 Ga (Tera et al., 1974; Bottke and Norman, 2017). These early publications paved the way for the LHB hypothesis: at $\sim$4 $\pm$ 0.1 Ga, the inner solar system purportedly experienced a distinctly elevated rate of bombardment by impactors, followed by a decline (Zellner, 2017; Hartmann, 2019). The dynamical basis of the LHB was posited to arise from the migration of the giant planets in our solar system after a lengthy period of relative quiescence (Gomes et al., 2005).

While the lunar cataclysm and LHB were historically popular, recent evidence increasingly favours a more-or-less smooth decline of the impactor rate over time, that is, the *accretion tail scenario* (Morbidelli et al., 2018). Support for the accretion tail scenario stems from numerical simulations of solar system dynamics, investigations of lunar samples with newer methods, and geochemical analyses of other worlds like the asteroid Vesta (Michael et al., 2018; Morbidelli et al., 2018; Cartwright et al., 2022; Raymond and Morbidelli, 2022). If the LHB was non-existent or overstated, then its ramifications for Earth's habitability and biosphere would likewise diminish (Zellner, 2017).

4.3.2 Pros and cons of large impacts

We round off our analysis with a synopsis of a select handful of benefits and detriments expected to ensue from large impacts.

We start by listing a select few negatives associated with large impacts.

1. Impacts are capable of triggering substantial atmospheric loss (consult de Pater and Lissauer 2015, Chapter 4.8).
2. After the impactors deposit their kinetic energy, they could deplete planets of their surface volatiles; in extreme cases, they can cause (transient) evaporation of the oceans (refer to Question 4.7).
3. Partly due to the preceding two reasons, impacts may pose impediments to pre-existing biospheres, and might drive large-scale biological extinctions, although they may not necessarily wipe out all life.
4. Impacts could strip planets of a fraction of their crusts and radioactive isotopes therein, thereby potentially shrinking the internal heat budget, as well as the inventory of certain elements.
5. Because of the dust produced in the aftermath of impacts in conjunction with the weathering of the ejecta leading to reduction in CO_2 levels, a sharp decrease in the temperature and entering an ice-dominated state is theoretically feasible (Kadoya et al., 2020).

Now, we turn our attention to a subset of the positives attributed to large impacts. Before doing so, note that giant impacts can engender a magma ocean, whose significance we have touched on in Sections 4.1.1 and 4.2.

1. Impacts could weaken the lithosphere and induce plate tectonics (Section 4.2.4), as well as promote the genesis of continents (Johnson et al., 2022).
2. Under some conditions, impacts might initiate or modulate the geodynamo (Cattaneo and Hughes, 2022; Lichtenberg et al., 2023).
3. In certain cases, impacts may *add* volatiles (including organics) to the atmosphere or surface instead of depleting them. Moreover, impacts can create transient reducing atmospheres (see Section 4.2.3).
4. These reducing atmospheres are conducive to the synthesis of prebiotic compounds; furthermore, the shock waves produced during impacts could also enable prebiotic synthesis (refer to Section 5.4).
5. Impacts excavate craters and give rise to *hydrothermal systems*, characterised by the circulation of heated fluids. Such sites are promising candidates for the origin of life, and might persist for $\sim$1 Myr for craters with diameters of $\sim$100 km (Osinski et al., 2020).

4.4   Problems

Question 4.1: By starting with (4.1) and using the mass-radius relation $M(r) = 4\pi r^3 \rho_p/3$ for a homogeneous object of radius r and density ρ_p, prove that (4.2) is obtained on integrating (4.1) between $r = 0$ and $r = R_p$.

Question 4.2: Calculate the value of ΔT_p from (4.3) for (1) Earth; (2) Mars; (3) Moon; and (4) Ceres. Which of these worlds, if any, exceeds the melting point of iron ($\sim$1,800 K) and/or rock ($\sim$1,200 K)? Look up the data for C_p or adopt $C_p \sim 10^3$ J kg^{-1} K^{-1} (Salyk and Lewis, 2020, Chapter 6.3.3).

Question 4.3: From Section 4.1 and the following definition of the gravitational energy E'_g, verify that (4.6) is the final result.

$$E'_g = \int_0^{R_c} \frac{G M_c \, dM_c}{r} + \int_{R_c}^{R_p} \frac{G M_m \, dM_m}{r}. \tag{4.12}$$

In (4.12), the first and second terms on the RHS are the contributions from the core and mantle, respectively. Next, compute $\delta_g = |E_g - E'_g|/E_g$ (measure of heat released via differentiation) for the Earth; you may use $\rho_c \approx 13$ g/cm^3, $\rho_m \approx 4.5$ g/cm^3, $\rho_p \approx 5.5$ g/cm^3, and $\alpha_c \approx 0.55$.

Question 4.4: Estimate ΔT_p for radiogenic heating using (4.8) for aluminium-26 ($\tau_i \approx 7.2 \times 10^5$ yr) and potassium-40 ($\tau_i \approx 1.25 \times 10^9$ yr), and compare the results with Question 4.2; you can assume that $C_p \sim 10^3$ J kg^{-1} K^{-1}.

Question 4.5: In the Earth's mantle, adopt $L \approx 3,000$ km, $\alpha \approx 10^{-5}$ K^{-1}, $\rho_0 \approx 3.5$ g/cm^3, $\eta \approx 10^{21}$ Pa s, $\kappa \approx 10^{-6}$ m^2/s, and $\Delta T = 1,400$ K (Douce, 2011, pg. 156). For these values, compute the Rayleigh number (Ra), the diffusion timescale (t_{cond}), and the convection timescale (t_{conv}).

Question 4.6: Earlier models for the formation of the Moon include the fission hypothesis and capture hypothesis. Briefly summarise their key aspects, and explain why they are not considered plausible; make sure to cite peer-reviewed sources. What is the minimum angular velocity required for a planet of mass M_p and radius R_p to eject material into space (i.e., exceed the escape velocity)? Compute this value for Earth, and the rotation period.

Question 4.7: Prove that the total energy E_i carried by an impactor of mass m_i when it strikes a planet of mass M_p and radius R_p is

$$E_i = \frac{1}{2} m_i v_\infty^2 + \frac{G m_i M_p}{R_p}, \tag{4.13}$$

where v_∞ is the initial velocity of the impactor far away from the planet. Next, show that the energy E_b needed to boil off the oceans completely is

$$E_b = M_{oc} C_w (373 \, \text{K} - T_{oc}) + M_{oc} L_w, \tag{4.14}$$

where M_{oc} and T_{oc} are the ocean mass and temperature, C_w and L_w are the specific heat capacity and latent heat of vaporisation for water. From the above two relations and setting $v_\infty \approx 15$ km/s, compute the minimum mass of the impactor that would be necessary to boil modern Earth's oceans. How does this mass compare with that of the asteroid Vesta?

5   Origin(s) of Life on Earth

In the preceding chapter, we have learnt that the Hadean eon was not as uninhabitable as once thought, and may have harboured liquid oceans and landmasses ~4.3–4.4 Ga, as well as perhaps plate tectonics and a global magnetic field, although the geological evidence is sparse and inherently subject to ambiguity. Even if we accept the reasonable notion that the Hadean and/or early Archean were indeed conducive to the emergence of life, we are still confronted with a multitude of unresolved mysteries, of which we shall highlight merely three of them:

1. *How* did the origin(s) of life from non-living matter (i.e., known as *abiogenesis*) transpire?
2. *Where* was abiogenesis engendered on Earth? In other words, what were the sites where the transition(s) from non-life to life occurred?
3. *When* did the origin(s) of life happen on Earth?

Collectively, this trio of questions (especially the first) embody one of the three fundamental questions in astrobiology elucidated at the start of this book (in Section 1.1): *where did we come from?*

It is easy to fathom why answering this question is daunting. First, we are stymied by the fact that Earth's geological record beyond ~4 Ga is patently scarce, consequently hampering our ability to gaze into the past. Second, pinning down the origin(s) of life is difficult because the boundary between 'life' and 'non-life' is challenging to determine precisely; as mentioned in Section 1.1.3, even defining life is far from straightforward. Third, the sheer complexity of just the smallest known microbes ought not be underestimated, thereby illuminating how hard it might be to construct even the 'simplest' cells from scratch in laboratory settings. Last but not least, understanding the origin(s) of life requires an integrated, multidisciplinary approach (as illustrated below) that is innately complicated, particularly when grappling with all the missing pieces of knowledge.

To build on the last theme, there are many perspectives or lenses through which the origin(s)-of-life conundrum can be tackled. For instance, we could focus exclusively on *chemistry*, and attempt to synthesise biological molecules and pathways universal to current living systems by commencing with simple inorganic feedstock molecules. Alternatively, we may foreground the *physics* of life, such as thermodynamic disequilibrium (Branscomb et al., 2017), and endeavour to employ appropriate branches (e.g., non-equilibrium statistical mechanics) for unravelling this phenomenon.

Continuing in this vein, we can appeal to the datum that abiogenesis did not manifest in a vacuum, but in real-world environments on Earth, thus necessitating a deep understanding of our planet's *geology* in that epoch. This emphasis is pertinent since the components of the Earth system are intertwined with each other, effectively rendering life a planetary phenomenon (Smith and Morowitz, 2016; Sasselov et al., 2020). Finally, living systems reveal great sophistication in the avenues whereby they acquire, store, transmit, and act upon *information*. Therefore, areas like information theory might shed light on the enigma of life's origin(s).

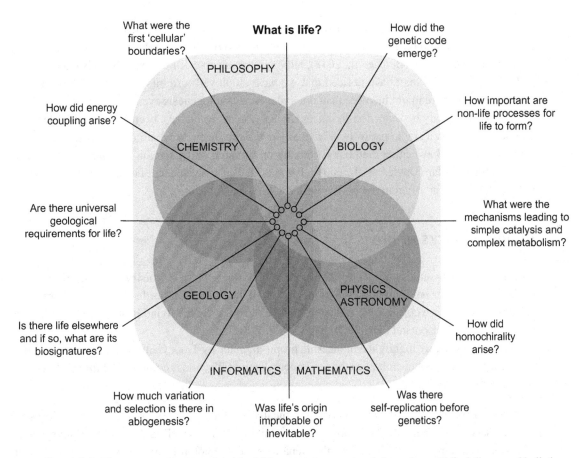

What were the first 'cellular' boundaries?

What is life?

How did the genetic code emerge?

PHILOSOPHY

How did energy coupling arise?

How important are non-life processes for life to form?

CHEMISTRY   BIOLOGY

Are there universal geological requirements for life?

What were the mechanisms leading to simple catalysis and complex metabolism?

GEOLOGY   PHYSICS ASTRONOMY

Is there life elsewhere and if so, what are its biosignatures?

How did homochirality arise?

INFORMATICS | MATHEMATICS

How much variation and selection is there in abiogenesis?

Was life's origin improbable or inevitable?

Was there self-replication before genetics?

Figure 5.1 The core questions in origin(s)-of-life research warrant an inherently multidisciplinary and holistic perspective built on the basis of synergy between outwardly disparate fields. (Credit: Preiner et al. 2020, Figure 1; CC-BY 4.0 license)

At this stage, it is tempting to enquire: which of the above perspectives/lenses is correct? It seems plausible that the suitable response is that all of these approaches are valid, but none of them in isolation may paint the full picture and tell the complete story. This situation is reminiscent of the insightful parable of the blind men and an elephant – which originated in ancient India more than 2,000 years ago – wherein several blind men feel different parts of this animal and arrive at contrasting, but (partly) correct, impressions of an elephant. By integrating these impressions, however, it should be possible to obtain a deeper comprehension of the object or phenomenon in question (i.e., more than the sum of the parts alone); this theme is embodied in Figure 5.1 for origin(s)-of-life research.

Taking our cue from the aforementioned parable, we will seek to address the *how* and *where* of the origin(s) of life from multiple viewpoints in this chapter. We shall not delve into the matter of *when* life emerged on Earth (see Question 5.1), as this important topic is briefly examined in the context of biological signatures of life (*biosignatures*) in Chapter 13. In a nutshell, it may safely be said that life has been documented on Earth by ∼3.4–3.5 Ga, that it was probably existent on Earth by ∼3.7–3.8 Ga, and that there are enigmatic and controversial traces of its presence in the Hadean eon. Some *molecular clock* studies, which investigate genetic differences between species to estimate when they diverged from a common ancestor, also favour the origin(s) of life in the Hadean $\lesssim 100$ Myr after the Earth became habitable (Betts et al., 2018), although the accuracy of these analyses is uncertain.

Finally, the reader may have noticed that we have employed 'origin(s)' instead of 'origin'. The rationale is that there might be more than one independent abiogenesis event that occurred on Earth. The (fuzzy) boundary between non-living and living systems could have been crossed multiple times, and not always via a single pathway (viz., many paths can birth many variants of life). These early forms of life may have either died out eventually or retreated into a *shadow biosphere* that we have not unearthed so far (Davies et al., 2009). For the sake of convenience, however, we will use the simpler term 'origin' often in the book.

5.1   Building blocks of life

As this textbook assumes only a basic familiarity with biology and chemistry, we will review some of the biomolecular building blocks documented in all known life on Earth. Our primer is brief and selective due to space constraints, and an in-depth exposition can be found in standard biochemistry textbooks (e.g., Nelson and Cox, 2004; Milo and Phillips, 2016).

Before doing so, we highlight a pioneering proposal, made by Tibor Gánti in 1971, dubbed *chemoton* theory. The chemoton embodies the fundamental unit of life, and is composed of three subsystems (Gánti, 2003, pg. 4):

1. *Metabolic* subsystem: A reaction network, typically involving simple substances, that produces the necessary (free) energy and compounds for sustaining itself, as well as the other two subsystems.
2. *Membrane* subsystem: A boundary that separates the chemoton from the external environment, and is sustained via compounds and energy generated by the metabolic subsystem.
3. *Genetic* subsystem: A polymeric substance that acts as the repository of information and facilitates self-reproduction based on a template, harnessing compounds and energy produced by the metabolic subsystem.

All three subsystems were postulated to be *autocatalytic*, that is, their constituent chemicals collectively catalyse their own synthesis. The chemoton idea contains the farsighted recognition of the essential subsystems of cells and their forerunners (*protocells*), of the interconnected nature of the subsystems, and of the importance of autocatalysis; it also emphasises metabolism, replication, and compartmentalisation, thereby motivating our study of the four major classes of *biomolecules* (i.e., molecules synthesised by cells capable of performing vital biological functions) hereafter.

We will now introduce the concept of *bioessential elements*, to wit, elements that are imperative for life-as-we-know-it, and are found in at least some of the biomolecules. The six canonical bioessential elements are carbon (C), hydrogen (H), nitrogen (N), oxygen (O), phosphorus (P), and sulfur (S); therefore, they are jointly called CHNOPS or SPONCH.

5.1.1   Proteins

It is not an exaggeration to argue that proteins are unmatched in terms of their functional versatility among all biomolecules. In the bacterium *Escherichia coli*, which is the most widely investigated microbe, proteins make up ~15% of the body mass, while the other three classes of biomolecules add up to merely ~12% (Nelson and Cox, 2004, Table 1.2).

Enzymes are one of the most crucial among the various types of proteins. The role of enzymes is to function as biological catalysts. The rate k of a generic chemical reaction may be crudely calculated from the Arrhenius equation, which is given by

$$k \propto \exp\left(-\frac{E_a}{k_B T}\right), \tag{5.1}$$

where k_B is the Boltzmann constant, T is the temperature, and E_a is the activation energy – roughly, the (minimum) energy needed to initiate the reaction. Catalysts (e.g., enzymes) often, albeit not always, boost reaction rates by lowering the activation energy. As seen from (5.1), k has an exponential dependence on E_a, implying that a modest decrease in the latter can substantially increase the former.

Besides their importance as enzymes, proteins are involved in many other biological functions. A few examples suffice to illustrate their diversity and centrality: (a) structural elements (e.g., actin) that confer support upon the cell; (b) messengers (e.g., growth hormone) that transmit signals or substances; (c) signal receptors (e.g., G-protein-coupled receptors) that receive signalling molecules; (d) antibodies (e.g., immunoglobulin G) that counteract pathogens; and (e) storage proteins (e.g., ovalbumin) that contain reserves of chemical compounds for future use by organisms.

Proteins are biopolymers composed of monomeric units called *amino acids*. The term amino acid is derived from the fact that these molecules consist of the amino group ($-NH_2$) and the carboxyl group ($-COOH$). Although the total number of documented amino acids is >500, only a small subset of them are commonly employed in life. In particular, only 22 of them are utilised in protein biosynthesis, of which 20 are more widespread; these amino acids are called *proteinogenic amino acids*. Of these 22 amino acids, 21 are depicted in Figure 5.2, along with some of their key properties.

A handful of salient characteristics of amino acids are worth highlighting. First, amino acids are *zwitterions* – molecules with zero net charge, but hosting an equal number of atoms with positive and negative charges in their *functional groups*;[1] the zwitterionic character of amino acids facilitates their solubility in water, thereby rendering them a good match with this solvent. Second, α-*amino acids* are those in which the amino group is attached to the α-carbon, that is, one carbon away from the carboxyl group (consult Figure 5.3 for a couple of generic examples). Likewise, in β-*amino acids*, γ-*amino acids*, and δ-*amino acids*, the amino group is present at the β-carbon (two carbon atoms away from the carboxyl group), γ-carbon (three carbon atoms away from the carboxyl group), and δ-carbon (four carbon atoms away from the carboxyl group), respectively.

Third, and most crucially, the proteinogenic amino acids (barring glycine) are *chiral*. A molecule is chiral when its mirror image cannot be superimposed on the original; this feature is also exhibited by human hands. The two distinct mirror images of this molecule are referred to as *enantiomers*. Naïvely, we may expect nearly equal abundances of the two enantiomers in general; such a mixture is called *racemic*. In actuality, however, living systems exhibit the opposite scenario of (near-)*homochirality*, that is, where one enantiomer is virtually absent and the other is overwhelmingly dominant. For instance, life on Earth almost exclusively consists of left-handed (*levorotatory*) amino acids (*L*-amino

---

[1] Functional groups are a collection of atoms that exhibit similar chemical properties whenever they appear in compounds.

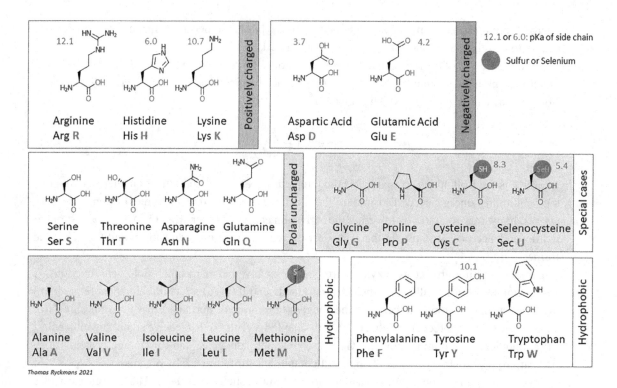

Thomas Ryckmans 2021

Figure 5.2 Twenty-one proteinogenic amino acids categorised based on their side chain properties. Note that 'polar' substances possess a finite dipole moment, while 'hydrophobic' (water-repelling) substances are non-polar. pKa is a measure of the acidity of the chemical species, with lower values indicative of higher acidity. (Credit: https://commons.wikimedia.org/wiki/File:Proteinogenic_Amino_Acid_Table.png; CC-BY-SA 4.0 license)

acids); similarly, biological sugars are chiefly found in their right-handed (*dextrorotatory*) form (*D*-sugars).[2]

Taking our cue from the fact that homochirality is a striking attribute of life on Earth, we may hypothesise that it is a signature of extraterrestrial life. We shall revisit this matter in Section 13.1.4, and explore some of the accompanying subtleties. The topic of homochirality continues to deal with unresolved major conundrums: Is homochirality required for extraterrestrial life? Does it need to resemble the setup on Earth or could it be the opposite (right-handed amino acids and left-handed sugars)? What are the mechanisms that drive the genesis of homochirality? The reader can consult Blackmond (2020) for a recent overview of homochirality.

The polymerisation of amino acids is a *condensation reaction*, where two molecules combine to yield a single molecule, with the loss of a small molecule; in this case, the latter is water (H_2O), owing to which the reaction constitutes a dehydration reaction. The two polymerised amino acids are linked through a *peptide bond*, as illustrated in Figure 5.3. The ensuing polymers are called *peptides*. Proteins are long peptides, which typically comprise $\gtrsim 50$ amino acids. However, proteins

[2] The concept of left- and right-handedness stems from the direction in which plane polarised light would be rotated when passed through a chiral sample.

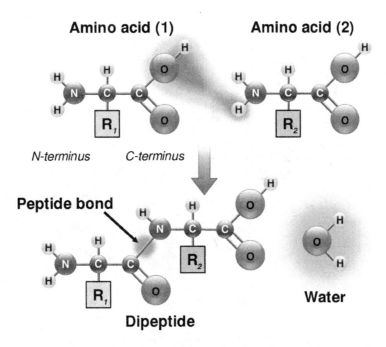

Figure 5.3 Polymerisation of two amino acids to produce a dipeptide via a peptide bond, and accompanied by the release of water. (Credit: https://commons.wikimedia.org/wiki/File:Peptidformationball.svg)

rarely occur as simple chains and instead fold into complex three-dimensional structures; the *tertiary structure* of a protein describes its 3D shape. The tertiary structure plays a vital role in determining the functionality of a given protein.

5.1.2 Nucleic acids

The primary function of nucleic acids is that they encode and store information in all life on Earth. The two classes of nucleic acids are *deoxyribonucleic acid* (DNA) and *ribonucleic acid* (RNA).

In the vast majority of organisms on Earth (including viruses under this umbrella for now), DNA constitutes the key repository of genetic information, one that is essential for their growth, development, reproduction, and functioning; loosely speaking, DNA contains the 'instructions' to synthesise other vital biomolecules such as RNA and proteins. A *gene* is a sequence of DNA, widely interpreted as the unit of heredity, that contributes to a certain function. Likewise, RNA is primarily known as a carrier of genetic information (involved in protein synthesis), but RNA molecules also execute several other functions in a cell – the most notable among them, with regard to this chapter, is catalysis, a role usually performed by enzymes.

The basic structure of DNA is as follows: it is predominantly a double-stranded macromolecule. DNA is a biopolymer whose repeating units are *nucleotides*. Each nucleotide consists of a *nucleobase* (nitrogenous base), the sugar *deoxyribose*, and a phosphate (PO_4^{3-}) group.[3] An *N-glycosidic bond* connects the sugar with the nucleobase, while a *phosphodiester bond* connects the sugar with the phosphate. There are four nucleobases in DNA: *adenine* [A] and *guanine* [G] (the *purine* bases), and

---

[3] *Nucleosides* are nucleotides sans the phosphate groups, that is, they are composed of a nucleobase bonded to a sugar. For example, *adenosine* is adenine bound to the sugar ribose, and is a part of *adenosine triphosphate* (ATP), whose significance is elucidated in Section 7.1.2.

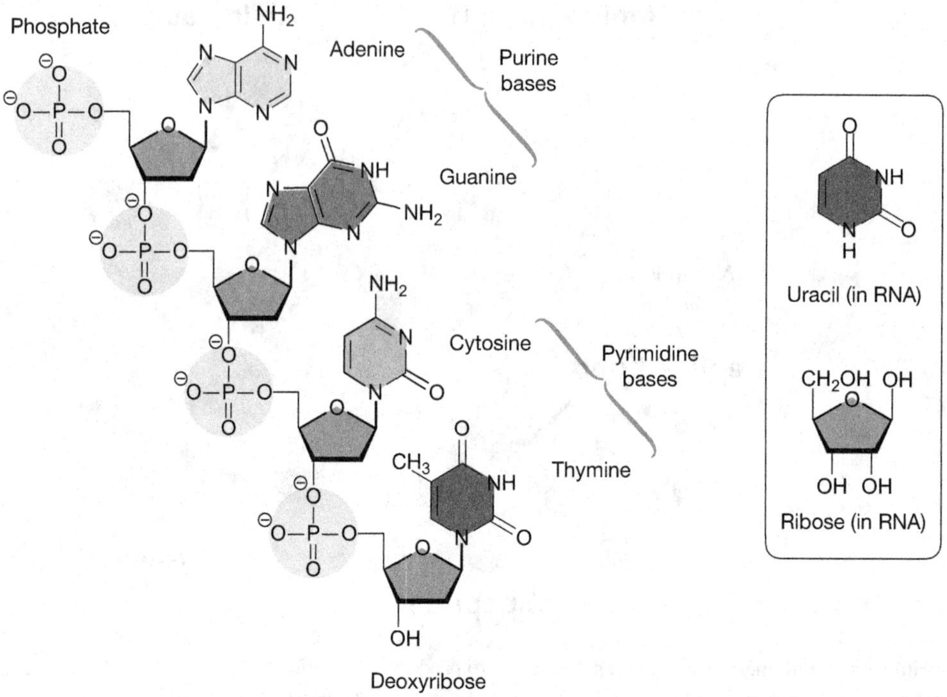

Figure 5.4 A segment of DNA (viz., polynucleotide chain), with the same features manifested in RNA except that thymine must be replaced with uracil and deoxyribose with ribose.

cytosine [C] and *thymine* [T] (the *pyrimidine* bases). The *Watson–Crick–Franklin base pairs* in DNA are A–T and G–C, with two and three hydrogen bonds, which allow the two strands to be linked and form the famous double helix structure.

Extensive research by Francis Crick (1916–2004), James Watson (1928–present), Rosalind Franklin (1920–1958), Maurice Wilkins (1916–2004), and their groups revealed DNA's structure in 1953. The features of one strand of DNA are presented in Figure 5.4. The structure of RNA is quite similar to that of DNA, albeit with three differences: (1) RNA is often single-stranded as opposed to the conventionally double-stranded DNA; (2) RNA has the sugar *ribose* in place of deoxyribose in DNA; and (3) the nucleobase thymine in DNA is replaced with *uracil* (U) in RNA. The second and third points of divergence are evident from inspecting Figure 5.4.

The *central dogma of molecular biology*, which was articulated by Francis Crick in 1957–1958, states that information flows from nucleic acids to nucleic acids or proteins, but that the converse (proteins to proteins or nucleic acids) is not feasible (Crick, 1970); as per our current knowledge, the central dogma remains valid. A further simplified and popular, but factually incorrect, version of the central dogma states that 'DNA makes RNA makes protein' or, alternatively, 'DNA → RNA → protein' (refer to Question 5.3).

5.1.3   Carbohydrates

Akin to how nucleic acids and proteins are polymers made up of monomeric units, *polysaccharides* are polymers of long chains of *monosaccharides* (i.e., simple sugars). The latter are usually characterised by the molecular formula $(CH_2O)_n$, but not all monosaccharides possess this formula and not all compounds with this formula are actually monosaccharides. The term *carbohydrates* includes monosaccharides, short chains of these monomers (called *oligosaccharides*), and polysaccharides. Carbohydrates are organic molecules composed of only carbon, hydrogen, and oxygen, typically possessing the molecular formula $C_m(H_2O)_n$.

Carbohydrates are documented to have multiple uses in life on Earth. Perhaps the most famous among them is their capacity to serve as energy sources in metabolism (e.g., oxidation of glucose in aerobic respiration) and energy storage (as starch in plants and glycogen in animals). Moreover, two sugars are essential components of nucleic acids: ribose in RNA and deoxyribose in DNA. Lastly, certain carbohydrates act as structural support, such as cellulose in cell walls and organs of plants, and chitin in the cell walls of fungi and exoskeletons of arthropods.

A few points concerning monosaccharides are worth mentioning. First, as indicated earlier in Section 5.1.1, sugars are chiral molecules with several chiral centres, implying that many *stereoisomers* (i.e., molecules with the same molecular formula, but distinct spatial configurations of atoms) are possible. Right-handed sugars (*D*-sugars) are much more common than their left-handed counterparts in biology, but the latter are not altogether absent. Second, in aqueous solution, many well-known monosaccharides (e.g., glucose) have a cyclic structure, as seen from Figure 5.4 for ribose and deoxyribose. Third, the number of carbon atoms in the sugars determine their nomenclature: monosaccharides with 4, 5, and 6 carbon atoms are tetroses, pentoses, and hexoses, respectively.

The synthesis of disaccharides and polysaccharides regularly entails the formation of O-glycosidic bonds between the hydroxyl (–OH) group of one monosaccharide and the so-called hemiacetal group (which also consists of an –OH group) of another; in other words, these glycosidic bonds link the sugars together. This reaction is a dehydration reaction, analogous to the formation of peptides from amino acids. Subtle differences in the structure of glycosidic bonds can lead to profound variations in the chemical properties of the ensuing polymers, which in turn might have major consequences at the level of ecosystems or even the biosphere.

This statement is best illustrated by comparing amylose, a constituent of starch (energy storage) versus cellulose (structural support). In the former, the –OH in the hemiacetal group is below the glucose ring, and the opposite is true for the latter. Although this difference may seem small, it has a substantial effect on the overall structure of these two polysaccharides, thereby influencing their characteristics and functions.

5.1.4   Lipids

The category of lipids is rather broad, as it encompasses diverse molecules united by their relative insolubility in water. Unlike the preceding trio of biomolecules, lipids are not macromolecules with a large number of monomeric units, even though they may be built from smaller molecules.

Lipids play a multitude of key roles in biology on Earth. In animals, *fatty acids* and *triglycerides* (shown in Figure 5.5) are both energy sources in metabolism and energy storage molecules. Crucially, *phospholipids* and *sterols* (refer to Figure 5.5) are structural components of cell membranes. In addition, lipids are also involved in functions such as electron carriers, chaperones for protein folding, and facilitators of enzyme activity, as reviewed in Nelson and Cox (2004, Chapter 10). Among these roles, we will highlight the propensity of certain lipids to assemble and form cell membranes, because the latter are a distinguishing feature of life on Earth.

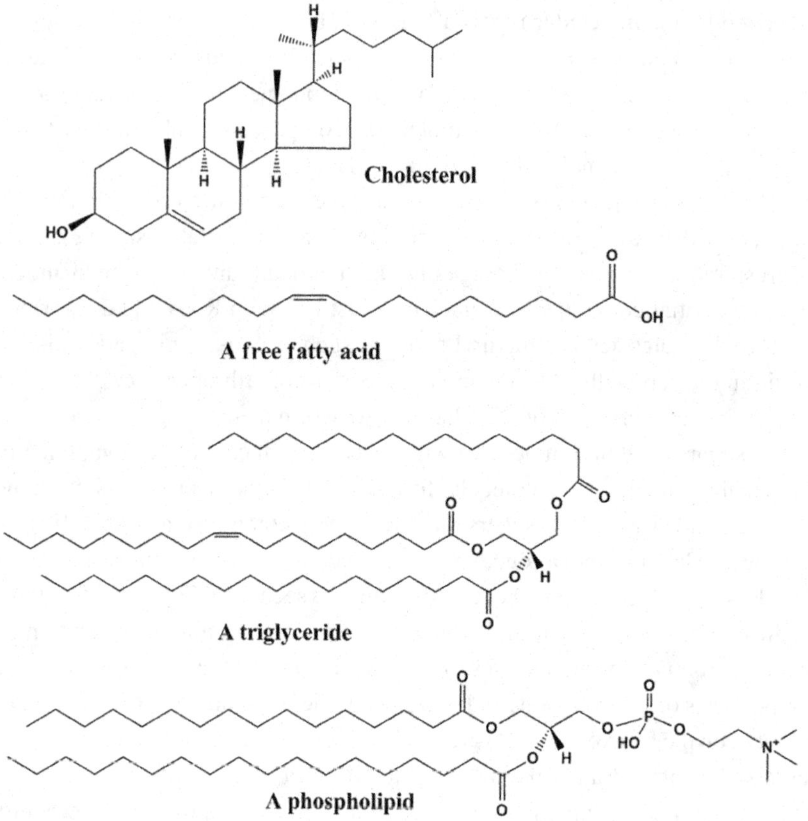

Figure 5.5 Fatty acids and triglycerides are energy storage molecules, while phospholipids and cholesterols bolster cell membranes. (Credit: https://commons.wikimedia.org/wiki/File:Common_lipids_lmaps.png; CC-BY-SA 3.0 license)

An important subset of lipids are endowed with one end that is *hydrophilic* (i.e., water-loving), whereas the other end is *hydrophobic* (i.e., water-repelling). The former often tends to be polar (i.e., with a finite electric dipole moment), while the latter is non-polar (zero dipole moment) in nature. For instance, as evident from Figure 5.5, phospholipids – the most abundant membrane lipids – have a hydrophilic phosphate group (polar 'head') and two hydrophobic 'tails' composed of fatty acid derivatives; note that fatty acids are essentially hydrocarbon chains with a carboxyl group (–COOH) at the end (i.e., they are examples of *carboxylic acids*).

A molecule with both hydrophilic and hydrophobic parts is *amphipathic* or *amphiphilic*; lipids belong to this category. When those lipids are therefore added to water, they can spontaneously organise such that the hydrophilic heads face towards the water and the hydrophobic tails face away from it. This organisation gives rise to the famous *lipid bilayer* structure encountered in cells, because the tails of the two layers face each other inward (away from the water) and the two heads are directed outward (towards the water).

Prokaryotes are single-celled organisms whose cells canonically lack internal membrane-bound compartments called *organelles*. The two kingdoms comprising prokaryotes are the renowned *bacteria* and the lesser known *archaea*. Some noteworthy differences do exist among the membrane lipids found in bacteria and archaea. The latter exhibit a higher degree of branching, and can be joined together, thus enhancing the robustness of archaea. It is not surprising that many archaea are well adapted to live in extreme environments; we delve into the topic of *extremophiles* in Section 7.2.

5.2 Minimum cell size based on biochemical constraints

Hitherto, we have witnessed how biomolecules are vital for life-as-we-know-it. Hence, if we posit that a minimum inventory of such molecules is necessary for life's functioning, we can attempt to derive a heuristic and simplified lower bound for the minimum cell size (or radius) of an organism. Naturally, as our analysis is based on life-as-we-know-it, the obtained limits are not necessarily valid for extraterrestrial life. A more sophisticated, and generalised, treatment is theoretically feasible by taking either energetic considerations (e.g., protein synthesis) or cell functions (e.g., information sensing) into account (Lingam, 2021b). This topic has attracted a great deal of attention for nearly a century, as summarised in Knoll et al. (1999).

Why is it valuable to determine the minimum cell size? The answer is apparent: if we unearth fossilised cell-like structures (i.e., potential microfossils), a reliable estimate of the minimum size could help us assess whether they are of biological origin. To put it another way, if the putative microfossils are noticeably below the lower bound, this result may diminish the likelihood that they are biogenic. To offer a striking real-world case, the discovery of cell-like structures in the Martian meteorite ALH84001 (sketched in Section 9.4.2) generated much excitement, but it is now thought that they lie below the lower bound for microbes (Cockell, 2020, Chapter 18).

We caution, however, that the lower bounds we calculate hereafter should not be regarded as definitive in view of the Earth-centric assumptions and simplifications that enter the model.

5.2.1 Protein-based bound

In Section 5.1.1, we have highlighted both the centrality and versatility of proteins in cell functioning, and that proteins are the most prevalent biomolecules in terms of the mass content. Thus, we can attempt to formulate a lower bound based on these points.

We consider a spherical cell of radius $\mathcal{R}_o$ and density ρ_o. The latter is close to the density of liquid water (around 10^3 kg/m^3), namely, the dominant component in the cell. Water is needed for many purposes, ranging from acting as a solvent for biochemical reactions to serving as the medium for diffusively transporting substances in and out of the cell. Of the total cell mass, let us suppose that a fraction f_{Pr} is composed of proteins; for both *Escherichia coli* and yeast, we may specify $f_{Pr} \approx 0.15$ (Milo, 2013, pg. 1053). Combining these pieces together, we find that the total protein mass in the cell is approximately equal to $4\pi f_{Pr}\rho_o \mathcal{R}_o^3/3$.

Next, the typical mass of a protein is taken to be $m_{Pr} \approx 5 \times 10^{-23}$ kg (Milo, 2013, pg. 1052). Finally, the minimum number of proteins required to sustain a cell is denoted by N_{Pr}. This quantity is not precisely determined, but an insightful analysis by Harold Morowitz (1927–2016) proposed a nominal value of 45 enzymes, with 3 copies of each enzyme (Morowitz, 1967, pg. 52), thereby yielding $N_{Pr} \approx 135$. The minimum total protein mass is expressed as $m_{Pr}N_{Pr}$. We emphasise, at this juncture, that we have focused solely on proteins, and not the manifold other components of the cell that are equally vital (e.g., nucleic acids that encode the synthesis of proteins); this is why our treatment is heuristic.

On equating the results from the above two paragraphs and simplifying, the minimum cell radius is found to be

$$\mathcal{R}_o \approx \left(\frac{3m_{Pr}N_{Pr}}{4\pi f_{Pr}\rho_o}\right)^{1/3}, \tag{5.2}$$

which reveals a relatively weak dependence on the parameters, thus suggesting that the findings may be partially tenable. On substituting the above values in this formula, along with $\rho_o \approx 1.1 \times 10^3$ kg/m^3, we end up with $\mathcal{R}_o \approx 21$ nm, consequently yielding a minimum cell size of about 42 nm. However, because we neglected other biomolecules, it is likely that this estimate is unrealistic and the actual lower bound is somewhat higher. The theoretical model by Morowitz (1967), which included these molecules, arrived at a minimum cell size of 92–104 nm.

5.2.2   Proton-based bound

Biomolecules are not the only crucial ingredients of a cell. A variety of salts and inorganic ions, among other chemical species, are also deemed necessary for life. Of this diverse group, we single out protons (H$^+$ ions), which are employed by cells to synthesise ATP, the 'energy currency' of the cell, through the process of chemiosmosis elaborated in Section 7.1.2; it may be said that proton gradients power cells (Lane, 2009). We will study how proton influx (from cell exterior to interior) constrains the cell size.

Consider a spherical cell of radius $\mathcal{R}_o$, whose volume is therefore $4\pi\mathcal{R}_o^3/3$. The maximum volume of water is also equal to this value. Now, suppose that the pH of the cell is known, from which we can roughly calculate the molar concentration of H$^+$ ions (denoted by [H$^+$]) as follows:

$$[H^+] \approx 10^{-pH}, \tag{5.3}$$

and a more accurate definition is described in Section 7.1.2. Cells canonically possess a near-neutral pH (i.e., close to 7), although this number is subject to some variability (Milo and Phillips, 2016, Chapter 2). As molar concentration is measured in units of mol/L, the maximum number of protons in the cell ($N_{p,max}$) is estimated from the above information to be

$$N_{p,max} \approx \left(\frac{4\pi\mathcal{R}_o^3}{3}\right) \times \left(1000\,N_A\,10^{-pH}\right), \tag{5.4}$$

where the factor of 1,000 stems from converting 1 m^3 to litre (L), and the Avogadro number N_A is introduced to transform the number of moles to the number of particles. It might be reasonable to postulate, albeit with caveats, that at least one proton must permeate into the cell at any given moment in time to execute its biological function(s). Hence, by setting $N_{p,max} = 1$, we can solve for the minimum cell radius, which leads us to

$$\mathcal{R}_o \approx \left(\frac{3}{4000\pi N_A}\right)^{1/3} (10)^{pH/3} \approx 0.73\,\text{nm}\,(10)^{pH/3}, \tag{5.5}$$

from which we see that this expression is sensitive to the pH (exponential dependence); an interesting feature of (5.5) is that the pH is the only variable, in contrast to (5.2). At least on Earth, biomolecules such as proteins exhibit high stability at pH of around 7 (e.g., Talley and Alexov, 2010), which may decrease in acidic or alkaline settings. On inputting this fiducial value in (5.5), we obtain $\mathcal{R}_o \approx 157$ nm.

The smallest documented microbes on Earth are ostensibly close to the above radius threshold, thus improving the credibility of our approach. Certain species from the free-living SAR11 clade (of bacteria) have effective radii of approximately 130–200 nm; the parasitic bacterium *Mycoplasma genitalium* has a radius of about 150 nm; and the phylum *Nanoarchaeota* may comprise species with radii of ∼100 nm (Lingam, 2021b).

5.3 Paradigms galore in the origin(s) of life

When readers consult the origin(s)-of-life literature, there are several theories or concepts that are commonly encountered (e.g., Fry, 2000; Luisi, 2016; Smith and Morowitz, 2016). We will outline some of them next, with the proviso that a number of subtleties are swept under the rug.

5.3.1 Metabolism-first versus replication-first

Two of the most distinguishing attributes of living systems are metabolism (namely, the set of chemical reactions that supply energy and resources for sustaining the organism) and self-reproduction. It is not currently known which of these two features developed earlier. Hence, it is unsurprising that the overarching camps of metabolism-first and replication-first exist.

To understand the replication-first paradigm,[4] let us take a step back and examine current life on Earth. DNA contains the information for making a cell, but its synthesis entails the catalytic action of enzymes (DNA polymerases). Conversely, the synthesis of proteins is dependent on the presence of DNA, which encodes the instructions for producing them. Thus, we are confronted by a classic chicken-and-egg conundrum, which can be resolved if one type of molecule could perform both functions: catalysis on the one hand, and information storage and transfer on the other. In the 1980s, a crucial breakthrough was achieved when certain RNA molecules called *ribozymes* (RNA enzymes) were demonstrated to operate as catalysts.

This discovery, which led to Sidney Altman (1939–2022) and Thomas Cech (1947–present) being awarded the 1989 Nobel Prize in Chemistry, augmented the appealing notion of the *RNA world* (Gilbert, 1986): a hypothetical epoch that commenced around 4.36 Ga wherein RNA, or a closely related molecule, constituted the basis of the earliest life forms (Benner et al., 2020). The RNA world is often dated to a 1962 paper by Alexander Rich (1924–2015), but its direct antecedents extend further back to at least the 1950s. The RNA world hypothesis is reviewed in Orgel (2004), Robertson and Joyce (2012), Higgs and Lehman (2015), and Benner et al. (2020). Several proposals have suggested that the RNA world was preceded by (simpler) pre-RNA worlds that fulfilled similar functions (Orgel, 2004). At this juncture, we emphasise that replication-first is *not* synonymous with the RNA world; in fact, an early conjecture of an enzyme-like self-replicating molecule was advanced by Leonard Troland (1889–1932) in 1914–1917 (Lazcano, 2010; Fry, 2011).

Despite the prior clarification, the RNA world is the dominant replication-first hypothesis because it draws on empirical evidence from life on Earth (e.g., ribozymes), on extensive laboratory investigations, on increasingly sophisticated theoretical models, and exhibits the twin virtues of parsimony

---

[4] This school of thought is conventionally, though inappropriately, equated with the gene(tics)-first camp, which postulates that polymers with analogues of genes (i.e., units of heredity) initiated the onset of biological evolution (Fry, 2011).

and elegance. However, this theory is not devoid of negatives. First and foremost, many critics have contended that the synthesis of such a large and complex molecule as RNA is extremely improbable (e.g., Shapiro, 2007). Additional issues evaluated include the initial absence of templates for aiding RNA polymerisation, the susceptibility of RNA to hydrolysis causing its breakdown, and hurdles in RNA replication (Szostak, 2012).

In contrast, the metabolism-first approach advocates that small molecules participated in chemical reaction networks capable of some form of evolution driven by an energy source (to wit, engendering protometabolism), after which the transition to information storage in polymers occurred. Shapiro (2007, pg. 50) proposed that five basic requirements must be satisfied by the diverse array of metabolism-first theories: (1) a compartment should separate the living system from its environment; (2) an energy source must be available; (3) this energy should sustain chemical reactions; (4) the chemical reaction network so created must have the capacity for adaptation and evolution of some kind; and (5) the network should accrue feedstock materials faster than their loss, in tandem with the self-reproduction of compartments.

The first major metabolism-first hypothesis was advanced by Alexander Oparin (1894–1980) in the 1920s and 1930s, although certain forerunners appear as early as the nineteenth century. Oparin posited that protometabolic networks were initiated and maintained in droplets, which had the capability to grow and divide. The offspring droplets could exhibit variations in their metabolic properties, which would result in some of them growing and reproducing more efficiently at the expense of others, thereby permitting a rudimentary form of evolution via natural selection (Fry, 2011). In the 1970s and 1980s, a host of metabolism-first proposals emerged to the forefront, due to which we restrict ourselves to sketching a handful of them.

From 1988 onward, Günter Wächtershäuser (1938–present) has published an innovative series of papers on the *iron–sulfur world* (e.g., Wächtershäuser, 1988, 2007), in which the energy source for protometabolism is the reaction of iron sulfide and hydrogen sulfide giving rise to pyrite (FeS_2):

$$FeS + H_2S \rightarrow FeS_2 + H_2. \tag{5.6}$$

An unusual feature of Wächtershäuser's original proposal(s) was that protolife did not necessitate clear-cut compartments, but rather was confined to the mineral surface of pyrite, on which it could grow and reproduce.

Wächtershäuser conjectured in some papers that the above putative protometabolism resembled the *reverse Krebs cycle* (which we shall revisit shortly) documented in certain bacteria. Likewise, using comprehensive analyses of possible protometabolic networks subject to geological constraints imposed by the conditions on early Earth, Morowitz and colleagues have argued since the 1990s that a version of the reverse Krebs cycle, also called the *reverse tricarboxylic acid (TCA) cycle*, constituted the first metabolic pathway (Smith and Morowitz, 2016).

A related, yet distinct, proposal was put forward by Michael Russell (1939–present), William Martin (1957–present), and colleagues: the *submarine alkaline hydrothermal vent theory* (e.g., Martin et al., 2008; Branscomb et al., 2017; Russell, 2023), whose origins may be traced to 1989. In a nutshell, submarine hydrothermal vents are cracks on the ocean floor that transport heated water and chemicals from Earth's interior, broadly analogous to hot springs and geysers on the surface. They evince high temperatures (at times exceeding 100°C) and high pressures. We will return to alkaline hydrothermal vents (AHVs) in Section 5.5.1, but the gist can be expressed as follows. Alkaline hydrothermal vents have steep chemical gradients that could be harnessed as a chemical energy source, inorganic catalysts that can substitute for enzymes, carbon sources for building organics, and inorganic compartments for localising metabolic networks. The protometabolism operational at AHVs was potentially akin to the *Wood–Ljungdahl pathway* (Sojo et al., 2016) – alternatively dubbed the *acetyl coenzyme-A pathway* – found in some bacteria and archaea, although other candidates have also been elucidated.

A great deal of work has arisen in parallel on the theoretical front. Two publications from the 1980s that stimulated this field merit mention. Freeman Dyson (1923–2020) developed a mathematical model for generic catalytic molecules, and established how the system could transition to self-organisation reminiscent of life when the population exceeded $\mathcal{O}(1000)$ units (Dyson, 1982). Likewise, ever since the 1970s, Stuart Kauffman (1939–present) has proposed that a network of peptides may eventually develop autocatalytic properties as the number of chemical species and reactions in the network increases (Kauffman, 1986, 1995, 2011), thereby facilitating self-replication despite the absence of replicators analogous to nucleic acids.

The metabolism-first approach is endowed with multiple advantages, but we will only review a select few. First, the synthesis of small molecules is comparatively easy, unlike complex informational molecules like RNA. Second, by drawing on the principle of congruity between protometabolic networks and extant metabolisms (Granick, 1957), laboratory experiments have demonstrated that simplified versions of the reverse Krebs cycle and the Wood–Ljungdahl pathway may have been feasible through the action of inorganic catalysts such as metals, that is, sans enzymes (Muchowska et al., 2020). Last, experimental and theoretical research on autocatalytic chemical systems has shown that they are roughly capable of Darwinian evolution (loosely speaking, evolution by means of natural selection), enabled by some form of collective metabolism (Adamski et al., 2020; Ameta et al., 2021).

Yet, the metabolism-first paradigm is not devoid of difficulties. A common argument, albeit one that is diminishing in significance, is that the spontaneous genesis of an intricate coordinated network (protometabolism) is unlikely (Pross, 2004), and that not many complex molecules have been synthesised from scratch in such a network (Westall et al., 2023, Section 5.4). Moreover, unlike DNA/RNA, where information is *digital* (i.e., composed of a string of discrete entities, the nucleotides), most metabolism-first theories invoke the notion of *analogue* information. It is conventionally held that digital information is better insofar as long-term fidelity and replication is concerned, but this stance is not universally accepted; if valid, it may pose issues for sustaining protometabolic pathways over long intervals.

5.3.2 Autotrophic origin versus heterotrophic origin

As described further in Section 7.1, the core distinction between *heterotrophs* and *autotrophs* is that the former derive the carbon needed for their functioning from organic compounds, while the latter employ inorganic compounds, specifically carbon dioxide, for this purpose.

The heterotrophic theory for abiogenesis has a long and distinguished history, arguably dating back to the writings of Charles Darwin (1809–1882). In an 1871 letter to Joseph Hooker, Darwin speculated that life might have originated in '*some warm little pond*' in which a '*protein compound was chemically formed*' that underwent subsequent complexification to give rise to life.[5] The heterotrophic theory is commonly associated with the renowned *primordial soup* concept that was

---

[5] www.darwinproject.ac.uk/letter/DCP-LETT-7471.xml

independently put forward by Alexander Oparin in 1924 and J. B. S. Haldane (1892–1964) in 1929. As per this hypothesis, the first living systems assembled together from a diverse pool of biomolecular building blocks (e.g., amino acids) that were synthesised, polymerised, and accumulated through abiotic channels (Lazcano, 2010).

In the heterotrophic theory, there is a clear emphasis on synthesising the biomolecular building blocks through abiotic avenues. As noted in Section 1.2, the first abiotic synthesis of amino acid was achieved over a century ago, by Löb and Baudisch in 1913. However, as these findings were not situated in the context of the origin(s) of life, and did not make explicit connections with the early Earth, the spark discharge experiments reported by Stanley Miller (1930–2007) in Miller (1953), called the *Miller–Urey experiment*, heralded the advent of *prebiotic chemistry*. A great deal of research has since been undertaken in this vein, and is reviewed in McCollom (2013), Luisi (2016), Sutherland (2017), and Kitadai and Maruyama (2018). We revisit this area in Section 5.4, which implicitly has a heterotrophic slant.

Moving onward to the realm of autotrophic theory, this paradigm posits that only low molecular weight molecules are necessary for life's origination, and not pre-existing complex building blocks. This category of proposals is often conflated with metabolism-first theories, but the two are not wholly synonymous (Fry, 2011). It is true, however, that many of them draw inspiration from extant metabolic pathways, and try to envision what their abiotic counterparts may have looked like. One of the first autotrophic hypotheses was outlined by Alfonso L. Herrera (1868–1942) in the early twentieth century (Lazcano, 2010), but this field truly accelerated from the 1970s.

Hartman (1975) suggested that the earliest metabolism was autotrophic, and corresponded to an archaic version of the reverse Krebs cycle. As we have seen in Section 5.3.1, the likes of Wächtershäuser, Morowitz, and others have also favoured this pathway. The interest in the reverse Krebs cycle is justified because this metabolism is an autocatalytic network that could sustain itself in principle, it produces precursors of all the classes of biomolecules, and it consists of relatively few reactions. Hence, recent research has sought to implement an abiotic version of the reverse Krebs cycle in the laboratory, with some promising breakthroughs (Muchowska et al., 2020).

The other major candidate for autotrophic origin is the Wood–Ljungdahl pathway. We have encountered a few of the researchers who support this pathway in Section 5.3.1. One appealing point in its favour is that a comprehensive analysis of 6 million genes by Weiss et al. (2016) came to the conclusion that the *Last Universal Common Ancestor* (LUCA) of all cells might have involved the Wood–Ljungdahl pathway and inhabited a submarine hydrothermal vent; note, however, that these results have been debated. Two other advantages of the Wood–Ljungdahl pathway are that it is simple and linear, and may be less susceptible to the problem of diminishing yields. As with the reverse Krebs cycle, ongoing research has witnessed some success in realising this network in the laboratory (Muchowska et al., 2020).

In closing, we remark that autotrophic chemical reaction networks that merge the desirable aspects of the Wood–Ljungdahl pathway and the reverse Krebs cycle have also been proposed.

5.3.3 Many-worlds interpretation of the origin of life

This title is a playful nod to the many-worlds interpretation of quantum mechanics, and alludes to the fact that many different molecules have been conjectured to lie at the heart of life's origin(s). We have already described the RNA world, owing to which we turn our attention to other contenders.

Peptide and RNA-peptide world: As we shall witness in Section 5.4, there are multiple abiotic pathways for synthesising amino acids and peptides. Partly motivated by this consideration, several authors suggested that peptides and proteins came first (i.e., before RNA and pre-RNA worlds) and facilitated early chemical evolution of life-like systems (e.g., Plankensteiner et al., 2005; Frenkel-Pinter et al., 2020). Peptides can fulfil manifold roles in prebiotic chemistry such as catalysis (like proteins), long-range charge transport, membrane stabilisation, and replication. With respect to the latter, autocatalytic networks comprising peptides have been demonstrated in laboratory settings (Kauffman, 2011), whose importance in metabolism-first models was sketched in Section 5.3.1.

Looking beyond peptides in isolation, one of their desirable traits is that they can interact with other (bio)molecules, and enhance the function of the latter and vice versa (Frenkel-Pinter et al., 2020). In particular, the *RNA-peptide world* is built on the premise that peptides/proteins and RNA co-evolved together (Fried et al., 2022). The former could have aided the genesis of the latter by providing scaffolding, compartmentalisation, catalysis, and so forth. The converse scenario is also feasible: for instance, non-standard RNA nucleobases are documented to enable the synthesis of peptides and create RNA-peptide chimeras (Müller et al., 2022).

Lipid world: The lipid world takes its cue from the significance of compartmentalisation in abiogenesis (Deamer et al., 2002), implying that protocells with amphipathic lipid membranes (typically bilayers) were an '*early evolutionary step in the emergence of cellular life on Earth*' (Segré et al., 2001). Over the years, the autocatalytic nature of certain lipid systems has been established, and so has their capacity to undergo chemical evolution, achieve self-reproduction, participate in chemical signalling, and manifest a form of composition-based inheritance (Lancet et al., 2019; Gözen et al., 2022); all of these discoveries have broadened the notion of the 'lipid world' itself.

Furthermore, protocells assembled from amphipathic lipids can interact with biomolecules situated inside them, and mutually enhance their function (Gözen et al., 2022). For example, laboratory experiments performed by Adamala and Szostak (2013) concluded that peptide synthesis and compartment growth could exert a positive influence on each other.

Miscellaneous worlds: *Coenzymes* are organic molecules that are required by enzymes to fulfil their catalytic activity. Coenzymes have a relatively simple structure, and share close connections with both nucleotides (in terms of their structure) and proteins (viz., enzymes). In the *coenzyme world*, these 'helper' molecules might have played the role of catalysts, while also maintaining some capacity for self-reproduction.

As opposed to (proto)cells, entities analogous to viruses might have been the earliest living systems, and may have originated from a pool of rudimentary genetic elements, thereby constituting the *virus world* scenario.

Finally, we reiterate a theme that was articulated previously, which applies to most (if not all) paradigms encountered in this section. The divisions between 'opposing' paradigms are not as deep or as fixed as once thought; these barriers are dissolving due to the progress made by ongoing laboratory experiments, molecular clock analyses, and mathematical/computational models (Domagal-Goldman et al., 2016; Preiner et al., 2020). This point is reinforced by Figure 5.6, illustrating how multiple pathways intertwined with one another can engender multiple forms of life.

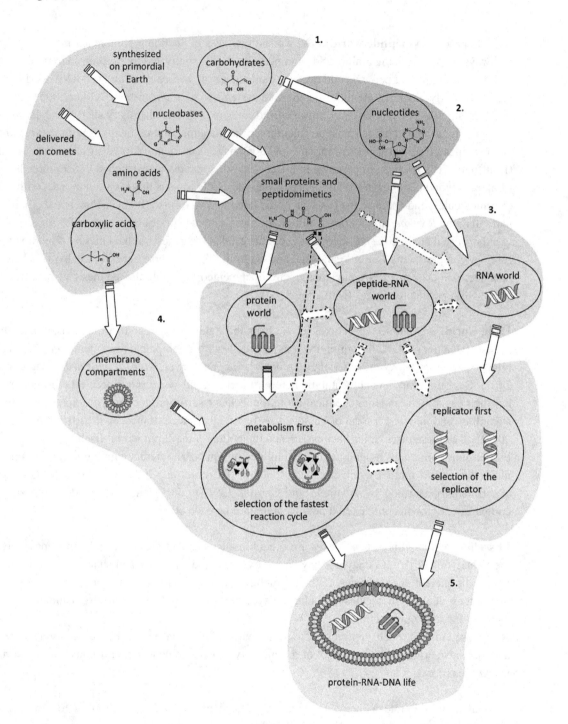

Figure 5.6 Possible paths and overall stages (1 to 5) starting with the synthesis of biomolecular building blocks and culminating in the origin(s) of life. (Credit: Domagal-Goldman et al. 2016, Figure 7; CC-BY-NC 4.0 license)

5.4   Synthesis of biomolecular building blocks on Earth

The origin(s)-of-life community has grown tremendously and encompasses many disciplines, as indicated at the start of the chapter. It is safe, however, to state that the question of how abiogenesis occurred has mostly been investigated from a chemistry perspective, with a particular emphasis on the synthesis of biomolecular building blocks and their polymers (Ruiz-Mirazo et al., 2014). Given the vastness of this field and the modest chemistry prerequisites assumed, we will barely scratch the surface. A selective timeline of major developments in the period 2009–2019, which is already becoming dated in some respects, is presented in Preiner et al. (2020, Figure 2).

5.4.1   Synthesis and delivery of monomers

Here, we will briefly examine the avenues through which biological monomers (amino acids, sugars, and nucleotides) are producible on the early (Hadean or Archean) Earth. We note in passing that many (free) energy sources would have existed on early Earth, ranging from electromagnetic radiation and chemical energy to electrical discharges (lightning) and radioactivity.

5.4.1.1   Amino acids

As remarked in Section 5.3.2, the Miller–Urey experiment can be regarded as a landmark event in the field of prebiotic chemistry because it demonstrated the synthesis of five amino acids, as well as other organic compounds (e.g., carboxylic acids) starting from simple atmospheric gases. The simplest proteinogenic amino acid, glycine, was detected at a fairly high yield of 2% (i.e., 2% of the available carbon was converted to glycine) (Miller and Urey, 1959). The classic Miller–Urey experimental setup, which supplied the energy through (electrical) spark discharge, is shown in Figure 5.7.

A few aspects are worth underscoring about the Miller–Urey experiment. First, the composition of gases consisted of highly reducing agents, whereas there is evidence that early Earth's atmosphere was quite neutral, albeit interspersed with transient reducing epochs (refer to Section 4.2.3). When the gaseous composition is made less reducing, it was originally found that the yields of amino acids dropped by more than two orders of magnitude (McCollom, 2013, pg. 212), although innovative and plausible workarounds have been identified (see Question 5.7). Second, the amino acids produced did not include any with sulfur in them, which is expected since none of the gases contained sulfur. When the Miller–Urey experiment was repeated with H_2S as one of the gases, sulfur-containing amino acids were produced.

The synthesis of amino acids may proceed through the *Strecker synthesis* identified in the nineteenth century. In the first step, an aldehyde (R–CHO) reacts with hydrogen cyanide (HCN) and ammonia (NH_3).

$$R - CHO + HCN + NH_3 \rightarrow R - CH(NH_2) - CN + H_2O. \tag{5.7}$$

The product on the RHS is then subjected to reaction with water (hydrolysis), finally giving rise to the amino acid.

$$R - CH(NH_2) - CN + H_2O \rightarrow R - CH(NH_2) - CONH_2, \tag{5.8}$$
$$R - CH(NH_2) - CONH_2 + H_2O \rightarrow R - CH(NH_2) - COOH + NH_3.$$

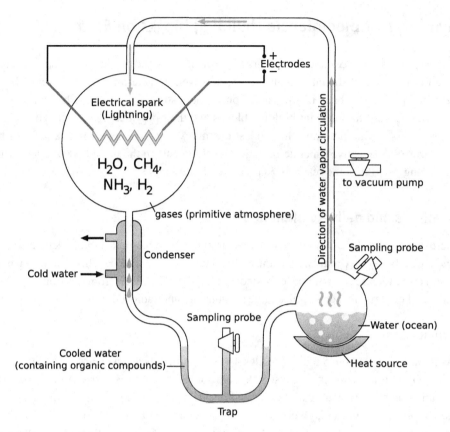

Figure 5.7 Schematic of the Miller–Urey experimental apparatus that paved the way for the field of prebiotic chemistry. (Credit: https://commons.wikimedia.org/wiki/File:Miller-Urey_experiment-en.svg; GNU Free Documentation License)

It is appropriate to take a brief detour at this juncture. Hydrogen cyanide is widely accepted to be a vital feedstock molecule in origin(s)-of-life research. This school of thought is remarkably traceable to Eduard Pflüger (1829–1910) in 1875. To offer a modern perspective, HCN constitutes an essential ingredient of the synthetic *cyanosulfidic protometabolism* network that creates the precursors of RNA, proteins, and lipids from a mixture of simple molecules possibly prevalent on early Earth (Patel et al., 2015).

We will now sketch a few other channels that can produce amino acids. First, ultraviolet (UV) radiation was the largest energy source (by orders of magnitude) on the young Earth. Hence, several studies investigated whether UV radiation could drive the synthesis of amino acids, and demonstrated that several proteinogenic amino acids (e.g., alanine, serine, cysteine) were formed (e.g., Sagan and Khare, 1971). Second, as indicated in Section 4.2, Hadean Earth experienced a much higher rate of impacts than today. These impacts triggered shock waves that propagated through the atmosphere, heating it in the process. Laboratory experiments mimicking shock wave conditions have generated amino acids with high efficiency (e.g., Bar-Nun et al., 1970). Third, Earth was frequently bombarded by high-energy particles from the Sun. Kobayashi et al. (2023) experimentally simulated these particles and showed that they yielded sizeable amounts of amino acids.

Next, we turn our gaze to submarine hydrothermal vents. From a theoretical perspective, the synthesis of amino acids is thermodynamically favourable at these sites (Amend and Shock, 1998). A modified version of the *Fischer–Tropsch* pathway (the canonical reaction chain is displayed below) may enable the synthesis of amino acids (McCollom and Seewald, 2007, pg. 383).

$$CO_2 + H_2 \rightarrow CH_4 + C_2H_6 + \ldots C_nH_{2n+2} + H_2O. \tag{5.9}$$

In (5.9), the molecular hydrogen is supplied through water–rock reactions (Schwander et al., 2023), discussed further in Section 7.1.3. From an experimental standpoint, several studies have synthesised amino acids in hydrothermal-like conditions (e.g., Huber and Wächtershäuser, 2006). However, one caveat is that the high temperatures typically render amino acids unstable, and cause their degradation (Aubrey et al., 2009).

We close with a couple of general statements. First, most laboratory pathways leading to amino acid synthesis produce nearly equal amounts of left- and right-handed amino acids, in sharp contrast to the (near-)homochirality observed in living systems (see Section 5.1.1). Second, while there are evidently manifold avenues for producing amino acids, the experiments should be carefully evaluated in terms of their plausibility on early Earth (e.g., concentrations of catalysts and reactants). Third, we have only focused on the generation of amino acids, but it must be appreciated that many channels facilitate their destruction as well.

5.4.1.2 Sugars

The best known pathway for producing sugars is the *formose reaction*, also called the *Butlerov reaction*, named after Aleksandr Butlerov (1828–1886), who discovered this in 1861. The formose reaction is crudely expressible as

$$n \, HCHO \xrightarrow[\text{H}_2\text{O}]{\text{Ca(OH)}_2} \text{sugars}, \tag{5.10}$$

where HCHO refers to formaldehyde. The formose reaction is canonically catalysed by calcium hydroxide, but other divalent metals can also perform this job. HCHO is a crucial feedstock molecule for prebiotic chemistry, which may be produced through multiple pathways, such as the photochemical reaction of carbon dioxide and water as shown in (5.11) (Cleaves, 2008).

$$CO_2 + H_2O \xrightarrow{\text{UV light}} HCHO + O_2. \tag{5.11}$$

However, in order to participate in subsequent prebiotic reactions involving HCHO, the accumulation of formaldehyde is necessary, which is possible only under relatively limited conditions (Cleaves, 2008).

The formose reaction is (in)famous for producing a wide variety of sugars; the significance of this class of molecules in metabolism (to wit, acting as energy sources) was outlined in Section 5.1.3. In addition, in conditions resembling alkaline hydrothermal vents, this pathway yields miscellaneous compounds associated with extant metabolic networks (e.g., carboxylic acids). Moreover, the formose reaction is autocatalytic, and is thus theoretically capable of sustaining itself. On account of these reasons, it is plausible to conclude that the formose reaction is of interest/relevance for the origin(s)-of-life from a metabolic standpoint (Omran et al., 2020).

The value of the formose reaction extends to the replication-first paradigm (specifically the RNA world). The rationale is that the formose reaction produces ribose, one of the constituents of RNA.

However, one drawback is that ribose is ostensibly only a small fraction ($\lesssim 1\%$) of the synthesised compounds. This hurdle may be overcome by including substances that can increase the yield of 5-carbon sugars (pentoses) in general and ribose in particular; of the candidates explored in this respect, boron-containing minerals are apparently promising (Benner et al., 2019).

5.4.1.3 Nucleotides

Unlike the building blocks examined so far, nucleotides (the monomers of nucleic acids) are comparatively complex, since they consist of a nucleobase, a sugar, and a phosphate group (refer to Section 5.1.2). Hence, we may anticipate that the synthesis of nucleotides ought to be challenging, which has indeed been historically the case, as reviewed in Yadav et al. (2020).

As we have already briefly delved into the synthesis of ribose, the next component we shall consider is the nucleobase. The synthesis of RNA/DNA nucleobases is over six decades old, which commenced with a famous experiment by Joan Oró (1923–2004) in 1960, wherein highly concentrated solutions of ammonium cyanide were heated to a temperature of $\leq 100°C$, and reported to comprise adenine (effectively a pentamer of HCN), among other compounds (Oró, 1960). The early studies based on HCN and its derivatives employed concentrations that are now thought to be incompatible with the general conditions on early Earth; however, in specialised settings (e.g., ice), attaining high abundances of HCN might be feasible.

This was the status quo in the first few decades of prebiotic chemistry, but a variety of experimental pathways are now confirmed to synthesise nucleobases (consult Question 5.8 for more information). We will sketch two of them. In place of pure water, solutions with formamide or urea have been demonstrated to produce nucleobases (Saladino et al., 2019). However, the question of whether such solutions were plausible on early Earth remains unsettled. Second, Fischer–Tropsch-type synthesis could give rise to nucleobases at submarine hydrothermal vents, because this reaction is thermodynamically favourable. One drawback, however, is that the stability of nucleobases declines sharply at high temperatures; for instance, the half-lives of adenine and guanine are only about one year at $100°C$ (Levy and Miller, 1998).

Equipped with the knowledge of how to synthesise ribose and nucleobases, the process for creating nucleosides appears straightforward, namely, combining them together directly. Afterwards, it would seem that nucleotides can be produced through the addition of the phosphate group (known as *phosphorylation*). However, this approach met with limited success,[6] owing to which innovative alternatives have been investigated (Sutherland, 2017). In a major breakthrough, Powner et al. (2009) were able to synthesise pyrimidine ribonucleotides (RNA nucleotides) from plausible feedstock molecules. A key ingredient of this pathway was the formation of 2-amino-oxazole, a chimeric molecule that is part ribose and part pyrimidine-nucleobase.

The synthesis of nucleotides has blossomed in the last decade, since new methods for producing nucleosides and subjecting them to phosphorylation have sprung up. The latter is summarised in Yadav et al. (2020, Section 4), and often entails the use of an *activating agent* that provides the energy input, loosely speaking, for the reaction to occur. To offer just one example: from small molecules (e.g., cyanoacetylene) and ribose, Becker et al. (2019) were able to form RNA nucleosides, which

---

[6] A recent successful example of this strategy is the synthesis of RNA nucleosides in the specialised environment of aqueous microdroplets by starting from ribose and RNA nucleobases in the presence of Mg^{2+} (Nam et al., 2018).

then reacted with phosphate minerals to synthesise ribonucleotides. Crucially, this synthesis was driven by alternating wet and dry conditions (i.e., wet–dry cycles).

5.4.1.4 Lipids

Amongst the diverse molecules belonging to the category of lipids, we will centre our attention on fatty acids (encountered in Section 5.1.4), because they are widely perceived as viable candidates for building protocell membranes (Szostak et al., 2001). Fatty acids are endowed with several advantages such as their high permeability to ions and molecules, capacity to form stable compartments, and ease of synthesis.

One of the customary avenues postulated for producing fatty acids is the Fischer–Tropsch-type pathway in hydrothermal vent settings, since their synthesis would be thermodynamically favoured in principle (McCollom and Seewald, 2007), although the predicted abundances are rather low (and decrease with increasing carbon chain length). Starting with the relatively simple formic acid (HCOOH) and oxalic acid (HOOC–COOH), the synthesis of fatty acids up to 22 carbons in length was reported in experiments performed at optimal temperatures of 150–250°C, but some of these findings are equivocal (Kitadai and Maruyama, 2018, Section 4.5). Current research suggests that, under the assumption that the production of long-chain fatty acids is realisable, these molecules can assemble to yield protocells in alkaline hydrothermal vents (Jordan et al., 2019).

Among other channels, the synthesis of fatty acids up to 12 carbon atoms in length was documented in Miller–Urey-type experiments (albeit involving reducing gases), although the mechanisms for their production remain poorly characterised. Fatty acids have also been synthesised in experiments based on the action of UV radiation or shock wave heating on atmospheric gases.

5.4.2 Exogenous delivery

Hitherto, we have operated under the implicit premise that all building blocks were synthesised endogenously on Earth. However, given the frequent rate of bombardment by impactors of myriad sizes in the Hadean (and Archean), an obvious avenue for enriching the abundances of organic molecules on Earth is exogenous delivery (Chyba and Sagan, 1992).

Revisiting Section 2.4.3, and Table 2.4 specifically, we see that the building blocks of life can be synthesised in interstellar ice grains through the action of UV radiation or high-energy particles. It is natural to then ask whether meteorites retrieved on Earth harbour these building blocks. The answer is in the affirmative, particularly in the case of the water- and organic-rich carbonaceous chondrites (introduced in Section 4.2.2). A few examples are furnished next (Kitadai and Maruyama, 2018, Section 4.1.3).

- Nearly 100 amino acids are confirmed, including at least 12 proteinogenic amino acids. Moreover, some of these meteorites have a distinct excess of left-handed amino acids (Glavin et al., 2019), which may have served as the seed, so to speak, for biological homochirality. The total abundance of amino acids in some carbonaceous chondrites is >1,000 ppm; comets might contain even higher concentrations.
- Adenine, guanine, and uracil have been detected in carbonaceous chondrites at abundances of ≲500 ppb. In addition, nucleobase analogues (e.g., xanthine), some of which are not naturally occurring on Earth, were recovered from these meteorites.

- Sugars such as arabinose, lyxose, and, most strikingly, ribose have been discovered in carbonaceous chondrites (Furukawa et al., 2019).
- A rich assortment of fatty acids and other lipids have been identified in carbonaceous chondrites at high abundances. It is well established that these extraterrestrial molecules can assemble into compartments.

Hitherto we have focused on meteorites, but in actuality Earth receives a substantial influx of *micrometeorites* (broadly interpreted as particles with sub-mm sizes). The micrometeorite flux may exceed that of its meteoritic counterpart by about three orders of magnitude (Westall et al., 2023, Section 5.3). Hence, even if a smaller fraction of their organic inventory survives passage through Earth's atmosphere compared to meteorites, they could still be a prominent source of biomolecular building blocks (see Question 5.9).

5.4.3   Polymerisation of monomers

Of the four categories of biomolecules, we single out peptides and *oligonucleotides* (short polymers of nucleotides). Lipids are not necessarily composed of strict monomeric units, and while carbohydrates (polysaccharides) are polymers, their solo prebiotic relevance is less characterised.

5.4.3.1   Formation of peptides

Just as with amino acids, there are different avenues for synthesising peptides, some of which are delineated below. Before tackling them, we highlight that peptide bond formation is a dehydration reaction (see Figure 5.3), due to which it is generally not thermodynamically favoured in water.

- As the temperature increases, the formation of peptides can become relatively less disfavoured. Hence, this feature has motivated models and experiments supporting the synthesis of peptides in submarine hydrothermal vents (e.g., Imai et al., 1999). We caution, however, that higher temperatures also cause rapid degradation, and the amino acids concentrations used in certain experiments appear to be unrealistically high.
- Mechanisms such as the *dry* heating of amino acids to temperatures $>150°C$, or modifying amino acids to incorporate polymerisation agents, do generate proteins, but they seem unlikely on early Earth.
- The notion that minerals could facilitate polymerisation and other vital roles in abiogenesis was crystallised in the treatise by Bernal (1951).[7] Silicate clays (e.g., kaolinite) and layered double hydroxides (e.g., fougerite) have been demonstrated to promote peptide formation in various settings (e.g., Cleaves et al., 2012; Erastova et al., 2017).
- In *salt-induced peptide formation*, the synthesis of peptides in the presence of sodium chloride (dehydrating agent) and Cu^{2+} (catalyst) was demonstrated after subjecting the mixture to wet–dry cycling (Rode, 1999).

---

[7] Although we address peptides, the significance of minerals is much broader and includes many functions (e.g., concentration and selection) and types (Cleaves et al., 2012).

- The volcanic gas, carbonyl sulfide, was shown to be an effective dehydrating agent, and yielded high peptide concentrations on short timescales (Leman et al., 2004). However, the abundance of this gas in volcanic emissions on early Earth remains poorly constrained.
- Several experiments mimicking impact-generated shocks and their aftermath have achieved the synthesis of short peptides.

To complement the endogeneous synthesis, the exogenous delivery of peptides may be viable since potential pathways for their production in interstellar ices have been proposed (Krasnokutski et al., 2022).

5.4.3.2 Formation of oligonucleotides

As polymers of nucleotides are the forerunners of nucleic acids, whose importance is readily apparent, a vast body of literature has sprung up around this subject. Like peptides, the polymerisation of nucleotides in water is often unfavourable on thermodynamic grounds, owing to which innovative workarounds have been identified. Another crucial issue is the tendency to form many structural isomers (i.e., molecules with identical chemical formula, but different structures) instead of the desired molecule(s) (Yadav et al., 2020, Section 5); to put it simply, an undesirable mixture is produced. Some common pathways for synthesising oligonucleotides are as follows.

- Initial attempts sought to heat nucleotides to temperatures of $\sim 160°C$, which yielded complex mixtures of short oligonucleotides.
- Short oligonucleotides (e.g., dimers to tetramers) can be synthesised in conditions akin to submarine hydrothermal vents, where polymerisation is rendered comparatively favourable (e.g., Burcar et al., 2015).
- A wide range of activating agents, notably phosphorimidazolides, mediate the formation of phosphodiester bonds and thus polymerisation. Polymers up to ~ 10 units in length have been synthesised in the presence of catalysts such as Pb^{2+}, Zn^{2+}, and UO_2^{2+}.
- The clay mineral, montmorillonite, is effective at catalysing the polymerisation of activated nucleotides (i.e., nucleotides combined with an activating agent) and can form polymers up to 50 units.
- In numerous publications, wet–dry cycles were applied to solutions of nucleotides and lipids. It was found that the synthesis of RNA-like polymers (whose structure is not precisely known) with ≤ 100 units occurred, which were further encapsulated in lipid compartments.
- As opposed to conventional nucleotides (see Figure 5.4), the phosphate group in cyclic nucleotides is linked to the sugar at two distinct sites. These cyclic nucleotides are more reactive, and have produced oligonucleotides in diverse experiments; the polymers generated have ≤ 40 units.
- Instead of proceeding from nucleotides, Gibard et al. (2018) began with nucleosides and added the phosphorylating agent diamidophosphate, which eventually led to the formation of short oligonucleotides (tetramers).

We close with the general observation that environments with strong non-equilibrium forces (corresponding to the existence of gradients in temperature, chemical composition, pH, and so forth) are well-suited for enabling the synthesis and replication of biomolecules (Ianeselli et al., 2023).

5.5   Potential sites for the origin(s) of life

As stated at the outset, the goal of this chapter was to explore two distinct questions, namely: the *how* and *where* of abiogenesis. Since we have tackled the former, it is now time to delve into the latter. This brings us to the critical question of what factors make a given *microenvironment* (i.e., localised setting) conducive for instantiating the origin(s) of life. However, answering this question is not easy because ambiguities still persist in the *how* of abiogenesis, which in turn affect the criteria for assessing the *where* of abiogenesis. We will summarise two sets of basic criteria in this context.

On combining Cockell (2020, Chapter 12.6) and Domagal-Goldman et al. (2016, Section 3.3.1), a desirable microenvironment for the origin(s) of life may need to exhibit the following characteristics.

1. Source(s) of energy for driving prebiotic synthesis.
2. Adequate supply of (in)organic building blocks.
3. Physical and chemical mechanisms for concentrating and/or enhancing yields (e.g., via catalysis) of (pre)biotic molecules.
4. Ambient conditions suitable for biomolecule formation and assembly.

An extended catalogue of factors necessary to allow the genesis of life was articulated in Deamer et al. (2022), which displays agreement with the preceding list, except for the inclusion of the following criteria: (5) wet–dry cycling for driving polymerisation, and (6) the presence of environmental selection pressures to facilitate the evolution of protocell populations.

An impressive array of microenvironments (dubbed *sites* henceforth) have been proposed, which span a broad range of parameters in physicochemical space, as reviewed in Saha et al. (2022). Owing to this multitude of putative sites, we will restrict ourselves to only a few examples hereafter. In Figure 5.8, the pros and cons attributable to some promising sites are depicted, but we caution that

		Origination			Complexification				Plausibility	
		Production of nutrients	Elemental complements	Availability of energy	Diversity of minerals	Concentration of organic molecules	Suitability of temperatures	Gradient in temperature, ph and redox conditions	Fluid dynamics propensity to fuel reactions	Widespread on the Hadean Earth?	Protection from exogenous threats
Environment	Submarine hydrothermal vents	5	5	5	1	5	5-1	5	3-1	5	5
	Vesicles in pumice clasts	0	5	3	5	5	5	1	3	1	3
	Subaerial hot springs and geysers	5	5	5	3	1	5-1	5	3-1	1	0
	Nuclear geyser	5	3	■	■	■	1	3	■	0	0
	Volcanic-hosted coastal splash pools	0	3	3	5	1	5	3	5	3	0
	Hydrothermal sedimentary reactors	3	5	5	5	5	5	5-3	5-3	5	5

Figure 5.8 Potential sites for the origin(s) of life (Westall et al., 2018, Figure 10), where *score* 5 and *score* 0 are the most and least favourable; black indicates insufficient knowledge. 'Origination' alludes to the capability of providing ingredients for prebiotic chemistry; 'Complexification' represents the ability to sustain directed reactions and molecular diversification; and 'Plausibility' is the likelihood of the site occurring on early Earth.

the list of evaluation criteria and the accompanying final scores evince a certain degree of ambiguity and subjectivity.

5.5.1 Submarine (alkaline) hydrothermal vents

Ever since the 1980s (Baross and Hoffman, 1985), it has been recognised that (deep sea) submarine hydrothermal vents (SHVs), defined in Section 5.3.1, are promising sites for initiating abiogenesis. The initial focus was directed in favour of SHVs at high temperatures and acidic pH, but the attention has distinctly shifted towards alkaline hydrothermal vents with temperatures $\lesssim 100°C$ (often in the 70–90°C range) and pH of 9–11. Succinct reviews of AHVs are furnished in Martin et al. (2008), Sojo et al. (2016), and Russell (2023). The variegated environs of AHVs are shown in Figure 5.9.

The AHV theory appears to have many advantages. First, it displays steep gradients in temperature ($\sim 100°C$ vent fluids versus $\sim 2°C$ deep ocean), pH (9–11 of vent fluids versus 6–7 of the ocean), and more. Furthermore, these gradients are separated by inorganic membranes comprising iron minerals like fougerite with sophisticated catalytic and concentration properties. The pH gradient across the membrane may create a *proton motive force* (summarised in Section 7.1.2) that could enable the reduction of CO_2 and the emergence of protometabolic networks such as the Wood–Ljungdahl pathway from Section 5.3.1 or the *denitrifying methanotrophic acetogenic pathway*, the latter of which is crudely expressible as (Russell, 2023, equation 2):

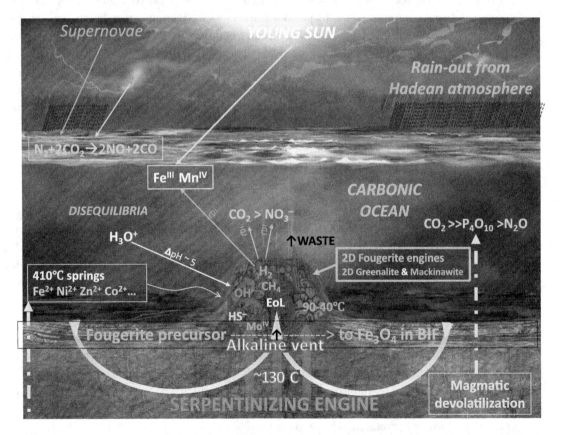

Figure 5.9 Physicochemical conditions in an alkaline hydrothermal vent conducive to the emergence of life. (Credit: Russell 2023, Figure 1; CC-BY 4.0 license)

$$\{4H^+ + CO_2 + N_2O\}_{ocean} + \{4H_2 + CH_4 + HS^- + OH^-\}_{hydrothermal}$$
$$\rightarrow \{CH_3COSH + 2NH_4^+\}_{metabolism} + \{3H_2O\}_{waste}. \qquad (5.12)$$

To reiterate, autotrophic protometabolic networks such as the Wood–Ljungdahl pathway might be feasible in AHVs as per laboratory experiments (Muchowska et al., 2020), especially since AHVs harbour a diverse assortment of metals and minerals that could act as inorganic catalysts.

From the related standpoint of the abiotic synthesis of biomolecular building blocks (consult Section 5.4), experimental studies and theoretical models suggest that the high temperatures and pressures of SHVs (which includes AHVs) are favourable, in the thermodynamic sense, for the formation of biological monomers and their polymerisation. Moreover, aside from these useful physical parameters, iron minerals at AHVs were empirically shown to drive the synthesis of amino acids from their precursors (Barge et al., 2019), which may be produced by a suitable protometabolism network.

The list of potential benefits does not end here. For instance, AHVs were likely prevalent on early Earth, and protected from the damaging effects of UV radiation. Next, AHVs possess a complex internal structure as well as strong non-equilibrium forces, both of which might permit the concentration, replication, and diversification of biomolecules (Ianeselli et al., 2023). Last but not least, the protometabolisms hypothesised in AHVs are congruent with extant metabolic pathways on Earth, thereby maintaining continuity and conceivably boosting their plausibility (Harrison et al., 2023).

However, AHVs are not devoid of downsides. Some crucial steps in the synthesis of biomolecules are not yet documented empirically, objections have been raised against the operation of the proton motive force (powered by the pH gradient) across the inorganic membrane, the high temperatures could destabilise and degrade biomolecular building blocks, and the abundances of certain bioessential elements and compounds are weakly constrained.

To round off, shallow water hydrothermal environments (depths of $\sim$10–100 m) are intriguing sites for abiogenesis, since they may evince several positives of AHVs, lack some possible pitfalls of AHVs (e.g., high temperature), and exhibit unique advantages. Two such candidates are hydrothermal sediments in proximity to volcanoes – advocated by Frances Westall (1955–present) and colleagues (Westall et al., 2018) – and shallow sea AHVs.

5.5.2   Hydrothermal fields

An alternative to SHVs (which are marine environments) is that life originated in land-based surface settings. Darwin's 'warm little pond' is a classic example (see Section 5.3.2) modelled in detail by Pearce et al. (2022). This paper concluded that the concentration of adenine in warm little ponds may not exceed the low value of $\sim 2 \times 10^{-8}$ mol/L at $\sim$4.4 Ga. Lakes are closely related water bodies which have also attracted research in the context of UV-driven prebiotic chemistry (e.g., Sasselov et al., 2020).

We will, instead, consider the surface environments of *hydrothermal fields* analogous to those in Kamchatka, Russia; Mount Lassen, California (USA); and Rotorua, New Zealand. These hot spring pools consist of freshwater, and are often associated with volcanic activity (Ross and Deamer, 2023). The positive facets linked with hydrothermal fields are reviewed in Damer and Deamer (2020), Van Kranendonk et al. (2021), and Ross and Deamer (2023), which we shall summarise in the next paragraph. In Figure 5.10, a putative path proceeding from early prebiotic chemistry to the origin(s) and evolution of life in hydrothermal fields is depicted.

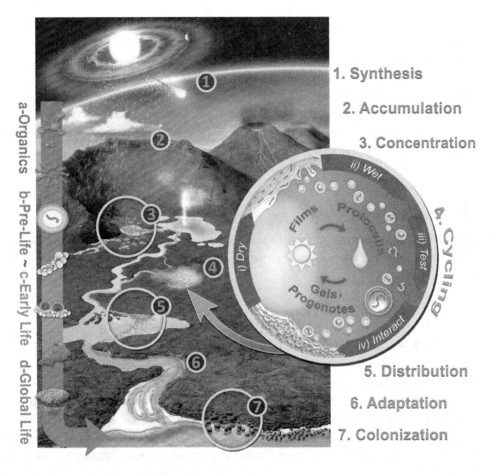

Figure 5.10 Steps leading from initial stages of prebiotic chemistry to early life in hydrothermal fields. (Credit: Damer and Deamer 2020, Figure 7; CC-BY-NC 4.0 license)

To begin with, surface and atmospheric pathways could have synthesised and/or delivered monomers of biomolecules (e.g., amino acids) to hydrothermal fields (refer to Section 5.4.1), in comparison to the somewhat sealed-off submarine hydrothermal vents. Hydrothermal fields are also endowed with diverse substances and circulation systems that enable the concentration of valuable trace elements (e.g., metals). Moreover, the action of wet–dry cycles, in conjunction with other non-equilibrium phenomena (gradients), can synthesise biologically relevant polymers (e.g., oligonucleotides) and encapsulate them within compartments generated from lipids, as briefly elucidated in Section 5.4. These protocells may be subsequently subjected to a form of evolution by natural selection, and eventually give rise to full-fledged biomolecules and cells, thus engendering life.

Many, albeit not all, of these steps have been reported in laboratory experiments, geological field studies, and mathematical models. Furthermore, extrapolating backwards from the ionic composition desirable for modern cells, Mulkidjanian et al. (2012) proposed that the attendant requirements exhibit a low likelihood in ocean water, but were realisable in hydrothermal fields. Thus, at first glimpse, these sites represent promising nurseries for the emergence of life, but we must now delineate and examine their weaknesses.

For starters, the UV flux, when suitably weighted by the damage caused to biomolecules, was nearly three orders of magnitude higher than modern Earth (Rugheimer et al., 2015, Table 3), which might be detrimental to early life; on the other hand, it can kickstart prebiotic chemistry. Second, the young Earth was exposed to frequent bombardment by impactors, which could have caused a disruptive and potentially hostile surface environment. Third, the fraction of early Earth covered by landmasses is still unresolved (consult Section 4.2.2), and it is conceivable that this fraction may be much lower than the present day. Last, the inventory of minerals and bioessential elements might be relatively limited (Westall et al., 2018).

5.5.3   Miscellaneous candidates

In analysing submarine hydrothermal vents and surface hydrothermal fields, we have addressed two common hypotheses, but many other possibilities exist. We will provide short synopses for a handful of them hereafter.

Beaches: The young Earth experienced far more intense tides than nowadays (work out Question 5.12). The alternating rapid action of high and low tides could have induced the repeated wetting and drying of beaches (and adjacent regions), thereby functioning as wet–dry cycles (Lingam and Loeb, 2018a), whose significance in the synthesis and polymerisation of biomolecular building blocks is apparent from Section 5.4.

In addition, beaches are characterised by other gradients (e.g., in salinity), a wide assortment of minerals, and radioactive sands conducive for the synthesis of prebiotic molecules. Finally, if (proto)life did evolve on beaches, it might have developed biological rhythms (akin to circadian rhythms) from the exposure to tidal cycles, which may be beneficial from an evolutionary perspective (Lingam and Loeb, 2018a). The stumbling blocks for beaches are presumably identical to those encountered in Section 5.5.2.

Semi-arid aquifers: Steven Benner (1954–present) and colleagues have formulated a pathway to the RNA world that unfolds in a semi-arid aquifer, that is, rock layer harbouring groundwater (e.g., Benner et al., 2019). The semi-arid nature of this site helps surmount the *water problem*, which alludes to the hurdle that many reactions and products pertaining to RNA are rendered unstable in water. The pathway involves the delivery of nucleobase precursors to the aquifer, formation of pentose sugars (e.g., ribose) from forerunners (mediated by boron and/or molybdenum minerals), the synthesis of nucleosides and nucleotides facilitated by phosphorus and boron compounds, and polymerisation through demonstrated channels.

As this process operates on land, it faces the obstacles attributed to hydrothermal fields in Section 5.5.2. Moreover, the availability of boron, molybdebum, and even phosphorus minerals on early Earth is poorly constrained.

Aerosols: Several papers have established that the air–water interface is markedly efficient at concentrating molecules and boosting reaction rates by orders of magnitude. Aerosols (i.e., tiny particles suspended in air/gas) corresponding to an air–water interface (e.g., sea spray) were probably commonplace on early Earth. Recent experiments mimicking these aerosols (i.e., utilising aqueous microdroplets with sizes of $\lesssim 1$ μm) have synthesised RNA nucleosides from ribose and RNA nucleobases (Nam et al., 2018), and peptides from amino acids (Holden et al., 2022). Thus, aerosols might be viable prebiotic reactors for enabling the emergence of life.

We caution, however, that these sites are not thoroughly explored. The effect of UV radiation and high-energy particles on these aerosols is poorly understood, as is their lifetime and subsequent fate of organic molecules after the destruction of the aerosols.

We reiterate that we have tackled only a small subset of candidates; we shall explore one other potential site for abiogenesis, namely ice and icy environments, in Section 10.4.2 dealing with subsurface ocean worlds.

5.5.4   Abiogenesis as planetary-scale process

Hitherto, we have singled out certain putative sites for abiogenesis, and life might have emerged on one or more of these microenvironments. To put it differently, life is interpreted to have originated in definite setting(s), that is, it should have arisen *somewhere* at some moment in time.[8] However, the vital (yet unspoken) point is that none of these sites were isolated from the entire Earth system, as they would have been receiving and/or transmitting flows of energy and raw materials from other locales.

Hence, when viewed in this light, even though abiogenesis on Earth is assumed to have occurred in particular microenvironments (i.e., it was localised), it could – and perhaps should – be concurrently perceived as a planetary-scale process. The notion of our planet as a global reactor for (pre)biotic chemistry has a long history; a modern summary of this theme is furnished in Stüeken et al. (2013). The network, loosely speaking, of potential sites for the emergence of life and the circulation systems (e.g., oceans, atmosphere) linking them is shown in Figure 5.11.

This global picture of abiogenesis is valuable from not just a conceptual standpoint (i.e., entailing a holistic vision) but also from a practical perspective. Analyses of early Earth and habitable worlds in our solar system (and in extrasolar planetary systems) should endeavour to investigate the existence of appropriate sites for abiogenesis in tandem with the circulation systems that transport energy and materials throughout the planet.

5.6   Final remarks: a physics coda

The motif of 'order' recurs in the book, from large-scale structure in the Universe (see Section 2.1.1) to the widely accepted premise that life is characterised by high internal order. In physics, order shares close connections with *entropy* S, whose statistical-mechanical (microscopic) definition is

$$S = -k_B \sum_{i}^{\mathcal{N}_{\text{tot}}} p_i \ln p_i, \tag{5.13}$$

where i stands for the i-th microstate (i.e., a microscopic configuration of the system), p_i is the probability associated with that state, and $\mathcal{N}_{\text{tot}}$ is the total number of microstates. When all the probabilities

---

[8] The alternative, which is that it emerged in multiple locations (virtually) simultaneously, appears highly improbable and may be dismissed.

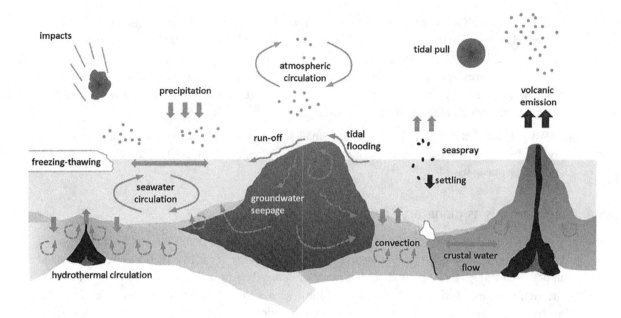

Figure 5.11 A selection of possible sites for the emergence of life and the circulation systems that may have distributed energy and materials amongst them. (Credit: Stüeken et al. 2013, Figure 5; reproduced with permission from John Wiley and Sons)

are equal, it can be shown that (5.13) reduces to $S = k_B \ln \mathcal{N}_{\text{tot}}$. In this scenario, the higher the number of microstates, the greater would be the entropy. When $\mathcal{N}_{\text{tot}}$ is increased, the probability of finding the system in a given configuration decreases, which may be very roughly interpreted as amounting to greater disorder. Thus, crudely speaking, disorder might be equated with higher entropy and order with lower entropy (e.g., Young and Freedman, 2018, Chapter 20.8).

Next, we turn our attention to the *Second Law of Thermodynamics*, which essentially states that $dS/dt \geq 0$ for an isolated system (e.g., Kondepudi and Prigogine, 2015, Chapter 3). When we naïvely apply this law to living systems, it indicates that their entropy, and therefore their disorder, must grow with time, seemingly contradicting the observation that life has a high degree of order. This conundrum is readily resolved since organisms are *not* isolated, but rather interact with their external environment frequently. Hence, the rules of physics entirely allow one subsystem (life) to undergo a decrease in entropy while another subsystem (environment) increases its entropy, with the net result being that the total system (which is isolated) experiences an increase in its entropy.

As seen from this viewpoint, life is a phenomenon that converts thermodynamic disequilibria (Schrödinger, 1944; Branscomb et al., 2017). The Universe as a whole, and the Earth–Sun system specifically, has considerable thermodynamic disequilibrium (consult Section 2.1.1), broadly manifested in the form of gradients. This disequilibrium is dissipated through a variety of avenues, which are thermodynamically favourable and accompanied by an increase in the entropy; these avenues may be loosely regarded as the energy inputs or sources on mapping to the language of this book.

A thermodynamically favourable (i.e., downhill) reaction of the above kind can be coupled together with an unfavourable (i.e., uphill) reaction, the latter of which creates thermodynamic disequilibrium and is characterised by a decrease in entropy. However, to maintain compatibility with the Second Law, the net reaction must still be thermodynamically favourable and entropy-increasing. On assembling the pieces, we may conclude that living systems are therefore intricate 'engines' that consolidate the downhill and uphill reactions into a single unified process, thus operating as sophisticated agents of disequilibria conversion.

When viewed from this humbling stance, the origin(s) of life – which can be envisioned as the emergence of the aforementioned engines – constitutes a triumph of fragile order over the ever-growing disorder in the Universe.

5.7   Problems

Question 5.1: We will seek to construct a simple mathematical model to gauge the timescale for the origin(s) of life on Earth. Let us model the growth in the total number of distinct species – the species richness ($N_\star$) in ecology – on Earth by an exponential function (which is not unreasonable).

$$N_\star = \exp\left(\frac{t}{\tau}\right) - 1, \tag{5.14}$$

where t is the time that has elapsed after the onset of habitable conditions on Earth; we will optimistically suppose that the Earth has been habitable since ~ 4.4 Ga (Gyr ago). Here, τ is an unknown parameter and represents the characteristic growth timescale.

  a. At $t = 0$, what is the value of $N_\star$? Justify why the result makes sense.
  b. Some papers have concluded that the Earth may be home to $N_\star \sim 10^{12}$ species currently, whereas other studies have yielded $N_\star \sim 10^6$ species. By setting $t = 4.4$ Gyr, estimate the values of τ for these choices of $N_\star$.
  c. Using the above two estimates for τ (derived from the two cases of $N_\star$), compute the timescale t_0 required for abiogenesis in each instance. Explain why t_0 should be calculated from the criterion $N_\star(t_0) = 1$.
  d. From the two estimates for t_0 obtained above, determine when the origin(s) of life would have transpired according to this simple model, by expressing your answer in terms of Ga.
  e. Discuss whether this result (i.e., the timing of abiogenesis) is compatible with the evidence introduced at the start of this chapter and in Section 5.3.1.

Question 5.2: As remarked in Section 5.1.1, a total of 22 amino acids are documented in proteins. Answer the following questions regarding them.

  a. Of the 22 amino acids, which is the 'odd one out' (i.e., missing amino acid) in Figure 5.2? Clarify why this amino acid was excluded.
  b. Determine how many of the 22 amino acids contain sulfur and list their names. Likewise, how many of this group have elements other than CHNOPS? Specify the names of these amino acids, and mention what element(s) they possess in addition to CHNOPS.
  c. How many of the 22 amino acids belong to the category of α-amino acids? Provide an example in biology where β- or γ- or δ-amino acids are used, with the appropriate peer-reviewed reference(s).

Question 5.3: We presented a simplified version of the central dogma of molecular biology in Section 5.1.2, namely: DNA $\rightarrow$ RNA $\rightarrow$ protein. By undertaking a literature survey of peer-reviewed references, explain why this picture is incorrect, and then furnish a couple of counterexamples.

Question 5.4: The theoretical model developed by Morowitz (1967) for a minimal self-reproducing entity yielded a cell radius $\mathcal{R}_o$ of

$$\mathcal{R}_o \approx \left[10\left(1.334\,\mathcal{N}_{Pr} + 29.7\right)^{1/3} + 7.5\right] \text{nm}, \tag{5.15}$$

where $\mathcal{N}_{Pr}$ is the number of *distinct* proteins in the cell. Plot (5.15) as a function of $\mathcal{N}_{Pr}$ on a log-log scale, with $\mathcal{N}_{Pr}$ ranging from 10 to 10^4. By setting $\mathcal{N}_{Pr} \approx 3000$ for *E. coli* (Nelson and Cox, 2004, Table 1-2), calculate the cell volume and compare this with the *E. coli* cell volume of 1.3 μm^3 (Milo and Phillips, 2016, pg. 10). Comment on the discrepancy, if applicable.

Question 5.5: The Martian meteorite ALH84001 hosts microfossil-like structures with an effective radius of $\sim$50 nm (McKay et al., 1996, Figure 6).

  a. As per (5.15), what is the number of distinct proteins that would be needed in a putative cell to create this size? Compare this estimate with common bacteria and viruses by citing peer-reviewed publications.

  b. By drawing on Section 5.2.2, determine the pH of the intracellular fluid that would support an organism with the above radius.

Question 5.6: In Section 5.3.3, many different 'worlds' for the origin(s) of life were outlined. By consulting and citing peer-reviewed papers, elaborate on the potential advantages of the coenzyme world and virus world. Next, identify and describe one world that was not covered in this section.

Question 5.7: The Miller–Urey experiment mimicked a reducing atmosphere, and the yields dropped substantially when a neutral atmosphere was used, as mentioned in Section 5.4.1. However, these studies did not take the surface geology into account, which may help offset this issue. Read and summarise the paper by Cleaves et al. (2008) pertaining to this subject, and provide quantitative details (e.g., yields) where possible.

Question 5.8: The synthesis of nucleobases has been demonstrated through the following experimental avenues: (a) spark discharge, (b) high-energy particles, and (c) shock waves. For each of these three sources, locate and summarise (with quantitative details) at least one peer-reviewed publication.

Question 5.9: By performing a literature search of peer-reviewed papers, describe three to five organic compounds reported in micrometeorites, along with their respective abundances. Have any of the biomolecular building blocks (e.g., nucleobases) been detected in micrometeorites (and if so, state them)?

Question 5.10: Aside from the pathways explicitly delineated in Section 5.4.3, describe two other avenues each for the synthesis of peptides and oligonucleotides after surveying the peer-reviewed literature. Based on our knowledge of early Earth, how viable would you deem these pathways?

Question 5.11: Read Section 5.5 before answering the following questions.

  a. Planets with only oceans on the surface have been dubbed *water worlds* or *ocean planets*, such as Kamino from the *Star Wars* franchise. For such worlds, identify which of the geological environments considered plausible sites for abiogenesis are potentially (1) rarer than on Earth (or even absent), and (2) more widespread relative to Earth. Explain your reasoning.

  b. Repeat the analysis specified above, but in this case for planets with minimal water bodies and mostly composed of deserts. These planets are known as *desert planets* or *Dune worlds*; the latter takes its title from the famous science fiction series by Frank Herbert bearing this name.[9]

---

[9] Strictly speaking, the desert planet in the *Dune* franchise is called Arrakis. As a bonus, name your favourite character from the books or the recent, critically acclaimed, movies.

Question 5.12: The equilibrium tidal height H_t induced by an object with mass M situated at distance a from the Earth is roughly expressible as

$$H_t \approx \frac{15}{8} \frac{M}{M_\oplus} \frac{R_\oplus^4}{a^3}, \tag{5.16}$$

where $M_\oplus$ and $R_\oplus$ are the mass and radius of Earth, respectively. Although inaccurate, assume that H_t indicates the amplitude of the oceanic tides.

  a. Given $M \approx 1.2 \times 10^{-2} M_\oplus$ and $a \approx 60 R_\oplus$ for the Moon currently, compute H_t for the Moon. Next, input $M = M_\odot$ and $a = 1$ AU for the Sun ($M_\odot$ is the Sun's mass) and calculate H_t for the Sun.

  b. The value of a has evolved significantly over time for the Moon, because it continues to recede away from the Earth since formation. In the mid-Hadean, the Earth–Moon distance might have been $a \approx 15 R_\oplus$. Compute the corresponding height H_t in this period.

  c. Compare the results obtained above for the present-day and past lunar distances, and verify that the tides were much stronger in the latter scenario. Elucidate one positive and one negative consequence of such strong tides in connection with the origin(s) of life.

Question 5.13: Aside from the microenvironments elaborated in Section 5.5, select two additional contenders for enabling abiogenesis and summarise three pros and three cons associated with each of them; cite relevant peer-reviewed publications to bolster your arguments.

Question 5.14: Suppose that the shortest ribozyme capable of self-replication would need to consist of ~40 nucleotides (Robertson and Joyce, 2012, Section 2.3). Given that RNA is constructed from four building blocks (nucleotides), how many RNA molecules of this length are mathematically possible? If each nucleotide has an average mass of ~5×10^{-25} kg, what is the total mass of all these 40-mers put together? What inference(s) can be drawn upon comparing this estimate with the mass of Earth?

Question 5.15: After a literature search, what is one 'school of thought' on abiogenesis not addressed herein that you would like to learn about?

6   Co-evolution of Life and Environment on Earth

In the previous chapter, we have witnessed how and where (and briefly when) the genesis of life (abiogenesis) on our planet may have transpired. The transition from an assortment of simple (inorganic) molecules to even the most rudimentary of cells is undoubtedly a profound one. Not only is there a substantial increase in the complexity of the molecules themselves but also a massive jump in the ways whereby they interact with one another, that is, a cell evinces considerable *relational* complexity. It is, however, evident that the '*tape of life*', to borrow Stephen Jay Gould's (1941–2002) famous phrase (Gould, 1989, Chapter 1), did not cease to play once the origin(s) of life had occurred. Countless myriad life forms emerged during the course of Earth's history via Darwinian evolution (refer to Section 1.1.3), thereby constituting and sculpting our planet's rich and intricate biosphere.

As we shall describe subsequently in Chapters 13 and 14, the search for extraterrestrial life – which is one of the fundamental questions in astrobiology (namely, '*Are we alone?*') – involves not just seeking microbes but also extends to complex multicellular life and 'intelligent' life. Thus, it is appropriate, and arguably necessary, to understand how the evolutionary trajectory of life has unfolded on planet Earth, as it might offer us some clues as to how evolution may operate on other worlds. However, as we merely have one data point, it is important to avoid the pitfall of overly extrapolating from this single instance by uncritically invoking a *principle of mediocrity* that perceives Earth and its biosphere as 'typical' (Balbi and Lingam, 2023); we will revisit and reinforce this theme in Section 14.2.3.

The evolution of life on our planet is deeply intertwined with the environment. Loosely speaking, natural selection acts as a filter, and preferentially disfavours individuals that are less adapted to their settings. It is, therefore, apparent that the environment exerts a major influence on life, but it is crucial to appreciate that the converse is also valid. On account of a panoply of biological processes (e.g., metabolism), organisms modify their habitats, sometimes to a dramatic degree, as we shall outline later in this chapter. In fact, the propensity of life to alter its settings is responsible for some of the biosignatures (i.e., markers of biological origin) in Chapter 13 and the technosignatures (i.e., signs of technology) in Chapter 14. Hence, it is accurate to state that life and its environment *coevolve* together.

Before doing so, we note that the *Gaia hypothesis* – which postulates that '*a coupled system of life on Earth and its abiotic environment self-regulates in a habitable state, despite destabilising influences*' (Lenton et al., 2018, pg. 633) – may be viewed as a strong version of the aforementioned planetary-scale co-evolution. Since the Gaia hypothesis was elucidated by James Lovelock (1919–2022) and Lynn Margulis (1938–2011) in their pioneering 1970s publications (Lovelock and Margulis, 1974), it has attracted its share of detractors (e.g., Tyrrell, 2013) and supporters (e.g., Lenton et al., 2018).

On the basis of the preceding discussion, we will broadly divide this chapter on co-evolution of life and its environment into two components. In the first, we will sketch the evolutionary timeline of life on Earth, and mention a few schemes that have spotlighted major evolutionary breakthroughs by employing a suitable framework. In the second, we will explore two specific examples of how life and its environment have interacted with one another: (1) the rise in oxygen levels and the ensuing consequences for the biosphere; and (2) the causes and implications of mass extinctions.

6.1   Earth's evolutionary timeline

Evolution on Earth has engendered 'endless forms most beautiful and most wonderful', as poetically expressed by Charles Darwin at the conclusion of *The Origin of Species* (1859). Our exposition will be manifestly brief and selective for various reasons, two of which are the absence of background information (only high school biology is assumed) and space constraints.

Of equal, if not greater, importance is the possibility that replaying the 'tape of life' (introduced earlier) may well lead to radically different results, in which case the Earth might not constitute a reliable proxy for envisioning and seeking extraterrestrial life, and attaching too much weight to the specific sequence of events on Earth could be counterproductive. The debates over whether the outcomes of replaying the tape of life are similar (convergent) or divergent, predictable or a matter of chance, have garnered significant attention in evolutionary biology (e.g., Gould, 1989). We refer the readers to the modern reviews by Blount et al. (2018) and Powell (2020), the latter of which displays an explicit astrobiology perspective.

Dozens of books have charted our knowledge of the evolutionary history of life on our planet – starting with the formation of Earth and ending their narrative with the growth of (industrial) human societies – such as Stanley (2005), Dawkins and Wong (2016), and Knoll (2021). We will chiefly parallel the discussion in Knoll and Nowak (2017), owing to its succinct character. Before reading the forthcoming (sub)sections, we recommend the reader to familiarise themselves with Figures 6.1 and 6.2, which will be valuable in clarifying technical nomenclature and painting a global picture.

6.1.1   Archean eon

We will skip the Hadean eon altogether, as it has already been tackled in Chapter 4. The remarkable transition from non-living systems to life may have been actualised in the Hadean, although the first unequivocal markers of life appear in the early Archean at 3.4–3.5 Ga, as remarked at the beginning of Chapter 5. It seems plausible, however, that life had emerged much earlier, say, in the Hadean, but that no robust traces exist.

The atmospheric composition in the Archean, which spanned 4.0–2.5 Ga, was clearly distinct compared to today. The abundance of N_2 was similar to (perhaps slightly less than) its current value, the concentration of O_2 was $\lesssim 10^{-6}$ times the present atmospheric level (PAL), and the greenhouse gases CO_2 and CH_4 occurred at potential abundances of $\sim$10–2,500 and $\sim 10^2$–10^4 PAL, respectively (Catling and Zahnle, 2020). The ocean temperature was conceivably $\sim$0–40°C, although higher values are not ruled out. These estimates have been obtained through a combination of interpreting geological proxies and state-of-the-art numerical models.

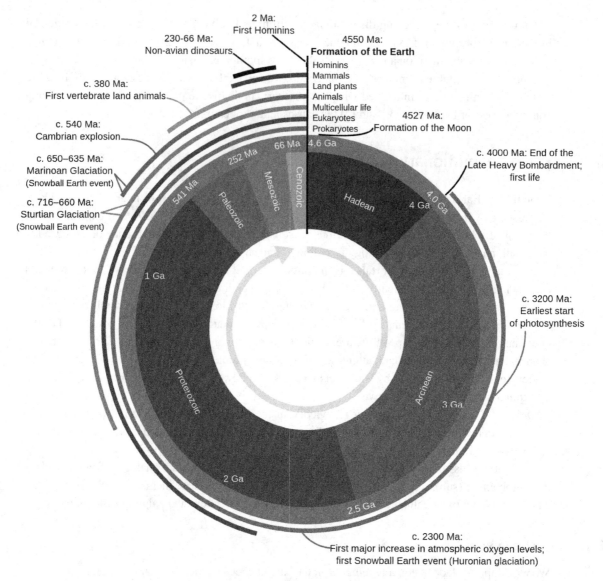

Figure 6.1 Chronicle of Earth's geological history, along with the major evolutionary events and their timing. In this 'clock', Ma is 1 million years ago, and Ga is 1 billion years ago. Note that some of the dates are subject to uncertainty. (Credit: https://commons.wikimedia.org/wiki/File:Geologic_Clock_with_events_and_periods.svg)

Some facets of Archean Earth remain uncertain to date. One of them is the fraction of the surface attributable to continental landmasses. Based on the analysis of oxygen isotope ratios determined from rock samples, Johnson and Wing (2020) concluded that Archean Earth mostly had limited continental coverage, and was nearly a water world, with such landmasses only arising during the late Archean (3.0–2.5 Ga). However, other publications support the presence of quite extensive continents throughout the Archean; a review of this subject is provided in Korenaga (2021).

The availability of dissolved phosphorus (in the form of phosphates) in the oceans is another debated matter. As seen in Section 5.1, phosphorus (P) is a bioessential element, and its abundance is

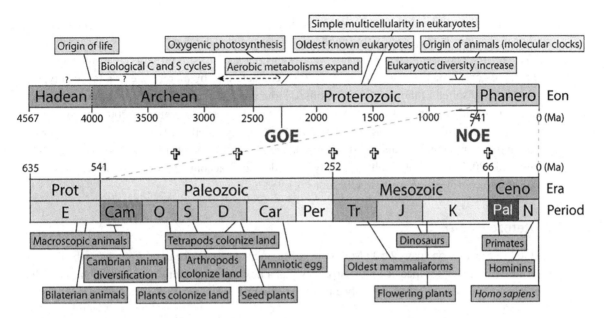

Figure 6.2 Timeline of major evolutionary events in Earth's history. Crosses signify the 'Big Five' Phanerozoic mass extinctions. Abbreviations are: Phanero (Phanerozoic); Prot (Proterozoic); Ceno (Cenozoic); E (Ediacaran); Cam (Cambrian); O (Ordovician); S (Silurian); D (Devonian); Car (Carboniferous); Per (Permian); Tr (Triassic); J (Jurassic); K (Cretaceous); Pal (Paleogene); and Neo (Neogene). (Credit: Knoll and Nowak 2017, Figure 1; reproduced with permission from AAAS)

presumed to regulate the biological productivity in the oceans. Many papers have used experiments, models, and geological proxies to conclude that the oceanic flux of P might have been limited in the Archean, thereby constraining biological productivity to <10% the modern value (e.g., Hao et al., 2020). However, this stance is not universally accepted, as a couple of studies have shown that phosphate-rich oceans in the Hadean and Archean are tenable.

Archean Earth was apparently composed exclusively of prokaryotes (see Section 5.1.4), the bacteria and the archaea. In the Archean, metabolic pathways associated with these microbes emerged and diversified. We will touch on several of these pathways such as methanogenesis and photosynthesis in Section 7.1. Empirical evidence for these metabolisms is derived from analyses of isotope ratios, because biology preferentially utilises certain isotopes over others, thus skewing the isotope ratios of samples; we elucidate this process of isotopic fractionation in Section 13.1.1. The pathway of nitrogen fixation, which involves the conversion of atmospheric N_2 into nitrogen species like ammonium (NH_4^+) that can be assimilated by organisms, evolved by at least 3.2 Ga as per measurements of the $^{15}N/^{14}N$ isotope ratio in rocks from the Soanesville Group, Australia (Stüeken et al., 2015).

Of greater significance was the evolution of *oxygenic photosynthesis*, whose importance will become apparent in Section 6.3.3. The enrichment of certain metals (e.g., molybdenum) >2.5 Ga has been interpreted as ensuing from the oxidation of sulfide minerals in the crust, which in turn was potentially due to O_2 produced by oxygenic photosynthesis (Lyons et al., 2014); however, this evidence is subtle and admits alternative explanations. Molecular clocks, which draw on genetic data, suggest an even deeper origin of oxygenic photosynthesis, possibly extending as far back as 3.6 –3.2

Ga (Sánchez-Baracaldo et al., 2022), although it is worth recognising that these clocks typically decline in accuracy with age, and some publications have yielded much younger ages (e.g., Shih et al., 2017). The first photosynthetic organisms were ostensibly ancestors of present-day *cyanobacteria*.

Last, the genesis of *simple multicellularity* (i.e., characterised by merely one or a few cell types) might have transpired in the Archean. Fossilised structures akin to filaments and mats may reflect simple multicellularity in prokaryotes by 3.5–3 Ga (Grosberg and Strathmann, 2007, pg. 622), but these findings are subject to ambiguity. Potential drivers of multicellularity are varied, ranging from mitigating predation and enhancing stress resistance to increased motility and resource utilisation (Tong et al., 2022). However, multicellularity is not devoid of accompanying costs, which must also be taken into consideration (refer to Questions 6.1 and 6.2).

6.1.2   Proterozoic eon

As the Proterozoic eon spans 2,500 Ma to ~539 Ma, it encompasses nearly 2 Gyr and is the longest eon in Earth's history. Hence, it is not surprising that a host of major events unfolded in this eon. We will restrict ourselves to sketching only a handful of them.

6.1.2.1   The Great Oxidation Event and Snowball Earth

From a geological standpoint, two prominent features heralded the Proterozoic eon. The first is the substantial increase in atmospheric O_2 levels beyond 10^{-5} PAL, which commenced about $2.5 - 2.4$ Ga, termed the *Great Oxidation Event* or *Great Oxygenation Event* (GOE) (Lyons et al., 2014); a detailed analysis of the GOE is deferred to Section 6.3. O_2 levels may, however, have fluctuated for ~200 Myr until a quasi-stable state was achieved at ~2.2 Ga (Poulton et al., 2021); some debated studies have proposed an even longer interval extending to ~1.7 Ga. Oxygen concentrations during much of the Proterozoic were possibly ~1% PAL, although the predicted range is ~0.1–10% PAL (see Figure 6.3). However, across an interval of order 100 Myr in the so-called *Lomagundi-Jatuli Excursion*, O_2 levels might have approached modern concentrations, as revealed by Figure 6.3.

Second, in the broad vicinity of 2.4 Ga (Gumsley et al., 2017), our planet may have experienced a period of global glaciations collectively constituting a *Snowball Earth* (Kirschvink et al., 2000),

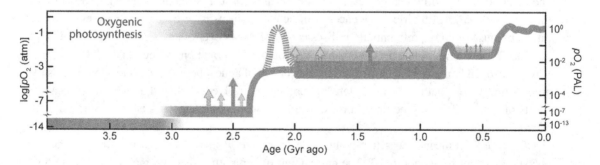

Figure 6.3 Earth's atmospheric oxygen levels as a function of time. The two different shaded regions from ~2.0 to ~0.5 Ga embody contrasting interpretations of O_2 evolution. The arrows indicate the presence of oscillatory behaviour in oxygen concentration. (Credit: Lyons et al. 2021, Figure 3; CC-BY-NC 4.0 license)

where virtually the entire planet was covered by ice sheets (even at low latitudes). Several lines of geological evidence appear to support the Snowball Earth hypothesis (Gumsley et al., 2017), such as glacially derived deposits (e.g., diamictite) and *cap carbonates* (i.e., carbonate rocks with distinctive textures). Since this Snowball Earth episode was roughly coeval with the GOE (i.e., an O_2 increase), potential connections between these two phenomena have attracted extensive research, some of which we outline in Section 6.3.4. We caution, however, that not all scientists accept the Snowball Earth state, with alternatives including the *waterbelt state*, wherein ice-free oceans persisted at latitudes of <10–$30°$.

6.1.2.2 Eukaryotes

The next major breakthrough is the origin of *eukaryotes*. This group of organisms (the domain *Eukaryota*) includes, among other categories, all *complex multicellular life* documented on Earth, which we will address shortly. The common textbook definitions of eukaryotes list several criteria that distinguish them from prokaryotes: genetic material encapsulated within a nucleus, and the existence of additional intracellular compartments (*organelles*) such as *mitochondria* (sites of energy production via metabolism), *plastids* (sites of photosynthesis), and the *endoplasmic reticulum* (crucial for protein and lipid synthesis). However, exceptions in the form of bacterial organelles are documented (e.g., in the gigantic cm-long *Thiomargarita magnifica*), rendering unambiguous definitions of eukaryotes difficult.

This definitional issue complicates the task of demarcating early eukaryotes from their prokaryotic ancestors (Donoghue et al., 2023). State-of-the-art phylogenetic analyses indicate that the pathway(s) from the *first eukaryotic common ancestor* (FECA) to the *last eukaryotic common ancestor* (LECA) – the latter of which is the most recent population or individual from which eukaryotes descended – was intricate and prolonged; it may have required as much as >500 Myr (Porter and Riedman, 2023, pg. 178). Despite the attendant uncertainties, it is (almost) universally acknowledged that the emergence of eukaryotes entailed *endosymbiosis*; this phenomenon can be succinctly expressed as (Martin et al., 2015, Section 1):

Endosymbiotic theories have it that cells unite, one inside the other, during evolution to give rise to novel lineages at the highest taxonomic levels, via combination.

It is widely held, barring a small minority, that both mitochondria and plastids originated through endosymbiosis (Martin et al., 2015; Eme et al., 2017). Endosymbiotic theories have a long and winding history, with Constantin Mereschkowsky's (1855–1921) 1905 scheme constituting a famous example: the plastids and nucleus were postulated to arise from free-living *cyanobacteria* and bacteria from the genus *Micrococcus*, respectively. An endosymbiotic origin for mitochondria was subsequently proposed by Paul Portier (1866–1962) and Ivan Wallin (1883–1969) in the 1910s and 1920s. Lynn Margulis wove these individual strands together with the available molecular, cellular, paleontological, and geological evidence in her 1967 landmark publication that breathed new life into endosymbiosis (Sagan, 1967).

The nature of the host into which the endosymbionts were assimilated is not fully understood: most (albeit not all) hypotheses assert that it was an archaeal cell (Eme et al., 2017; Mills et al., 2022). In particular, molecular clock studies support a close relationship between the remarkable *Asgard*

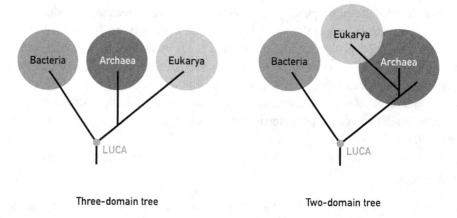

Three-domain tree   Two-domain tree

Figure 6.4 Schematic of three- versus two-domain Tree of Life; the former represents the canonical picture, which is being challenged by the latter. LUCA represents the last universal common ancestor.

archaea (Spang et al., 2015) and the provenance of eukaryotes (Liu et al., 2021); the former group exhibits eukaryotic-like aspects such as compartmentalised DNA and elaborate *cytoskeleton* (i.e., a network of filaments that bolsters cell structure) (Donoghue et al., 2023). This breakthrough lends credence to an ongoing profound shift from the classic three-domain *Tree of Life* (ToL) comprising bacteria, archaea, and eukaryota (Woese et al., 1990) to a two-domain ToL composed exclusively of bacteria and archaea (e.g., Liu et al., 2021), as depicted in Figure 6.4. However, the latter scheme has garnered critiques, and is not accepted by everyone.

Notwithstanding the rapid pace of advances, many mysteries surround eukaryogenesis. One of the most significant among them concerns the role and timing of mitochondria: endosymbionts that share deep connections with *Alphaproteobacteria* (Martin et al., 2015; Eme et al., 2017). A prominent class of hypotheses contend that the acquisition of mitochondria exerted strong selection pressure on the host cell and this endosymbiont, thereby sparking the birth of eukaryotic traits such as certain organelles (e.g., endoplasmic reticulum), sexual reproduction (reportedly present in LECA), and enhanced genome and cell size (Lane and Martin, 2010; Raval et al., 2022); however, some of the bioenergetic arguments have been disputed. Other hypotheses have suggested that the incorporation of mitochondria happened at an intermediate or late stage in eukaryotic evolution, and that it was not of critical importance (see Donoghue et al. 2023 for a review).

The reason we have highlighted eukaryotes and their origin(s) is because of their numerous advantages over prokaryotes. For starters, the existence of internal compartments (i.e., organelles) allows eukaryotes to maintain distinct physicochemical conditions within them (e.g., lysosomes have acidic pH to break down biopolymers), and thus execute diverse and specialised functions. Second, eukaryotes are capable of *phagocytosis*, namely: the engulfment of particles with sizes ≥ 0.5 μm through folding motions of the cell membrane. Phagocytosis is linked with several benefits such as driving major evolutionary and ecological changes due to predation, improving nutrition, and aiding the immune system; we will encounter it again shortly.

Last, eukaryotes are equipped with novel avenues to sense and interact with their environments, which in turn facilitated complex behaviours (e.g., predation, sophisticated movement) (Wan and

Jékely, 2021). We will outline a simple model to demonstrate the relevance of cell size in this regard. The Péclet number (Pe) is defined as

$$\text{Pe} = \frac{\mathcal{U}_o \mathcal{L}_o}{D}, \tag{6.1}$$

where D is the diffusion constant, while $\mathcal{U}_o$ and $\mathcal{L}_o$ are the typical cell speed and size, respectively. The Péclet number crucially embodies the ratio of the advective and diffusive timescales, such that Pe $\gtrsim 1$ enables a significant shift towards organisms with the capacity to traverse and harness sizeable volumes (of liquid). A biophysical model by Meyer-Vernet and Rospars (2016) concluded that $\mathcal{U}_o \propto \mathcal{L}_o$, and obtained a proportionality constant of $\sim 10 \, \text{s}^{-1}$. On solving for $\mathcal{L}_o$ after imposing the criterion Pe $\gtrsim 1$, we arrive at

$$\mathcal{L}_o \gtrsim 10 \, \mu m \left(\frac{D}{10^{-9} \, \text{m}^2/\text{s}} \right)^{1/2}, \tag{6.2}$$

where the normalisation for D is based on its value for small molecules in water (Wan and Jékely, 2021, Section 3). Interestingly, eukaryotic cells have characteristic sizes of $\gtrsim 10 \, \mu m$, whereas pro-karyotes are often $\lesssim 1 \, \mu m$ in size (although exceptions do exist). This calculation illustrates how eukaryotes might access locomotion regimes that are generally ruled out for prokaryotes.

Moving on to the earliest evidence for eukaryotes, we refer the reader to Porter and Riedman (2023). Fossils of large organisms with eukaryotic-like structural complexity have been identified in rocks dating from 2.2 to 1.8 Ga, but the evidence is decidedly ambiguous. The Changcheng Group (North China), with an age of 1673–1638 Ma, harbours some of the earliest fossils acknowledged as eukaryotic organisms on the basis of careful examination of their morphologies. From a phylogenetic perspective, the ranges for FECA and LECA are 3.0–2.3 Ga and 1.9–1.1 Ga, respectively.

It is apparent, therefore, that the first eukaryotes might have emerged shortly after the post-GOE stabilisation of atmospheric O_2, which may have occurred at ~ 2.2 Ga (Poulton et al., 2021). Hence, this feature has stimulated conjectures, both classic and modern, that the rise in O_2 levels paved the way for eukaryogenesis (Sagan, 1967; Knoll and Nowak, 2017). However, this stance remains speculative, as there is seemingly a substantial gap separating the earliest attested fossils of eukaryotic provenance and the aforementioned O_2 increase (Mills et al., 2022). Additional data are thus needed to resolve the connection(s) between atmospheric O_2 and eukaryogenesis.

6.1.2.3 Complex multicellularity

If we envision complex life on Earth, two groups that spring to mind are animals (*Animalia*) and land plants (*Embryophyta*). Thus, the next milestone we will tackle in this evolutionary tale is *complex multicellularity*, overviews of which are provided in Knoll (2011) and Chaigne and Brunet (2022).

Several features are conventionally invoked for distinguishing complex multicellularity from simple multicellularity introduced in Section 6.1.1. Notable attributes include (Knoll, 2011): (1) cellular differentiation (equivalent to multiple cell types); (2) cell–cell adhesion mediated by spe-cialised proteins; (3) inter-cellular communication involving diverse signalling molecules; and (4) three-dimensional morphology, wherein some cells never experience direct contact with the external environment over their lifetime.

Complex multicellularity is documented in five lineages: animals, land plants, fungi, brown algae, and red algae (Chaigne and Brunet, 2022). In contrast, simple multicellularity has evolved >25 times in eukaryotes (Grosberg and Strathmann, 2007), thus revealing the rarity of complex multicellularity. The steps leading to the emergence of complex multicellularity are numerous, some of which are not well understood, owing to which we skip this exposition. It suffices to say that the evolution of proteins for binding and signalling, extracellular matrix (which provides scaffolding for the cellular components), division of labour among cells, and cytoplasmic bridges joining cells are crucial in this context (e.g., Brunet and King, 2017).

The first lineage that we single out is the land plants, which constitute around 80% of the biomass on our planet (Bar-On et al., 2018). As per the available empirical evidence (e.g., spores and cuticles), land plants evolved from green algae (specifically the class *Charophyceae*) at least by ~470 Ma (Bowman, 2022), although a few molecular clock studies have controversially predicted that the land plants originated in the *Neoproterozoic* (1,000–539 Ma). It is obvious that plants impact the biosphere tremendously, ranging from sustaining other complex multicellular organisms (e.g., animals) to modifying atmospheric composition and hydrological cycles.

The second lineage of interest to us is the animals: the only instance of complex multicellularity arising from eukaryotes capable of phagocytosis and retaining this trait (Cole et al., 2020). As opposed to the multiple origins of complex multicellularity in algae (which encompasses plants) and fungi, animal multicellularity ostensibly evolved just once, thereby highlighting the difficulties underpinning this pathway. In particular, a high degree of coordination and cooperation might have been necessary for the transition from unicellularity to animal multicellularity in the realms of feeding, locomotion, and reproduction (Brunet and King, 2017). Some of the oldest credible animal fossils are from the Drook Formation (Newfoundland, Canada), with an age of ~574 Ma. In contrast, several phylogenetic studies have determined an origination time of roughly 800 Ma, which is broadly when eukaryotes appear to have diversified (see Cole et al. 2020).[1]

Animals – with their unique combination of complex multicellularity, macroscopic size, motility, and heterotrophy (Butterfield, 2011) – are endowed with the capacity to alter, maintain, and create habitats through their activities, owing to which they are termed *ecosystem engineers* (Jones et al., 1994), and this process of environmental modification corresponds to *niche construction* (e.g., Odling-Smee et al., 2013). Through this property of ecosystem engineering, animals have exerted a powerful influence on the cycling of nutrients, biological productivity, and the nature of their habitats. Animals have been linked with enhancing diversification of prokaryotes and eukaryotes (e.g., land plants); shifts in body size; production of biologically derived minerals; and much more (Butterfield, 2011; Judson, 2017).

6.1.2.4 The Neoproterozoic Oxidation Event and Snowball Earths

Geological evidence, ranging from shifts in carbon isotope ratios to trace metal abundances, potentially indicates large-scale changes on Earth starting ~800 Ma (Lyons et al., 2021, Section 5), loosely when animals evolved.

[1] The period of ~1 Gyr between ~1.8 Ga (when LECA may have existed) and ~0.8 Ga (when animals might have emerged) is called the *Boring Billion*, which is considered a misnomer.

Atmospheric O_2 levels may have begun to ascend $\sim$800 Ma (refer to Figure 6.3), thus signifying the onset of the *Neoproterozoic Oxidation Event* (NOE) (Och and Shields-Zhou, 2012), roughly assigned a window of $\sim$800–600 Ma (Lyons et al., 2021, Section 5). It is, however, important to recognise that this rise in O_2 was protracted and likely non-monotonic; deep ocean oxygenation might have taken several 100 Myr, with certain geochemical proxies supporting a timing as late as $\sim$400 Ma in the Paleozoic era (Tostevin and Mills, 2020). Therefore, the notion of a clear-cut NOE is unsettled, with some studies proposing substantial oscillations in atmospheric O_2 ($\sim$1–50% PAL) in place of a directional increase (e.g., Krause et al., 2022).

The *Cryogenian period* ($\sim$720–635 Ma) in the Neoproterozoic is conventionally believed to have experienced two global glaciation events, that is, Snowball Earth episodes (Hoffman et al., 1998, 2017). The geological evidence cited in their favour includes cap carbonates (introduced previously) and various glaciogenic sedimentary deposits (e.g., the so-called dropstones and rhythmites). The first glaciation, called the *Sturtian glaciation*, was long lived, since it approximately spanned 717–659 Ma. The second, the *Marinoan glaciation*, was relatively short in duration, as it existed between 639 Ma and 635.5 Ma. As with the aforementioned putative $\sim$2.4 Ga global glaciation event, alternative interpretations (e.g., waterbelt states) of the Sturtian and Marinoan glaciations have been explored.

Last, given the conceivable atmospheric and oceanic oxygenation that unfolded in the second half of the Neoproterozoic, as well as the Snowball Earth episodes and their aftermath, this trend raises a vital question: could an increase in O_2 levels have enabled the emergence and/or diversification of animals? This school of thought has thrived for more than half a century (Nursall, 1959; Mills et al., 2023), though state-of-the-art evidence and modelling has challenged the premise of a definite causal connection between these two phenomena (Cole et al., 2020). Some publications have advanced alternative scenarios – where oxygenation was driven by animal evolution, or they coevolved together (Lenton et al., 2014) – but they remain debated.

6.1.3 Phanerozoic eon

The *Phanerozoic eon* (539 Ma to present) commenced with the *Cambrian period* (539–485 Ma), which is renowned for hosting the *Cambrian explosion*, which is canonically represented as a '*unique episode in Earth history, when essentially all the animal phyla first appear in the fossil record*' (Marshall, 2006, pg. 355). The Cambrian was preceded by the *Ediacaran period* (635–539 Ma), which was marked by profound environmental upheavals and evolutionary change (Droser et al., 2017).

The Ediacaran life forms were predominantly soft-bodied, with the majority of body plans seemingly distinct from modern animals. In the Cambrian, the proliferation of body forms (e.g., the major animal group *Bilateria*, with bilateral symmetry) and species akin to current fauna is strikingly documented in the fossil record (Marshall, 2006), which spawned initial hypotheses of an 'explosion' of biota (see Gould 1989). However, recent fossil and geochemical evidence jointly suggests that the Cambrian explosion was merely one stage in a series of evolutionary radiations from the Ediacaran onward (Wood et al., 2019). Potential causes for the Cambrian explosion are myriad, ranging from the development and refinement of genetic toolkits to the moderate increase of O_2 levels (to $\sim$10% PAL) that may have facilitated certain behaviours (e.g., carnivory), thereby sparking an evolutionary 'arms race'. On the whole, a multifaceted array of abiotic and biological factors engendered the Cambrian explosion (Knoll and Carroll, 1999).

Due to the wealth of fossil and geochemical data, reconstructing and dating the major evolutionary innovations of the Phanerozoic has been feasible. Instead of an exhaustive treatment (e.g., Stanley, 2005; Dawkins and Wong, 2016; Knoll, 2021), which is obviously unrealistic, we will provide extremely brief and selective synopses of the geological periods next.

1. Cambrian (539–485 Ma): In the Cambrian explosion, major animal groups (e.g., vertebrates like fish) conspicuously manifested in the fossil record. Phylogenetic studies and tentative biological markers indicate that land plants might have evolved in the second half of the Cambrian.

2. Ordovician (485–444 Ma): Life, especially marine organisms (e.g., fish), is widely held to have diversified in a series of radiations called the *Great Ordovician Biodiversification Event* (GOBE). The *Late Ordovician mass extinction* closing this period may have witnessed the extinction of ∼85% of marine species; we address mass extinctions in Section 6.4.

3. Silurian (444–419 Ma): Jawed vertebrates underwent diversification, and the first traces of *vascular plants* (i.e., with specialised tissues for internally transporting materials) appear in this period.

4. Devonian (419–359 Ma): In this period, life in the oceans and on land diversified considerably; notable examples of the latter are the spreading of vascular plants and the evolution of tetrapods (i.e., four-limbed vertebrates). A series of extinctions (dubbed the *Late Devonian mass extinction*) may have collectively eliminated ∼70% of animal species.

5. Carboniferous (359–299 Ma): The atmospheric O_2 concentration (of >25%) was perhaps higher than its modern value (Mills et al., 2023). Terrestrial animals expanded in multiple ways, ranging from the emergence of reptiles and the putative ancestors of mammals to the radiation of amphibians and insects (which grew to great sizes). The supercontinent *Pangaea* formed in this period, and lasted for >100 Myr.

6. Permian (299–252 Ma): The diversification of synapsids (ancestors of modern mammals) and reptiles happened in this period. Three or four sizeable extinctions occurred, with the *Permian–Triassic mass extinction* at the end of this period constituting the most severe event in the entire Phanerozoic. It is estimated that ∼80–95% of all marine species went extinct, with recovery taking ≳10 Myr (Chen and Benton, 2012).

7. Triassic (252–201 Ma): Reptiles were the dominant terrestrial animals. The first dinosaurs evolved in the Triassic, and so did the early mammals. The *Triassic–Jurassic mass extinction*, a series of extinctions that closed this period, might have eliminated as much as ∼76% of species.

8. Jurassic (201–∼145 Ma): The breakup of Pangaea occurred during the early Jurassic. The dinosaurs were the dominant group of fauna on land. In this period, the first birds may have evolved from the dinosaurs. Flowering plants (*angiosperms*) might have emerged in the Jurassic.

9. Cretaceous (∼145–66 Ma): While dinosaurs were still prominent, the diversification of mammals, birds, and insects has been reported. Flowering plants radiated, and perhaps originated, in the Cretaceous. The famous *Cretaceous–Paleogene (K-Pg) mass extinction* that terminated this period was characterised by the extinction of ≲75% of all species, which included the non-avian dinosaurs.

10. Paleogene (66–23 Ma): In the aftermath of the K-Pg extinction event, mammals spread into many of the niches occupied by non-avian dinosaurs. The earliest primates may have evolved in an epoch not far removed from the Cretaceous–Paleogene boundary, as per fossils and phylogenetics. This

period also contains the *Paleocene–Eocene thermal maximum* at ∼56 Ma, when the planet warmed by 5–8°C.

11. Neogene (23–2.6 Ma): The diversification of mammals and birds continued in this period. The split between chimpanzees and *hominins* (i.e., members of the human lineage) was lengthy and intricate, owing to which it has a relatively broad range of ∼5–9 Ma (Bobe and Wood, 2022). With regard to hominin evolution, the earliest fossil specimens tentatively ascribed to our genus (*Homo*) date from ∼2.8 Ma.

12. Quaternary (2.6 Ma to present): This period has evinced regular fluctuations in environmental conditions (e.g., ice ages). Hominin evolution in the Quaternary was chiefly operational in Africa (Bergström et al., 2021), with modern *Homo sapiens* having emerged ≲0.3 Ma. Our species has engineered massive transformations of Earth's biosphere, particularly since the Industrial Revolution(s), perhaps warranting the introduction of a new geological epoch: the *Anthropocene* (Lewis and Maslin, 2018).

6.2 Paradigms for understanding major evolutionary events

In view of the wealth of information in Section 6.1, it is challenging to devise frameworks that can pinpoint the major evolutionary events that transpired in our planet's geological history by unearthing common or similar features linking them. Identifying these major evolutionary events could, in turn, enable us to speculate about the likelihood of their functionally analogous counterparts arising on other worlds, with the strong caveat that the same trajectories are not guaranteed to manifest in extraterrestrial environments.

We will briefly summarise three such frameworks herein, with a fourth one (entailing energy sources) deferred to Question 6.7. Yet another scheme (composed of ten 'inventions') is furnished in Lane (2009).

6.2.1 Major transitions in evolution

In 1995, John Maynard Smith (1920–2004) and Eörs Szathmáry (1959–present) authored an influential book: *The Major Transitions in Evolution* (Maynard Smith and Szathmáry, 1995). The paradigm of *major evolutionary transitions* (METs) introduced by the authors has sparked substantial research in multiple disciplines (e.g., Calcott and Sterelny, 2011; Szathmáry, 2015).

In the publication by Szathmáry and Maynard Smith (1995), the unifying aspects of the METs were proposed to be:

- Formation of more 'complex' entities from 'simpler' units. A classic example in this respect is eukaryogenesis, given that a couple of organelles (e.g., mitochondria) were once independent organisms.

- However, in some scenarios, these simpler units could hinder or even nullify the formation of the complex entities.

- The aforementioned simpler units once capable of self-reproduction are now restricted to reproduction as components of the complex entity. In the case of eukaryotes, certain organelles were free-living organisms that reproduced independently before they became part of the eukaryotic cell.

- The division of labour among the simpler units enabled task specialisation and boosted functionality. For instance, in complex multicellular organisms, the variegated cell types fulfil distinct purposes.
- The complex entities are endowed with improved or novel modes of information storage and transmission. To offer an example, sexual reproduction (observed in eukaryotes) allows for genetic 'shuffling'.

In place of the original METs (Maynard Smith and Szathmáry, 1995), we will outline the updated list presented by Szathmáry (2015, Table 1):

1. Origin(s) of protocells with autocatalytic networks in compartments.
2. Origin of full-fledged prokaryotic cells endowed with the *genetic code* (viz., 'instructions' in the nucleic acids for synthesising proteins) and translation (i.e., protein assembly based on messenger RNA).
3. Origin of eukaryotic cells (eukaryogenesis).
4. Origins of plastids (which enable eukaryotic photosynthesis).
5. Origins of complex multicellularity.
6. Origins of eusociality, characterised by living in groups (with multiple generations), cooperative rearing of juveniles, and division of labour.
7. Origin of human societies with language.

In closing, we caution that METs have garnered widespread usage (and acceptance), but also some critiques (refer to Question 6.8).

6.2.2   Megatrajectories

As mentioned in Section 6.1, it is unclear whether evolution is deterministic and/or directional; or unpredictable and/or random; or some combination thereof. Knoll and Bambach (2000) postulated a path that straddled the first two perspectives, where evolutionary diversification is both diffusive (i.e., characterised by a random walk, so to speak) and directional to an extent, that is, within certain organismal groups.

To capture the directional facet, Knoll and Bambach (2000) proposed the existence of six broad megatrajectories. They were argued to constitute a logical sequence, in the sense that the $(N + 1)$-th level could evolve only if the N-th level had already arisen. A distinguishing trait of megatrajectories is that the ecological dimension was taken into consideration: each megatrajectory introduces novel avenues whereby the emergent organisms may expand into novel ecological modes of life (i.e., unoccupied regions of *ecospace*) by acquiring resources via new channels and engendering more complex ecosystems. We will next sketch the six megatrajectories.

1. The steps from the origin(s) of the earliest living systems to the relatively sophisticated Last Universal Common Ancestor (LUCA).
2. The metabolic diversification(s) of bacteria and archaea, which caused significant changes in the cycling of chemical elements (e.g., nitrogen).
3. The genesis of eukaryotes.
4. The origin(s) of (complex) multicellularity.

5. The spreading of complex multicellular life on land, particularly land plants and animals, which facilitated the emergence of novel ecoystems.

6. The advent of intelligence and technology in humans, which has allowed them to profoundly modify the biosphere.

In closing, Knoll and Bambach (2000) adopted an explicitly astrobiological slant in discussing the extraterrestrial signatures of each megatrajectory, if they were to evolve on other worlds. The authors suggested that solar system exploration is well-suited for detecting megatrajectories #1 and #2 (see Section 13.1), remote sensing of exoplanets for seeking megatrajectories #2 and beyond (refer to Section 13.2), and the search for extraterrestrial intelligence for unearthing megatrajectory #6 (see Chapter 14).

6.2.3 Evolutionary singularities

By examining Earth's history, Christian de Duve (1917–2013), who was awarded the 1974 Nobel Prize in Physiology or Medicine, introduced a series of 'singularities' that were interpreted as *'events or properties that have the quality of singleness, uniqueness'* (de Duve, 2005, pg. viii). These singularities might have helped surmount certain evolutionary bottlenecks, and paved the way for the emergence of greater biological complexity.

In the same vein, Lingam and Loeb (2021, Chapter 3.9.4) highlighted five singularities that potentially exhibited the following attributes in common:

- Each of the singularities appears to have evolved only once on Earth. While this datum does not necessarily render them improbable, it does suggest that their likelihood of emergence is not high.
- Akin to the megatrajectories from Section 6.2.2, all the singularities caused major ecological shifts (e.g., by altering environmental conditions), which in turn shaped the pathways of evolution.
- Each singularity permitted bottlenecks on biological complexity to be overcome, by introducing novel evolutionary innovations.
- The singularities might have followed a sequence analogous to megatrajectories, in which the N-th singularity represented a prerequisite (of sorts) for the advent of the $(N + 1)$-th singularity.

The five singularities of Lingam and Loeb (2021, Chapter 3.9.4) are as follows:

1. Origin of life, from which LUCA and its descendants evolved.
2. Origin of oxygenic photosynthesis, whose importance will become apparent shortly in Section 6.3.3.
3. Origin of eukaryotes, which was explored in Section 6.1.2.
4. Origin of animals, whose significance for the Earth system was briefly documented in Sections 6.1.2 and 6.1.3.
5. Origin of humans (with the capacity for technological intelligence).

6.3 Rise in atmospheric oxygen: causes and consequences

We have already surveyed the levels of atmospheric oxygen, and their ramifications, in Section 6.1.2. Given, however, that the interplay of oxygen (an environmental parameter) and organismal evolution (a biological phenomenon) offers a compelling example of how life and its environment may coevolve together, we will examine this topic in more detail.

6.3.1 Evidence for oxygenation

Let us first take a closer look at a select few lines of geochemical evidence favouring a conspicuous rise in atmospheric O_2 around 2.4 Ga, to wit, the Great Oxidation Event (GOE); our exposition parallels that of Catling and Kasting (2017, Chapter 10). We bypass a similar analysis for the Neoproterozoic Oxidation Event (NOE), as the magnitude of the rise (expressed in logarithmic terms) was smaller (refer to Figure 6.3) and possibly more ambiguous (Tostevin and Mills, 2020; Krause et al., 2022).

Paleosols are ancient soils reported to be deficient in iron before the GOE, which has been explained by invoking the fact that the prevalent ferrous (Fe^{2+}) iron was soluble in rainwater and could get washed away, whereas modern ferric (Fe^{3+}) iron documented in oxygenated settings is relatively insoluble. *Detrital minerals* (which are deposited in sediments) prior to the GOE contain certain reduced minerals such as pyrite (FeS_2) and *uraninite* (UO_2) that would have formed in low-oxygen environments. *Red beds* are sedimentary rocks with a reddish hue arising from iron oxides (e.g., *hematite*, namely, Fe_2O_3), which are thought to have formed in oxygenated settings.

Banded iron formations (BIFs) appear in the geological record at $\sim$3.7–3.8 Ga, but decline and subsequently vanish in the *Paleoproterozoic* (2.5–1.6 Ga), although they reappear briefly in the Neoproterozoic. The formation of BIFs has been linked to the advent of (an)oxygenic photosynthesis, which oxidised the soluble Fe^{2+} to the insoluble Fe^{3+}, thereby causing precipitation of the latter. We note, however, that there are additional subtleties concerning the distribution of BIFs that we shall not address here.

We have sketched the concept of isotopic fractionation in Section 3.1.7, which corresponds to processes that induce changes in isotopic ratios. In most of these processes, the magnitude of fractionation is proportional to the mass difference between isotopes, justifying the nomenclature of *mass dependent fractionation* (MDF); the converse is true for *mass independent fractionation* (MIF). Photochemistry involving UV photons with wavelengths <230 nm is the only known mechanism for producing sulfur mass independent fractionation (S-MIF), via the UV photolysis of sulfur oxide gases such as SO_2. However, in an oxygenated atmosphere comprising non-negligible O_2 and ozone (O_3), S-MIF is suppressed, as those gases screen the required UV photons. Hence, the presence of S-MIF $\gtrsim$2.4 Ga and its disappearance $\lesssim$2.4 Ga is interpreted to reflect an increase in atmospheric O_2.

Other data cited in favour of the GOE include variations in the abundances of elements whose physicochemical properties are based on oxidation state (e.g., molybdenum is highly soluble in oxygenated conditions), and isotopes of certain elements such as transition metals, nitrogen, and carbon.

6.3.2 Mechanisms for increasing atmospheric oxygen

We offer a brief account of the potential mechanisms underpinning the rise in O_2 during the GOE by mirroring Catling and Kasting (2017, Chapter 10.7); it is plausible that some of them also triggered the NOE.

We introduce a dimensionless parameter Δ_{oxy} as follows:

$$\Delta_{oxy} \equiv \frac{O_2 \text{ source flux}}{O_2 \text{ sink flux}}, \tag{6.3}$$

where continental weathering should be excluded from the list of sinks (i.e., the denominator), because it is significant only when O_2 has already accumulated to a sizeable degree. Broadly speaking, $\Delta_{oxy} > 1$ permits the eventual build up of atmospheric O_2 (viz., an oxic atmosphere), whereas $\Delta_{oxy} < 1$ would favour the opposite outcome. In the case of modern Earth, Catling and Kasting (2017, pg. 282) estimated that $\Delta_{oxy} \approx 1.8$.

There are two ways of enhancing Δ_{oxy}, as seen from (6.3). Either the source flux of O_2 could increase (Case I) or the sink flux of O_2 can decline (Case II). We sketch a couple of hypotheses from these camps.

- Continental weathering elevated by the availability of new rocks (e.g., during continental growth or breakup) could have supplied phosphorus (a bioessential element), and thereby boosted biological productivity. The potential increase in buried dead organisms (which would otherwise consume O_2) might have thus raised the O_2 source flux (Case I).
- Oxygenic photosynthetic organisms may have been in competition with their anoxygenic counterparts, before overwhelming the latter. If this scenario based on ecological dynamics was applicable (Knoll and Nowak, 2017), it would be tantamount to a net increase in O_2 source flux (Case I).
- O_2 is currently depleted through reactions with reducing gases supplied through volcanic activity. If the inventory of volcanic gases was substantially transformed over time by becoming less reducing, this trend would translate to a decline in the O_2 sink flux (Case II).
- The process of serpentinisation (tackled in Section 7.1.3) produces H_2, a reducing gas that can react with O_2 and diminish its concentration. It is conceivable that the rate of serpentinisation decreased with time, consequently lowering the O_2 sink flux accordingly (Case II).
- Reducing gases generated by varied metamorphic activity (i.e., at high temperature and pressure) may have declined over time, equivalent to decreasing the O_2 sink flux (Case II). For instance, hydrogen escape to space (elucidated in Section 8.2.3) could have caused net oxidation of the crust, in turn constraining reducing gases and O_2 sink flux.

We have only barely scratched the surface of this topic, and many alternative phenomena remain under active investigation.

6.3.3 Ramifications of higher oxygen levels

It is well established that the advent of oxygenic photosynthesis had profound environmental, ecological, and evolutionary consequences, since it facilitated the accumulation of atmospheric O_2 that galvanised many dramatic changes (Lane, 2002; Judson, 2017), some of which are delineated hereafter.

For starters, the build up of O_2 allowed the synthesis of ozone in the presence of UV radiation at wavelengths <240 nm. As the ozone layer was formed, the UV flux reaching the surface of Earth in the wavelength range of 280–315 nm (called *UV-B radiation*) diminished conspicuously. Numerical simulations indicate that the UV-B flux at the Earth's surface declined from 3.4 W m^{-2} at 4.0 Ga to 2.2 W m^{-2} at 2.0 Ga to 0.8 W m^{-2} currently (Rugheimer et al., 2015, Table 5). The ensuing decrease in UV radiation would have lessened the damage to biomolecules, and may have allowed organisms to occupy new niches that were hitherto inaccessible to them.

Second, the diversity of minerals expanded considerably after the GOE, with the inventory roughly doubling compared to its prior value (Hazen and Morrison, 2022, Table 1). As minerals fulfil myriad functions in chemical and biological processes, ranging from catalysis to protection (e.g., UV screening), the GOE may have played a vital role in this respect. Third, the accumulation of atmospheric O_2 could have aided the formation of oxidised compounds (e.g., nitrates and sulfates), which were then utilised by microbes in assorted metabolic pathways (some of them are summarised in Section 7.1.3); in other words, the GOE might have enabled the expansion of the metabolic repertoire (Fischer et al., 2016).

Fourth, the existence of oxygenated settings would have represented both a boon and bane for organisms; we tackle the former shortly. Insofar as the latter is concerned, oxygen and the *reactive oxygen species* (ROS) – such as hydrogen peroxide (H_2O_2) and superoxide (O_2^-) – derived from O_2 are markedly toxic. In particular, ROS can cause damage to lipid membranes and DNA, impede enzyme operation, inhibit cell growth, and so forth. Therefore, it is plausible that the GOE might have induced a severe mass extinction of anaerobic and low-O_2 tolerant organisms, while simultaneously creating new niches for life forms better adapted to withstand oxidative stress.

Last, we mentioned the potential connections between O_2 and complex life in Section 6.1.2, both when discussing eukaryotes and animals. We caution at the outset that the evidence for a causal link between O_2 and complex life displays some ambiguity, and cannot be regarded as settled. With this caveat out of the way, many authors have implied or stated that sufficiently high O_2 levels are essential for complex life (e.g., Nursall, 1959; Fischer et al., 2016; Knoll and Nowak, 2017; Mills et al., 2023). We will parallel the analysis and arguments formulated by Catling et al. (2005) in this context.

One crucial benefit of O_2 is that *aerobic respiration* yields around an order of magnitude more energy compared to anaerobic respiration for a fixed amount of food (refer to Section 7.1.3). This is expected because the reduction (i.e., gain of electrons) of O_2 releases higher energy per electron transfer than any other element except fluorine and chlorine, by virtue of its strong electron affinity. However, fluorine and chlorine are relatively uncommon in the Universe, and thus unlikely to accumulate to significant levels. If this reasoning is correct, then O_2 might be desirable, and perhaps even strictly necessary, for the emergence of complex life (especially macroscopic, motile, complex multicellularity) on other worlds.

Catling et al. (2005, Table 2) derived the maximal sizes of aerobic organisms that transport O_2 via simple diffusion and circulation systems, denoted by $\mathcal{L}_{\text{diff}}$ and $\mathcal{L}_{\text{circ}}$, respectively. After further simplification, we end up with

$$\mathcal{L}_{\text{diff}} \sim 4\,\text{mm} \sqrt{\frac{P_{O_2}}{1\,\text{bar}}}, \qquad (6.4)$$

$$\mathcal{L}_{\text{circ}} \sim 100\,\text{mm} \left(\frac{P_{O_2}}{1\,\text{bar}} \right), \qquad (6.5)$$

where P_{O_2} is the external partial pressure of O_2. From these scalings, we notice that the maximal size increases monotonically with P_{O_2}.

Last but not least, initiating fire and combustion on Earth requires the atmospheric O_2 concentration to exceed $\sim 16\%$. Given the importance of these phenomena for the advancement of human technology, Balbi and Frank (2024) speculated that crossing this threshold is likewise critical for the emergence of (extraterrestrial) technological intelligence.

6.3.4   Snowball Earth episodes and rises in oxygen

In Section 6.1.2, we noted that Snowball Earth episodes and rises in atmospheric O_2 may have been broadly coeval; we briefly explore this further.

To begin with, suppose that an increase in O_2 levels preceded a Snowball Earth. This ordering of events raises the possibility that the resultant O_2 could have reacted with reducing greenhouse gases like methane (e.g., Kopp et al., 2005), and thus caused a drop in temperature and increased ice coverage. The onset of glaciation could have initiated the *ice-albedo feedback*, whereby the high albedo of ice would reflect more sunlight and drive additional cooling and glaciation, consequently exerting a positive feedback.

Now, consider the opposite scenario wherein the Snowball Earth preceded a rise in O_2. Among other routes, the former may be initiated, in principle, via enhanced weathering of newly available rocks during continental reorganisation (e.g., Pu et al., 2022), thereupon leading to CO_2 sequestration, cooling, and the ice-albedo feedback. Our planet might have eventually exited the Snowball Earth phase because of the gradual build up of CO_2 in the atmosphere – since it cannot react with rocks (under ice sheets), undergo weathering, and get depleted – and the accumulation of dust, which could raise albedo and enable warming (Hoffman et al., 2017).

In the above ordering, some models have proposed that the Snowball Earth might have triggered an O_2 increase (e.g., Laakso and Schrag, 2017). During the Snowball Earth stage, the sinks of O_2 may have become decoupled (i.e., effectively shut down), allowing atmospheric oxygen to increase, as outlined in Section 6.3.2. Furthermore, high rates of weathering during the deglaciation phase could have released large quantities of the nutrient phosphorus (P), consequently boosting biological productivity and elevating O_2 levels (see Section 6.3.2). In support of this hypothesis, there is tentative geochemical evidence that substantial changes in the P cycle (Reinhard et al., 2017), as well as the diversity and density of food webs (Brocks et al., 2017), occurred in proximity to the Neoproterozoic Snowball Earth episodes.

Thus, it is plausible that there may be deep connections between shifts in Earth's continental distribution and volcanism, increase of atmospheric oxygen, cycling of phosphorus (and/or other bioessential elements), and Snowball Earth episodes, although some of these links are not well understood.

6.4   Mass extinctions

We will sketch some of the basics of mass extinctions, as they offer another compelling illustration of the interplay of life and its environment. At the outset, we must clarify the meaning of a mass extinction event. We will adopt the general definition delineated in Hallam and Wignall (1997, pg. 1):

…a mass extinction is an extinction of a significant proportion of the world's biota in a geologically insignificant period of time.

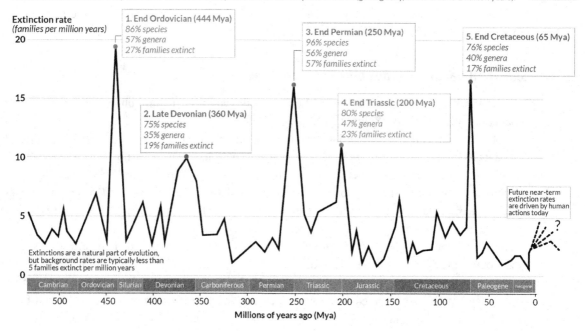

Figure 6.5 Approximate record of extinction rates in the Phanerozoic. (Credit: Hannah Ritchie courtesy of https://ourworldindata.org/mass-extinctions; CC-BY 4.0 license)

If we envision a mass extinction engendered by some abiotic trigger (e.g., asteroid impact or large-scale volcanism), as discussed shortly, it is clear how the environment is able to shape the evolution of life in this instance. However, we may now ask: what about the converse situation for mass extinctions? Although notable uncertainties persist, it seems feasible that certain evolutionary innovations (e.g., rise in atmospheric O_2) sparked substantial environmental changes, and aided the onset of mass extinctions (Algeo and Shen, 2023). Thus, in this scenario, life can not only alter its environment, but also modulate its own subsequent evolutionary trajectory. In the aftermath of many, albeit not all, mass extinctions, proliferation of species into recently vacated niches is documented.

We shall focus exclusively on mass extinctions in the Phanerozoic, owing to its better fossil records. In Section 6.1.3, we already introduced the *Big Five* mass extinctions: (1) Late Ordovician mass extinction; (2) Late Devonian mass extinction(s); (3) Permian–Triassic mass extinction; (4) Triassic–Jurassic mass extinction(s); and (5) Cretaceous–Paleogene mass extinction. The extinction rates in the Phanerozoic, along with the Big Five, are depicted in Figure 6.5. However, a couple of caveats merit highlighting. Crucially, growing evidence indicates that other large-scale extinctions of comparable or even higher magnitude existed, and/or that the Big Five extinctions were composed of multiple phases (Bambach, 2006). Moreover, the data in Figure 6.5 is only approximate and may get updated.

The drivers of mass extinctions are still not fully resolved, and continue to be debated even with respect to the Big Five. The reader may consult the reviews by Hallam and Wignall (1997), Bambach (2006), Bond and Grasby (2017), and Algeo and Shen (2023) for summaries of the immediate (proximal) and ultimate causes of mass extinctions. *Large igneous province* (LIP) events – which correspond to volcanic eruptions of $\gtrsim 10^5$ km^3 of magma, that is, around 4 orders of magnitude higher than volcanoes in attested human history (Grasby et al., 2019) – may be regarded as the current front runners for explaining several large-scale Phanerozoic extinctions. We provide brief overviews of the underlying mechanisms for the Big Five, but caution that our abbreviated treatment does not reflect all the complexities.

Late Ordovician mass extinction (LOME):   LOME consisted of two phases: the first characterised by widespread glaciation and the second by *anoxia* (oxygen depletion) in the oceans. The causes of the glaciation or anoxia are still not understood, but may stem from a confluence of factors such as tectonic shifts, volcanic activity, and/or the evolution of land plants, which might have accelerated weathering and sequestered enough CO_2 to produce cooling (Lenton et al., 2012). Another intriguing possibility is that LOME was triggered by a *gamma ray burst* (Melott et al., 2004), a high-energy astrophysical phenomenon outlined in Section 8.4.3.

Late Devonian mass extinction (LDME):   The mechanisms underpinning the series of extinctions comprising the LDME remain rather uncertain. The Late Devonian climactic perturbations potentially responsible for extinction, which include large-scale ocean anoxia, may have arisen from volcanism (e.g., LIP event), *bolide impact* (i.e., an extraterrestrial object colliding with the Earth), the broadening evolution of land plants, or even supernovae (high-energy astrophysical processes covered in Section 8.4.2).

Permian–Triassic mass extinction (PTME):   Two major extinctions unfolded in the Permian, namely: the less-known *Capitanian mass extinction* and the famous PTME (colloquially termed the *Great Dying*). A bevy of proximal factors seem to have engendered the PTME – the greatest biotic crisis in the Phanerozoic – such as global warming, ocean acidification, and anoxia. The consensus for the ultimate cause of the PTME is a LIP event, to wit, the formation of the *Siberian Traps*, which could have generated the above factors (Dal Corso et al., 2022). For instance, this volcanism may have added massive amounts of CO_2 (boosting atmospheric abundance by a factor of ~ 6), enabling global warming and ocean acidification.

Triassic–Jurassic mass extinction (TJME):   The extinctions at the end of the Triassic (i.e., constituting the TJME) exhibit distinct similarities with the above PTME in multiple aspects (Wignall, 2015). Global warming of several degrees Celsius, and especially ocean acidification, are considered putative drivers of the TJME. Both these phenomena are believed to have ensued from a LIP event: volcanic eruptions in the *Central Atlantic Magmatic Province*; the latter is associated with the breakup of Pangaea, consequently illustrating the relevance of (super)continents in Earth's history.

Cretaceous–Paleogene (K-Pg) mass extinction (CPME):   The celebrated paper by Alvarez et al. (1980) concluded that an observed excess of iridium at the K-Pg boundary could be explained by a bolide impact (*Chicxulub impact*) that triggered the CPME. A wealth of geological data, such as abundances of platinum group elements and certain minerals (e.g., nickel-rich spinels), indicate that a large asteroid (~ 10 km in diameter) struck the Earth. The impact is thought to have generated heat pulses, tsunamis, and shock waves in its aftermath, and inputted soot, dust, and sulfate aerosols into the atmosphere (Morgan et al., 2022), all of which might have promoted surface cooling of $\lesssim 15°C$ and fostered the CPME.

However, given that a LIP event, the *Deccan Traps* eruption(s), occurred close to the Chicxulub impact (separated by $\lesssim 0.1$ Myr), a number of studies have postulated that the Deccan Traps produced similar effects and initiated the CPME instead. State-of-the-art analyses suggest that this volcanism subsided before the impact, and that the extinction record coincides with the latter (Hull et al., 2020). While the case for the Deccan Traps appears to be weakened at the moment, its role ought not be discounted altogether.

Finally, we round off our narrative on a sobering note. Anthropogenic actions are increasingly responsible, among other detriments, for global warming, ocean acidification, and deforestation, with no signs of cessation shortly. In the past 500 years, Cowie et al. (2022) estimated that as much as $\sim 10\%$ of documented species might have become extinct, at least partly due to human activities. Hence, unless humanity implements avenues for mitigating and halting anthropogenic threats to the biosphere, the rapidly accelerating *Anthropocene extinction* may rank alongside the Big Five extinctions of the Phanerozoic, ushering in the tragic demise of countless species.

6.5   Problems

Question 6.1: Suppose that N_u unicellular (spherical) life forms, each of radius $\mathcal{R}_o$, assemble to yield a simple multicellular (spherical) organism of radius $\mathcal{R}_m$. Assuming mass conservation, and that density stays constant, calculate $\mathcal{R}_m$ as a function of $\mathcal{R}_o$ and N_u. On dimensional grounds, show that the diffusion timescale for this multicellular organism is $\mathcal{R}_m^2/D$, where D is the diffusion constant. From this expression, what is one of the potential drawbacks of multicellularity, that is, when N_u is raised?

Question 6.2: In the example from Question 6.1, assume that the basal power required to maintain the simple multicellular organism scales linearly with its mass (Hoehler et al., 2023, equation 1), and that the rate of acquisition and transport (via diffusion) of resources (e.g., food) scales linearly with its area. By taking the ratio of the latter to the former, does this quantity grow or decline with N_u? Discuss the anticipated positive or negative implications of this scaling for the origin(s) of simple multicellularity.

Question 6.3: By consulting the reviews cited in Section 6.1.2 on eukaryotes, briefly summarise the strengths and limitations of three pathways of eukaryogenesis (e.g., *phagocytosing archaeon model*); from your perspective, which one of this trio do you find the most compelling and why?

Question 6.4: By harnessing (6.2), determine the organismal size when advection becomes important on Saturn's moon Titan (potentially allowing for a wider range of behaviours) by substituting the appropriate value of D.

Question 6.5: By surveying the peer-reviewed literature, elucidate three examples of ecosystem engineering and niche construction. Situate these cases in an astrobiological context by speculating how these processes may be detectable by telescopes and missions; you can draw on Chapter 13.

Question 6.6: In each geological period of the Phanerozoic, briefly describe one major evolutionary and environmental event not mentioned in Section 6.1.3, along with the appropriate references.

Question 6.7: Carefully read the paper by Judson (2017), which attempts to trace the major evolutionary shifts in Earth's history using energy availability as the criterion, and then answer the following questions.

   a. Describe the five energy expansions, and their underlying rationale.

   b. Provide two reasons why you deem this classification advantageous and/or robust, and two potential drawbacks of this scheme. Outline whether you find the paper convincing overall or not, and justify your statement(s).

   c. In an ocean planet (sans landmasses), do you expect the five energy expansions to evolve, and in the same temporal order? Your response should be backed by peer-reviewed references wherever possible.

Question 6.8: After consulting the peer-reviewed literature, elucidate two putative weaknesses of major evolutionary transitions and megatrajectories.

Question 6.9: Based on Figure 6.3 and Section 6.3.3, calculate the maximal organismal sizes with diffusion and circulation at the following times: (a) prior to the GOE; (b) at ~ 2.0 Ga; and (c) at ~ 0.5 Ga. Compare your results with the sizes of the smallest known eukaryotes and animals (cite appropriate references), and comment on the ramifications.

Question 6.10: Moderately high O_2 levels might have been achieved during the Lomagundi-Jatuli Excursion (refer to Section 6.1.2), owing to which, in principle, commensurately large and complex organisms could have existed at that time; see Question 6.9(b). Hypothesise (with peer-reviewed citations) as to whether and why the fossil record seems to be devoid of such organisms.

Question 6.11: Using the data tabulated in Rampino (2020) for large bolides, we can crudely adopt the following power-law scaling between the extinction percentage of species (denoted by $\mathcal{P}_E$) and impactor kinetic energy (KE):

$$\mathcal{P}_E \sim 88\% \sqrt{\frac{\text{KE}}{10^{24}\,\text{J}}}. \tag{6.6}$$

Determine the kinetic energy, mass, and radius of a spherical impactor that may cause $\mathcal{P}_E \approx 100\%$. Assume that the impactor velocity is 18 km/s when it strikes the surface of the Earth, and that it has a mean density of 2.7 g/cm^3.

Part III

Habitability

7   Instantaneous Habitability

What are the necessary and sufficient conditions for an environment to support life (i.e., to be deemed *habitable*)? The answer is patently obvious: we do not truly know, which is the upshot of the multiple unknowns and gaps in our current understanding and database of living systems. For instance, as elucidated in Chapters 1, 5 and 6, we have not yet resolved: (1) what is life? (2) how, where, when, and possibly why does life originate? (3) what are the plausible evolutionary trajectories for early life?

Even if we restrict ourselves to life-as-we-know-it, these ambiguities remain, to say nothing of 'exotic' life that will be investigated in Chapter 11. Yet, despite these hurdles, several authors have attempted to tackle the central question posed at the beginning; one such example is Benner et al. (2004), who provide a wide-ranging analysis. In this chapter, we will adopt the criteria for *instantaneous habitability* encapsulated in Cockell et al. (2016) and Domagal-Goldman et al. (2016); the relevance of instantaneous habitability was sketched in Section 1.1.2.

1. The presence of a solvent: liquid water for life-as-we-know-it.
2. Physical and chemical factors (e.g., temperature) that permit the existence of organisms and their biomolecules, as well as the solvent.
3. Free energy sources for powering essential biological functions like maintenance, reproduction, and growth.
4. Sufficient inventories of bioessential elements: carbon, hydrogen, nitrogen, oxygen, phosphorus, and sulfur (CHNOPS) for life-as-we-know-it.

Let us now examine these criteria in succession to gauge their implications and motivate our decision to focus on criteria #2 and #3.

For much of the book, we do not explicitly evaluate criterion #1. The reason is because, when contemplating putative astrobiological targets, these environments have been shortlisted in the first place since they are confirmed or believed to host liquid water at some point in their history. To put it another way, environments of relevance to astrobiology already fulfil #1 in a sense. The search for planets in the *habitable zone* – which we encounter in Section 8.1 – mirrors the 'follow the water' strategy adopted by NASA. Furthermore, unlike the other criteria, inferring the existence of past or current water is not very hard in principle; even in the case of exoplanets, signatures of water vapour are discernible via spectroscopy of their atmospheres, as elaborated in Section 12.2.

We can ask ourselves why criterion #1 is necessary in the first place, that is, why does life require a solvent? In order for chemical reactions underpinning living systems to operate, the liquid phase is well-suited because it straddles two extremes represented by solids and gases. In the former, diffusion and transport are comparatively minimal, posing difficulties for acquiring reactants and expelling waste. The opposite situation impedes the gaseous phase: the containment of useful atoms and molecules is rendered challenging. This explanation is admittedly rather facile, and we cannot dismiss the prospects for exotic life sans solvents altogether.[1]

If we accept the need for a solvent, the emphasis on water has compelling justifications. Water offers a panoply of potential benefits to biology due to its chemical and physical properties, which are summarised in Chapter 11. However, theoretically viable alternatives to water have been identified (also covered in Chapter 11), such as ammonia and sulfuric acid. The focus on water, setting aside its physicochemical benefits, is actually (largely) pragmatic. Comprehending and searching for life-as-we-know-it is tremendously intricate and sweeping in its own right, let alone going beyond the standard paradigm. Hence, to philosophise briefly, the overarching ethos of astrobiology is perhaps expressible as follows: seek out the familiar, as that is relatively easier to accomplish, while not discounting the unexpected and keeping both eyes open for genuine anomalies.

We jump over criteria #2 and #3 for now, and turn our attention to criterion #4. The significance of having adequate abundances of bioessential elements is self-evident for sustaining biospheres, owing to their centrality in biological functions (see Section 5.1). For example, nitrogen is one of the requisite elements for synthesising amino acids, which are then assembled to yield proteins. Likewise, phosphorus comprises the backbone of nucleic acids, it is an indispensable part of metabolic activities mediated by certain 'energy carriers', and so forth. Biology on Earth utilises not just CHNOPS but also a sizeable fraction of all naturally occurring elements (Fraústo Da Silva and Williams, 2001; Wackett et al., 2004), especially metals.

Therefore, the omission of #4 may appear all the more surprising. We do not exclude this criterion on grounds of its importance, but rather because of the anticipated scarcity of empirical and/or theoretical constraints. To elaborate, if we consider the Earth, clearly the best characterised world, crucial holes in our knowledge of how the abundances of bioessential elements like phosphorus transformed over time still persist. When we turn our gaze to other worlds, the uncertainties are accordingly amplified. For instance, current data favours dissolved phosphorus concentrations of $\sim 10^{-3}$ to $\sim 10^{-2}$ M in the subsurface ocean of Enceladus (Postberg et al., 2023), thoroughly updating prior predictions of $< 10^{-10}$ M.[2]

As illustrated in the preceding paragraph, ascertaining the abundances of bioessential elements is an uphill task. This problem is compounded by the fact that we wish to know the inventories of these elements specifically in the form of compounds that are suitable for uptake by organisms (e.g., revisiting phosphorus, the dissolved concentration of P is what matters). From an empirical perspective, we are hampered by the sparseness of data from Earth's geological record, and the paucity of astrobiology-themed missions to other worlds. With that being said, there are worlds in the solar system where we may possess better information about the abundances of bioessential elements than the full repertoire of energy sources. In the realm of theory, estimating molar concentrations demands sophisticated (bio)geochemical models – an emerging line of enquiry that is at a rudimentary stage in connection with environments outside Earth.

On the basis of our choices so far, we are consequently left with criteria #2 and #3, which duly constitute the rest of the chapter. As both these topics are vast and are witnessing substantial progress, the

---

[1] The astronomer Sir Fred Hoyle (1915–2001) wrote an intriguing science fiction novel *The Black Cloud* (1957) about a gaseous life form.

[2] Note that 1 M is the unit for molar concentration, and indicates that 1 mole of the substance is found in 1 L of the solvent.

exposition will be selective, and will strive to highlight the salient physical and chemical processes. We reiterate that criteria #1 and #4 are just as vital, and the decision to bypass them in this chapter is partly motivated by length constraints and partly on account of the aforementioned reasons.

7.1 Energy sources for life

Metabolism involves the transmutation of inputs (e.g., carbon, energy, and nutrients) into outputs of biological relevance (e.g., biomolecular building blocks). As far as we can tell, life on Earth predominantly utilises two sources of energy: chemical energy and electromagnetic energy (light). We shall, therefore, restrict our analysis to these two sources. In principle, 'exotic' forms of energy such as kinetic energy, thermal energy, gravitational energy, and magnetic energy might be suitable for metabolism, all of which are explored by Schulze-Makuch and Irwin (2018, Chapter 5).

We will adopt a classification scheme in which organisms are divided into four categories as per how they derive their carbon and energy.

1. *Photoautotrophs* tap electromagnetic energy and their carbon source is inorganic in nature (canonically carbon dioxide).
2. *Photoheterotrophs* use electromagnetic energy and obtain their carbon from organic compounds.
3. *Chemoautotrophs* – often loosely labelled as *chemolithoautotrophs* and *chemolithotrophs* – extract their energy from redox reactions (distinguished by electron transfer from one chemical species to the other) and utilise inorganic carbon (carbon dioxide) to fulfil carbon requirements.
4. *Chemoheterotrophs* (also dubbed *chemoorganotrophs*) harness redox reactions for energy and organic compounds for their carbon.

It is apparent from the descriptions that autotrophs are characterised by inorganic sources of carbon (carbon dioxide to be precise) and heterotrophs get their carbon from organic compounds.[3] This scheme is depicted in Figure 7.1, along with examples of life forms from each of the four groups.

Before tackling chemotrophs and phototrophs, which respectively employ chemical energy derived from redox reactions and electromagnetic energy, clarifying some of the underlying (bio)physical and (bio)chemical concepts is helpful, which is addressed in Sections 7.1.1 and 7.1.2.

7.1.1 A primer on thermodynamics

Thermodynamics is a vast field spanning both science and engineering, in which many excellent textbooks have been published. A clear exposition of the subject is provided in Schroeder (2020), while the treatise by Kondepudi and Prigogine (2015) is tailored towards more advanced students.

Our starting point is the internal energy of the system, denoted by U, which may be roughly envisioned as the sum of the microscopic kinetic (e.g., vibrational) and potential (e.g., chemical bonding) energies. The *enthalpy H* of the system is defined to be

$$H = U + PV, \tag{7.1}$$

---

[3] The versatile mixotrophs, as hinted by their name, possess the capacity to obtain carbon from CO_2 as well as organic compounds.

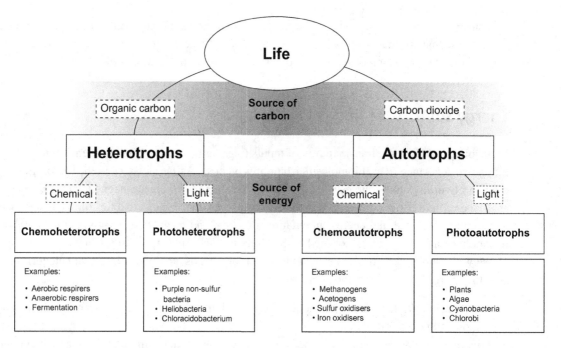

Figure 7.1 A schematic of the metabolic pathways found on Earth, organised according to the diverse strategies employed by life forms in acquiring both energy and carbon essential for their sustenance.

where P and V are the pressure and volume of the system; the second term on the RHS is a measure of the work done. The change in enthalpy ΔH attributable to a chemical reaction determines the 'heat' associated with it. If $\Delta H > 0$, the reaction takes up thermal energy from the environment, that is, it absorbs heat and is called *endothermic*. The situation is exactly the opposite for *exothermic* reactions with $\Delta H < 0$, characterised by the liberation of heat into the environment. The *Gibbs free energy*, G, can be obtained once H and the following thermodynamic variables are specified.

$$G = H - TS, \tag{7.2}$$

where T is the temperature of the system and S is the entropy. The latter is a subtle property of the system – often crudely interpreted as a measure of disorder – that we briefly summarised in Section 5.6. The change in Gibbs free energy ΔG tells us, in essence, the maximal work that may be performed at constant pressure and temperature, and is given by

$$\Delta G = \Delta H - T\Delta S. \tag{7.3}$$

The sign of ΔG is critical for determining the direction in which a reaction would proceed under specific physical conditions. If $\Delta G < 0$, the reaction is thermodynamically favoured and would spontaneously occur in the forward direction; this reaction is called *exergonic*. In contrast, when $\Delta G > 0$, the backward reaction would spontaneously take place and the forward reaction is thermodynamically disfavoured; those reactions are *endergonic* in nature. If $\Delta G = 0$, the system exists in a state of equilibrium.

There are many equivalent formulae for ΔG in the so-called 'standard state', which is denoted by $\Delta G°$ and is not the same as ΔG. The simplest version is to take the sum of the standard Gibbs free energy of formation for the reactants and subtract it from the sum of the standard Gibbs free energy of formation for the products. The second formula, namely (7.5), is furnished for the particular reaction below, but can be generalised appropriately.

$$r_1 R_1 + r_2 R_2 \leftrightarrow p_1 P_1 + p_2 P_2, \tag{7.4}$$

where R_1 and R_2 are the reactants, and P_1 and P_2 are the products. The labels r_1, r_2, p_1, and p_2 signify the abundances of the corresponding chemical species (e.g., r_1 moles of R_1). For (7.4), the standard Gibbs free energy is defined as follows:

$$\Delta G^\circ = -RT \ln \left(\frac{[P_1]^{p_1} [P_2]^{p_2}}{[R_1]^{r_1} [R_2]^{r_2}} \right) \equiv -RT \ln K_{eq}, \tag{7.5}$$

where the square brackets represent the concentrations of the given substances, $R \approx 8.3$ J/(K mol) denotes the gas constant, and K_{eq} is known as the reaction equilibrium constant.

Last, and perhaps most important, the standard Gibbs free energy evinces a compact form for electrochemical reactions, which entail the flow of electrons. Oxidation–reduction reactions (redox reactions) constitute the basis of electrochemistry, and are ubiquitous in biochemistry, as we shall witness in Section 7.1.3. In a nutshell, oxidation and reduction involve the loss and gain of electrons (i.e., an increase and decrease in oxidation state), respectively. In essence, oxidising agents are electron acceptors, whereas reducing agents are electron donors. Note that, somewhat counter-intuitively, reducing agents get oxidised and oxidising agents are reduced in redox reactions. The standard Gibbs free energy for redox reactions turns out to be

$$\Delta G^\circ = -n_e F \Delta E^\circ, \tag{7.6}$$

where n_e represents the number of electrons transferred (units of moles), $F \approx 9.65 \times 10^4$ J/(V mol) is the Faraday constant, and ΔE° embodies the redox potential *difference* in the standard state, since there are two half reactions (redox couples) effectively at play in any redox reaction, which are none other than oxidation and reduction.

Broadly speaking, the standard *redox potential* (or dubbed the electrode potential), denoted by E° (with units of V), quantifies the inclination of a chemical species in the standard state to lose or gain electrons (with respect to an electrode). Strong reducing agents are distinguished by negative values of E°, and the opposite is valid for potent oxidising agents. The standard redox potentials are often arranged in an electron tower, to wit, a visualisation tool in which powerful reducing agents populate the top and strong oxidising agents occupy the bottom. As already indicated, taking the potential difference is essential, and requires us to keep close track of the signs. For example, imagine that we have a reducing agent with $E^\circ = -0.3$ V and the oxidising agent has $E^\circ = +0.5$ V. In this instance, we find that ΔE° would be equal to 0.5 V $- (-0.3$ V$) = 0.8$ V.

In closing, the significance of the change in Gibbs free energy ΔG is manifest from our discussion below (7.3). If we were to compute ΔG for a potential metabolic pathway in a specific geochemical environment, it would permit us to ascertain whether the reaction is favoured on thermodynamic grounds – in other words, $\Delta G < 0$ in this setting would imply that the reaction can occur spontaneously. However, we caution that thermodynamics represents only one half of the story, and kinetics is the other, that is, reactions may unfold too quickly or too slowly. We will return to kinetics when exploring the roles of temperature in Section 7.2.1.

7.1.2   ATP and the proton motive force

A commonly encountered truism is that *adenosine triphosphate* (ATP) functions as the 'energy currency' of the cell. The molecular structure of ATP consists of a nucleobase (adenine), sugar (ribose), and three phosphate groups linked together.

The hydrolysis of ATP is spontaneous under the typical conditions of a cell, with ΔG approximately ranging between -50 and -70 kJ/mol (Milo and Phillips, 2016, pg. 183), and yields adenosine diphosphate (ADP) and inorganic phosphate (P_i), as shown here.

$$\text{ATP} + \text{H}_2\text{O} \leftrightarrow \text{ADP} + \text{P}_i. \tag{7.7}$$

By virtue of the relatively small size and sizeable energy released on hydrolysis, among other factors, ATP is ideal for supplying (chemical) energy to the appropriate components of the cell. However, it must be synthesised in the first place to fulfil this purpose. The converse of (7.7) is *phosphorylation*, which involves adding inorganic phosphate to ADP for synthesising ATP. As phosphorylation in this context is endergonic, an external source of free energy would be necessary to drive it.

This free energy is derived ultimately from redox reactions: as we have remarked in Section 7.1.1, electrons are transferred from an electron donor to an electron acceptor in those reactions. This process of electron transfer is facilitated in cells by means of an intricate series of molecules termed an *electron transport chain* (ETC). The ETC is composed of protein complexes such as iron–sulfur proteins, cytochromes, flavoproteins, and hydrogenases. We will not delve into the details of how the ETC operates, as the underlying biochemistry prerequisites are not mandated for this book; the interested reader can peruse Nicholls and Ferguson (2002) and Nelson and Cox (2004).

The free energy extracted from this sequence of redox reactions is employed to pump protons across the cell membrane from the interior (inside) to the exterior (outside). The ensuing excess of protons on the outside is tantamount to two available sources of energy. First, on account of the unequal distribution of protons, an electrochemical gradient is generated, and the associated (chemical) potential energy is proportional to the pH difference, given by $\Delta(\text{pH}) = \text{pH}_{\text{in}} - \text{pH}_{\text{out}}$. Note that the pH is the logarithm of the inverse of the hydrogen ion activity (a_{H^+}),[4] or in mathematical parlance, $\text{pH} = -\log(a_{\text{H}^+})$; the activity is approximately equal to the (normalised) concentration in the case of dilute solutions. The second contribution stems from the electrostatic potential difference $\Delta\psi$ that is engendered through the buildup of protons in the exterior.

Thus, the pH and charge gradients enable the movement of protons across the membrane back into the interior to reach equilibrium. The 'force' promoting this tendency goes by the name of the *proton motive force* (PMF), although this quantity has units of V (and not N) owing to its connections with chemical/electrostatic potential. The expression for the PMF is

$$\text{PMF} = \Delta\psi - \left(\frac{\ln 10\, RT}{F} \right) \Delta(\text{pH}). \tag{7.8}$$

The PMF has multiple uses in cell biology (e.g., uptake of organic molecules and ions), which are not described here. The most pertinent among them, however, is the synthesis of ATP by a remarkable protein known as *ATP synthase*. This complex protein, comprising multiple subunits, is analogous to a

---

[4] It is more accurate to speak of the hydronium ion (H_3O^+) instead of H^+, but we use the latter as a proxy because it is better known.

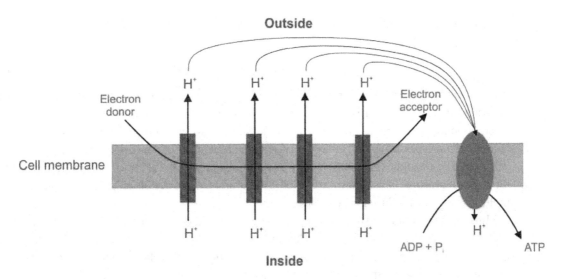

Figure 7.2 A depiction of how chemiosmosis operates in a generic cell. The electrons released from an electron donor pass through an electron transport chain and are taken up by an electron acceptor. The free energy liberated in the process is used to pump protons across the cell membrane, which are later utilised to synthesise ATP from ADP and inorganic phosphate P_i via the protein ATP synthase. (Credit: Cockell, 2020, Figure 6.5; reproduced with permission from John Wiley and Sons)

turbine and has consequently been labelled a 'molecular turbine' in some quarters. A turbine exploits the kinetic energy of flowing water and converts it to electrical energy. If we replace flowing water with the proton gradient, the turbine with ATP synthase, and electrical energy with ATP production, the analogy is complete. The minimum energy 'quantum' for driving ATP synthesis might be around -10 kJ/mol (Müller and Hess, 2017).

The mechanism whereby the ETC powers the pumping of protons across the cell membrane, which are subsequently harnessed to synthesise ATP, is called *chemiosmosis*. The theory was proposed by Peter Mitchell (1920–1992) in a seminal publication (Mitchell, 1961), who went on to win the 1978 Nobel Prize for Chemistry for this discovery. An illustration of the salient steps constituting chemiosmosis is provided in Figure 7.2.

7.1.3   Chemical energy

The domain of chemotrophs is undoubtedly broad, as evident from Figure 7.1. It encompasses both chemoheterotrophs and chemoautotrophs, which obtain their carbon from organic and inorganic (carbon dioxide) sources, respectively. Chemotrophs inhabit, among other locations, the deep subsurface (i.e., at depths of km) that may harbour the majority of microbial life on Earth (Magnabosco et al., 2018). The metabolic pathways are complex and characterised by the intricate orchestration of multiple reactions. Hence, we will restrict ourselves to sketching a bird's-eye view of these pathways.

Aerobic respiration is a natural starting point in our voyage. This pathway is documented in many microbes and multicellular organisms; as the life forms increase in complexity, the prevalence of such respiration is duly enhanced. For example, aerobic respiration is documented in the vast majority of animals, albeit *not* all of them (e.g., *Henneguya salminicola*). The simplified net reaction of aerobic respiration is expressible as

$$C_6H_{12}O_6 + 6O_2 \rightarrow 6CO_2 + 6H_2O, \tag{7.9}$$

and the standard Gibbs free energy is $\Delta G° = -2880$ kJ/mol (Atkins and De Paula, 2006, pg. 173), indicating that the reaction is highly spontaneous and strongly favoured. As hinted by (7.9), the electron donors are organic molecules while the (terminal) electron acceptor coupled to the ETC is molecular oxygen (O_2). The energy released in aerobic respiration is partly harnessed to synthesise ATP. The theoretical prediction cited in textbooks is that 38 ATP molecules are generated per molecule of glucose (Nelson and Cox, 2004; Atkins and De Paula, 2006); in reality, around 29 or 30 molecules may be produced instead (Rich, 2003).

Hitherto, the implicit assumption concerning respiration was that O_2 is the terminal electron acceptor in the ETC. If we were to relax this premise, other candidates can step in and perform the job of O_2; the ensuing pathway is *anaerobic respiration*. One key difference, however, is that most of these chemical species have a weaker redox potential than O_2, and do not yield as much energy for biological functions. As a rough rule of thumb, the free energy associated with anaerobic respiration is about an order of magnitude smaller relative to its aerobic counterpart, in which event the amount of ATP molecules produced must be scaled downward commensurately; we caution that this heuristic in not always valid, and must be employed carefully.

Aside from anaerobic respiration, a variety of *fermentation* mechanisms exist. The latter resemble the former in the sense that both can operate without O_2. However, there are notable divergences that stand out – for instance, fermentation is distinguished by the absence of the ETC and the deployment of chemiosmosis to synthesise ATP, owing to which only 2 ATP molecules are typically produced. The terminal electron acceptor is neither oxygen nor another inorganic species, but rather an organic species such as pyruvate (CH_3COCOO^-) or acetaldehyde (CH_3CHO). A collection of respiration and fermentation pathways are presented in Table 7.1. The appearance of acetogenesis and methanogenesis, as well as some other metabolisms in Table 7.1, may seem confusing since we allude to them later on as chemoautotrophs. This ambiguity arises because certain acetogens and methanogens derive their carbon from CO_2, and others in this group from organic compounds (see Schuchmann and Müller, 2016; Mand and Metcalf, 2019).

Our discussion so far dealt with chemoheterotrophs, as the metabolisms in question obtained the carbon from organics. Moving on, some of the most well-known chemoautotrophic pathways are summarised in Table 7.2. Instead of getting into the specifics of each pathway, it is worth understanding the general trends that underlie chemoautotrophs.

1. In chemoheterotrophs, the electron donors are predominantly organic molecules, whereas nearly all pathways that utilise inorganic electron donors (for redox reactions) are autotrophic. The presence of methanotrophy in Table 7.2 might seem odd at first glimpse because methane is ostensibly an organic molecule, unlike other pathways where the electron donors are inorganic. However, some methanotrophs apparently source their carbon from CO_2, justifying their inclusion, although there are attendant subtleties not addressed here (De Marco, 2004).
2. Chemoautotrophy is comparatively difficult to actualise since the inorganic electron donors are often not as strong reducing agents as organics, and the process of converting CO_2 (the carbon source) into biomolecules is energetically expensive. Despite these obstacles, some species may

Table 7.1 Select metabolic pathways reported for chemoheterotrophs

Pathway	Electron acceptor	Product(s)	Example(s)
Aerobic respiration	O_2	H_2O, CO_2	Virtually all animals
Nitrate reduction	NO_3^-	NO_2^- or NH_3 or N_2 or NO or N_2O	Members of genus *Pseudomonas*
Sulfate reduction	SO_4^{2-}	sulfides (S^{2-})	Members of genus *Desulfobacterales*
Iron reduction	Fe^{3+}	Fe^{2+}	Members of genus *Geobacter*
(Per)chlorate reduction	ClO_4^-, ClO_3^-	Cl^-, O_2	*Dechloromonas hortensis*
Uranium reduction	U^{6+}	U^{4+}	*Shewanella oneidensis*
Fumarate respiration	Fumarate	Succinate	*Wolinella succinogenes*
Organohalide respiration	Halocarbons	Halide ions	Members of genus *Dehalococcoides*
Acetogenesis	CO_2	CH_3COO^-	*Acetobacterium woodii*
Methanogenesis	CO_2	CH_4	Members of genus *Methanosarcina*
Fermentation	organics (e.g., pyruvate)	organics (e.g., lactic acid or ethanol)	Lactic acid bacteria and yeast

Note: In the third and fourth columns, only a subset of options are highlighted. Aside from aerobic respiration and fermentation (i.e., first and last rows), all the other entries correspond to anaerobic respiration. Fermentation encompasses a multitude of pathways that are not elaborated herein. It must be understood that not all species in the taxa listed above are capable of the metabolic pathway in question.

grow quickly in conducive environments. For example, the sulfur-oxidising bacterium *Thiomicrospira crunogena*, which dwells near deep-sea hydrothermal vents, can double its population in ∼1 hour.

3. Table 7.2 illustrates how *syntrophy* is viable in microbial metabolisms. In syntrophy, the waste products of one pathway constitute the food for another. For instance, the sulfide arising from sulfate reduction could serve as the electron donor for sulfur-oxidising microbes.

4. Just as we had done with chemoheterotrophs, we may divide chemoautotrophs into aerobic and anaerobic life forms. In both categories, a rich palette of metabolic pathways is documented, at times involving the same electron donor (encapsulated in Table 7.2).

Before moving to phototrophy, we emphasise one pathway: *hydrogenotrophic methanogenesis*. Sacrificing some accuracy, we use 'methanogenesis' as shorthand for hydrogenotrophic methanogenesis hereafter, unless stated otherwise. The simplified chemical reaction for this type of methanogenesis is

$$4H_2 + CO_2 \rightarrow CH_4 + 2H_2O. \tag{7.10}$$

Table 7.2 Select metabolic pathways reported for chemoautotrophs

Pathway	Donor	Acceptor	Product(s)	Example(s)
Hydrogenotrophic methanogenesis	H_2	CO_2	CH_4	*Methanopyrus kandleri*
Acetogenesis	H_2	CO_2	CH_3COO^-	*Sporomusa ovata*
Methanotrophy	CH_4	O_2	CO_2	*Methylococcus capsulatus*
Iron oxidation	Fe^{2+}	O_2, NO_3^-	Fe^{3+}	*Acidithiobacillus ferrooxidans*
Phosphite oxidation	HPO_3^{2-}	SO_4^{2-}	HPO_4^{2-}, HS^-	*Desulfotignum phosphitoxidans*
Sulfur oxidation	H_2S, S	O_2, NO_3^-	S, SO_4^{2-}	Members of genus *Thiobacillus*
Nitrification	NH_3	O_2	NO_2^-, NO_3^-	Members of genus *Nitrosomonas*
Anaerobic ammonium oxidation	NH_4^+	NO_2^-	N_2	Members of genus *Candidatus Brocadia*
Manganese oxidation	Mn^{2+}	O_2	Mn^{4+}	*Candidatus Manganitrophus noduliformans*
Sulfate reduction	H_2	SO_4^{2-}	S^{2-}	*Desulfobacterium autotrophicum*

Note: In the second to fifth columns, only a small subset of possibilities are highlighted. The terms 'Donor' and 'Acceptor' stand for electron donor and electron acceptor, respectively. It must be understood that not all species in the taxa listed above are capable of the metabolic pathway in question.

A number of compelling reasons motivate the focus on (hydrogenotrophic) methanogenesis, many of which are revisited in subsequent chapters.

1. Hydrogenotrophic methanogenesis is potentially the most widespread version of methanogenesis (Thauer et al., 2008). Moreover, the core pathway is ancient and may share deep connections with the origin(s) of protometabolism and life, as reviewed in Schwander et al. (2023).
2. Methanogenesis requires neither molecular oxygen nor access to electromagnetic radiation, making it well-suited for deep subsurface ecosystems, which might be ubiquitous in the Universe (Lingam and Loeb, 2020a).
3. The feedstock molecules were widely available on early Earth. Atmospheric CO_2 abundance was, in all likelihood, higher by orders of magnitude relative to modern Earth, and so was the dissolved CO_2 concentration in the ocean. Proposed sources of H_2 include remnants of the H/He primordial atmosphere; radiolysis (splitting molecules by radiation) of water; and serpentinisation, which is a water–rock reaction. In simplified form, serpentinisation is expressible as (Sleep et al., 2004, equation 2):

$$3Fe_2SiO_4 + 2H_2O \rightarrow 2Fe_3O_4 + 3SiO_2 + 2H_2, \tag{7.11}$$

where Fe_2SiO_4 on the LHS is fayalite, the iron end-member of the mineral olivine (a magnesium-iron silicate); Fe_3O_4 (magnetite) on the RHS is composed of iron oxides; and SiO_2 is silica (silicon dioxide).

4. Turning our gaze to other worlds, the aforementioned trio of H_2 production mechanisms could have operated on early Mars. Molecular hydrogen has been detected on Saturn's moon, Enceladus, by the *Cassini–Huygens* mission (Waite et al., 2017). Additionally, laboratory experiments suggest that worlds with iron-rich olivine may yield much higher fluxes of H_2 via serpentinisation than on Earth (McCollom et al., 2022).

5. Intriguingly, whiffs of methane were discovered in the Martian atmosphere, but abiotic causes are likely or plausible.[5] The situation is even more promising for Enceladus. Laboratory experiments have demonstrated that some methanogens (e.g., *Methanothermococcus okinawensis*) might be able to thrive in this moon's ocean (Taubner et al., 2018), and the high abundance of methane detected by *Cassini* is statistically compatible with the existence of methanogens (Affholder et al., 2021).

6. Last but not least, methane is synthesised predominantly by biological activities on Earth (Krissansen-Totton et al., 2022), suggesting that this gas may represent a robust indicator of life (consult Section 13.2.1).

7.1.4 Electromagnetic energy

The most dominant source of energy on Earth is electromagnetic radiation from the Sun. It should come as no surprise, therefore, that *phototrophy* (i.e., harvesting of electromagnetic energy) is a fundamental aspect of Earth's rich biosphere. In fact, phototrophs comprise at least approximately 80% of Earth's biomass (Bar-On et al., 2018). On paring down the net reaction for photosynthesis to the bare essentials, we roughly have

$$CO_2 + 2H_2X + \gamma \xrightarrow{\text{pigments}} CH_2O + H_2O + 2X, \tag{7.12}$$

where CH_2O is, somewhat confusingly, a stand-in for the biologically synthesised organic compounds, H_2X signifies the reducing agent (electron donor), X is the metabolic product, and γ represents photons. Oxygenic photosynthesis, which we return to shortly, relies on photosynthetically active radiation (PAR) in the canonical wavelength range of $\sim$400–700 nm (Blankenship, 2014, Chapter 1.2); the exact range is slightly broader.

Various chemical species can serve as the electron donor such as molecular hydrogen, ferrous iron (Fe^{2+}), hydrogen sulfide, and nitrite (NO_2^-) on Earth and elsewhere. However, if the inventories of these species were limited on Earth, then it would have been hard for those variants of photosynthesis to spread globally. In contrast, oxygenic photosynthesis has water as the electron donor, owing to which (7.12) takes the form

$$CO_2 + 2H_2O + \gamma \xrightarrow{\text{pigments}} CH_2O + H_2O + O_2. \tag{7.13}$$

Oxygenic photosynthesis possesses a number of intrinsic benefits, which are further explored in Question 7.3, as well as Section 6.3.3. Before getting into the functional features of this pathway, it

---

[5] Methanogenesis in the subsurface of Mars was theoretically explored in Boston et al. (1992), and laboratory experiments have assessed its practicality on Mars (e.g., Sinha et al., 2017).

is worth bringing up one of the prominent advantages: the molecular oxygen in Earth's atmosphere is attributable to oxygenic photosynthesis. The rise in O_2 levels had a profound impact on Earth's biosphere, as witnessed in Section 6.3.3. Oxygenic photosynthesis may be decomposed into two overarching components.

In the *light reactions* (the first part), the electromagnetic energy of the star is transducted (i.e., harnessed and converted) into chemical energy, and in the *dark reactions* (the second phase), the biosynthesis of organic carbon compounds from CO_2 (to wit, carbon fixation) takes place. As the terminology suggests, the latter can occur in the absence of light, as opposed to the former. As both these phases are intricate and involve complex (bio)molecules and chemical reactions, we aim to capture a bird's-eye view, akin to Section 7.1.3; comprehensive reviews of this vast subject are provided in Falkowski and Raven (2007) and Blankenship (2014).

The light-harvesting pigments in oxygenic photosynthesis are chiefly *chlorophylls*, of which the most notable is chlorophyll *a* (Chl *a*). This molecule exhibits absorption maxima at $\sim$435 nm, $\sim$680 nm in photosystem II, and $\sim$700 nm in photosystem I. When electromagnetic radiation is incident on chlorophylls, it can cause excitation of an electron, crudely resembling the behaviour of semiconductors. If the photon has sufficiently high energy (around 1.9 eV for Chl *a*), electron detachment would be feasible, in loose analogy to the *photoelectric effect* from quantum physics.[6] This electron, which has absorbed energy $\propto 1/\lambda$ (where λ is the photon wavelength), is taken up by an electron acceptor after shuttling along an ETC.

In oxygenic photosynthesis, the aforementioned steps occur in the reaction centre of photosystem II, whose (figurative) heart is the Chl *a* dimer P680. After electronic excitation and ejection, thereby serving as the primary electron donor of photosystem II, the cationic P680$^+$ (with an electron hole) is an extremely powerful oxidising agent, with a redox potential of $\sim$1.2 V. This characteristic is crucial because water is not a particularly effective reducing agent, with a redox potential of $\sim$0.82 V, implying that a strong oxidising agent is necessary to oxidise it. The electrons derived from the oxidation of water serve to replace the electrons conveyed to the ETC. The 'splitting' of water to release the electrons is expressible as

$$2H_2O \rightarrow O_2 + 4H^+ + 4e^-. \tag{7.14}$$

This reaction is accomplished in photosystem II via the oxygen-evolving complex (OEC) that consists of a tetramanganese cluster at its core. While the excitation (and ejection) of an electron unfolds step-by-step, (7.14) demonstrates that four electrons must be extracted to produce a single molecule of O_2. This discrepancy is handled ingeniously by the OEC, which is reviewed in Blankenship (2014, Chapter 7) and Fischer et al. (2016).

Returning to our main narrative, the high-energy electron liberated from P680 travels down the ETC and drives the synthesis of ATP by generating the PMF, as described in Section 7.1.2. This electron is ultimately inputted to the reaction centre of photosystem I, which consists of the Chl *a* dimer P700. An electron in P700 is excited by the incident radiation to a higher energy level, such that P700* (i.e., the excited state) is rendered a strong reducing agent with redox potential of about -1.2 V. The electron thus donated is transferred to an ETC (its replacement is obtained from photosystem II),

---

[6] The underlying theory (refer to Young and Freedman, 2018, Chapter 38) was formulated by Albert Einstein in one of his *Annus mirabilis* papers from 1905. Einstein was awarded the 1921 Nobel Prize in Physics for the photoelectric effect, ironically not for special/general relativity as commonly assumed.

and is utilised – at the end of the chain – by the electron acceptor *nicotinamide adenine dinucleotide phosphate* ($NADP^+$) to produce its reduced version, NADPH. The latter molecule plays a vital role in the dark reactions for synthesising organic compounds.

The mechanisms of light reactions are encapsulated in Figure 7.3. Now, we turn our attention to the dark reactions wherein organics are synthesised by oxygenic photoautotrophs. This phase is essentially as complex as the light-dependent reactions, and arguably ventures even deeper into biochemical details, on account of which we merely sketch the bare-bones outline. The biosynthesis of organic molecules operates through an intricate pathway known as the Calvin–Benson–Bassham (CBB) cycle, or the *Calvin cycle* for short. The net reaction of the Calvin cycle has the form

$$3CO_2 + 6NADPH + 9ATP \Rightarrow G3P + 6NADP^+ + 9ADP + 8P_i, \tag{7.15}$$

where P_i is inorganic phosphate and G3P denotes glyceraldehyde 3-phosphate, a three-carbon molecule that is subsequently employed to synthesise molecules such as glucose ($C_6H_{12}O_6$). As one starts with a one-carbon molecule and ends up with a 3-carbon product, it is apparent that three turns of this cycle are needed for generating G3P. If we suppose that two G3P combine to yield glucose, (7.15) suggests that 18 ATP are consumed to produce one molecule of glucose. The central enzyme of this cycle is ribulose 1,5-bisphosphate carboxylase/oxygenase (abbreviated as *RuBisCO*). RuBisCO is presumably the most abundant enzyme on Earth, although it is famously regarded as rather inefficient in fulfilling its function.

With our summary of oxygenic photosynthesis wrapped up, we now delve into anoxygenic photosynthesis, whose major attributes are summed up in Table 7.3. In some publications (e.g., Hamilton, 2019), the latter implicitly embraces both autotrophs and heterotrophs, while others subscribe to the stance that photosynthesis should instead be considered synonymous with photoautotrophy (e.g., Bryant and Frigaard, 2006, Box 1). We will adopt the former perspective, for the simple reason

Table 7.3 Select anoxygenic photosynthetic taxa and their properties

Taxon	RC	Carbon fixation	Electron donor
Chloracidobacterium	Type I	No	–
Chloroflexus	Type II	3-hydroxypropionate bicycle/No	H_2, H_2S
Oscillochloris	Type II	Calvin cycle	H_2, H_2S
Chlorobi	Type I	Reverse Krebs cycle/No	H_2S, S, Fe^{2+}
Heliobacteria	Type I	No	–
Gemmatimonas	Type II	No	–
Purple sulfur bacteria	Type II	Calvin cycle/No	H_2, H_2S, Fe^{2+}, NO_2^-, $S_2O_3^{2-}$
Purple non-sulfur bacteria	Type II	No/Calvin cycle	H_2, H_2S, Fe^{2+}, $S_2O_3^{2-}$

Note: In the first column, not all members of that particular taxon are guaranteed to be capable of photosynthesis. Likewise, not all potential electron donors are listed in the fourth column (e.g., they are excluded in the absence of carbon fixation). The third column specifies the carbon fixation pathway, in which all avenues (or lack thereof) are not included. 'RC' (second column) stands for reaction centre. The table is adapted from the data in Hamilton (2019, Figure 1).

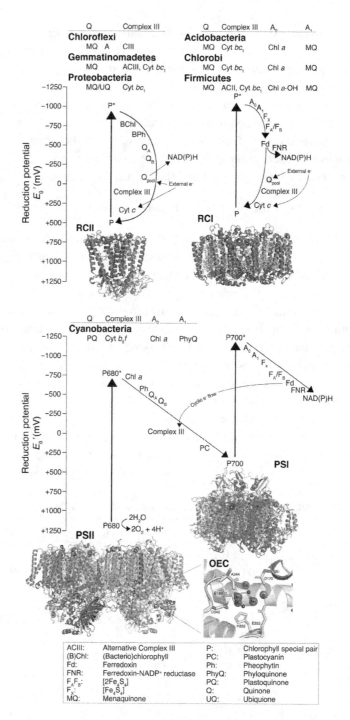

Figure 7.3 A schematic of the biochemistry of light-mediated reactions in oxygenic (photosystem II [PSII] and photosystem I [PSI]) and anoxygenic (Type I and Type II reaction centres) photosynthesis: mechanisms of electron transfer, along with the constituent molecules and their approximate redox potentials, are shown. (Credit: Hamilton, 2019, Figure 2; reproduced with permission from Elsevier)

that doing so enables us to avail ourselves of certain commonalities among the diverse groups of anoxygenic photosynthesisers, and accordingly formulate a more compact treatment.

Anoxygenic photosynthesisers diverge from their oxygenic counterparts in a few crucial respects. First, they feature only a single reaction centre, either Type I or Type II, as depicted in Figure 7.3, whereas oxygenic photosynthesisers have two reaction centres (incorporated in photosystems I and II). Second, these anoxygenic organisms do not harness water as electron donor and do not produce O_2 in the process; other electron donors such as H_2S, Fe^{2+}, and H_2 are utilised. Third, their key light-harvesting pigments, the *bacteriochlorophylls* (BChls), are closely related to chlorophylls (Chls), but differ in molecular structure and absorption properties. The maximum absorbances of BChls manifest at infrared wavelengths in certain instances (e.g., BChl *b* has an absorption peak at ∼1015–1040 nm).

Type I reaction centres are usually characterised by strong reductants and weak oxidants, while the converse often holds true for Type II reaction centres (Bryant and Frigaard, 2006). Broadly speaking, in Type I reaction centres, the flow of electrons is crudely linear because there is a directional movement from the source (excited pigment) to ferredoxins, which are proteins containing iron and sulfur. The electrons derived in turn from ferredoxins facilitate the synthesis of NADPH, whose importance in carbon fixation was previously encountered. In contrast, organisms with Type II reaction centres demonstrate roughly cyclic electron flow – excited electrons from the reaction centre pass through an ETC, and are ultimately returned to the reaction centre. As a fraction of the electrons are consumed by non-cyclic processes, they must be replenished by an external electron donor.

Until now, we have explored (an)oxygenic photosynthesis, which is distinguished by the excitation of electrons in (bacterio)chlorophylls followed by electron transfer by means of an ETC (thereby permitting the synthesis of ATP via the PMF), irrespective of the exact pathway. However, photosynthesis is not the only version of phototrophy. A plethora of microbes, drawn from manifold lineages, possess light-sensitive proteins known as *rhodopsins*, wherein the chromophore retinal is a central component.[7] Rhodopsins are of much interest in metabolism because some of them function as light-driven pumps, whereby ions are pumped across the cell membrane to set up electrochemical gradients (Rozenberg et al., 2021).

The connection with Section 7.1.2 on chemiosmosis and ATP synthesis is obvious. Rhodopsins that serve as ion pumps could synthesise ATP through the PMF. However, rhodopsin-based phototrophy is reportedly rather inefficient. For example, it is estimated that about three to four molecules of bacteriorhodopsin (found in archaea) must each absorb a photon and pump a proton (across the cell membrane) to generate sufficient PMF for producing one molecule of ATP (Bryant and Frigaard, 2006). This negative is counterbalanced by the fact that rhodopsins may easily spread from one organism to another through the mechanism of horizontal gene transfer (since only a few genes would need to be transported), and are abundant in consequence. Around 50–70% of all microbes in the sunlight zone of Earth's oceans appear to possess rhodopsins (Rozenberg et al., 2021).

To sum up, we have introduced two separate modes of phototrophy that are viable. The first is photosynthesis in which (bacterio)chlorophylls constitute the bedrock of energy conversion; as stated earlier, photosynthesis is interpreted to span both autotrophs and heterotrophs. The second entails ion pumping by rhodopsins, which is solely heterotrophic in nature insofar as our current knowledge is concerned. However, a host of unresolved questions linger: What wavelengths could oxygenic

---

[7] Chromophores are chemical groups that absorb light at particular wavelengths and bestow molecules with colour.

photosynthesis tap into on other worlds? How old is (an)oxygenic photosynthesis? Does photosynthesis engender any signatures detectable by our instruments? Are there any pigments apart from (bacterio)chlorophylls and rhodopsins that comprise the basis of energy transduction on Earth (and elsewhere)?

Some of these questions are tackled in previous and forthcoming chapters, but others lack answers so far.

7.2   Physical and chemical limits of life

The field of extreme microbiology has grown by leaps and bounds in the past couple of decades. Organisms inhabiting environments that are extreme in some fashion (viewed from an explicitly anthropocentric perspective) are loosely classified as *extremophiles*. The reader can consult the state-of-the-art reviews by Merino et al. (2019) and Carré et al. (2022) for further information, as well as the earlier treatises by Rothschild and Mancinelli (2001), Seckbach et al. (2013), and McKay (2014).

Before proceeding onward, there is a subtle distinction that deserves to be spelt out (see Rampelotto, 2013). Extremophiles are, strictly speaking, adapted to live in particular extreme conditions, and therefore require those settings in order to grow. On the other hand, extremotolerant life forms have the capacity to tolerate extreme environments (typically for modest timescales), but otherwise grow in normal conditions. While we endeavour to distinguish between these two scenarios, we caution that we use the generic term 'extremophiles' at times to encompass both categories.

The study of extremophiles confers many benefits upon astrobiology, some of which are delineated next.

1. Earth's biosphere is home to millions to trillions of species (the number is weakly constrained). A systematic investigation of extremophiles may help us constrain and/or infer the limits of Earth's biosphere, and thus construct the boundaries of the 'biospace', which represents the region (hypervolume) wherein life-as-we-know-it could operate; the axes in this multidimensional parameter space correspond to different physicochemical variables (in the manner of Figure 7.4).
2. Mapping this biospace from Earth-based extremophiles may provide a valuable benchmark and heuristic guide for exploring targets beyond Earth. If we discover that the conditions in some environment are far outside the confines of Earth's biospace, this feature might indicate that the setting pragmatically warrants a low priority in searching for life.
3. Alternatively, if we find extremophiles elsewhere that take us much beyond Earth's biospace, that would compel us to revisit our expectations of the frontiers of life, and thereby arrive at a deeper understanding of life's thresholds. Moreover, such a discovery can help initiate life-detection surveys in analogous environments on Earth (if they exist).
4. Last but not least, extremophiles are the natural testing grounds for theoretical (bio)physical and (bio)chemical models that aim to quantify limits of life. If a good match is found between these models and empirical data on extremophiles, it would lend credence to the former, and simultaneously advance biophysics and biochemistry.

Our upcoming discussion will make it apparent that the boundaries of Earth's biospace are primarily set by microorganisms. This fact is not surprising, given that coordinating and adapting the multiple cell types of complex multicellular organisms in extreme environments is innately challenging.

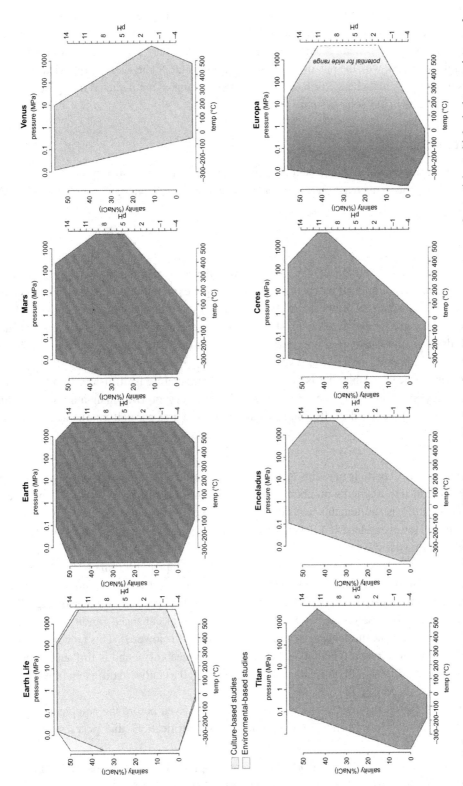

Figure 7.4 The physicochemical boundaries of habitats and life forms on Earth compared with the parameter spaces of potential astrobiological targets in our solar system. Each edge demarcates the range of permitted values for a given environmental variable (e.g., temperature), and its absence conveys the lack of data. (Credit: Merino et al., 2019, Figure 2; CC-BY 4.0 license)

However, some animals and fungi are remarkably resilient in spite of their complexity – a famous example are the tardigrades (viz., species of phylum *Tardigrada*), colloquially called 'water bears'.

In keeping with the general theme of this chapter and the book, the thrust is towards developing a panoramic overview, with the risk of losing some accuracy along the way. The physical and chemical factors we shall evaluate henceforth (and their possible ranges) are depicted in Figure 7.4 for Earth and other objects in our solar system.

7.2.1   Temperature

Perhaps the most crucial aspect to bear in mind is the impact of the temperature (T) on reaction rates. With some provisos,[8] the reaction rate constant k can be crudely modelled by means of the Arrhenius equation:

$$k = \mathcal{A} \exp\left(-\frac{E_a}{k_B T}\right), \tag{7.16}$$

where k_B is the Boltzmann constant, E_a is the activation energy, and $\mathcal{A}$ is the Arrhenius factor that has a relatively modest temperature dependence. This simple expression for k tells us a great deal about the pervasive influence of temperature. For starters, observe that $1/k$ has an exponential dependence on the inverse temperature ($1/T$). Hence, if the temperature is lowered, the reaction rates can decline dramatically. The opposite scenario is manifested when the temperature is raised. Both these underlying trends could give rise to severe issues, as we will witness shortly.

Another physical variable that is governed by the temperature is diffusion. In one of his *Annus mirabilis* papers from 1905, Einstein elucidated the theory of Brownian motion, including the mathematical formula for the diffusion constant D (units of area per unit time),[9] which has the form

$$D = \frac{k_B T}{6\pi\eta R_0}, \tag{7.17}$$

for a small spherical particle of radius R_0 travelling in a liquid with dynamic viscosity η; this equation is applicable at low Reynolds number. As the diffusion timescale is $\tau_D \sim L^2/D$ on dimensional grounds, where L is the length scale of the system, we see that it also scales inversely with the temperature. However, since the dependence of diffusion on T is rather weak, this facet may be somewhat unimportant (see Question 7.5).

Let us first consider the effect(s) of cold temperature. It is apparent from (7.16) that reaction rates will decrease, sometimes significantly. For instance, the rates of catalysis and protein folding are slowed down; the latter is problematic because the 3D structure of proteins (after folding) determines their properties. To these obstacles, we must add the issue of enhanced rigidity and compactness, which occurs due to the diminished motion of atoms when T is lowered. This reduction in flexibility acts against cell membrane functions (e.g., impeding permeability), and is further compounded by the lower inward diffusion of nutrients and reactants, as well as outward diffusion of waste products; this is demonstrated by (7.17).

Last but not least, the danger of freezing looms large. Even before the completely frozen state, intracellular fluids can transition to a glass-like state (vitrification) that poses severe challenges

---

[8]   In actuality, the thermal responses of organisms are progressively impaired after crossing a certain threshold, on account of which biological rates do not obey the Arrhenius equation at all temperatures (Schulte et al., 2011).

[9]   The result was previously derived by William Sutherland (1859–1911) in 1904.

to metabolism (Clarke, 2014). Consider a spherical organism of radius $\mathcal{R}_o$, much like the fabled 'spherical cow' in theoretical physics. The body mass, which remains constant, is proportional to $\rho_o \mathcal{V}_o$, where ρ_o is the mass density; note that the spherical volume $\mathcal{V}_o$ scales as $\mathcal{R}_o^3$. By imposing mass conservation, we end up with

$$\mathcal{R}_o \propto \rho_o^{-1/3}. \tag{7.18}$$

Given that $\gtrsim 70\%$ of the cell (by mass) consists of water, we may approximate ρ_o by ρ_w, the density of H_2O. Hence, when water transitions from the liquid phase ($\rho_w \approx 1$ g/cm^3) to the solid ice phase ($\rho_w \approx 0.916$ g/cm^3), the radius is estimated to increase by $\sim 3\%$ as per the above equation. The increase in area, which is proportional to $\mathcal{R}_o^2$, would then be $\sim 6\%$. A classic experiment by Evans et al. (1976), albeit specifically for human red blood cells, discovered that lysis (disruption) occurred for an area expansion of 2–4%. Subsequent investigations have raised this bound under certain conditions, but our simple calculation nevertheless captures the salient (bio)physics and the potential hazard posed by freezing.

A number of ingenious strategies are adopted by extremophiles that grow at cold temperatures (called *psychrophiles* or *cryophiles*); cold temperatures of $<5°C$ comprise a sizeable fraction of Earth's biosphere. Their enzymes are distinguished by weaker intermolecular interactions, which serve to boost flexibility. These organisms produce antifreeze proteins that inhibit ice crystallisation, and they also synthesise promoters of protein folding. Psychrophiles possess cell membranes with unsaturated fatty acids, which introduce bending and indirectly enhance flexibility. They are often found in high-salinity environments, since salts lower the melting point of water.

Next, let us consider the opposite regime of hot temperature. Many of the above issues associated with cold temperature are inverted, as are the evolutionary adaptations. Vital intermediaries in metabolism like ATP and NADPH have short half-lives at high T. The same is true for peptides (e.g., Aubrey et al., 2009) and RNA; the latter explains why some scientists who subscribe to the RNA world paradigm in origin(s)-of-life debates (consult Section 5.3.1) disfavour a high-temperature genesis (e.g., Levy and Miller, 1998). When the temperature is raised, atomic motions are increased, thereby triggering a boost in flexibility and weakening of chemical bonds, to the point where protein unfolding could be enhanced. In the case of DNA, high T can initiate unbinding of the Watson–Crick–Franklin base pairs and the ensuing single-strand version is more susceptible to damage (e.g., by hydrolysis). Last of all, the permeability of cell membranes may get elevated to a degree where maintaining internal stability is difficult.

Organisms that grow optimally at temperatures of $>60°C$ and $>80°C$ are respectively known as *thermophiles* and *hyperthermophiles*. Broadly speaking, the biomolecules of these organisms exhibit higher rigidity to cope with the hot temperatures. Thermophilic proteins are distinguished by their thermostability, often achieved by the incorporation of additional chemical (e.g., ionic) bonds. The reverse gyrase enzyme expressed in hyperthermophiles causes the DNA to wind more tightly (dubbed positive supercoiling), which can prevent or mitigate thermal alteration of DNA's structure. Cell membranes of hyperthermophilic archaea exhibit robust features such as ether linkages in place of ester linkages (enhances stability) and lipid monolayer scaffolding instead of bilayers (boosts rigidity).

The minimum and maximum temperatures at which organisms survive on Earth are $-25°C$ (a strain of *Deinococcus geothermalis*) to $130°C$ (a strain of *Geogemma barossii*). On the other hand, the minimum and maximum temperatures at which metabolic activity is sustained are $-20°C$

(a microbial culture extracted from Siberia) and 122°C (a strain of *Methanopyrus kandleri*). These limits will almost certainly change as new discoveries are made. From a theoretical standpoint, the lower bound might be −40°C (refer to Question 7.7), and the upper limit may extend to 150°C; the latter is plausibly dictated by the stability of biomolecules, as well as metabolic products and intermediaries (Clarke, 2014; Schulze-Makuch et al., 2017).

7.2.2   Pressure

We will not delve into the low-pressure limit here. Barely above the triple point of water (with pressure of 6.1×10^2 Pa) – where water is stable in the liquid state – there is ample evidence for growth of multiple bacteria species (Verseux, 2020). In Low Earth Orbit (LEO) and space vacuum, the pressures decline to $\sim 10^{-7}$–10^{-4} Pa. Yet, a diverse array of fungi, lichens, and prokaryotes are documented to have survived in these harsh conditions after exposure times of $\sim$0.1–6 yr (Horneck et al., 2010). The fraction of viable organisms declines substantially due to the twin effects of desiccation (i.e., removal/loss of water) and DNA degradation, among others, but theoretical calculations suggest that survival timescales of order $10^5 - 10^6$ yr are possible if sufficient shielding, either via inorganic compounds (e.g., salts) or biofilms, is available (Mileikowsky et al., 2000).

At high pressure, large assemblies of proteins and/or nucleic acids are disrupted, particularly at values $\gtrsim 2 \times 10^8$ Pa. The pressure-induced packing also contributes to protein unfolding at $\sim 4 \times 10^8$ Pa, and stabilises double-stranded nucleic acids (at least up to pressures of $\sim 10^9$ Pa). While the latter may seem like a positive, unwinding of DNA is needed for some biological functions, which would be hampered by over-stabilisation. If we consider cell membranes, the effects of high pressure are essentially analogous to those attributable to low temperature: enhanced rigidity and compactness alongside reduced permeability and flexibility.

Organisms that display optimal growth at high pressures are referred to as *piezophiles*. In view of the preceding paragraph, it is not surprising that piezophiles have some traits in common with psychrophiles. Piezophiles are composed of loosely packed protein assemblages, flexible individual proteins, and unsaturated cell membranes with branched lipids. This example illustrates how polyextremophiles (namely, organisms that live in multiple extremes) do not necessarily compensate for (or adapt to) every environmental parameter separately. To put it differently, the various environmental extremes are non-additive: in certain cases, the deleterious effects amount to more than the sum of the parts, and at other times, they add up to less – as in the current situation of cold temperatures and high pressures.

The archaeon *Thermococcus piezophilus* achieves optimal growth at 5×10^7 Pa, and can survive at the pressure of 1.25×10^8 Pa. However, in light of the vast number of uncultured microbes, the true record holder remains unknown in not just this category, but most (if not all) others. On short timescales, it has been demonstrated that model organisms like *Escherichia coli* may evince pressure resistance up to $\sim 2 \times 10^9$ Pa in controlled settings.

7.2.3   Salinity and water activity

A variety of environments on Earth harbour high concentrations of salts. Soda lakes are highly alkaline environments with salinity (expressed as mass fraction) confirmed to reach as much as 37%. We will not focus on the lower bound of salinity, since organisms can survive and grow in aquatic settings with 0% salinity. We will, therefore, explore the difficulties posed by high salinity. Organisms that grow in high-salinity settings are known as *halophiles*. A strain of the bacterium *Halarsenatibacter silvermanii* was isolated from a lake with 35% salinity, and might be the record holder.

The general issue in saline environments is osmotic stress. In osmosis, the solvent molecules move from the region with lower concentration of a solute to another with higher concentration to attain equilibrium. When applied to our scenario, we see that water would tend to exit the cell into the salty surroundings. This loss of liquid water would impact cellular functions, as it represents the solvent for life. The second obstacle we wish to highlight is that ions sourced from salts may damage proteins and other biomolecules because they restrict or suppress intermolecular interactions. High ionic strength, which roughly captures the overall concentration of ions in a solution, apparently stymies the growth of microbes in brines (high-concentration salt solutions) akin to those that could exist on Mars (Fox-Powell et al., 2016).

It is important to differentiate between chaotropic and kosmotropic ions, which may be crudely envisioned as those ions that stabilise and destabilise macromolecules (e.g., proteins), respectively (reviewed in Ball and Hallsworth, 2015). More specifically, chaotropic ions can expose the protein backbone to water and interact with side chains of amino acids, both of which could adversely affect protein functions. Examples of chaotropic cations are Ca^{2+} and Mg^{2+}, while NH_4^+ and K^+ are kosmotropic. Likewise, perchlorate (ClO_4^-) and thiocyanate (SCN^-) are chaotropic anions, whereas hydrogen phosphate (HPO_4^{2-}) and sulfate (SO_4^{2-}) are kosmotropic.

To deal with the challenges of high salinity, organisms have evolved two broad strategies. In the 'salt-in' approach, the ion concentration inside the cell is high and maintained at osmotic equilibrium with the exterior; this process often entails accumulating K^+ within the cell. As mentioned in the prior paragraph, this cation is kosmotrophic and may boost protein stability. In addition, the proteins themselves possess characteristics that are conducive to preserving their activity and functionality. They are enriched in acidic sites that promote protein stability and flexibility, and also help prevent protein aggregation. The cell membranes of the 'salt-in' microbes have special features (e.g., carotenoid pigments) to selectively reduce permeability and regulate the transport of K^+, Na^+, and H^+ ions.

The second strategy is the 'salt-out' pathway, where the salts (or their constituent ions) are pumped out into the cell exterior for mitigating damage to biomolecules. As a result, however, the osmostic stress alluded to earlier is further enhanced. To combat this problem, the uptake or production of compatible solutes is commensurately increased. Compatible solutes are small molecules (e.g., amino acids and sugars) that balance osmostic pressure, are non-toxic, and foster protein stability. The salt-out method is more prevalent, but might not be as effective as the salt-in strategy insofar as the most extreme saline habitats on Earth are concerned.

A physicochemical parameter connected to salinity is *water activity* (a_w), which is interpreted as the effective mole fraction of water (or the relative humidity divided by 100). A low value of a_w implies that the environment has limited water availability and vice versa. Note that $a_w = 1$ for distilled (pure) water and $a_w = 0.755$ for saturated sodium chloride solution. Microbes can regulate a_w to an extent by storing or attracting water by means of the synthesis of some metabolic products. However, beyond a certain stage, water availability would plausibly become a fundamental bottleneck. Organisms that dwell in water-limited conditions are termed *xerophiles*.

The xerophilic fungus *Aspergillus penicillioides* is capable of growth at $a_w = 0.585$ (experiments) and $a_w = 0.565$ (theory) (Stevenson et al., 2017), and recent experiments support microbial viability at $a_w = 0.540$. If this threshold stands the test of time, then water activity may exclude life-as-we-know-it in the clouds of Venus (possibly with $a_w \sim 0.004$) and perhaps Mars, but not those of Jupiter (Hallsworth et al., 2021).

7.2.4   pH

As stated in Section 7.1.2, the pH signifies the concentration of protons. Earth hosts remarkably diverse environments vis-à-vis pH. Water from Richmond Mine (Iron Mountain, USA) has pH of -3.6 (extremely acidic), while Gorka Lake (Chrzanow, Poland) has pH of 13.3 (extremely alkaline).

Cells have a pH not far removed from 7 in most settings; the pH of the cytoplasm (the liquid inside the cell) is typically (but not always) $\sim 7 \pm 1$. Maintaining this value is desirable for the proper functioning of the cell. In Section 7.1.2, we witnessed how proton pumps are crucial for ATP synthesis, and similar conclusions apply to other ion pumps. In either highly acidic or alkaline environments, the pH gradient between the interior and exterior is substantial, which favours a tendency for H^+ to diffuse inward or outward; either scenario would tend to disrupt the (quasi-)equilibrium of the cell.

If we consider highly acidic environments, protons are liable to diffuse into the cell from the outside. The ensuing increase in proton abundance can induce several negative consequences such as protein unfolding due to the disruption of chemical bonds among side chains of amino acids. A number of strategies enable organisms to handle acidic environments. First, plentiful production of proton pumps and transporters enable excess H^+ to be expelled. Second, cell membranes could evolve lower-than-average permeability towards protons. Third, the synthesis of acid-stable proteins (with an excess of glutamic acid and aspartic acid) would partly offset the buildup of positive charges. Last, organisms may secrete metabolic products into their immediate surroundings that aid in regulating external pH.

The mechanisms facilitating the survival and growth of microbes in alkaline environments are not as well understood, but some of them are the opposite of the above paragraph. For instance, the low proton concentrations in the cell exterior indicate that high efficiency of proton transport across the cell membrane could be beneficial. Furthermore, cell membranes with negative ionic species impede the diffusion of the hydroxide ion OH^- (associated with alkaline settings), which can damage the cell membrane and disrupt the double-stranded structure of DNA.

Organisms that thrive at low and high pH are respectively called *acidophiles* and *alkaliphiles*. The hyperacidophilic archaea *Picrophilus oshimae* and *Picrophilus torridus* survive at pH of approximately 0, and display optimal growth at pH ≈ 0.7; these archaea are also thermophilic. Ultra-small archaea (ostensibly *Nanohaloarchaea*) inhabit the Dallol Hot Springs of Ethiopia, which not only have pH ~ 0 but are also hot (temperature approaching 100°C) and saline to boot (Gómez et al., 2019). On the other end of the spectrum, some bacteria from the genus *Serpentinomonas* exhibit optimal growth at pH ≈ 11, and reportedly survive at maximum pH of 12.5.

7.2.5   Radiation

High-energy 'radiation' encompasses not just electromagnetic radiation such as ultraviolet (UV), X-rays, and gamma rays but also high-energy particles (e.g., protons). This radiation may inactivate or kill organisms via manifold avenues. To offer an example, UV radiation can trigger the formation of reactive oxygen species, which in turn could destroy biomolecules, cause potentially fatal single-strand and double-strand DNA breaks, and impair Watson–Crick–Franklin base pairs in DNA. Many of these dangers are likewise induced by the above types of radiation aside from UV light.

Fortunately, prokaryotes and eukaryotes are equipped with an impressive array of adaptations (whose provenance is partly unclear) to combat the aforementioned stressors. Some of the noteworthy ones include repair mechanisms to restore DNA structure, multiple copies of the genome to boost redundancy, production of screening compounds (e.g., pigments), incorporation of manganese (Mn^{2+}) since it can scavenge reactive oxygen species, and cooperation among microbes (via extracellular processes). It is important to recognise that most (perhaps all) radiation-resistant organisms are extremotolerant rather than extremophiles in the strict sense; the subtle distinction between this duo was spelt out at the start of Section 7.2.

Unlike before, we bypass precisely quantifying empirical radiation limits because this domain is broad, and the thresholds are sensitive to multiple factors (e.g., duration of exposure, access to shielding, and nature of radiation). The hyperthermophilic archaeon *Thermococcus gammatolerans* can tolerate an integrated gamma radiation flux of 3×10^4 Gy (where 1 Gy = 1J/kg), which is three to four orders of magnitude higher than the limit for most mammals. The well-known extremophile and model organism *Deinococcus radiodurans* has survived exposure to gamma radiation flux of 50 Gy/h. Certain mosses are documented to withstand high UV fluxes: the survival fractions remained high after exposure to UV fluences (time-integrated fluxes) of $\sim 10^8$ J/m^2 in the wavelength range of 200–400 nm (Huwe et al., 2019).

In passing, we mention a specific limit on photosynthesisers imposed by radiation. If the levels of PAR are too low, then photosynthesis would not be able to operate, although life could still exist by harnessing chemical energy from redox reactions. Based on a combination of (bio)chemical constraints such as protein turnover, oxidation of proteins in photosystem II, and pumping efficiency of H^+, Raven et al. (2000) suggested that the lower bound for the PAR flux is ~ 2.4–6×10^{15} photons m^{-2} s^{-1}, that is, about five orders of magnitude lower than direct sunlight at Earth's surface (McKay, 2014).

7.3   Final remarks: polyextremophiles and transfer of life

In this chapter, we dove into the fundamental criteria for life-as-we-know-it, with an emphasis on energy sources and physicochemical conditions. We divided the former into chemical energy (obtained from redox reactions) and electromagnetic energy. Multiple avenues are conceivable for harvesting energy from chemical sources, and at least two for electromagnetic radiation, which augurs well for putative extraterrestrial life. With regard to environmental parameters, we uncovered how life on Earth has expanded into myriad habitable niches via extremophiles, as illustrated in Figure 7.4. Instead of reiterating prior themes and findings, we will round off with a brief exposition of some crucial facets not highlighted so far.

To begin with, when contemplating electron transfer reactions (yielding chemical energy), it is appealing to visualise the Earth as one gigantic electronic circuit characterised by the flow of electrons. The oxidants and reductants utilised by biology for maintaining this circuit must not only be available for consumption but also transported on substantial scales. The latter is realisable through the two major circulation engines: the atmosphere and the oceans. In view of the centrality of metabolic pathways as sources and sinks for bioessential elements,[10] it is not surprising that life itself is a fundamental component and driver of the cycling of these elements (i.e., biogeochemical cycles), as reviewed in Jelen et al. (2016).

Next, the purpose of Section 7.2 was to summarise various environmental stressors and elucidate how extremophiles have evolved ingenious adaptations to counter them. However, we looked at these stressors in isolation, and thus did not examine what would happen when more than one of them operate simultaneously, as is the case in most extreme environments on Earth. As hinted in Section 7.2.2, the effects are often non-additive, to wit, the sum is more or less than the individual parts. A careful treatment necessitates the study of polyextremophiles, which goes beyond the scope of this book. The majority of extremophiles are actually polyextremophiles (Merino et al., 2019), whose properties are quite underexplored. Reviews of this topic can be found in Seckbach et al. (2013) and Carré et al. (2022).

Finally, it is apparent that space is an incredibly extreme environment from the perspective of temperature, pressure, water availability, radiation, and more. It is natural to presume that organisms cannot survive in this setting for prolonged periods of time. What if, however, they are embedded in some type of protective cocoon? This question directly ties in with a deeper question: could life be transported from one world to another? This conjecture goes by the name of *panspermia*, and has an unusually rich history stretching back to at least the pre-Socratic Greek philosopher Anaxagoras (500–428 BCE) in the fifth century BCE. Panspermia is subject to a multitude of severe challenges, some of which might be insurmountable.

The most promising version of panspermia is *lithopanspermia*, wherein: (1) life-bearing rocks are ejected into space from the donor world (often triggered by a large impact); (2) survive the lengthy and harsh voyage through space; and (3) subsequently land and microbes replicate on the recipient world. It is believed that lithopanspermia between Mars and Earth (i.e., interplanetary panspermia) may be possible in theory (Mileikowsky et al., 2000), but empirical evidence is lacking. The prospects for interstellar panspermia are much dimmer on account of the great distances involved, although a minority of scientists maintain otherwise (e.g., Wickramasinghe et al., 2019).

7.4   Problems

Question 7.1: You are given the information that the electric potential difference due to the charge accumulation outside the cell membrane is $\Delta\psi = -0.1$ V. Moreover, the pH outside the cell is 6 and inside the cell is 8. Find the proton motive force (PMF) at: (i) $T = 280$ K and (ii) $T = 380$ K. Does the PMF increase or decline with T, with all other parameters held fixed?

Question 7.2: We will attempt to gauge the photon flux that could be harnessed by oxygenic photosynthesis. In Section 12.2.3, the blackbody spectral radiance B_λ is presented in (12.31). The photon spectral flux density n_λ (i.e., number of photons per unit area per unit time per unit wavelength) is found by dividing B_λ by the photon energy hc/λ.

---

[10] For example, the oxidation of sulfides/sulfur and the reduction of sulfates correspond to previously encountered metabolic pathways, and contribute to sulfur cycling.

a. We can determine the stellar photon production rate in the band $\lambda_{min} < \lambda < \lambda_{max}$ by integrating n_λ over this wavelength range, and multiplying it with the area of the star, whose radius is $R_\star$. Justify this procedure and obtain the formula for this quantity.

b. For a given star–planet distance (denoted by a), the stellar photon flux $\mathcal{F}$ at the top of its atmosphere may be estimated from part **(a)**. Complete this task and write down the integral expression.

c. For the Earth–Sun system, set $\lambda_{min} = 400$ nm and $\lambda_{max} = 700$ nm, and numerically solve for $\mathcal{F}$. How does this value match against the photon fluxes of 1.05×10^{21} photons m^{-2} s^{-1} and 4×10^{20} photons m^{-2} s^{-1} incident at sea level and required to support Earth's photoautotrophs, respectively?

Question 7.3: Several advantages are attributable to oxygenic photosynthesis, a few of which can be deduced systematically, as demonstrated here.

a. The photon spectral flux density n_λ is defined in Question 7.2. Compute the wavelength at which n_λ is maximised (a calculator is needed), and then compare this wavelength at $T = 5,780$ K – the blackbody temperature of the Sun – with the absorption peaks of chlorophylls (see Section 7.1.4). What do you find, and what is the inference that may be drawn?

b. What is the electron donor in oxygenic photosynthesis? Instead of thinking in terms of redox potential, if global abundance is used as the yardstick, why is this electron donor rendered appealing?

c. What is the significance of the gaseous metabolic product of oxygenic photosynthesis for life? You can peruse Section 6.3.3 and references therein.

Question 7.4: How many independent avenues for phototrophy are documented on Earth? How likely is it that other pathways for phototrophy do exist, but have not been discovered yet? To help justify your answer, you are encouraged to read up on the expected percentages of uncultured microbes, the rapid development of –omics methods (e.g., metagenomics), and so forth; peer-reviewed sources must be consulted and referenced.

Question 7.5: Calculate the diffusion constant of a particle of radius 1 nm (typical of small molecules) on Earth and Titan; for the latter, you will need to look up the temperature and viscosity of liquid methane (the ambient liquid). How long would this particle take to diffuse across a length of 1 µm, taken to be the characteristic microbe size, on Earth and Titan?

Question 7.6: Mars has an *average* surface temperature of 215 K, and possesses a characteristic geothermal gradient of $\sim$10 K/km; in actuality, this value is uncertain by a factor of $\sim$2. Assume that the thermal limits for putative Martian life forms are exactly the same as those on Earth. You are required to calculate the following quantities.

a. The depth H_1 (in km) beneath the surface of Mars where the lower thermal bound of life is attained. The depth H_2 (in km) below the Martian surface where the upper thermal bound of life is reached. The width of the subsurface habitable region ($H_2 - H_1$) on Mars.

b. The pressure at depth H_2 if the mean rock density is taken to be $\sim$2.6 g/cm^3 and $g = 3.7$ m/s^2 for Mars. Is this pressure high enough to rule out piezophiles introduced in Section 7.2.2?

Question 7.7: A famous study by Price and Sowers (2004) estimated that the (ensemble-averaged) metabolic rate of microbes surviving in extreme cold environments obeys the Arrhenius equation, with $E_a \approx 110$ kJ/mol, and that the rate at $-12°$C is $\sim 10^{-9}$ h^{-1}. Convert E_a from kJ/mol to eV/particle since the latter is necessary. If we select a temperature of $-40°$C, what is the value of the metabolic rate? What is the ballpark lifetime of the organism (crudely the inverse of the metabolic rate)? Is this timescale compatible with alleged microbial lifespans of $\sim$100 Myr?

Question 7.8: Employ the high-pressure threshold from Section 7.2.2 here.

    a. The internal structure of Europa (surface gravity of 1.3 m/s^2) may consist of an icy crust of thickness $\sim$25 km and a subsurface water ocean of depth $\sim$100 km underneath; for simplicity, the density of both layers is set equal to $\sim$1 g/cm^3. Calculate the pressure at the bottom of Europa's ocean. Is this value higher or lower than the pressure limit(s) of piezophiles?

    b. Some exoplanets may harbour oceans much deeper than the Earth, and we wish to know whether piezophiles can live at the bottom of the oceans. Consider an exoplanet with radius R_p and utilise the scaling $M_p \propto R_p^{3.7}$ from numerical simulations (M_p is the planet's mass) to derive the surface gravity as a function of R_p. The density of liquid water is chosen to be 1 g/cm^3. Obtain the expression for the maximum ocean depth H_{max} at which piezophiles could grow as a function of the above variables. By setting $R_p = 0.5\,R_\oplus$ (i.e., a Mars-sized planet), determine H_{max} for this world.

Question 7.9: Building on Question 7.6, life in the deep subsurface faces manifold challenges. On the basis of what you have learnt in the current chapter, as well as a judicious literature survey, list four parameters that pose serious impediments to life in this environment. Justify your reasoning, and outline possible pathways whereby these obstacles might be overcome.

Question 7.10: In Section 7.2, we encountered many different kinds of extremophiles. We did not, however, attempt to catalogue every such '-phile'. By carrying out a literature survey of peer-reviewed publications, discuss two examples of -philes not mentioned in this chapter. Explain what extreme environmental parameters are prevalent, and how the ensuing hurdles are overcome through specialised adaptations.

8   Continuous Habitability

In the previous chapter, we expounded the requirements that are canonically deemed necessary for an environment to sustain life at any moment in time (i.e., instantaneous habitability). We now delve into a different, but related, question: what makes that environment conducive to life over an extended period, compatible with the long timescales of biological evolution? In other words, we will investigate the theme of *continuous habitability*.

This is complex topic, as a multitude of stellar and planetary characteristics, and their temporal variations and mutual interplay, can affect the conditions on a planet's surface. We will divide our discussion into distinct parts, but it should be kept in mind that it is not always possible to separate the various factors involved, and the final outcome will be determined by an often non-trivial interdependence among them. To begin with, we will introduce the crucial concept of the circumstellar habitable zone, and discuss some basics of planetary climates. Next, we will explain and quantify how the host star(s) may regulate habitability via its electromagnetic spectrum (e.g., UV and optical fluxes), winds, flares, and space weather.

Moving onward to the planetary facets, we will delineate the significance of planetary mass, orbital properties, plate tectonics, atmospheric composition, and other factors. The chapter will conclude by sketching how high-energy astrophysical phenomena, such as supernovae and supermassive black hole activity, could modulate habitability on galactic scales.

8.1   The circumstellar habitable zone

As remarked in Chapter 7, the availability of liquid water is one of the essential requirements for life-as-we-know-it. This datum suggests that the conditions for its possible presence can be employed as a crude criterion to identify potentially habitable environments outside Earth. A useful tool, in this respect, is the concept of the *circumstellar habitable zone*, or simply the *habitable zone* (HZ). The HZ is the annular region surrounding a star where the physical conditions are compatible with the existence of liquid water on the surface of a terrestrial (i.e., rocky) planet. Although the notion of the HZ remarkably dates as far back as Sir Isaac Newton, its 'modern' usage apparently emerged in the 1880s (Lingam, 2021a). State-of-the-art observations appear to indicate that the frequency of terrestrial planets in the HZ is loosely of order 0.1 per star.

Needless to say, a planet situated in the HZ is neither guaranteed to have liquid water bodies, nor to be habitable. The use of this terminology, to some extent, may be misleading: a more neutral and appropriate denomination might be the *temperate zone*.[1] However, despite its inherent limitations,

---

[1] A commonly adopted colloquial term is the *Goldilocks zone*, a reference to the famous fairy tale 'Goldilocks and the Three Bears', wherein the eponymous character settles for the middle choice between two extremes, which is 'just right'.

the notion of the HZ can render helpful guidance, especially in identifying and selecting interesting targets for detailed follow-up observations.

Most calculations of the HZ start from the assumption that the planet under consideration has geological, atmospheric, and physical properties broadly similar to the modern Earth: such a world is usually referred to as an *Earth analogue*. If this is the case, we can define the inner and outer edges of the HZ as the distance from the host star where the planet's average surface temperature (T_s) falls within 273 K and 373 K, which is none other than the range for liquid water at a pressure of 1 atm.

We will now explore how to derive an accurate estimate of the location of the circumstellar HZ for Earth analogues around generic stars.

8.1.1   Radiative balance

To compute the average temperature of a planet, we need to know how the energy radiated from the host star interacts with its surface and atmosphere. The first useful quantity we introduce is the *albedo*; this parameter accounts for the fact that the stellar energy flux is not entirely absorbed by a planet, as part of it is reflected into space. Various definitions of albedo are possible, depending on the context: for our purposes, we will adopt the *Bond albedo* A_b – this is simply the ratio of total reflected or scattered radiation power (at all frequencies) divided by the total incident radiation power (at all frequencies). By definition, the Bond albedo is always $A_b \leq 1$. Table 8.1 shows the albedo values for some notable objects in the solar system.

The next step is to compute the total flux that is incident on a planet from its star. If we label the stellar luminosity (measured in W) as $L_\star$, then the total flux at orbital distance a from the star is typically expressible as

$$F_\star = \frac{L_\star}{4\pi a^2}. \tag{8.1}$$

Table 8.1 Bond albedo values of solar system objects

Object	Bond albedo
Mercury	0.09
Venus	0.76
Earth	0.30
Moon	0.11
Mars	0.25
Jupiter	0.50
Saturn	0.34
Uranus	0.30
Neptune	0.29
Pluto	0.41

Note: Data is tabulated from Mallama et al. (2017).

The energy intercepted by the cross-sectional area of the planet is $\pi R_p^2 F_\star$, where R_p is the planetary radius: by spreading this flux over the entire planetary surface, $4\pi R_p^2$, we obtain the total averaged flux, $F_\star/4$.

We estimate the planetary equilibrium temperature T_{eq} by approximating the planet as a blackbody, with a corresponding flux of $\sigma_{SB} T_{eq}^4$. At equilibrium, the incoming and outgoing flux must balance out, and therefore,

$$T_{eq} = \left[\frac{(1 - A_b)L_\star}{16\pi\sigma_{SB}a^2} \right]^{1/4}, \tag{8.2}$$

where we accounted for the fact that only a fraction $(1 - A_b)$ of the incident stellar flux is actually absorbed by the planet. By inverting this equation, we can easily find the distance from the host star where the planet has a given equilibrium temperature. The range of distances for the HZ (encapsulated by a_{HZ}) are then estimated to be

$$a_{HZ} = \left[\frac{(1 - A_b)L_\star}{16\pi\sigma_{SB}T_{H_2O}^4} \right]^{1/2}. \tag{8.3}$$

where $T_{H_2O} = 273\text{--}373$ K is the thermal range for liquid water at 1 atm.

To check the reliability of these calculations, we use them to gauge the width of the habitable zone around the Sun. By adopting the values for the solar luminosity, $L_\odot = 3.83 \times 10^{26}$ W, and the Earth's albedo, $A_b = 0.3$, the inner and outer edges of the HZ ensuing from (8.3) are $a_{HZ} = 0.47\text{--}0.87$ AU, so that the Earth would fall outside the habitable zone in the solar system. In fact, upon invoking (8.2), we determine that the equilibrium temperature at 1 AU ought to be 255 K, that is, well below freezing point, and 33 K short of Earth's average surface temperature of 288 K.

Clearly, the equilibrium temperature is not necessarily an accurate representation of a planet's surface temperature. The previous treatment neglected several factors that have a sizeable impact on the final result, and that shall be elucidated in the following sections.

8.1.2   The greenhouse effect

For many planets, the most significant influence on their average surface temperature arises from the so-called *greenhouse effect*, whereby heat radiated from the planet is partially trapped by its atmosphere, raising T_s of that world. This phenomenon occurs via molecular species that absorb infrared thermal radiation, thus reducing the amount of energy that can escape into space. Direct or indirect greenhouse gases include water vapour (H_2O), carbon dioxide (CO_2), ozone (O_3), methane (CH_4), and nitrous oxide (N_2O). On modern Earth, H_2O and CO_2 are the two most important greenhouse gases, contributing $\sim 2/3$ and $\sim 1/3$ of the total warming, respectively. The role of water vapour, however, is subject to the influence of the other greenhouse gases: if the Earth warms, the evaporation of surface water boosts the atmospheric H_2O pressure, amplifying the greenhouse effect.

Clouds are also vital in regulating the average temperature, as they affect both the planetary albedo and the absorption of infrared radiation, and could therefore either warm or cool the planet, depending on the predominant mechanisms. For modern Earth, the net effect of clouds reportedly lowers the average temperature. Modelling the overall impact of clouds on climate is difficult, and can introduce considerable uncertainties if solid observational constraints are lacking, which is the case for planets around other stars. It is worth highlighting here that both greenhouse gases and cloud coverage may be modulated by biological and technological activities.

A compact means of quantifying the greenhouse effect is by defining an optical depth τ_{IR} of the planetary atmosphere to infrared radiation. When $\tau_{IR} \leq 1$, the atmosphere is partially transparent to thermal radiation, while for $\tau_{IR} \gg 1$, it may be considered opaque. If $\tau_{IR} \neq 0$, the surface temperature is elevated with respect to the equilibrium value in (8.2), as carefully derived in Chamberlain and Hunten (1987, pg. 12):

$$T_s = T_{eq} \left(1 + \frac{3}{4}\tau_{IR} \right)^{1/4}. \tag{8.4}$$

For Earth, $\tau_{IR} \approx 0.8$ drives a 33 K increase relative to the equilibrium temperature. Venus' atmosphere is essentially opaque to the infrared, with $\tau_{IR} \approx 142$ and a surface temperature of $T_s \approx 737$ K, which is dramatically higher than the equilibrium temperature $T_{eq} \approx 229$ K. The atmosphere of Mars, on the contrary, is largely transparent to thermal radiation, with $\tau_{IR} \approx 0.13$ and $T_s \approx 215$ K compared to $T_{eq} \approx 210$ K.

8.1.3 Climate modelling

The average surface temperature of a planet with an atmosphere depends on the combined action of stellar flux, albedo, and greenhouse effect (Figure 8.1). Yet, this is not to say that understanding variations in planetary temperature is an easy task. In fact, the existence of *feedbacks* may alter the evolution of climate in complicated ways. Feedbacks can be either *positive* or *negative*: in the former, a change in T_s of the planet – either an increase or a decrease – gets amplified in the same direction, potentially leading to a runaway effect; whereas in the latter case, the change in temperature is counteracted by the system as a whole. The existence of negative feedbacks is crucial for the long-term stability of planetary climates.

Four major feedbacks are operational in the Earth's climate, and potentially on Earth analogues around other stars (Catling and Kasting, 2017):

- *Positive feedback from atmospheric water vapour*: Water vapour in the atmosphere is close to its condensation temperature, and therefore its concentration is quite sensitive to the global temperature. If the climate is subject to warming, a higher amount of vapour builds up in the atmosphere, thereupon increasing the greenhouse effect and further amplifying warming. The opposite trend is expected if climate cools.
- *Positive feedback from ice albedo*: As ice has higher albedo than the ground, decrease in ice coverage due to climate warming boosts the fraction of stellar flux absorbed by the planetary surface, which in turn raises the global temperature; the opposite happens when climate cools and ice coverage grows, which can cause runaway glaciation (see Section 6.3.4).
- *Negative feedback from outgoing infrared radiation*: The flux emitted by a planet depends on the surface temperature as $F_p = \sigma_{SB} T_s^4$. Hence, somewhat counter-intuitively, a temperature increase is accompanied by higher emission of infrared radiation, and therefore enhanced cooling. This feedback contributes to climate stability on short timescales.
- *Negative feedback from the carbonate–silicate cycle*: On longer geological timescales ($\sim 10^6$ yr), another negative feedback plays a key role in climate stability: the amount of CO_2 in the atmosphere is controlled by the so-called *carbonate–silicate cycle* (Walker et al., 1981). This mechanism

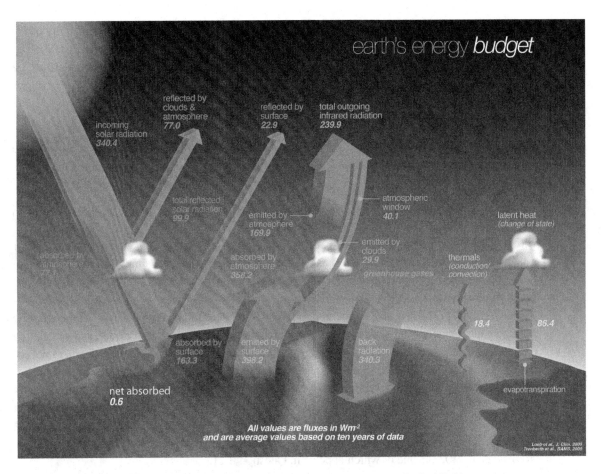

Figure 8.1 The Earth system's energy budget and its myriad components. (Credit: NASA; https://mynasadata.larc.nasa.gov/basic-page/earths-energy-budget)

is so important for the habitability of the Earth (and possibly for 'Earth-like' planets) that we tackle it separately in Section 8.3.2. Here, we just mention that the cycle can remove CO_2 from the atmosphere when temperature rises, thus cooling the climate; likewise, when the temperature drops, more CO_2 is released, thereby increasing the greenhouse effect.

We may now ask what is the change in surface temperature (ΔT_s) ensuing from a variation in *radiative forcing* (also called *climate forcing*), denoted by ΔQ; the latter (units of energy flux) measures the imbalance between incoming and outgoing radiation due to a change in stellar flux, greenhouse effect, or albedo. A positive ΔQ signifies that the planet receives more energy than it radiates into space, and vice versa for $\Delta Q < 0$. The above question can be addressed by evaluating the *climate sensitivity* (λ_c) as

$$\lambda_c = \frac{\Delta T_s}{\Delta Q}. \tag{8.5}$$

The apparent simplicity of the formula veils the feature that this climate sensitivity innately depends on the complex interplay of physical, chemical, geological, and biological processes and feedbacks regulating planetary climates. For illustrative purposes, let us compute the climate sensitivity of the Earth to a variation in the mean solar flux reaching the top of the Earth's atmosphere (or *solar constant*), which is given by $S_0 = 1361$ W m^{-2}. By utilising (8.2), it can be shown in Question 8.2 that

$$\lambda_c = \frac{T_{eq,\oplus}}{4S_0}, \tag{8.6}$$

which yields $\lambda_c = 0.047$ K W^{-1} m^2 for $T_{eq,\oplus} \approx 255$ K. This result would imply a temperature variation $\Delta T_s \approx 0.64$ K for a 1% change in solar flux. However, a more realistic model of the Earth's climate leads to

$$\lambda_c = \frac{1 - A_b}{4b}, \tag{8.7}$$

with $b = 2.2$ W m^{-2} K^{-1} (Catling and Kasting, 2017, pg. 38). This equation yields $\Delta T_s \approx 1.1$ K for a 1% change in solar flux, almost twice as much as the estimate from the simplified calculation.

A reliable estimate of the average planetary surface temperature necessitates the adoption of full-fledged climate modelling. The simplest approach is based on the *one-dimensional radiative–convective model* (Manabe and Wetherald, 1967),[2] which determines the planetary temperature by averaging the incident solar radiation and surface temperature over the entire planet, and calculating the vertical variation in atmospheric temperature. More sophisticated models, called *general circulation models*, account for the dynamics of oceans and atmosphere by using a discretised grid of cells to solve the salient differential equations for motion and energy transfer over time. Initially developed for Earth's climate and weather forecasts, they are increasingly being harnessed to predict the thermal properties of exoplanets (covered in Chapter 12). Several numerical codes for computing exoplanetary climates are freely available at NASA's Exoplanet Modelling and Analysis Centre (Renaud et al., 2022),[3] one of which appears in Question 8.11.

8.1.4   Habitable zone calculations

The first realistic calculations of the circumstellar habitable zone for Earth analogues around main-sequence stars – based on one-dimensional numerical climate models – were performed by Kasting et al. (1993), and subsequently refined by Kopparapu et al. (2013). These publications adopted the following criteria to demarcate the boundaries of the HZ:

- The inner edge of the HZ is defined by the onset of the *moist greenhouse* effect; this phenomenon occurs when substantial water vapour accumulates in the upper atmosphere, becoming subject to photodissociation by stellar radiation, which leads to hydrogen escape into space. In this water-loss limit, the process will gradually deprive the planet of all its surface water. A more extreme limit is the *runaway greenhouse* threshold, reached when the positive feedback from atmospheric water vapour surpasses the negative feedback from outgoing infrared radiation (see Section 8.1.3). As a result, temperatures escalate rapidly and the planet dries up completely. Conservative estimates set the current inner edge of the HZ for the solar system at $\sim$0.95 AU (Ramirez, 2018).

---

[2] Syukuro Manabe (1931–present) received the 2021 Nobel Prize in Physics partly for this work.
[3] https://emac.gsfc.nasa.gov/.

- The outer edge of the HZ arises from the *maximum greenhouse limit*, when the greenhouse effect cannot warm the planet further. For an Earth analogue, this occurs when the climate is too cold for the carbonate–silicate cycle to operate, resulting in the condensation of carbon dioxide built up in the atmosphere, to the point of preventing additional greenhouse warming. For the present solar system, a conservative estimate of the outer boundary of the HZ is $\sim$1.7 AU (Ramirez, 2018).

To a first approximation, simple scaling criteria dictate how the HZ shifts when the host star luminosity varies. As per (8.2), a planet with the same albedo of Earth has the same equilibrium temperature at an orbital radius

$$a_\star = 1\,\mathrm{AU} \left(\frac{L_\star}{L_\odot} \right)^{1/2} = 1\,\mathrm{AU} \left(\frac{M_\star}{M_\odot} \right)^{7/4}, \tag{8.8}$$

where we invoked the stellar mass–luminosity relation $L_\star \propto M_\star^{3.5}$. However, as stated earlier, the star's luminosity is not the sole determinant of planetary climates. When the stellar spectral type changes, non-trivial interactions are expected to occur in the planetary atmosphere, which can only be assessed properly by a radiative transfer model. For example, the atmosphere should scatter radiation from a bluer star with greater efficiency, due to an increased Rayleigh scattering cross section, and will typically absorb red and infrared light more easily. The wavelength changes also affect albedo (ice, for example, has higher reflectivity at shorter wavelengths). Effects of this sort will slightly alter the location of the HZ around main-sequence stars, compared to a simplistic prediction based solely on their luminosity.

Therefore, numerical climate simulations are a better way of demarcating the edges of the HZ. This is usually done in terms of an effective flux, $S_{\mathrm{HZ}} = S/S_0$, where S is a generic flux and S_0 is the solar flux at 1 AU (Ramirez, 2018). If $S_{\mathrm{HZ},\odot}$ represents the effective flux for the HZ around the Sun, a good fit to the effective flux for the HZs of other main-sequence stars has the form (Kopparapu et al., 2013):

$$S_{\mathrm{HZ}} = S_{\mathrm{HZ},\odot} + A\mathcal{T}_\star + B\mathcal{T}_\star^2 + C\mathcal{T}_\star^3 + D\mathcal{T}_\star^4, \tag{8.9}$$

where $\mathcal{T}_\star = T_\star - 5{,}780$ K, the star's effective temperature is $T_\star$, and the numerical constants are calibrated from climate models. As a reference, the effective fluxes for the inner and outer edges of the HZ for the solar system are $\sim$1.1 and $\sim$0.35, respectively (Catling and Kasting, 2017, pg. 425). Once the effective fluxes for a given boundary are known, the corresponding orbital distances can easily be computed via the inverse square law:

$$a_{\mathrm{HZ}} = 1\,\mathrm{AU} \left(\frac{L_\star/L_\odot}{S_{\mathrm{HZ}}} \right)^{1/2}. \tag{8.10}$$

Figure 8.2 depicts the result for the classic HZ calculations described so far; this plot is not definitive for reasons explained shortly. The current solar system has two rocky planets in the HZ, to wit, Earth and Mars, with the Earth in close proximity to the moist greenhouse limit. The fact that Mars has no stable liquid water on the surface is a clear indication that the distance from the star has only some bearing on a planet's habitability, as numerous other factors may prove to be more important.

It is worth recognising that the boundaries of the HZ, in principle, become much broader if one allows for significant departures from Earth-like physical properties and atmospheric compositions. For example, the inner edge of the HZ depends on surface water distribution, and may be pushed

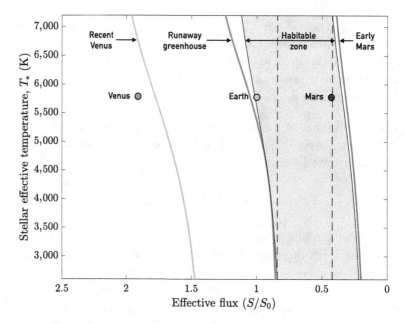

Figure 8.2 The circumstellar habitable zone for cloud-free Earth analogues around main-sequence stars, constructed from Kopparapu et al. (2013). The shaded region of the habitable zone is limited by the moist and maximum greenhouse limits (see text). The vertical dashed lines represent regions that are in the habitable zone for any stellar type.

closer to the star by decreasing the water content. Arid planets might remain in the HZ even for effective stellar fluxes as high as ~1.55 (Kodama et al., 2019). Similarly, the outer edge could shift farther away from the star if the greenhouse effect is enhanced by altering the chemical composition of the atmosphere. Notable extensions of the HZ might be obtained by adding H_2 to the mix, or increasing CH_4 abundance (Ramirez, 2018). In fact, planets with thick H_2 atmospheres of $\mathcal{O}(100)$ bar may harbour liquid water far away from the star (Madhusudhan et al., 2021). These planets, dubbed *Hycean worlds*, are regarded as promising candidates for seeking gaseous biosignatures, which are elucidated in Section 13.2.1.

An additional caveat concerns the dependence of the HZ location on stellar evolution. Since the luminosity of a main-sequence star increases with time, if all other factors are held fixed, the HZ will move outward, and its width would evolve over the course of a star's life (Kasting et al., 1993). As a reference, the Sun's luminosity varies approximately as (Gough, 1981):

$$L_\odot(t) \approx \left[1 + \frac{2}{5} \left(1 - \frac{t}{t_0} \right) \right]^{-1} L_\odot(t_0), \tag{8.11}$$

where $t_0 \approx 4.6$ Gyr is the Sun's current age. Therefore, if a planet exists in the HZ at some moment in time, it will not necessarily remain in the HZ at all other times. We may define a *continuously habitable zone*, that is, the region around a star wherein a planet could maintain liquid surface water for a given duration (which can, at most, equal the star's main-sequence lifetime). Detailed estimates of the interval spent in the HZ by Earth analogues orbiting stars of different mass were computed by Rushby et al. (2013), ranging from ~42.2 Gyr for a $0.2\,M_\odot$ M-dwarf to ~4.6 Gyr for a $1.2\,M_\odot$

F-type star. The increase in the HZ lifetime with decreasing stellar mass is along expected lines because smaller stars possess longer lifespans.

Remarkably, empirical data from the solar system indicate that the HZ calculations from climate models are not too far removed from reality (Catling and Kasting, 2017, Chapter 15). One noteworthy constraint comes from the past history of Venus. Evidence from its cratering record suggests that the planet had already lost all its water through the runaway greenhouse by ~0.7 Gyr ago. When the lower luminosity of the Sun in that epoch is duly accounted, the current inner edge of the HZ is estimated to be ~0.75 AU (i.e., close to the minimum distance allowed for dry planets). This so-called *recent Venus* inner limit translates to an effective flux of $S_{HZ} \approx 1.8$. Likewise, the geological history of Mars implies that surface liquid water was prevalent ~3.8 Gyr ago, which would set the *early Mars* limit at 1.77 AU (or $S_{HZ} \approx 0.32$), nearly equal to the maximum greenhouse edge.

As we have repeatedly emphasised, the concept of circumstellar habitable zone is only an initial heuristic for singling out potentially habitable locations. We will now delve more deeply into the various factors that regulate planetary habitability, starting with the influences of the host star.

8.2   Stellar factors

Stars are obviously the main source of free energy for planets – and conceivably for their biospheres – and have a direct impact on their climate. However, the role of stars in regulating planetary habitability is more diverse. We will provide some insight into this issue by focusing our discussion on main-sequence stars, using stellar mass as a convenient parameter to encapsulate the ensemble of their physical properties. Special emphasis is devoted to M-dwarf stars, whose masses span $0.075 < M_\star/M_\odot < 0.6$, not only because they are the most abundant (~75% of all stars) and long-lived stars in the Milky Way, but also because their planets are well-suited for characterisation by current and future telescopes (refer to Section 12.2).

A general factor to consider is the energy spectra produced by stars, which differ in the relative intensity of radiation emitted at various wavelengths. It is difficult to constrain the range of biologically relevant wavelengths, but some tentative criteria may be adopted. Haqq-Misra (2019) proposed a lower limit of $\lambda_{min} \approx 200$ nm based on the negative interactions of such photons with biomolecules (e.g., DNA), and an upper limit of $\lambda_{max} \approx 1,200$ nm for an incident photon to excite an electron from a substrate, as witnessed in photosynthesis. If this range is indeed credible, planets around F- and G-type stars might be more conducive for supporting biospheres than those orbiting K- and M-dwarfs (see Question 8.3).

8.2.1   Ultraviolet radiation

The above concept can be refined by focusing on the role played by ultraviolet (UV) radiation for prebiotic chemistry. There is persuasive – albeit not yet conclusive – evidence that UV radiation was a prerequisite for the origin(s) of life on Earth, providing the necessary energy input for the synthesis of the major building blocks of biomolecules such as proteins, lipids, and nucleic acids (see Section 5.4.1). If planets do not receive enough UV photons from their star, they might be disadvantaged in terms of habitability. We may anticipate this issue to become pertinent for planets around M-dwarfs, which are cooler and emit at longer wavelengths.

Rimmer et al. (2018) conjectured the existence of an *abiogenesis zone*, where planets intercept sufficient UV flux in the 200–280 nm wavelength band to mediate prebiotic chemistry as per laboratory experiments. This putative zone was estimated to overlap with the HZ only when $T_\star \gtrsim 4{,}400$ K; in fact, M-dwarf exoplanets in the HZ receive up to $\sim 10^3$ times less UV radiation in the above range compared to early Earth (Rugheimer et al., 2015). On such planets, the paucity of UV radiation may suppress prebiotic reactions or significantly diminish their rates (Ranjan et al., 2017).

However, we caution that UV radiation is simultaneously known to be detrimental to living organisms, engendering damage to DNA and other biological molecules, especially in the UV-C range ($200 < \lambda < 280$ nm). A commonly used metric to express the drawbacks of UV radiation is the *biologically effective irradiance* (BEI), defined as follows:

$$\mathrm{BEI} = \int_{\lambda_1}^{\lambda_2} \mathcal{F}_\lambda^{\mathrm{surf}} \mathcal{S}(\lambda) d\lambda, \tag{8.12}$$

where $\mathcal{F}_\lambda^{\mathrm{surf}}$ is the spectral flux density on the planetary surface, and $\mathcal{S}(\lambda)$ is the *action spectrum* that quantifies the relative biological response (and the extent of damage) for a given biomolecule or organism.

For a given stellar flux, the BEI can vary substantially due to the interaction of UV radiation with the atmosphere. In particular, the presence (or lack thereof) of an ozone layer would strongly alter the amount of UV radiation reaching the ground. Effective shielding could also be supplied by organic-rich hazes in the atmosphere, and even modest amounts of ocean water containing organic and inorganic compounds. Therefore, the actual danger of UV radiation to living organisms should not be overstated, especially considering the capacity of organisms adapted to cope with UV-induced damage, by evolving DNA repair mechanisms for example. Exposure to UV radiation may, instead, be instrumental in driving mutations, therefore enhancing the rates of molecular evolution and the emergence of new species. For a wide-ranging exposition of the role of UV radiation in biology, we refer the readers to Cockell and Blaustein (2001).

8.2.2   Photosynthetically active radiation

On Earth, oxygenic photosynthesis canonically relies on electromagnetic radiation in the $400 \lesssim \lambda \lesssim 700$ nm band. Whether this range can be regarded as a universal property of photopigments that evolved elsewhere is unclear (Wolstencroft and Raven, 2002; Kiang et al., 2007a). The theoretical lower limit might be set by the constraint of mitigating ionisation and cellular damage, and is therefore potentially not far removed from that of Earth's photopigments. The upper limit is more uncertain, but theoretical and laboratory studies suggest that a value of $\sim 1{,}100$ nm is compatible with facilitating electronic excitation (Lingam et al., 2021), unless exotic pigments distinct from the group of chlorophylls were to exist.

In light of the extant unknowns, we can consider the preceding interval for *photosynthetically active radiation* (PAR) as a working hypothesis to investigate the feasibility and efficiency of oxygenic photosynthesis around stars of different spectral types. As indicated in Sections 6.3 and 7.1.4, oxygenic photosynthesis is not only a major metabolic pathway but is also indirectly responsible for the oxygenation of Earth's atmosphere. It could also create prospective biological signatures, as reviewed in Section 13.2. Hence, exploring oxygenic photosynthesis would pave the way for comprehending how habitability may be influenced by the star's energy spectrum.

By paralleling the approach in Lehmer et al. (2018) and Lingam and Loeb (2019b), we can take $F_\oplus \approx 4 \times 10^{20}$ photons m^{-2} s^{-1} to represent the flux of PAR that could sustain a net rate of biomass synthesis (via oxygenic photosynthesis) comparable to Earth. These two papers, along with Covone et al. (2021), concluded that late M-dwarfs (e.g., TRAPPIST-1) with masses of $M_\star \lesssim 0.2\,M_\odot$ might be disfavoured for habitability, insofar as the availability of PAR is concerned. It should be noted, however, that frequent stellar flares associated with young and/or active M-dwarfs stars may enhance the time-averaged PAR flux to such an extent that Earth-like photosynthetic biospheres are tenable on their planets in the HZ.

8.2.3 Atmospheric escape

The host star plays a critical role in regulating the rate of planetary atmospheric loss, thus impacting habitability. Various mechanisms involving stellar energy can impart enough velocity to atmospheric particles to overcome the gravitational pull of the planet. Perhaps the most intuitive is *thermal escape*, where loss occurs because of atmospheric heating (Figure 8.3).

Two main processes constitute drivers of thermal escape. The first is *Jeans escape*, which simply arises from temperature-dependent particle velocities. In the lower strata of the atmosphere, energy exchange is dictated by particle collisions, and velocities follow a Maxwell–Boltzmann distribution.

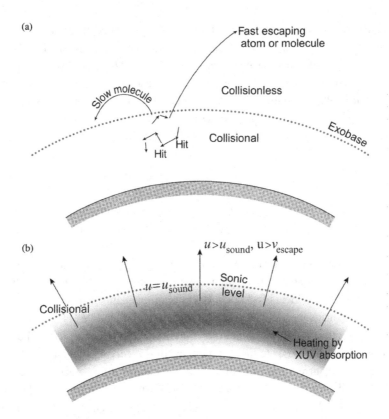

Figure 8.3 Thermal escape processes: (a) Jeans escape (b) Hydrodynamic escape. (Credit: Catling and Kasting 2017, Figure 5.2; reproduced with permission from Cambridge University Press)

At each temperature, a certain number of particles in the tail of this distribution possess kinetic energies greater than the gravitational potential energy, such that they can reach the *exobase* (the collisionless region of the atmosphere) and escape to space. This criterion is quantified by the *Jeans escape parameter*, $\lambda_J = (v_e/v_s)^2$, where $v_e = \sqrt{2GM_p/R_p}$ is the escape velocity for a planet with mass M_p and radius R_p, and $v_s = \sqrt{2k_BT/m}$ is the most likely Maxwell–Boltzmann velocity for a particle of mass m. On integrating the velocity distribution, the escape flux at the exobase is

$$\Phi_{\text{escape}} = \frac{1}{2\sqrt{\pi}} n_{\text{exo}} v_{s,\text{exo}} (1 + \lambda_{J,\text{exo}}) \exp(-\lambda_{J,\text{exo}}), \qquad (8.13)$$

where n_{exo} is the exobase's particle number density. We have $T_{\text{exo}} \sim 800\text{--}1250$ K, and $n_{\text{exo}} \sim 10^{11} - 10^{12}$ m^{-3} for Earth's hydrogen (H) at this location. Jeans escape is only significant for small λ_J, and therefore favours the loss of low-mass particles such as hydrogen, while it is relatively negligible for heavier species such as nitrogen and oxygen. Earth might be losing $\sim$10–40% of its hydrogen via Jeans escape (Catling and Kasting, 2017, pg. 143).

The second key mechanism for thermal escape is *hydrodynamic escape*. This occurs when heating induces an upward pressure gradient in the atmosphere. The result is that the whole upper atmosphere expands as a fluid into space, and can attain escape velocity. An interesting aspect of hydrodynamic escape is that strong flows of lighter particles (such as hydrogen) are capable of dragging along heavier species through collisions, enabling the loss of gases that would otherwise be too heavy to undergo Jeans escape.

The maximum hydrodynamic escape rate may be estimated in the *energy-limited* regime. The relevant energy flux of stellar radiation intercepting the planet's cross-sectional area (πR_p^2) is equated to the kinetic power of escaping particles, $\dot{M}_a v_e^2/2 = GM_a M_p/R_p$, where $\dot{M}_a$ is the atmospheric mass-loss rate. The appropriate energy range for H escape lies above the threshold for atomic H ionisation, which selects radiation <91.2 nm. This is satisfied by extreme UV and X rays, collectively termed *XUV radiation*, which encompasses photons with wavelengths of $\sim$0.1–90 nm (Linsky, 2019, Table 1.1). The resulting expression for energy-limited mass loss rate is

$$\dot{M}_a \sim \varepsilon \frac{\pi R_p^3}{GM_p} F_{\text{XUV}}, \qquad (8.14)$$

where ε is a factor that quantifies the efficiency of the mechanism and can be calibrated from numerical modelling, and F_{XUV} is the energy flux incident on the planet in the aforementioned XUV range.

Although low-mass stars such as M-dwarfs possess lower temperatures than Sun-like stars, this does not automatically translate to smaller fluxes in some UV wavelength bands, as one would conventionally expect for an ideal blackbody. Because of mechanisms occurring in the layers exterior to the visible stellar surface, certain M-dwarfs evince ratios of far-UV (115–180 nm) to near-UV (180–320 nm) fluxes up to $\sim$10^3 times higher than that of the Sun (Linsky, 2019). Furthermore, planets in the HZ of M-dwarfs are closer to the star, perhaps thus receiving comparatively higher UV fluxes.

For the Earth, the current rate of hydrogen loss computed by the energy-limited escape formula of (8.14) is $\sim$3 $\times$ 10^{15} H atoms m^{-2} s^{-1} (Catling and Kasting, 2017, pg. 157). This is merely a fiducial value, as modern Earth's atmosphere is hydrogen-poor, and the energy-limited regime does not apply. However, hydrogen escape was likely active on the early Earth, where its rate was significantly higher due to the higher XUV flux from the young Sun. This process is pertinent for gauging the amount of water that an Earth-like planet could lose over its history. As water vapour accumulates in the atmosphere (e.g., from the greenhouse effect), it is subjected to photolysis into hydrogen and oxygen, with the former susceptible to escape into space. The column density of hydrogen atoms in Earth's oceans today is 2×10^{32} H atoms m^{-2}, indicating that the lifetime of this hydrogen at the current energy-limited escape rate would be ~ 2 Gyr, which implies that twice the present water content might have been lost during Earth's entire history (Catling and Kasting, 2017, pg. 157).

In addition to thermal processes, a number of *non-thermal* mechanisms can power atmospheric escape. The most important, in connection with the host star's properties, involve interactions with *stellar winds* – namely, streams of energetic, charged particles (plasma) that propagate from the outermost regions of the star into interplanetary space. Various processes may trigger escape of atmospheric particles due to the energy transfer from stellar winds: these are generally hard to treat analytically and are best investigated by means of sophisticated multi-species magnetohydrodynamic (MHD) numerical simulations. A further complication stems from the planetary magnetic field, which could partially shield the atmosphere from stellar winds; we will touch on this specific aspect in Section 8.3.3.

For unmagnetised planets, it seems reasonable to assume that the particle loss per unit time via the above mode of escape (denoted by $\dot{N}_a$) is proportional to the number of stellar wind protons incident on the atmosphere. The latter can be estimated from the rate of mass lost by the star through winds ($\dot{M}_\star$), which yields a proton flux impinging on the planet equal to $\dot{M}_\star/(4\pi m_H a^2)$ at distance a, where m_H is the proton mass. Multiplying this with the planetary cross-sectional area of πR_p^2, we end up with

$$\dot{N}_a = \frac{\epsilon}{4} \left(\frac{R_p}{a} \right)^2 \frac{\dot{M}_\star}{m_H}, \tag{8.15}$$

where $\epsilon \sim 0.1$ is an efficiency factor derived from numerical simulations (Zendejas et al., 2010; Lingam and Loeb, 2019a).

A plausible consequence of this discussion is that planets with atmospheres comparable in mass to that of modern Earth, in the HZ of very low-mass M-dwarfs ($\sim 0.1 M_\odot$), might be fully depleted of their gas envelopes by the action of stellar winds on timescales of order 100 Myr.

8.2.4 Stellar flares and space weather

In addition to the steady flux of radiation and particles, stars can unpredictably emit short and intense bursts of energy known as *stellar flares*. These flares are probably produced by the conversion of magnetic energy via the breaking and re-connection of field lines. They are characterised by an emission of radiation in multiple bands (optical, ultraviolet, X-rays, and gamma rays), as well as energetic charged particles.

The most intense solar flares in modern history have energies of order 10^{25} J, but flares with energies larger than $\sim 10^{26}$ J (or *superflares*) have been documented for other stars (Airapetian et al., 2020). The frequency distribution of flares is proportional to an inverse power law of the flare energy (E_f): $\propto E_f^{-\gamma}$, with $\gamma \approx 1.5$ (Feinstein et al., 2020); the most energetic flares are the rarest along intuitive lines. This rate depends on the stellar type, age, rotation, and so forth. M-dwarfs are relatively more active and can exhibit much higher flaring rates than G-type stars like the modern Sun.

Flares (both on the Sun and other stars) are often accompanied by the expulsion of large masses of magnetised plasma, dubbed *coronal mass ejections* (CME). Coronal mass ejections may transport kinetic energies of $\sim 10^{28}$ J and masses of $\sim 10^{18}$ kg in active stars (Argiroffi et al., 2019). The majority of energetic stellar flares are also correlated with the production of *stellar energetic particles* (SEPs); the corresponding phenomena are called *stellar proton events* (SPEs). Collectively, the multifarious events attributable to flares – and, more generally, arising from stellar variability – are collectively designated as *space weather*, especially in the solar system context. The manifold impacts of space weather on exoplanet habitability are surveyed in Lingam and Loeb (2019a) and Airapetian et al. (2020).

The most obvious consequence of flares for planetary habitability is the enhanced flux of high-energy electromagnetic radiation reaching the surface during the event. This should be evaluated in conjunction with the effect of SPEs, which have long been known to interfere with the ozone layer (e.g., Crutzen et al., 1975), driving its reduction and therefore potentially increasing the surface UV flux to high levels. A SPE associated with a particular superflare of energy E_f might trigger depletion of the ozone shield according to the following semi-empirical formula (Lingam and Loeb, 2017b):

$$\Delta O_3 \sim 2.8\% \left(\frac{E_f}{10^{25}\,\text{J}}\right)^{9/25} \left(\frac{a}{1\,\text{AU}}\right)^{-18/25}, \tag{8.16}$$

where ΔO_3 is the percentage of ozone that is depleted. Repeated flares of $\sim 10^{25}$ J (and their SPEs) erupting once per week are predicted to induce complete ozone destruction on a rapid timescale of ~ 10 yr for an Earth analogue orbiting the M-dwarf GJ1243 (Tilley et al., 2019). Broadly speaking, planets around Sun-like stars seem less susceptible to ozone depletion than their counterparts around K- and M-dwarfs (Chen et al., 2021a).

We have already explored how the damage engendered from UV radiation may be quantified via the BEI in (8.12). This damage should be compared against a suitable *critical threshold* for different organisms, and is often defined by the limit at which only 10% of the initial population survive. The critical UV fluence for the model microorganism *Escherichia coli* at ~ 254 nm is 22.6 J/m^2, whereas the extremely radiotolerant *Deinococcus radiodurans* has a much higher threshold of 553 J/m^2 (Estrela and Valio, 2018).

The impact of flares on habitability may be assessed quantitatively by considering a planet in the HZ of an M-dwarf subjected to a superflare of $\sim 10^{27}$ J. Assuming no ozone layer is present, the overall UV radiation at the peak of the flare is about 50 times larger than the background value, but merely somewhat higher than the normal UV fluxes received on modern Earth (Segura et al., 2010). Therefore, the direct hazard posed by such an event to the biosphere might be negligible. In contrast, on a hypothetical Earth-like planet at 1 AU from a Sun-like star, the effective fluence (the BEI multiplied by the event duration) from a superflare of $\sim 10^{28}$ J lasting 580 s could be $\sim 2 \times 10^4$ J/m^2 for *E. coli* and $\sim 1.3 \times 10^4$ J/m^2 for *D. radiodurans* (Estrela and Valio, 2018), well above the critical threshold for both species, and lethal for most microorganisms on Earth. Similar conclusions potentially apply to Proxima b, the planet in the HZ of the closest star to our Sun, which was exposed to a flare of $10^{26.5}$ J in 2018 (Howard et al., 2018).

These results should, however, be interpreted with caution. For example, even without an ozone layer, the effective biological fluence would be drastically reduced if the organisms are underwater. The critical dose is also not a robust metric of the overall survivability of a given species to severe irradiation, as even a minute fraction of residual individuals could suffice to guarantee its persistence.

Moreover, as stated previously, the exposure to UV might have beneficial effects. Interestingly, the fact that M-dwarfs are more prone to flares and superflares may compensate for their relative paucity of UV (or even optical) radiation under normal conditions (Rimmer et al., 2018; Lingam and Loeb, 2019b). Likewise, SPEs can supply the requisite energy for prebiotic synthesis on M-dwarf exoplanets (Lingam et al., 2018), as well as on early Mars and on early Earth.

8.3 Planetary factors

Having delineated the main stellar factors that can impact planetary habitability, we will now move onward to investigate those factors that originate directly from the physical properties of the planet itself.

8.3.1 Mass

Mass has a direct bearing in regulating multiple planetary processes. The standard template for a habitable planet – to wit, a rocky world with surface water and sufficient atmosphere – is only to be expected in a certain mass interval. An obvious reason is that planetary mass defines surface gravity via $g = GM_p/R_p^2 \propto R_p^{1.7}$, where we have used the mass-radius scaling $M_p \propto R_p^{3.7}$. All things being equal, this may imply that low-mass planets require less energy for eroding their atmospheres, because the escape velocity obeys $v_e = \sqrt{2GM_p/R_p} \propto M_p^{0.36}$, regardless of particle mass.

Planetary mass governs a number of other factors, including internal energy dissipation and cooling (refer to Section 4.1). These processes may result in the early cessation of tectonic activity and volcanism, as well as the shutdown of the magnetic field. As we shall indicate in Section 9.3.1, a low mass (and the aforementioned consequences) was almost surely one of the dominant contributors to the rapid decline of Martian environmental conditions. At the other end of the spectrum, based on models and observational data, planets with radii larger than $\sim 1.5R_\oplus$ are unlikely to be rocky (Cloutier and Menou, 2020). This limit constrains the maximum mass of terrestrial planets through the prior mass-radius dependence.

Kopparapu et al. (2014) proposed that more massive planets with Earth-like atmospheres might have wider HZs than smaller ones. Less massive planets will conceivably possess thinner atmospheres and higher H_2O column depths, which causes the runaway greenhouse limit to be exceeded at an effective flux lower than modern Earth's, so that the inner edge of the HZ moves outward. The opposite behaviour is to be expected for larger planets: they would reach the runaway greenhouse limit only when the flux is higher than Earth. On the other hand, the change in the outer HZ edge is minimal, as an increase in albedo partly compensates for the higher greenhouse effect.

8.3.2 The carbonate–silicate cycle and plate tectonics

In Section 8.1.3, we alluded to a crucial negative feedback that stabilised Earth's climate on geological timescales, namely, the *carbonate–silicate cycle*.

Figure 8.4 shows a schematic representation of the cycle. To understand how it works, we can start with the formation of carbonic acid (H_2CO_3) from the dissolution of atmospheric CO_2 in rainwater:

$$H_2O + CO_2 \leftrightarrow H_2CO_3. \tag{8.17}$$

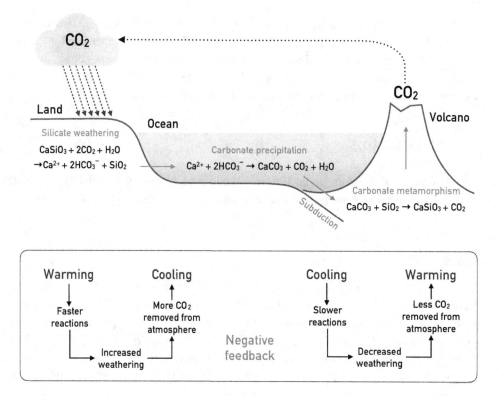

Figure 8.4 The components of the stabilising carbonate–silicate cycle.

The resultant mild *acid rain* is capable of dissolving continental silicate rocks. Conveniently, we can symbolise all these rocks through *wollastonite* ($CaSiO_3$) and write the reaction for *silicate weathering* as

$$CaSiO_3 + 2CO_2 + H_2O \rightarrow Ca^{2+} + 2HCO_3^- + SiO_2. \tag{8.18}$$

Subsequently, rivers and other channels transport the products of weathering to the oceans, where organisms like plankton can produce calcium carbonate shells ($CaCO_3$) via *carbonate precipitation*:

$$Ca^{2+} + 2HCO_3^- \rightarrow CaCO_3 + CO_2 + H_2O. \tag{8.19}$$

Most calcium carbonate is released when the organisms die and sink into the ocean, but some of it forms sediments on the seafloor. Adding the previous two equations together, we end up with the net result of silicate weathering and carbonate precipitation:

$$CaSiO_3 + CO_2 \rightarrow CaCO_3 + SiO_2. \tag{8.20}$$

To complete the cycle, CO_2 has to find a way to return back to the atmosphere. This is made possible by plate tectonics that enables the seafloor to move and sink down in the mantle (the process of *subduction*), reaching high temperatures and pressures; in these conditions, calcium carbonate combines with quartz (SiO_2) during the process of *carbonate metamorphism*:

$$CaCO_3 + SiO_2 \rightarrow CaSiO_3 + CO_2, \tag{8.21}$$

with the resulting carbon dioxide being eventually released into the atmosphere (i.e., outgassed) by way of volcanic activity.

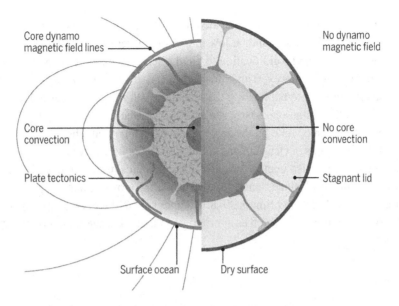

Figure 8.5 Plate tectonics versus stagnant-lid convection in terrestrial planets. (Credit: Shahar et al., 2019; reproduced with permission from AAAS)

This cycle induces a negative feedback on Earth's climate that stabilises temperature on timescales of ~ 1 Myr, but on longer timescales its status is less established (Arnscheidt and Rothman, 2022). The feedback could be envisioned as a thermostat that responds to changes in the climate forcing (consult the box in Figure 8.4). Planetary warming produces an increase in the weathering rates because of the speeding up of the relevant reactions (see Question 8.5), with the result being an enhanced removal of CO_2 from the atmosphere, and a consequent reduction of the greenhouse effect: the final outcome is a cooling of the climate. The opposite trend manifests when the planet starts to cool down, leading to less weathering, and therefore an increased greenhouse effect followed by warming.

It should be recognised that the functioning of the carbonate–silicate cycle is directly reliant on the existence of plate tectonics, that is, the relative movement of portions of the rigid outer shell of the planet comprising the crust and upper mantle (viz., the lithosphere); this topic is summarised in Korenaga (2013). In the solar system, the Earth appears to be unique among the rocky planets in this respect – although Venus may have (had) some plate tectonic activity – and it is thus important to ponder: what is the likelihood that plate tectonics is active on planets around other stars?

Plate tectonics broadly involves the convective dissipation of internal heat in the mantle (refer to Section 4.1.2). The latter can occur via an alternative path: *stagnant-lid convection*, wherein the entire external shell remains rigid (see Figure 8.5). Empirically, stagnant-lid convection is more common than plate tectonics in our solar system. This trend may be ascribed to the strong temperature dependence of mantle viscosity, which increases as one moves from the hotter internal layers towards the colder external layers, thus favouring the formation of a single rigid lid. To counteract this tendency, water could play a crucial role in weakening the lithosphere, and in providing the necessary lubrication for subduction to occur (Korenaga, 2013).

Numerical models also suggest that planetary mass might influence the sustenance of plate tectonics, with rocky worlds in the range $M_p \approx 1$–$5\,M_\oplus$ potentially favoured in this respect (Cockell et al., 2016). This may be qualitatively understood as follows. At high masses, the mantle viscosity could increase due to the elevated interior pressure, which in turn is expected to suppress plate tectonics. On the other hand, at low masses, the internal heat budget is typically lower, which may promote cooling and solidification to yield a stagnant lid (refer to Section 4.1.2). However, we caution that multiple factors exert an influence, such as the inventory of water, the planet's age (which affects the amount of internal heating that declines with time), the thermal and mechanical properties of the mantle, and so on. Lastly, stagnant-lid convection and plate tectonics are not mutually exclusive, as they can operate on the same planet at different epochs.

For all these reasons, a clear-cut answer to whether plate tectonics could be operative on another world is lacking. It is also unclear whether plate tectonics itself would be indispensable for climate stability. A number of studies analysed the habitability of stagnant-lid worlds on Gyr timescales (see, for example, Foley and Smye, 2018; Godolt et al., 2019), concluding that it might be possible if certain conditions are met, such as appropriate radiogenic heating and inventory of carbon dioxide and water.

8.3.3   Magnetic field

Another unique feature of the Earth, compared to other terrestrial planets in the solar system, is the presence of a strong intrinsic magnetic field. This is thought to be produced by a dynamo mechanism, whereby charged particles are transported by convective motions occurring within a partly molten metal core (Figure 8.5). Therefore, the presence of a magnetic field may be related to the amount of residual internal heat and its dissipation rate to some extent. If this hypothesis is correct, inferring plate tectonics (linked to internal heat) in exoplanets might be feasible through measuring their magnetic field strength, although this evidence would be ambiguous.

Several methods may be used for detecting magnetic fields; the best known of them entails observing *radio auroral emission* from the planet. This emission stems from a plasma instability known as the *electron cyclotron maser instability* (ECMI), with emission occurring at a frequency f_{ECMI}:

$$f_{ECMI} \approx 2.8\,\mathrm{MHz} \left(\frac{B_p}{10^{-4}\,\mathrm{T}} \right), \tag{8.22}$$

where B_p is the planetary magnetic field strength. Photons with wavelengths <10 MHz are reflected by the Earth's ionosphere; hence, space- or lunar-based observatories are ideal for detecting ECMI. The spectral flux density received at Earth depends on the power emitted at radio wavelengths, which is itself proportional to the stellar wind power and B_p (Zarka, 2007).

When planets possess an intrinsic magnetic field, the stellar wind plasma will be deflected (as it consists of charged particles), thereby creating a cavity-like structure called the *magnetosphere*. In general, although not always, it is thought that planets orbiting in the HZ of low-mass stars ought to have smaller magnetospheres, because of the increased stellar wind pressure (Vidotto, 2018), and also possibly due to the anticipated weaker intrinsic magnetic field. The latter is believed to arise from slower planetary rotation; we will tackle this aspect in the next section.

The presence of a magnetic field modulates the way a planet interacts with charged particles in the stellar wind and, consequently, has some impact on atmospheric escape. At first, it seems straightforward to conclude that a magnetised planet (e.g., Earth) would be more protected from the action of stellar winds, and therefore exhibit reduced atmospheric escape compared to unmagnetised planets. However, detailed theoretical models indicate that this conclusion is too simplistic (Gunell et al., 2018; Lingam and Loeb, 2019a). In fact, the atmospheric escape rate is not monotonic with the planetary magnetic field, and it may be higher for magnetised planets in certain cases, owing to the escape of ions from the polar cap regions. All in all, the presence of a planetary magnetic field is not sufficient (and perhaps not necessary) to mitigate atmospheric loss. Detailed numerical simulations of the interaction between the stellar wind, the magnetosphere, and the atmosphere are needed to assess the habitability of exoplanets in this context.

The planet's magnetic field also regulates the amount of charged particles hitting the surface, and the associated biological consequences. The fluxes of galactic cosmic rays and SEPs penetrating to the ground are indeed higher sans a magnetic field, but the result is heavily influenced by the atmospheric column density, which is likely the decisive factor for the protection of a planetary biosphere (Atri et al., 2013). Indeed, for Earth-like atmospheres, even setting the magnetic field to zero does not make a big difference; the surface flux merely changes by a factor of $\sim$2.

8.3.4 Tidal locking and orbital factors

A number of orbital factors can affect planetary habitability. The most conspicuous, dating back to the 1960s, is the possibility of *tidal locking* arising from star–planet gravitational interaction; once tidal locking happens, the rotation rate of the world essentially stays constant. The variation of gravitational acceleration a_g along the radial direction at distance r from a star of mass $M_\star$ is computed as:

$$\left| \frac{da_g(r)}{dr} \right| = \left| \frac{d}{dr} \left(\frac{GM_\star}{r^2} \right) \right| = 2\frac{GM_\star}{r^3}. \tag{8.23}$$

Therefore, for a planet of radius R_p at distance r (with $R_p \ll r$), opposite hemispheres will experience a differential acceleration a_{tide} given by

$$|a_{\text{tide}}| = \left| \frac{da_g(r)}{dr} \right| R_p = \frac{2GM_\star R_p}{r^3}, \tag{8.24}$$

resulting in a *tidal force* that acts along the radial direction and distorts the planet into an ellipsoid. If the spin and orbital angular velocities of the planet differ from each other, the friction ensuing from tidal oscillations along successive orbits dissipates spin energy. This process occurs until the spin and orbital angular velocities become equal to each other, corresponding to a *1:1 resonance*. This tidally locked state is also known as *synchronous rotation*. The Earth–Moon system is an example: as a consequence of tidal locking, the Moon always shows the same hemisphere to the Earth, because its rotational and orbital periods are synchronised.

The time necessary for a planet to become tidally locked to its host star depends on a number of parameters, including the former's elastic and mechanical properties, which are not tightly constrained. If we consider Earth-analogues in the HZ, planets around low-mass stars will enter synchronous rotation earlier, since they orbit closer to the star (refer to Section 8.1), which duly decreases the tidal locking timescale. Planets in the HZ of M-dwarfs with $M_\star \lesssim 0.4 M_\odot$ might be tidally locked on timescales of $\lesssim$1 Gyr (Barnes, 2017; Lingam and Loeb, 2021). Another consequence of star–planet tidal interactions is that internal energy dissipation causes the planetary eccentricity to decline with time, thus leading to *orbital circularisation*. The timescale for circularisation may be potentially $\lesssim$1 Gyr for planets in the HZ of stars with $M_\star \lesssim 0.15 M_\odot$ (Barnes, 2017).

The onset of synchronous rotation may affect planetary habitability via various avenues. The most obvious effect is that one side of the planet will be permanently irradiated (dayside), while the other will remain forever in the dark (nightside). In principle, this could create an extreme temperature gradient, with the permanent dayside too hot for life and the permanent nightside too cold. It should be noted, however, that this drastic outcome is only expected for planets without an atmosphere, whereas even modest atmospheres containing a sufficient amount of CO_2 and/or oceans can efficiently redistribute heat, mitigating the temperature contrast. Detailed numerical simulations of the climates of tidally locked planets have shown that the dayside–nightside temperature gradient is not necessarily a severe concern for habitability (Pierrehumbert and Hammond, 2019).

A more concrete issue may be the absence of starlight on the nightside, as this would make photosynthesis impossible on that hemisphere. While not a showstopper for habitability in general, this outcome can nevertheless reduce by about one half the amount of planetary surface available to host a complex biosphere. A further repercussion of tidal locking stems from the slowing down of planetary rotation. Theoretical studies suggest that a lower spin angular velocity might result in a weaker planetary magnetic field, since many dynamo models predict that the magnetic field increases monotonically with the rotation rate (Christensen, 2010). As we have seen in the preceding section, the latter could affect habitability in turn.

Besides tidal locking, other orbital factors may have a direct impact on habitability. For example, on planets with highly eccentric orbits, the incident stellar fluxes and resultant equilibrium temperatures are anticipated to periodically drop below freezing. The orbit-averaged equilibrium temperature $\langle T_{eq} \rangle$ is estimated to be (Méndez and Rivera-Valentín, 2017):

$$\langle T_{eq} \rangle = T_{e=0} \frac{2\sqrt{1+e}}{\pi} E\left(\sqrt{\frac{2e}{1+e}} \right), \qquad (8.25)$$

where $0 \leq e < 1$ is the orbital eccentricity, $T_{e=0}$ is T_{eq} for a perfectly circular orbit, and $E(\dots)$ is the complete elliptic function of the second kind.[4]

Obliquity, that is, the tilt of the rotation axis with respect to the orbital plane, also shapes planetary climate and, as a by-product, habitability. It is well known that the 23.4° obliquity of the Earth dictates its seasonal variability. High obliquity and/or large variations of this parameter would result in extreme and chaotic seasons, both of which seem inimical to habitability. On the other hand, large-amplitude fluctuations extend the HZ and exert additional positive effects according to some models (e.g., Colose et al., 2019). If the obliquity is always nearly zero, this might have negative consequences for habitability (e.g., reduced natural selection pressure). M-dwarf exoplanets experience strong tidal forces, and modelling indicates that this may lead to small (or zero) obliquities, except in special cases.

Finally, the possible role of the Moon in stabilising Earth's spin axis merits mention. Earth's obliquity currently varies slowly between 22.1° and 24.5°; the present obliquity is roughly midway between these two values. The timescale attributable to oscillations of the obliquity is around 4×10^4 years: one of the famous *Milankovitch cycles*. Early theoretical work suggested that, in the absence of the Moon, the obliquity of Earth would vary chaotically from 0° to 85° (Laskar et al., 1993). Such results are regularly invoked to argue that a large moon is necessary to maintain a habitable climate on an Earth-like planet (Ward and Brownlee, 2000). However, subsequent studies have diverged from

---

[4] https://mathworld.wolfram.com/CompleteEllipticIntegraloftheSecondKind.html.

this conclusion, demonstrating instead that Earth's obliquity may remain within a restricted interval of $\sim$20–25° over timescales of $\mathcal{O}(100)$ Myr, even sans the Moon (Lissauer et al., 2012).

8.3.5   Atmospheric composition

Last but not least, atmospheric composition can influence planetary habitability in many profound ways. As we have already explored some of these gases in other chapters of the book, we will restrict ourselves to sketching a few of the most relevant impacts on habitability.

- *Carbon dioxide:* CO_2 is the basis for all modes of carbon fixation (biological synthesis of organic compounds), such as photosynthesis. Moreover, the outer edge of the HZ is defined by the maximum greenhouse effect from CO_2, so that habitable planets farther from the star require higher levels of carbon dioxide. Because this gas is known to be toxic for terrestrial animals above certain limits, it has been speculated that a narrower 'habitable zone for complex life' might exist, only $\sim$30–50% the width of the standard HZ (Schwieterman et al., 2019). The same study points out that CO generated from CO_2 is toxic to many known organisms, and may rule out M-dwarf exoplanets as hosts for complex life, due to high CO levels. These conclusions, however, are undoubtedly too reliant on terrestrial life to set general limits on putative extraterrestrial biospheres.
- *Methane and molecular hydrogen:* Each of them are quite potent greenhouse gases and may extend the outer limits of the HZ, as discussed earlier. Both gases are, moreover, valuable from the standpoint of potential microbial metabolisms (see Section 7.1.3). For example, H_2 could be used by hydrogenotrophic methanogens for reducing carbon dioxide to methane. Likewise, methane can be subjected to oxidation by aerobic methanotrophs to generate products like formaldehyde.
- *Molecular oxygen:* In Section 6.3.3, we have elucidated the remarkable environmental, ecological, and evolutionary changes engendered by increase of O_2 on Earth. Some of the notable consequences include the following: (1) formation of the ozone layer, thus greatly reducing the surface flux of UV-B radiation; (2) production of novel minerals and (oxic) habitats for life; and (3) the debated emergence of complex multicellular motile life, because aerobic respiration yields a substantial amount of energy.
- *Molecular nitrogen:* Atmospheric nitrogen (N_2) on Earth is fixed into compounds such as ammonia (NH_3) through abiotic and biological avenues, which in turn enables its assimilation into living organisms (as N is a bioessential element). Ammonia is also a powerful greenhouse gas, although it may not survive for long in the atmosphere.
- *Hazes:* They form in hydrocarbon-rich atmospheres, and are very effective as shielding agents of UV light, thereby mitigating radiation damage.

We conclude this exposition of the stellar and planetary factors modulating continuous habitability by underscoring the close interconnections between most of them, as partly illustrated in Figure 1.2 of Section 1.1.2. This figure exemplifies the difficulties in resolving whether a planet is really suitable for life, when limited physical information is available.

8.4   Galactic factors

Having examined some stellar and planetary factors that can regulate habitability, we may now ask whether additional processes arising in the broader astrophysical milieu of the Milky Way could also play a role. This question has received some attention over the last two decades, with various studies attempting to identify locations in the Milky Way where the chance of finding habitable worlds is higher. In analogy with the notion of circumstellar habitable zone, this putative region was termed the *Galactic habitable zone* (GHZ) (Gonzalez et al., 2001; Lineweaver et al., 2004; Prantzos, 2008).

Quantitative estimates of the GHZ have attempted to analyse the relative importance of several high-energy astrophysical phenomena that may influence the development of habitable planets at varying galactic locations and epochs. We will briefly discuss the main ones hereafter.

8.4.1   Metallicity

The mass fraction of all elements heavier than H and He is, rather confusingly, dubbed the *metallicity* in astrophysics. A low metal abundance might hinder the formation of habitable rocky planets, as terrestrial planets might form only above a metallicity threshold of order 0.1 of the solar value as per theoretical models (Johnson and Li, 2012). However, increased metallicity may correlate with over-production of Hot Jupiters (Osborn and Bayliss, 2020), with consequent disruption of inner planetary orbits in the circumstellar habitable zone (Lineweaver et al., 2004). Moreover, stars with high metallicity are predicted to emit high fluxes of UV radiation, which could be detrimental for habitability (Shapiro et al., 2023). Based on these arguments, it is conventionally expected that habitable planets should preferentially exist at intermediate metallicities, that is, closer to the solar value.

The metallicity of protoplanetary discs is presumed to reflect that of the parent star, owing to which the latter can embody the accessible material for planetary formation. Observations imply that warm rocky planets occur at loosely constant frequency around stars in the range of 0.25–2.5 times the solar metallicity, whereas hot Jupiters tend to form preferentially at higher metallicities, with a probability distribution of roughly $P_J \propto 10^{3.4 Z_g}$, where Z_g is the gas-phase metallicity (Petigura et al., 2018). Note, however, that these findings may get revised by future surveys.

In the thin disc of the Milky Way, the metallicity increases over time as successive generations of stars are formed and, at any given epoch, decreases upon moving away from the centre; the relation is crudely proportional to the number density of stars (which are the primary birth sites for metals). This has led some authors to regard the habitability of the inner and outer Milky Way unfavourably, the former because of the excessive abundance of heavier elements, and the latter for the opposite reason.

8.4.2   Supernovae

As we have witnessed, an important factor affecting planetary habitability is the incident flux of high-energy radiation and particles capable of causing ionisation. This ionising radiation is not only produced by the host star but may also originate from other astrophysical phenomena; a review of the main sources can be found in Melott and Thomas (2011).

By far, the most studied deterrent to the habitability of the Galaxy is the risk of nearby supernovae (SNe) explosions, which is directly connected to the local density of stars. The temporal dependence is also important, as the SNe rate is currently a declining function of the cosmic time. The most credible impediment is expected to stem from Type II supernovae; these phenomena are the endpoint of stars with masses above $8 M_\odot$. While the so-called Type Ia supernovae could be individually more luminous and, therefore, more dangerous, they are less frequent as well, because their progenitors are white dwarfs in binary systems, which only represent 1% of the stars in the mass range $0.08M_\odot < M_\star < 8M_\odot$ (Gowanlock et al., 2011).

SNe rates in the Milky Way are subject to some uncertainty. The mean frequency is estimated to lie somewhere between 0.1 and 0.01 per year in the whole Galaxy (Branch and Wheeler, 2017, Chapter 1). An average rate can be obtained by taking the total number of events per year in the Galaxy (hence formulating a rate 'per star') and multiplying it by the mean stellar density at a given Galactic location. For example, the average long-term SNe rate within a spherical volume of radius d smaller than 100 pc around the Sun ($P_{SN,\odot}$) is estimated as (Melott and Thomas, 2011):

$$P_{SN,\odot}(< d) = 2 \times 10^{-6} \mathrm{yr}^{-1} \left(\frac{d}{100\ \mathrm{pc}} \right)^3 . \tag{8.26}$$

The SNe rate within a spherical volume at any other location is found by rescaling the above value with the appropriate local stellar density $n_\star$:

$$P_{SN}(< d) = P_{SN,\odot}(< d) \frac{n_\star}{n_\odot}, \tag{8.27}$$

where $n_\odot \approx 0.1\ \mathrm{pc}^{-3}$ is the star density at the Sun's location.

SNe explosions are violent enough to cause potentially catastrophic consequences well beyond the surrounding planetary system and the immediate stellar neighbourhood. The net energy release from a supernova (SN) explosion is typically around 10^{44} J, with ionising radiation arising from both neutral (electromagnetic) and charged (cosmic rays) components, potentially triggering climatic and biological consequences broadly akin to what we had sketched for stellar flares, even at considerable interstellar distances.

A SN at the distance of $d \leq 8$ pc can significantly affect the habitability of an Earth-like planet (Gehrels et al., 2003). This estimate is based on a reference value of 30% atmospheric ozone depletion from the SN blast, which basically causes a doubling of UV radiation from the host star reaching the surface, in turn conceivably triggering mass extinctions of complex land-based life (Melott and Thomas, 2011). State-of-the-art numerical models have, however, revised the distance for such a 'lethal' SN event to even larger distances of around 20 pc (Thomas and Yelland, 2023). A 'sterilising' event – amounting to the complete eradication of life from an Earth-like planet – would demand the full evaporation of oceans along with high-energy fluences well beyond the survivability threshold of extreme radiation-tolerant organisms. This scenario would probably only occur during a much closer event at distance $d \leq 0.04$ pc (Sloan et al., 2017).

Empirically assessing the influence of SNe explosions on mass extinctions on Earth is difficult, because stellar motions may cause their remnants to drift as much as 10 kpc in 100 Myr. However, based on estimated rates, it is plausible that a few events closer than 10 pc (i.e., powerful enough to cause mass extinctions) transpired in the past 500 Myr. Furthermore, there is solid evidence of at least one SN explosion within $\sim$50–100 pc having occurred during the Pleistocene (i.e., in the past 2.58 Myr), which is unlikely to have severely affected the biosphere (Thomas and Ratterman, 2020).

8.4.3 Gamma ray bursts

Gamma ray bursts (GRBs) are rare, extremely energetic and transient ($\lesssim$1 min) events, with a non-thermal power spectrum peaking at photon energies of $\sim$10 keV to $\sim$10 MeV. The total energy released during GRBs is $\sim 10^{42}$–10^{48} J (Kumar and Zhang, 2015). As a consequence of this massive energy output, GRBs can adversely impact planetary environments at vast distances from their source location. A 'lethal' GRB radiation dose for Earth-based life is conventionally envisioned as a fluence of 10^5 J/m^2 (Piran and Jimenez, 2014). Akin to SNe explosions, GRBs could engender ozone depletion and increased surface fluxes of ionising radiation.

Gamma ray bursts have only been observed on extragalactic scales, with a detected rate of ~ 1 per day. Scaling this frequency to the Milky Way is subject to substantial error bars. The GRB rate depends on both the star formation history and metallicity, among other factors, but the extent to which chemical evolution has reduced the GRB rate over time is not yet precisely settled (Gowanlock, 2016). Using available estimates of the GRB rate in the Milky Way, Piran and Jimenez (2014) calculated a 50% probability that a lethal GRB swept over the Earth in the past 500 Myr, possibly triggering one of the recorded mass extinctions. According to the same study, there is a much higher chance that such an event transpired in the inner Milky Way (95% probability at radii <4 kpc from the Galactic centre), thereby diminishing its habitability with respect to the outer regions.

However, due to the above uncertainties, GRB formation might instead be restricted to low metallicity settings, implying that GRBs preferentially occur in metal-poor outskirts of galaxies at recent epochs (Gowanlock, 2016).

8.4.4   Supermassive black hole

A crucial factor that could affect the habitability of the Milky Way is the presence of a *supermassive black hole* (SMBH) at its centre – known as Sgr A* – with an estimated mass of $4.1 \times 10^6 M_\odot$. The role of the SMBH in governing Galactic habitability has gained prominence in the last decade (e.g., Balbi and Tombesi, 2017). Based on observational constraints, the heated disc of material surrounding Sgr A* was almost surely an important source of ionising radiation during its accretion peak (the *active phase*), which ostensibly occurred ~ 8 Gyr ago, and might have lasted for $\sim 10^7$–10^9 yr (Marconi et al., 2004). A number of negative (and some positive) ramifications of Sgr A* during the active phase have been proposed:

1. Cumulative high-energy irradiation from the disc could have driven energy-limited hydrodynamic escape (see Section 8.2.3). Assuming a plausible value of order 10^{36} J s^{-1} for the XUV luminosity, the atmospheric mass loss can be found from (8.14). Balbi and Tombesi (2017) estimated that terrestrial planets at distances of 0.2–1 kpc from the Galactic centre may have lost an atmospheric mass comparable to that of Earth.
2. The biological consequences of the irradiation from Sgr A* during its active phase are hard to evaluate, as they depend on atmospheric and magnetospheric screening, as well as the radiation tolerance of the salient organisms. For an Earth-like world, radiation damage may be an issue only up to short distances of $\lesssim 10$ pc (Lingam et al., 2019), although this length scale could extend up to ~ 1 kpc for planets with very tenuous atmospheres (Balbi and Tombesi, 2017).
3. Besides the electromagnetic radiation emitted during the active phase, the production of high-speed particle winds and outflows is feasible. Ambrifi et al. (2022) explored the impact of these processes on atmospheric heating and escape, and triggering ozone depletion (akin to SPEs). This study concluded that these negative effects might extend to $\lesssim 1$ kpc.

Table 8.2 Comparison of select analyses of Galactic habitability

Study	Effect	Result
Gonzalez et al. (2001)	Metallicity	Solar neighbourhood
Lineweaver et al. (2004)	SNe & Metallicity	$R = 7$–9 kpc
Prantzos (2008)	SNe & Metallicity	Entire disc
Gowanlock et al. (2011)	SNe & Metallicity	$R \approx 2.5$ kpc
Spitoni et al. (2014)	SNe & Metallicity	$R \approx 8$ kpc
Morrison and Gowanlock (2015)	SNe & Metallicity	$R \approx 2.5$ kpc
Vukotic et al. (2016)	SNe & star formation	$R \gtrsim 16$ kpc
Forgan et al. (2017)	SNe & Metallicity	$R = 2$–13 kpc
Piran and Jimenez (2014)	GRBs	$R \gtrsim 4$ kpc
Gowanlock (2016)	GRBs	Entire disc
Balbi and Tombesi (2017)	Active SMBH	$R \gtrsim 1$ kpc
Pacetti et al. (2020)	TDEs	$R \gtrsim 1$ kpc
Ambrifi et al. (2022)	SMBH winds	$R \gtrsim 1$ kpc

Note: The second column encapsulates the main detrimental effects evaluated in the analyses, while the third column signifies the region (measured by the distance R from the Galactic centre) that may be regarded as most habitable in connection with the specific effect(s).

Even after the active phase ends, Sgr A* could still influence Galactic habitability. This may occur, for example, due to *tidal disruption events* (TDEs), namely: the breakup of stars that pass close by the SMBH. Tidal disruption events have a typical duration of ~ 1 yr, an estimated rate of $\sim 10^{-4}$–10^{-5} yr, and are accompanied by a burst of high-energy radiation. Pacetti et al. (2020) analysed the cumulative impact of TDEs on Galactic habitability and demonstrated that it is potentially comparable to the active phase of Sgr A*. In particular, planets within distances of ~ 0.1–1 kpc from the Galactic centre can lose Earth-like atmospheres over a period of ~ 4.5 Gyr, and might experience large-scale biological damage once every TDE.

Admittedly, many factors that may modulate habitability on the Galactic scale are subject to uncertainty. Hence, there is no consensus on what, if any, are the region(s) of the Milky Way that would be relatively conducive to life. Table 8.2 summarises a sample of key publications in this vein, along with their findings. Although the GHZ concept is still rather nascent, it is nonetheless important to investigate this topic in detail, as the GHZ could provide us with a better theoretical understanding of the possible distribution of habitable planets in the Milky Way, and it might help guide future observational surveys and interpret the outcomes.

8.5   Problems

Question 8.1: Around 3.8 Gyr ago, geological evidence supports liquid water on Mars' surface, as expounded in Section 9.2.1. In this epoch, based on (8.11), what was the luminosity of the Sun? Next, for this luminosity and the same Bond albedo as modern Mars, calculate the equilibrium temperature of Mars (in that past period). In order to host liquid water on its surface for this value of T_{eq}, estimate the required optical depth τ_{IR} in the infrared. Comment on whether τ_{IR} is close to unity.

Question 8.2: For blackbody radiation, write down the energy flux Q in terms of the equilibrium temperature T_{eq}. By perturbing this expression, determine the relationship between ΔQ and ΔT_s, as outlined in Section 8.1.3. Show that you recover (8.6) for the Earth's parameters.

Question 8.3: In Section 8.2.1, the importance of the stellar energy spectrum was highlighted. Planck's law describes the spectral radiance, which is proportional to the power per unit wavelength, and is delineated in (12.31) of Section 12.2.3. Plot this equation as a function of the wavelength λ between 200 and 1,200 nm for the following temperatures: (1) 2,500 K (late M-dwarf), (2) 3,500 K (early M-dwarf), (3) 4,500 K (K-dwarf), (4) 5,500 K (G-type star), and (5) 6,500 K (F-type star). From this plot, which stars emit more radiation in the above range, and why? What implications may this have for the habitability of rocky planets in their HZs?

Question 8.4: Estimate the flare energies that may be sufficient to cause ozone depletion of $\Delta O_3 \approx$ 50% and $\Delta O_3 \approx 100\%$ at orbital distances of $a = 0.1$ AU and $a = 0.5$ AU. Outline at least two positive and two negative consequences of stellar flares, along with the accompanying reasons.

Question 8.5: The classic paper by Walker et al. (1981) on the carbonate–silicate cycle posited $k_W \propto$ $\exp\left(\delta T_s / 13.7\,\text{K}\right)$ for the weathering rate k_W, where $\delta T_s \approx T_s - 285\,\text{K}$. If the average temperature of the Earth were to increase or decrease by 10 K relative to today, by what factor would k_W change? By citing peer-reviewed sources, discuss why this simple formula is not accurate, and provide an updated version of the weathering rate; also mention the dependence on other variables (e.g., pressure).

Question 8.6: In (8.25), select $T_{e=0} = 255$ K (i.e., Earth's equilibrium temperature), and plot the orbit-averaged equilibrium temperature $\langle T_{eq} \rangle$ as a function of the eccentricity for $0 \leq e < 1$. Compute and tabulate $\langle T_{eq} \rangle$ for: (1) $e = 0$, (2) $e = 0.3$, (3) $e = 0.6$, and (4) $e = 0.9$. Does $\langle T_{eq} \rangle$ increase or decrease monotonically as a function of e?

Question 8.7: Obliquity can strongly influence planetary climate (see Section 8.3.4). In George R. R. Martin's *A Song of Ice and Fire* fantasy novels, adapted into the (in)famous TV franchise *Game of Thrones*, the home planet exhibits chaotic seasons, with lengthy and unpredictable summers and winters. A couple of scientific groups sought to model this climate on April Fools' Day (e.g., Paradise et al., 2019). Read this publication, and briefly summarise the potential orbits, obliquity, and major findings.

If you are familiar with the series/books, you may comment on whether you find the paper a convincing in-universe explanation for the climate.

Question 8.8: Over a timescale of Δt, compute the total number of SNe that have occurred in a spherical volume of radius d by using (8.26), and solve for the value of Δt at which the number of these SNe in this volume equals unity. Next, by setting $\Delta t = 500$ Myr (nearly the length of the Phanerozoic eon), estimate the corresponding d. Is this distance sufficient to trigger a mass extinction, in light of the exposition in Section 8.4.2?

Question 8.9: In Section 8.1.4, we learnt that the HZ in the solar system is situated at $\sim$0.95–1.7 AU. Employ the VPL code for computing the HZ (Kopparapu et al., 2013, 2014),[5] and verify that the inner and outer limits are reproduced for an Earth-mass planet around the Sun. Next, specify $T_* = 2,600$ K, which is close to that of the well-known star TRAPPIST-1, encountered in Chapter 12, and repeat the calculations. Report your results, and also verify that three of the TRAPPIST-1 planets fall within the HZ.

---

[5] https://live-vpl-test.pantheonsite.io/calculation-of-habitable-zones/.

Question 8.10: Besides single stars, binary (and multiple) stars can also host HZs. Hence, you will utilise a HZ calculator developed by Müller and Haghighipour (2014).[6] Assume that the binary star system is approximated by Alpha Centauri A and B (merely 1.3 pc from the Sun). By inputting the relevant stellar parameters, and holding all other variables fixed at their default values, determine the HZ limit(s) and depict the output plot(s).

Even if a planet dwells within the HZ of a binary, it might experience some issues with long-term habitability. Describe two such potential drawbacks: make sure to cite peer-reviewed sources and justify your reasoning.

Question 8.11: The field of modelling exoplanetary climates has grown tremendously. You will use the LAPS Project (Turbet et al., 2016),[7] a simple 1D climate model, to perform some open-ended investigations. Choose a couple of the 'Initial Parameters' (e.g., surface pressure) and the 'Parameters' (e.g., flux), and then assign three discrete values to each of them. Hold all other parameters fixed at their standard values. Run the simulations over a single orbit, and stop them afterwards. Write a short report of your primary conclusions, and try to explain or motivate your results, where possible.

Question 8.12: The duration of the active phase of a SMBH is taken to be the *Salpeter time*, that is, the time needed to approximately double the black hole mass, $t_{BH} = M_{BH}/\dot{M}_{BH}$. You may assume that a fraction $\eta_{BH} \sim 0.1$ of the accreted mass is converted into energy and radiated away, with a luminosity given by $L_{BH} = \eta_{BH}\dot{M}_{acc}c^2$ ($\dot{M}_{acc}$ is the mass accretion rate), while the remaining falls into the black hole, thereby increasing its mass as $\dot{M}_{BH} = (1 - \eta_{BH})\dot{M}_{acc}$. If the SMBH luminosity is close to the *Eddington limit*, $L_{Edd} \simeq 1.26 \times 10^{31} (M_{BH}/M_\odot)$ W, estimate t_{BH} for Sgr A* ($M_{BH} \approx 4.1 \times 10^6 M_\odot$).

Question 8.13: Use the Eddington limit from Question 8.12, and suppose that the luminosity of Sgr A* was $0.1 L_{Edd}$ during the active phase. At what distance from the black hole would a starless planet receive a total effective flux of $S/S_0 = 1$? Could such a planet be considered potentially habitable? Elaborate on your answer using what you have learnt in this chapter.

Question 8.14: Aside from the factors modulating habitability addressed in this chapter, identify at least two others by conducting a (peer-reviewed) literature search, and briefly write about their impact(s) on habitability.

---

[6] http://astro.twam.info/hz/.

[7] http://laps.lmd.jussieu.fr/index.html.

Part IV

Astrobiological Targets

9　Mars

Mars is the most extensively studied planet in the solar system besides Earth. After the first successful flyby performed by NASA's *Mariner 4* mission in 1965, dozens of space missions[1] have targeted the Red Planet for direct observation, by deploying orbiters, landers, and rovers. Many are currently active, and more are expected to join them in the future.

The interest in Mars is multifaceted, but questions concerning its possible habitability and the presence of life have always been at the forefront. The excitement and controversy surrounding the (in)famous Martian 'canals', at the end of the nineteenth century, have left a lasting mark in popular culture, to the extent where 'Martian' has become synonymous with extraterrestrial life. Even after the dust settled on the premature enthusiasm for such early and unsubstantiated claims of Martians, the scientific community still perceived this planet as a potential abode for microbial life. Thus, in the 1960s, Mars unsurprisingly became a prime objective for closer investigations.

Many vital reasons render Mars an important astrobiological target. The planet sits squarely in the habitable zone of the solar system (see Section 8.1), and it was birthed from broadly similar initial conditions as Earth. Early Mars was conceivably akin to early Earth: whether it hosted independent abiogenesis is an open question whose importance cannot be overemphasised. The evolution of Mars later diverged radically from Earth, with a severe 'degradation' of its environmental conditions. This makes it a remarkable case study in comparative planetology, with major consequences for our understanding of habitability within and beyond the solar system.

In this chapter, we review the main results of the long history of Mars exploration, and the wealth of data accrued concerning its past and present geology, climate, and atmosphere. Based on this information, we delineate the pros for life on current Mars (availability of bioessential elements, energy sources, and water) as well as the cons (aridity, low temperature, radiation, salts); we also discuss some refugia that could harbour life on present-day Mars. Lastly, we sketch the past, present, and future of life detection on Mars, including the debated results from the *Viking* biological experiments and from the analysis of the ALH84001 meteorite.

9.1　Basic physical properties

The salient features of present-day Mars are summarised in Table 9.1. With a mean radius of 3,389.5 km, Mars is the second smallest planet in the solar system. As we shall see, this diminutive size had some bearing on the evolution of its environmental conditions, which changed dramatically from its formation – essentially complete by 4.5 Ga, as per multiple lines of isotopic evidence (Lammer

---

[1]　For an up-to-date list, consult: https://en.wikipedia.org/wiki/List_of_missions_to_Mars.

Table 9.1 Mars physical properties

Parameter	Value
Mean radius	3,389.5 km ($0.532R_\oplus$)
Mass	6.4171×10^{23} kg ($0.107M_\oplus$)
Surface gravity	3.73 m/s^2 ($0.38\ g_\oplus$)
Semimajor axis	1.524 AU
Eccentricity	0.0935
Solar irradiance	586.2 W/m^2 ($0.431S_0$)
Mean orbital period	686.98 Earth days (1.88 Earth years)
Mean solar day	88,775.2 s (24.66 hr, 1.027 Earth days)
Axial tilt	25.19°
Mean surface temperature	215 K
Surface pressure	0.00636 bar
Main atmospheric constituents	Carbon dioxide (CO_2) 95.1%
(by volume)	Nitrogen (N_2) 2.59%
	Argon (Ar) 1.94%
	Oxygen (O_2) 0.16%
	Carbon monoxide (CO) 0.06%

Note: Data primarily adopted from https://nssdc.gsfc.nasa.gov/planetary/factsheet/marsfact.html.

et al., 2021) – to the present epoch. The habitability of Mars throughout its history might have actually hinged more on its modest size than its distance from the Sun.

Perhaps most notably, the relatively low surface gravity (38% that of Earth) played a role in the loss of Mars' atmosphere, whose surface pressure today is about 0.6% of Earth's value, equivalent to the pressure at 35 km above the surface of Earth. The current lack of a global intrinsic magnetic field on Mars (Acuna et al., 1998) can also be partly traced back to the small size (refer to Section 4.1), which caused Mars to potentially cool down rapidly after its formation, shutting down the internal dynamo mechanism. Likewise, volcanic activity and plate tectonics are apparently absent today, possibly on account of this reason.

With a semimajor axis of 1.52 AU, Mars receives an effective solar flux about 43% that of the Earth. Mars' orbital parameters, including eccentricity and obliquity, have been subject to larger variability compared to Earth, with periods of $\sim 10^5$ years, driving climate patterns akin to *Milankovitch cycles* on Earth, albeit with larger oscillations (Armstrong et al., 2004). The atmosphere is predominantly composed of carbon dioxide, but it is too thin and dry to induce a significant greenhouse effect: the corresponding estimated warming is only ~ 5 K. The mean Martian surface emission temperature is around 215 K or -58°C (Haberle, 2013), similar to winter in Antarctica; because of the minimal heat capacity (i.e., thermal buffering) of the atmosphere, diurnal variations may exceed 80 K (Martínez et al., 2017). The pressure and temperature of Mars are very close to the triple point of water, as verifiable from Figure 11.1, thereby making it essentially impossible to host long-lived (pure) surface liquid water on modern Mars.

From a geological point of view, Mars' northern and southern hemispheres differ significantly, with the former displaying lower crater density, elevation, and crustal thickness (see Figure 9.1a). Proposed explanations for the origin of this so-called *Martian crustal dichotomy* include internal (endogenous) processes, a single mega-impact, or several overlapping large impacts; of this trio, the

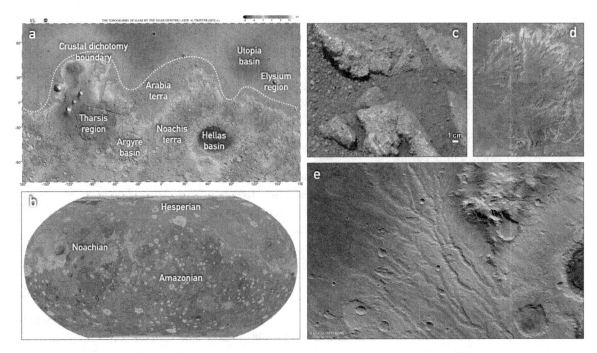

Figure 9.1 (a) Mars topography map from the Mars Orbiter Laser Altimeter (MOLA) data set from NASA's *Mars Global Surveyor*; (b) Distribution of terrain and related geological epochs; (c) Sedimentary deposits at the ≳3.6 Gyr old Gale Crater observed by NASA's *Curiosity* rover; (d) Deltaic lake deposits in Eberswalde Crater; (e) Dendritic valley networks in Paraná Valles (Credits: (a) NASA/MOLA Science Team; (b) USGS; (c) JPL-Caltech; (d) NASA/JPL/Malin Space Science Systems; (e) ESA/DLR/FU Berlin).

mega-impact hypothesis has garnered much attention (e.g., Andrews-Hanna et al., 2008). Because of the drastic evolution of Mars' properties over the course of the past ~4.5 Gyr, it is useful to demarcate the major geological periods (Meadows et al., 2020, pp. 155–156):

1. *Pre-Noachian*, from formation until ~4.1 Ga
2. *Noachian*, from ~4.1 Ga until ~3.7 Ga
3. *Hesperian*, from ~3.7 Ga unti ~3.0 Ga
4. *Amazonian*, from ~3.0 Ga to present.

Note, however, that these boundaries are subject to considerable uncertainty. The Martian surface is assigned the relevant period based on the impact record, especially the size-frequency distribution of craters, with the oldest terrain having the highest density of craters (Figure 9.1b). Besides the aforementioned crustal dichotomy, other large-scale features of Martian topography include the Tharsis bulge (a large region of elevated terrain near the equator), and the Hellas and Argyre impact craters in the southern hemisphere (see Figure 9.1a), all of which formed during the Noachian.

9.2   Instantaneous habitability

We will now review the various aspects pertaining to the instantaneous habitability of Mars, following the theme of Chapter 7.

9.2.1   Liquid water

As already mentioned, low temperatures and a thin atmosphere (i.e., low pressures) hinder stable surface liquid water on present-day Mars, but transient water is feasible (Wray, 2021). However, there is now conclusive evidence that Mars hosted large amounts of surface liquid water in the past, particularly in the Noachian and Hesperian (Wordsworth, 2016; McLennan et al., 2019). Most data are derived from surface geology and morphology, both from orbital observations and in situ analyses by robotic probes.

The existence of *dendritic valley networks* (Figure 9.1e) is regarded as compelling evidence in favour of surface water on early Mars. These are branching systems of channels that are fairly common in Noachian terrain, and resemble drainage basins found on Earth. Most of them are located in the equatorial regions and some extend for thousands of kilometres. The estimated minimal formation timescale for such networks is 10^5–10^7 years (Carr, 2012), and their morphology is strongly suggestive of a hydrological cycle driven by precipitation operating on ancient Mars (Schon et al., 2012). Geological data supports the presence of ancient Martian rivers until <3 Gyr ago, and conceivably as recently as <1 Gyr ago (Kite et al., 2019).

A second line of evidence for past liquid water on Mars is the existence of several hundreds of *crater lakes* connected to the valley network (Cabrol and Grin, 1999). The lifespan of individual lakes is estimated to be 10^4–10^6 years (Fassett and Head, 2008). The general scenario for their appearance involves the simultaneous presence of a cratered terrain and water flow through valley networks, enabling formation of lakes and ponds in the craters. If flow rates are high, water level will exceed the crater rim height, resulting in open lakes. Therefore, the relative frequency of closed-basin to open-basin lakes can constrain the hydrological cycle and precipitation rate. The predominance of closed-basin crater lakes observed on Mars may suggest that these features are not attributable to catastrophic floods or periods of intense warming driven by impacts (Fassett and Head, 2008).

The possible evidence for a past ocean covering part of the Northern hemisphere is more contentious: it was initially motivated by observations of the apparent remnants of ancient shorelines (Head et al., 1999) but this finding was questioned by other studies (Malin and Edgett, 1999). In general, the existence of a substantive lost ocean is not firmly established, and is perhaps at odds with the estimated global surface water inventory throughout Mars' history (Carr and Head, 2015). What is known of early Martian climate (discussed later in this chapter) seems to indicate that such a large body of water may not remain stable for long periods of time on the Martian surface.

In addition to morphological analysis of Martian terrain, geochemical observations from rovers and orbital probes support the conclusion that surface liquid water existed during the Noachian (for a review of Martian mineralogy, see Ehlmann and Edwards, 2014). The *Curiosity* rover discovered sediment deposits at Gale Crater (Figure 9.1c) that are interpreted as evidence of fluvial (i.e., river-based) erosion, albeit the chemical alteration caused by water is reportedly minimal and compatible with flows of relatively short duration (Williams et al., 2013); such conclusions are corroborated by orbital observations of open-basin lakes (Goudge et al., 2012). More generally, Noachian terrains contain iron- and magnesium-rich clays, or *phyllosilicates* (whose formation generally entails the presence of liquid water), as well as other aqueous minerals comprising sulfates, chlorides, and silicas (Poulet et al., 2005). It should be noted, however, that phyllosilicates may have also formed in water-poor conditions underneath the surface, and later transported to their current sites via crustal erosion (Ehlmann et al., 2011).

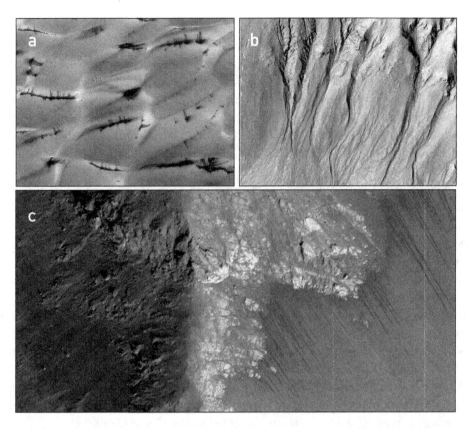

Figure 9.2 (a) Dark dune streaks; (b) Gullies; (c) Recurring slope lineae. (Credit: HiRISE images courtesy of NASA/JPL/University of Arizona)

Given the evidence for past liquid water on Mars, one may wonder what happened to this water inventory. As per the estimate by Jakosky and Treiman (2023, Table 2), the initial amount of surface H_2O was 685–1970 m, expressed in terms of the *global equivalent layer* (i.e., average water depth that would cover the surface of Mars), whereas the current surficial value might be 34 m (Carr and Head, 2015). The majority of H_2O was ostensibly either lost to space or sequestered in the crust (Scheller et al., 2021). Most H_2O on Mars is currently bound in minerals or frozen beneath the surface, but transient aqueous fluids have been invoked to explain active surface processes such as gullies and slope streaks, observed by orbital probes (see Figure 9.2). A review of contemporary liquid water on Mars can be found in Wray (2021). The abundance of (frozen) H_2O near the Martian surface ranges from a few per cent to tens of per cent by weight (Feldman et al., 2004).

The first notable class of variable surface features was observed by Mars orbiters around the polar regions (Figure 9.2a). Aptly christened *polar dark dune streaks*, these features consist of meter-size spots that appear when temperature rises after the winter season, and were interpreted by some authors as evidence for liquid brines, namely: high-concentration solutions of water and salts (both of which are known to exist on Mars) with lower melting points than pure water (Kereszturi et al., 2011). However, the observations of these streaks are also compatible with the downhill movement of solid debris triggered by CO_2 defrosting (Hansen et al., 2011).

Martian *gullies* (Figure 9.2b) are km-size elongated patterns on steep slopes on the walls of craters, pits, and valleys. They are observed at variegated terrain and locations, with prevalence at middle to high latitudes. Following their discovery, they were attributed to seepage and runoff from sources of liquid water at shallow depths beneath the surface (Malin and Edgett, 2000). This interpretation was subsequently questioned, as further data were consistent with dry flows of dust and silt, akin to snow avalanches on Earth (Treiman, 2003). Progress towards a resolution of the debate has been hampered by the lack of a terrestrial analogue to test the hypothesis of sediment transport by sublimating CO_2. A comprehensive review of Martian gullies and proposed explanations can be found in Conway et al. (2019).

Similar uncertainties surround the interpretation of *recurring slope lineae* (Figure 9.2c): these are dark streaks a few meters wide and up to a few hundred meters long, which appear during the warmest months at diverse latitudes and locations, including dunes, craters, and canyons. Partial seasonal temperature dependence suggests that they may be caused by melting of ground ice with briny composition (Chevrier and Rivera-Valentin, 2012), but conclusive spectroscopic evidence in support of this hypothesis is apparently missing, and alternative explanations based on the downhill motion of dry material are feasible (Dundas et al., 2017).

Perhaps the most intriguing scenario of liquid water on current Mars is the existence of subsurface reservoirs. In principle, they are identifiable via radar observations performed from orbit, since liquid water has a high dielectric constant that can differentiate it from rock or ice. Data from the MARSIS (Mars Advanced Radar for Subsurface and Ionosphere Sounding) instrument on board the *Mars Express* probe may constitute evidence for a stable body of liquid water over a $\sim$20 km wide area at $\sim$1.5 km depth beneath the Martian south pole (Orosei et al., 2018). The global temperature $T(z)$ at depth z (units of km) is approximately (refer to Question 7.6 of Chapter 7):

$$T(z) \approx 215\,\text{K} + z(\text{km}) \times 10\,\text{K/km}, \tag{9.1}$$

from which we discern that the predicted temperature at 1.5 km is $-43°$C. In reality, the temperature of the putative liquid inferred from the MARSIS data is likely around $-70°$C, which is close to the minimum melting point for perchlorate (ClO_4^-) brines. Should underground liquid water on current Mars be unambiguously confirmed and elaborated upon by future observations (as seems to be the case), this would reinforce the notion that subsurface environments might be promising targets in the quest for Martian life (Michalski et al., 2018).

9.2.2   Energy sources

Martian life based on chemotrophy (summarised previously in Section 7.1.3) has/had access to multiple electron donors and acceptors, as reviewed in Westall et al. (2015) and Clark et al. (2021). Even if both electron acceptors and donors are individually available on Mars, their co-location might be challenging (Cockell, 2014) except for select environments such as hydrothermal vents (Rucker et al., 2023), which would, in turn, limit the energy accessible to putative Martian microorganisms.

Examples of electron donors are H_2 from serpentinisation, hydrothermal activity, or radiolysis; Fe^{2+} and Mg^{2+} produced by alteration of magmatic rocks; and methane. Potential and confirmed electron acceptors include perchlorates, CO_2, SO_4^{2-}, Fe^{3+}, and organics, which could be reduced by the above molecular hydrogen. The available H_2 might also combine with atmospheric CO_2 in the well-known pathways of methanogenesis and acetogenesis (Section 7.1.3). Detection of nitric oxide (NO) by the *Curiosity* rover and nitrate in Martian meteorites like EETA 79001 (Stern et al., 2015;

Ansari, 2023) support adding nitrates – which has the capacity to facilitate anaerobic oxidation of iron, sulfur, methane, and carbon monoxide – to the list of available electron acceptors. Fermentation is another conceivable chemoheterotrophic pathway, because complex organics have been lately detected on the Martian surface (Ansari, 2023).

Moving on to phototrophy, various pathways are possible in principle, as elucidated in Section 7.1.4. For instance, anoxygenic photosynthesis with sulfur as electron donor may have been feasible. Oxygenic photosynthesis necessitates liquid water, owing to which its likelihood could be relatively low broadly since the end of the Noachian period.

9.2.3 Bioessential elements

We can now examine the Martian inventory of the basic elements (CHNOPS) needed to construct biomolecules (Cockell, 2014).

We begin our analysis with carbon, the backbone of biochemistry on Earth. Although multiple avenues for the synthesis of organic compounds exist (see Section 5.4), these molecules are also susceptible to destruction by UV radiation, cosmic rays, and reactions with oxidants in the soil (e.g., perchlorates). This complicates the task of estimating the actual concentration of organic material available on Mars; the abundance may be as high as $\sim$200 ppm (by weight) in mudstones from the Gale Crater region (Stern et al., 2022). A plethora of organics are attested on Mars such as hydrocarbons, halocarbons, organosulfur compounds, and carboxylic acids (Ansari, 2023, Table 1). Aside from organic molecules, two ready sources of carbon are atmospheric carbon dioxide and carbonates in Martian rocks.

Molecular hydrogen is a trace gas in the Martian atmosphere, and may be produced through several channels mentioned earlier. While molecular oxygen (O_2) is scarce in the Martian atmosphere, oxygen itself is accessible via various chemical species like CO_2, H_2O, and perchlorates. Previously, the availability of nitrogen in a form conducive for organismal uptake was considered a bottleneck, but results from the *Curiosity* rover support a nitrate abundance of $\sim$100–1000 ppm (by weight) at Gale Crater (Stern et al., 2015), suggesting that early Mars was not lacking in nitrogen for potential microbial ecosystems (Shen et al., 2019).

Finally, both phosphorus and sulfur are documented in various Martian minerals (Cockell, 2014). In particular, sulfate-dominated sedimentary deposits are prevalent on the surface of Mars, and the sulfur cycle exerts a major influence on the Martian geosphere, playing a role loosely reminiscent of the carbon cycle on Earth (Gaillard et al., 2013).

9.2.4 Physicochemical conditions

Next, we evaluate the numerous physical and chemical parameters that may have a direct negative impact on the possible survival of living organisms. These include high-energy radiation and charged particles, high salinity in brines, extreme dryness, low atmospheric pressure and surface temperature, acidic conditions, and oxidants in the soil. While such parameters would classify the Martian environment as extremely harsh and even lethal to many terrestrial organisms, it cannot be considered outright inhabitable or biocidal for *all* life, especially for extremophiles (Hallsworth, 2021). In fact, a vast array of laboratory and field studies conducted over the past decades demonstrated that some microbes might remain viable or in a dormant state in conditions analogous to that of present-day Mars.

For example, while UV radiation is lethal to most life, spore-forming microbes could survive prolonged UV irradiation when covered by either additional layers of spores or a thin film ($\sim$1 mm) of shielding (Mancinelli and Klovstad, 2000). Organisms isolated from rock-based microbial communities in Antarctica can apparently survive in a radiation environment similar to that of the Martian surface (Horneck et al., 2010; Onofri et al., 2018). In particular, pigmented microbes exhibit enhanced survival, because melanin confers protection against ionisation, and may even favour metabolic activity and growth by converting radiation to heat (Changela et al., 2021).

The UV radiation environment on the surface of Mars translates to a biologically effective irradiance for DNA damage (defined in Section 8.2.1) roughly three orders of magnitude higher than on the surface of Earth; however, this flux is fully suppressed at millimetre depths below the surface (Cockell et al., 2000). Accounting for the more penetrating solar energetic particles and galactic cosmic rays, the total measured dose rate of ionising radiation at the Martian surface is 76 mGy/year (Hassler et al., 2014), which decreases substantially at subsurface depths of $\gtrsim$1 m; note that 1 Gy is the dose corresponding to 1 J of ionising radiation absorbed by 1 kg of biological material. This flux is ostensibly tolerable by radiation-resistant microorganisms such as *Deinococcus radiodurans* (McKay, 2014).

Likewise, desiccation and low temperatures are not necessarily sterilising, and might even function as a preservation mechanism for microbial cells, although it is not clear at the moment whether metabolism could be sustained under such conditions (Hallsworth, 2021). While many Martian environments are either neutral or alkaline, certain local acidic conditions (pH 2–5) are implied by the existence of sulfate minerals, indicative of SO_2 outgassing from past volcanic activity (Gaillard et al., 2013). Even the lowest or highest inferred pH, however, does not lie outside the boundaries of known life.

Due to the fact that brines may constitute the only liquids viable on present-day Mars,[2] the survival of life in such high-salinity fluids acquires extra relevance. As stated in Section 7.2.3, microbial survival is sensitive to the specific salts involved, that is, their chemical properties and concentrations. For example, while brines containing sulfate salts tend to be kosmotropic, other salts like $MgCl_2$ are chaotropic (see Section 7.2.3). Thus, drawing general conclusions about the habitability of putative Martian brines is difficult. With that said, the low water activities and temperatures of certain surface and near-surface (i.e., a few centimetres deep) brines are seemingly beyond the tolerance levels of Earth-based life (Rivera-Valentín et al., 2020).

To sum up, a general assessment of Martian instantaneous habitability is challenging. Most, if not all, of the Martian surface is likely not habitable today. Yet, it is conceivable that local habitable settings may exist even in the current harsh conditions of Mars, especially in subsurface environments with access to aqueous fluids. Regions of Mars where terrestrial organisms could replicate, as well as regions interpreted to have a high potential for harbouring extant Martian life, have been designated as *Special Regions*. The identification of Special Regions is invaluable, not only as targets for future surveys of past and present Martian life, but also to prevent contamination from terrestrial microorganisms. The Special Regions of Mars are extensively described in Rummel et al. (2014) and Space Studies Board (2018).

---

[2] We caution that these brines are not necessarily capable of remaining stable for extended temporal intervals (Chevrier et al., 2022).

9.3   Continuous habitability

As already highlighted, there is plenty of evidence that the Martian environment underwent radical changes over the course of the past ∼4.5 Gyr. A *heuristic* timeline of some relevant events is depicted in Figure 9.3. It is, therefore, important to ask: was Mars more habitable in the past than it is today? If this was indeed the case, for how long did the favourable conditions last? Detailed observations support a positive answer to the first question, but the second is relatively difficult to address. To illustrate and unpack the problem, we will highlight the salient factors involved.

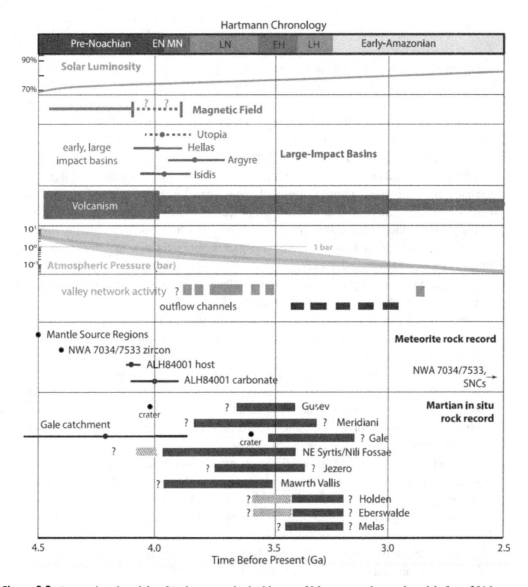

Figure 9.3 A tentative chronicle of major events in the history of Mars as per data and models from 2016. (Credit: Ehlmann et al., 2016, Figure 3b; reproduced with permission from John Wiley and Sons)

9.3.1   The evolution of Martian atmosphere

As seen earlier, multiple lines of evidence signify surface liquid water during the Noachian (and Hesperian) period. However, it is difficult to ascertain whether these data imply that past Mars was warm and wet/semi-arid over an extended period of time (Ramirez and Craddock, 2018), or rather that the planet was cold (i.e., much below freezing) and dry for the majority of its history, with short and episodic windows conducive for liquid water, perhaps linked to sporadic events such as impacts or increased volcanic activity (Scherf and Lammer, 2021; Wordsworth et al., 2021). The dichotomy between a warm-wet and cold-dry Mars may, however, be simplistic, thereby warranting a nuanced picture (McLennan et al., 2019).

This uncertainty chiefly stems from hurdles in reconstructing the evolution of Mars atmosphere, and trying to reconcile some aspects of the geological record (e.g., morphological evidence of large-scale liquid water bodies) with climate models and geochemical proxies. In particular, direct clues offer sparse insights into the state of the Martian atmosphere during the enigmatic pre-Noachian period (Scherf and Lammer, 2021). However, growing evidence from cratering history, Martian meteorites, geochemical data, and numerical simulations indicates that Mars had a denser atmosphere during the Noachian, with total pressure $\sim$0.5–2 bar, of which CO_2 might have approached $\lesssim$1 bar (Kurokawa et al., 2018; Kite, 2019; Thomas et al., 2023).

Higher X-ray and extreme ultraviolet radiation ($\sim$10–90 nm) and solar wind fluxes produced by the early Sun were jointly primary drivers of atmospheric loss across most of Mars' history, causing hydrodynamic escape as well as non-thermal processes (outlined in Section 8.2.3). The small mass of the planet explicitly facilitated the action of these mechanisms. Furthermore, impacts by sufficiently large asteroids and comets could have also played a key role in stripping away a portion of the atmosphere (Melosh and Vickery, 1989), while at the same time contributing a source of volatiles. Gauging the atmospheric evolution is challenging, because neither the sinks nor the sources (e.g., volcanic outgassing) are precisely understood.

Giant impacts potentially dominated loss at earlier times (i.e., in the pre-Noachian), when bombardment was much higher, whereas thermal and non-thermal escape were perhaps paramount from the Noachian onward (Jakosky and Phillips, 2001). Impacts deplete the atmosphere by producing a hot vapour plume that expands faster than the escape velocity (Melosh and Vickery, 1989). This process expels gases into space from the entire air column, affecting all species regardless of their mass, and is thus not expected to modify isotopic ratios. On the contrary, hydrodynamic escape and *sputtering* (i.e., ejection of particles after bombardment by solar wind ions) alter the isotopic ratios, since lighter species are preferentially lost to space.

This distinction allows us to differentiate atmospheric loss due to impacts from that driven by solar wind and radiation. Isotopic fractionation is documented for multiple atmospheric species on Mars (e.g., D/H, $^{13}C/^{12}C$, $^{15}N/^{14}N$, $^{18}O/^{16}O$ and $^{38}Ar/^{36}Ar$). Collectively, the evidence suggests that $\gtrsim$50% of the initial atmosphere became depleted by thermal and non-thermal processes (Jakosky et al., 2017; Lichtenegger et al., 2022). Theoretical models based on crater densities estimate an additional loss of $\lesssim$50% of the atmosphere due to impacts (Jakosky, 2019). Hence, the combined effects of impacts, solar radiation, and solar wind might have enabled the loss of $\gtrsim$90% of early Martian atmosphere to space (Brain and Jakosky, 1998).

Our current understanding of Martian loss rates are derived from state-of-the-art observations by the *MAVEN* spacecraft and sophisticated multi-species plasma modelling (Dong et al., 2018;

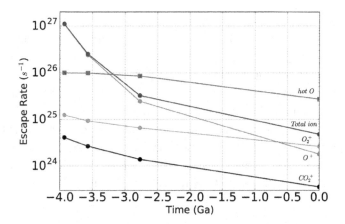

Figure 9.4 Estimated ion escape rates and photochemical escape rate (for hot oxygen) over Martian history. (Credit: Dong et al. 2018, Figure 4; ©AAS, reproduced with permission)

Jakosky, 2019). The inferred ion escape rates, which are the dominant channel for atmospheric loss (Lichtenegger et al., 2022), on Mars have varied notably over time, ranging from $\sim 10^{27}$ s^{-1} at ~ 4 Ga to $\sim 5 \times 10^{24}$ s^{-1} in the current epoch, as illustrated in Figure 9.4. An analytical model proposed by Dong et al. (2018) for the total ion escape rate $\dot{N}_{\mathrm{ion}}$ displays the power-law scaling,

$$\dot{N}_{\mathrm{ion}} \sim 5 \times 10^{24}\,\mathrm{s}^{-1} \left(\frac{t_M}{4.5\,\mathrm{Gyr}} \right)^{-2.33}, \qquad (9.2)$$

where t_M is the instantaneous age, that is, time lapsed after Mars formed by ~ 4.5 Ga. While hydrodynamic escape can function efficiently in a H$_2$-dominated atmosphere, this mechanism is thought to have become subdominant since at least the end of Noachian, with non-thermal processes (e.g., ion escape) collectively taking the lead thereon (Scherf and Lammer, 2021).

It has been estimated that the total atmospheric loss engendered by the aforementioned mechanisms, when integrated over the entire history of Mars, was perhaps $\lesssim 1.5$ bar (Jakosky, 2019; Lichtenegger et al., 2022).

9.3.2   Magnetic field

The evolution of the atmosphere is linked in complex ways to the presence (or absence) of a global magnetic field, as sketched in Section 8.3.3. A substantial magnetosphere could protect the atmosphere from solar wind particles, and the consequent erosion by ion-mediated processes. However, magnetic field lines can also concentrate a larger flux of the solar wind into the auroral region around the poles, driving enhanced atmospheric loss compared to the case sans magnetosphere. If we postulate that a global magnetic field is beneficial (which is contested), its absence on Mars might have caused: (1) elevated atmospheric escape and (2) higher surface flux of radiation and charged particles; the latter could promote the production of oxidants such as perchlorates and the destruction of organics.

Pinpointing the exact timing of the magnetic field cessation is arguably critical for deciphering the evolution of the Martian atmosphere and climate, and therefore for identifying the habitability interval. Some constraints can be obtained from the analysis of residual crustal magnetic fields and

Martian meteorites. The majority of publications favour a Martian dynamo shutdown by the end of the Noachian, although modern studies appear to indicate otherwise; this topic is reviewed in Mittelholz and Johnson (2022). The termination of the dynamo beyond the Noachian is loosely consistent with the overall degradation of its habitability potential thereafter.

Progress in understanding the genesis and decline of the Martian magnetic field necessitates a deeper understanding of the internal dynamo mechanism. During the first few hundred Myr after Mars' formation, the dynamo was probably driven by internal heat sources (refer to Section 4.1.2). The subsequent evolution, however, would have been sensitive to factors like core size and composition. In the case of present-day Mars, recent analyses of seismic data from the *InSight* mission suggest a mostly or fully liquid metal core, with a radius of $1{,}835 \pm 55$ km (Le Maistre et al., 2023).

9.3.3 Plate tectonics

We have discussed in Section 8.3.2 how plate tectonics are essential in regulating Earth's climate, and we highlighted its possible influence on planetary habitability. However, we reiterate that the relationship between tectonics and habitability is complex, and it is far from clear that maintaining plate motions and subduction is imperative to ensure suitable conditions for life. It is, nonetheless, important to ask whether this mechanism was ever active on Mars. Geological formations exhibiting features of fault systems and plate boundaries do exist (Changela et al., 2021), but there is no conclusive evidence that plate tectonics operated at any time (Lapôtre et al., 2022). Due to its small size, Mars may have cooled down rapidly via stagnant lid convection (refer to Section 4.1.2). Absence of plate tectonics is also congruous with the combination of Mars' thick crust and large molten core.

The lack of plate tectonics could have deprived Mars of a major mechanism for exchanging bioessential elements such as hydrogen, carbon, sulfur, and nitrogen between the atmosphere and the interior. Geochemical cycles would conceivably operate chiefly in one direction, with volatiles outgassed via volcanism either permanently trapped in the crust after weathering or lost into space. One implication of the limited circulation of material from the Martian interior to the exterior might be a diminished global redox gradient; currently, the surface of Mars is highly oxidised, as attested by its striking red colour. This trend may affect habitability, as life on Earth harnesses redox gradients as an energy source (see Section 7.1). However, alternative routes can circumvent this limitation – for example, the action of UV radiation on the first $\sim$10 cm of the surface (Changela et al., 2021).

Another key consequence of the absence of plate tectonics and its related geochemical cycles may be the absence of a feedback mechanism for mitigating long-term climate fluctuations: CO_2 outgassed from the interior via volcanism was not recycled by subduction, and was eventually lost to space. Therefore, the thermal buffering provided by the greenhouse effect would be missing. In turn, this aspect poses a conundrum: how was the climate of early Mars warm enough to be compatible with the evidence of past liquid water on its surface? We examine this question in the following section.

9.3.4 Climate of early Mars

Understanding the past climate of Mars is challenging: the subject is thoroughly reviewed by Wordsworth (2016) and Kite (2019). A major open conundrum can be expressed as follows (refer to Question 9.3): while the geological record unambiguously supports the presence of liquid water on the Martian surface during the late Noachian, it is unclear how a warmer climate could be maintained, especially given that the luminosity of the young Sun was merely around 75% of its present value at ~ 3.8 Ga. Even if all the incident solar radiation was absorbed by the Martian surface (i.e., zero albedo), the equilibrium temperature at the time would have been ~ 65 K below the melting point of water at 1 bar. Thus, the greenhouse effect to sustain liquid water on the surface of Noachian Mars had to be about twice that of present-day Earth, and ~ 10 times that of present-day Mars.

As witnessed earlier in this chapter, models and isotopic proxies indicate that early Mars had a thicker atmosphere. Hence, an appealing solution to the problem delineated above would be to assume that a denser atmosphere was responsible for the requisite greenhouse. At first glimpse, this solution does not seem unrealistic. It is estimated that $\lesssim 2$–3 bar of CO_2 might have entered the atmosphere of early Mars via outgassing and impacts, whereas a similar amount ($\lesssim 1.5$–3 bar) of CO_2 could have been lost over time through sequestration in the crust, impact ejection, and erosion by solar wind and radiation (Jakosky, 2019, Figure 3). The partial pressure of CO_2 (P_{CO_2}) as a function of time t_M may obey (Wordsworth et al., 2021, equation 17):

$$P_{CO_2} \approx 2.1 \, \text{bar} \left[1 - \tanh\left(\frac{t_M - 0.8 \, \text{Gyr}}{0.67 \, \text{Gyr}} \right) \right]. \tag{9.3}$$

However, the above numbers should be interpreted with due caution on account of the accompanying uncertainties.

Yet, even a ~ 1 bar CO_2-H_2O cloudless atmosphere may not have sufficed for warming early Mars (Kasting, 1991; Ramirez and Craddock, 2018). The reasons are as follows: (1) a high CO_2 concentration increases the albedo of the planet; and (2) CO_2 condenses into dry ice clouds at low temperatures and, at increased pressures, this phenomenon could result in the complete collapse of the atmosphere. Alternative gases, such as methane or hydrogen, have been proposed to provide the desired greenhouse effect. These gases cannot be discounted, especially because giant impacts might have created transiently reducing conditions by injecting H_2 into the atmosphere (Steakley et al., 2023), along the lines described in Section 4.2.3. Besides atmospheric gases, the putative existence of water ice clouds is expected to have contributed additional greenhouse warming (Kite et al., 2021).

On the other hand, geological evidence constrains the duration of liquid water flows and bodies in the Noachian and post-Noachian periods (Kite, 2019, Table 1), which is apparently compatible with a cold and arid climate interspersed with episodes of surface liquid water. This episodic warming might be linked to volcanic outgassing or impacts, among other factors, as demonstrated in a broad stochastic model developed by Wordsworth et al. (2021). For example, sporadic volcanic emissions of sulfur-bearing greenhouse gases such as SO_2 or H_2S could, in principle, warm the climate enough to permit transient liquid water (Halevy and Head, 2014). The extent to which this scenario affected Martian habitability is unresolved, although the diverse array of psychrophiles on Earth, as well as the potential global glaciation events in Earth's climate record, may leave room for optimism.

In summary, assessing the evolution of Martian habitability remains a daunting task filled with gaps in our knowledge, and different trajectories are compatible with the existing data. The classification scheme outlined by Cockell (2014) comprises five possible scenarios (see Question 9.7). One such example is that Mars was originally inhabited, life went extinct, and the planet became uninhabitable. Each of these tracks may be experimentally testable, and a major goal of astrobiological research entails deciding which paths are the most feasible. Advances in the last decade have lowered the likelihood of some trajectories, and boosted others.

We will now discuss the endeavours to unearth signatures of life on Mars, whether it be extant or extinct (see also Chapter 13).

9.4   The quest for Martian life

Based on what we learned so far, it seems tenable that Martian habitability was not a long-term and global planetary feature, but rather a transient occurrence limited to specific locales. Our quest for evidence of extant and/or extinct life on Mars, therefore, has to begin by identifying plausible past and present habitats, and then branch out into life-detection studies.

9.4.1   Past and current Martian habitats

A promising window for the potential emergence of life on Mars was early in its history: from the pre-Noachian at $\lesssim 4.2$ Ga (Goodwin et al., 2022) until at least the end of the Noachian, when surface liquid water was relatively common, and somewhat conducive settings for abiogenesis persisted on Myr timescales. Viable environments for the assemblage of protocells (requiring simultaneous existence of liquid water, prebiotic molecules, energy sources, etc.) were presumably limited in number and spatially separated, but the habitats that could host already established life may have been more numerous, as they involve less stringent constraints both spatially (regions as small as few microns) and temporally (minimal intervals of hours), a condition termed *punctuated habitability* (Westall et al., 2015).

Given the limitations and the heterogeneous nature of putative Martian habitats, life would have probably faced severe evolutionary constraints, and thus remained chemotrophic and anaerobic, akin to early organisms in similar abodes on Earth (Westall et al., 2015). The best places to uncover traces of past Martian life could be those regions that once had plenty of liquid water, and availability of prebiotic chemicals and free energy, such as ancient hydrothermal systems (see Figure 9.5). The modest geological activity of the planet might preserve some imprints of bygone life in rocks.

With the degradation of surface conditions, putative Martian microbes could have become restricted to narrow niches in the (sub)surface. For instance, chemoautotrophs adapted to rock-based environments on Earth such as those in hot and cold deserts (e.g., the McMurdo Dry Valleys of Antarctica) potentially constitute a credible analogue for microbial communities that might have persisted on Mars, even possibly until now (Horneck et al., 2010; Vago et al., 2017; Onofri et al., 2018). Environments that may currently host Martian life include (Hays et al., 2017; Carrier et al., 2020):

1. *Caves*: documented on Mars, they can provide direct access to the subsurface, as well as stable temperature and humidity, while at the same time protecting organisms from solar radiation and desiccation.
2. *Deep subsurface*: this region is conceivably the most widespread potentially habitable environment on Mars (Michalski et al., 2018), having existed at least since the Noachian. According to models, liquid water reservoirs are feasible at depths of a few kilometres (refer to Question 7.6 of Chapter 7). Subsurface life on Earth is found at depths of >4 km in the continental crust, isolated from surface waters for $\sim 10^7 - 10^9$ yr.
3. *Ice*: concentrated liquid water veins exist in ice and permafrost at temperatures as low as $-60°C$, offering a possible refuge for organisms that may survive by deriving energy and chemicals from

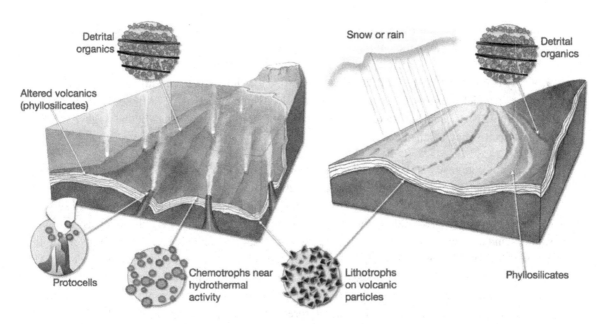

Figure 9.5 A schematic depicting putative habitats on early Mars based on the minimal requirements for habitability. (Credit: Vago et al., 2017, Figure 2; CC-BY-NC 4.0 license)

ions and salts (also see Section 10.4.2). Ice is almost ubiquitous in the near-surface of Mars, perhaps with the exception of equatorial regions.

4. *Salts*: brines with dissolved nutrients support extremophile communities on Earth, and could provide a viable habitat on Mars.

In theory, if not in practice, all these habitats are accessible to direct investigation by automated probes in future explorations.

9.4.2   The state of life detection on Mars

The only experiments explicitly designed to seek evidence of microbial life on Mars were performed in 1976, as part of the first successful landing mission on the planet. Two identical probes, the *Viking* 1 and 2, collected soil samples (at depths of a few cm) at separate locations via a robotic arm, and subjected them to four distinct tests. These are succinctly delineated below, mirroring the order in which they were executed (Klein, 1979):

- *Gas chromatograph/mass spectrometer (GCMS):* This separated the various chemical components of untreated Martian soil and analysed their molecular weight. Although not a biological experiment, the GCMS measurements were crucial for interpreting the results of the other three tests.
- *Gas exchange (GEX):* The soil sample was incubated for days in an inert helium atmosphere, after exposure to a mixture of organic and inorganic nutrients (as well as pure water). The experiment then looked for relevant gases released by the sample, including O_2, CO_2, N_2, H_2, and CH_4.
- *Labelled release (LR):* The soil sample was incubated over a timescale of days after mixing with a nutrient solution containing seven organics (formate, glycolate, glycine, D-alanine, L-alanine,

D-lactate and L-lactate) that were all labelled with the radioactive isotope ^{14}C. The release of radioactive carbon from the sample was subsequently sought as an indicator of potential metabolic activity taking place in the soil.

- *Pyrolytic release (PR):* The soil sample was exposed to water, light and a simulated Martian atmosphere, in which the gases CO and CO_2 were labelled with radioactive ^{14}C. After days of incubation, the sample was heated to 625°C, and then analysed for radioactive carbon-14 that might have been fixed due to photosynthetic activity.

The general consensus is that the *Viking* mission did not yield conclusive evidence of biological activity on Mars (see, e.g., Schuerger and Clark, 2008).

Despite the fact that all three biological experiments initially produced positive results (a surge of oxygen observed in GEX, radioactive carbon detected in both LR and PR), the analysis of control samples and additional contextual considerations suggest that a chemical explanation involving the action of oxidants in the soil is preferred over a biological one. The failure of the concurrent GCMS experiment to detect organic compounds in the samples at a level of a few parts per billion (well below the limit expected if any active, or even dead, organisms were present) was, at that time, a major factor in favouring a non-biological interpretation. It should be noted that the harsh properties of the Martian soil, as well as its unusual chemistry (such as hosting strong oxidants), were unknown or poorly understood when the *Viking* experiments were designed. This issue has complicated the evaluation of the results, and left some room for ambiguities that are not fully resolved to this day (see, e.g. Levin and Straat, 2016).

Two decades after the *Viking* mission, the next epochal event pertaining to putative Martian life stemmed from the analysis of the meteorite Allan Hills 84001 (ALH84001) (McKay et al., 1996; Treiman, 2021). This 1.9 kg rock, unearthed in Antarctica's Allan Hills in 1984, is one of ~300 meteorites retrieved on our planet identified as having a Martian origin.[3] Collectively termed *SNC meteorites*, they are currently the only samples from Mars available for direct analysis in laboratories on Earth. ALH84001 has been dated to $\gtrsim$4 Ga, and thus originated close to the threshold of the pre-Noachian period. It was likely ejected from Mars ~15 Myr ago, and landed on Earth ~1.3×10^4 yr ago, remaining in the Antarctic ice ever since.

After studying samples from ALH84001, McKay et al. (1996) highlighted several aspects that they interpreted as cumulative evidence in favour of possible relic biogenic activity (i.e., past Martian life), as described below.

1. the presence of carbonate globules;
2. an age of the globules younger than that of the igneous rock, indicating formation in an aqueous environment rather than from melting;
3. the presence of polycyclic aromatic hydrocarbons (encountered in Section 2.4.2) associated with the globules;
4. the presence of magnetite and iron sulfides, with structures similar to those found in magnetotactic bacteria on Earth (i.e., which are capable of orienting themselves along the Earth's magnetic field);
5. the presence of morphological features, revealed by microscope images, resembling those of fossilised terrestrial microorganisms.

[3] An up-to-date list can be generated by specifying 'Martian meteorites' in: www.lpi.usra.edu/meteor/.

A thorough review of the arguments for and against the hypothesis of past microbial life in ALH84001 is furnished in Martel et al. (2012). Although all the empirical data delineated above are compatible with a biological origin, they can be adequately explained by mechanisms involving known abiotic phenomena. Akin to what happened with the *Viking* experiments, the overall scientific community favours the non-biological interpretation of the ALH84001 data, as this hypothesis is arguably the simplest. Furthermore, these 'nanofossils' may be too small to support life (see Question 5.5).

Both the *Viking* and ALH84001 sagas offer us a useful lesson for the future: discovering proof of Martian life (or, for that matter, any extraterrestrial life) will probably be a gradual and painstaking process that will hinge on garnering multiple strands of evidence by complementary methods, rather than an unequivocal one-shot endeavour (Westall et al., 2015). For instance, potential Martian microfossils are likely to evince substantial ambiguity, and great care must be taken to unravel their origin(s) (McMahon and Cosmidis, 2022). In Section 13.3, we briefly explore the complicated matter of establishing rigorous criteria for life detection.

9.4.3 The future of life detection on Mars

Taking stock of what we have learned so far, we can conclude that the most promising avenue to seek out signatures of extant or extinct life on Mars is the direct investigation of past water-rich environments (sediments and deposits found in craters, lakes, deltas, hydrothermal systems, etc.) and the deep subsurface ($\gtrsim 2$ km depths). The second option is technically more difficult than the first, but it might also be the one with the highest chances of revealing currently active life (Michalski et al., 2018).

At this juncture, it is worth mentioning the detection, by the *Mars Science Laboratory* mission, of variable levels of methane in the Martian atmosphere, with a strong, repeatable seasonal pattern. Explaining these measurements has proven to be challenging, as they are ostensibly at odds with known atmospheric chemistry on Mars, and suggest the release of methane from localised surface or subsurface reservoirs; this subject is reviewed in Yung et al. (2018). Conceivable sources of methane include various geochemical processes but also, intriguingly, a putative subterranean biosphere.[4] Discriminating between abiotic and biotic origins for methane on Mars is among the key objectives of present and future explorations.

Current and upcoming astrobiologically relevant missions to Mars include:

- *Mars Science Laboratory*[5] (NASA, ongoing): This mission's primary goal is the exploration and characterisation of surface regions that are plausible abodes for extinct/extant life. The *Curiosity* rover is examining the Gale Crater region composed of stratified rocks and mineral signatures.
- *Mars 2020*[6] (NASA, ongoing): The *Perseverance* rover is exploring the Jezero impact crater, a region once flooded with water and situated near deltas of ancient rivers, that is, this area is rich in sedimentary material and eroded rocks (detritus). The objective is to analyse putative habitats, and seek traces of possible past microbial life. It has also collected and cached soil and rock samples for anticipated future return to Earth.

---

[4] On Earth, the biological production of methane by chemoautotrophs exceeds abiotic pathways by a few orders of magnitude (Thompson et al., 2022).

[5] https://mars.nasa.gov/msl.

[6] https://mars.nasa.gov/mars2020.

- *Tianwen-1*[7] (CNSA, ongoing): This mission has an orbiter, a lander, and a rover. The *Zhurong* rover is surveying the environment in the Utopia Planitia region, an impact basin, with a focus on volatiles and water.

- *ExoMars*[8] (ESA, planned): This mission aims to search for signatures of extinct Martian life, and to store samples for eventual return to Earth. The *Rosalind Franklin* rover, named after one of DNA's pioneers, will be equipped with a drill that could penetrate up to $\sim$2 m depth, in order to access potentially well-preserved biomolecules. The mission timeline is currently uncertain, but launch is not expected prior to 2028.

Therefore, the future of Mars exploration appears bright. In fact, ambitious plans to land humans on this planet in the 2030s have been advanced, although both their feasibility and desirability remain contested. Hence, in the most optimistic scenario, whether it be via robotic or crewed missions, it is conceivable in the coming decades that we might be able to resolve whether the Red Planet has harboured life at some point(s) in its history.

9.5   Problems

Question 9.1: A viable explanation for the small size of Mars is given by the 'Grand Tack' model of planetary migration, which we described in Section 3.3. After reading this segment, briefly summarise how this explanation works, by using quantitative arguments and additional references.

Question 9.2: Refer to Section 4.1.2, and compute the Rayleigh number (Ra), the diffusion timescale (t_{cond}), and the convection timescale (t_{conv}) for Mars, adopting the following parameters: $g \approx 3.7$ m s^{-2}, $L \approx 1,600$ km, $\alpha \approx 10^{-5}$ K^{-1}, $\rho_0 \approx 3.5$ g/cm^3, $\eta \approx 10^{21}$ Pa s, $\kappa \approx 10^{-6}$ m^2/s, and $\Delta T = 1,700$ K (Douce, 2011, pg. 177). How do your results compare with that of Earth in Question 4.5 of Chapter 4? In particular, does the calculated Rayleigh number suggest that mantle convection is active on Mars?

Question 9.3: At 3.8 Ga, determine the solar luminosity using (8.11). From this value, along with assuming zero albedo, calculate the equilibrium temperature of Noachian Mars by employing (8.2). If the surface temperature is taken to be 273 K for this equilibrium temperature, compute the necessary greenhouse effect in terms of the optical depth τ_{IR} by utilising (8.4). How many times larger is this τ_{IR} for Noachian Mars compared to present-day Earth and Mars listed in Section 8.1.2? How do all of your above results compare with the statements in the first paragraph of Section 9.3.4?

Question 9.4: Plot the partial pressure of CO_2 given by (9.3) as a function of the age of Mars. Calculate the partial pressure at the start of the Noachian ($\sim$4.1 Ga), and in the current epoch. By taking the difference of these two values, comment on whether the ensuing result is consistent with the picture painted in this chapter, particularly Sections 9.3.1 and 9.3.4.

Question 9.5: Atmospheric mass-loss rates ensuing from thermal (specifically hydrodynamic) and non-thermal escape are denoted by $\dot{M}_t$ and $\dot{M}_{nt}$, respectively. After perusing Section 8.2.3, prove that their ratio obeys

$$\frac{\dot{M}_t}{\dot{M}_{nt}} \propto \frac{R_p}{M_p} \propto R_p^{-2.7}, \tag{9.4}$$

---

[7] www.eoportal.org/satellite-missions/tianwen-1.

[8] www.esa.int/ExoMars.

where M_p and R_p are the planetary mass and radius; the second equality follows from the mass-radius scaling $M_p \propto R_p^{3.7}$. If all other factors are held fixed, what does the above equation imply for smaller rocky planets like Mars? Discuss how the conclusions stack up against Section 9.3.1.

Question 9.6: Integrate (9.2) from the start of the Noachian to the present day, and estimate the total number of ions (mostly O^+) that have escaped from Mars. Using the mass of a single O atom, convert your answer to bar (of atmosphere lost), and compare your results with Section 9.3.1.

Question 9.7: The classification scheme of Cockell (2014) comprises five evolutionary trajectories for Martian habitability and life. Briefly summarise all these outcomes, and how they may be empirically tested by future data. Drawing on this chapter, in conjunction with your own literature review, which of the five trajectories do you find the most plausible, and why?

Question 9.8: In Section 7.3, we briefly touched on the panspermia hypothesis. As per the contents of this chapter, would you deem Mars-to-Earth panspermia feasible in the (pre-)Noachian? Justify your answer with quantitative reasoning and citing appropriate peer-reviewed references.

Question 9.9: After performing a peer-reviewed literature search, sketch (with quantitative details) at least two proposed crewed missions to Mars and/or settlements on the planet. List at least five hurdles that such missions may encounter en route and after reaching Mars. On the basis of these constraints, do you find a crewed mission to Mars in the 2030s realistic?

10 Icy Worlds

Until now, whether it be Mars or Earth at various epochs, we have implicitly devoted our attention to rocky worlds with atmospheres situated within the habitable zone (HZ); the latter concept was defined in Section 8.1. It is worth recalling that the HZ is constructed based on the potential for hosting *surface* liquid water. This constraint is derived from one of the chief requirements for enabling habitability, namely: the existence of liquid water, as described at the beginning of Chapter 7. We emphasise, however, that the presence of liquid water is only partly synonymous with habitability, as the latter is inherently a multifaceted property.

Notwithstanding this caveat, if the availability of liquid water (often termed 'water' henceforth) is adopted as a guiding principle in the search for life in the Universe (a paradigm implicitly adopted by NASA, under the tagline 'follow the water'), it is natural to wonder whether the primary abodes for water in planetary systems are truly rocky worlds in the HZ. To put it differently, would water-bearing (and thus potentially habitable) worlds loosely resemble the Earth? Or could we conceive worlds that are distinct from Earth and still sustain water?

These questions are clearly of theoretical value, because they permit us to gain a deeper understanding of what exactly constitutes a habitable world, insofar as harbouring water is concerned. However, these questions are simultaneously of immense practical value, given that they can tangibly shape and guide our search for viable astrobiological targets (i.e., worlds that may host life) in our solar system and beyond. The latter avenue has already borne fruit within our solar system, as elucidated shortly hereafter.

In this chapter, we will take a deep dive into this subject by demonstrating that an entire category of worlds with *subsurface* oceans of water underneath shells made of icy materials – and typically comprising negligible atmospheres – are not just plausible but also impressively documented in our solar system. As these objects are quite divergent from Earth or Mars, we will revisit many of the topics already tackled in this book – such as the origin(s) of life and habitability – in the specific context of these subsurface ocean worlds, which are dubbed *icy worlds* hereon.

As we shall witness, several objects with subsurface oceans are confirmed in the solar system. Hence, unlike previous chapters on Earth or Mars that focused on a single world, we will instead carry out a generalised treatment of icy worlds and only delve occasionally into the properties and data of specific objects. The reasons underlying our strategy are threefold: (a) the sheer expected number of icy worlds makes a case-by-case analysis cumbersome; (b) sparse empirical data are available for the oceans of most of these worlds; and (c) we can highlight the commonalities linking this category of objects. It is important, however, to bear in mind the fact that icy worlds are a diverse group, and differ therefore in size, ocean composition, and so forth.

10.1 On the commonality of icy worlds

In Section 3.1.3, it was shown that the (midplane) disc temperature T_d drops off with the radial distance $R_d \equiv \sqrt{x^2 + y^2}$ in a protoplanetary disc. The crudest temperature profile is a power law that scales as $R_d^{-3/4}$, but this corresponds to a highly idealised scenario. A simplified, yet pedagogically relevant, profile is given by $T_d \propto R_d^{-1/2}$ (Weaver et al., 2018). We adopt a rough normalisation of $T_d = 4{,}000$ K at $R_d = 2R_\odot$ for Sun-like forming stars (Armitage, 2020, pg. 66),[1] thereby allowing us to express the profile as

$$T_d = 386\,\mathrm{K} \left(\frac{R_d}{1\,\mathrm{AU}} \right)^{-1/2}. \tag{10.1}$$

At temperatures greater than 150–170 K, water is predicted to exist only in the vapour phase. Hence, when $T_d \sim 170$ K, we can suppose that H_2O may theoretically exist in the ice phase past this radial location, which is termed the *snow line* or the *ice line* (refer to Section 3.1.5). On substituting this value of T_d into (10.1), we end up with $R_d \approx 5.2$ AU. Despite the numerous simplifications involved, this result is not far removed from empirical evidence for a snow line at ~ 2.7 AU in the early solar system.

A corollary is that objects beyond the snow line should contain substantial amounts of H_2O ice. In the solar system, this inference is validated by comets. This broad class of objects could consist of $\sim 10\%$ of their total mass in the form of water ice (e.g., Raymond and Morbidelli, 2022), whereas in comparison, the fraction of modern Earth's oceans is merely $\sim 2 \times 10^{-4}$ of the total mass. Classical Kuiper belt objects, situated at distances of ~ 30–50 AU from the Sun, are also rich in water ice (Brown, 2012).

Next, we can ask ourselves how many worlds in a particular size range occur beyond the snow line in our solar system. The Oort cloud, at distances between $\sim 10^3$ AU and $\sim 10^5$ AU from the Sun, is viewed as the repository of comets, and is presumed to host the largest collection of icy planetesimals. The number of worlds with diameters $\geq D$ (denoted by $n_>$) is roughly

$$n_>(D) \sim 1.5 \times 10^{12} \left(\frac{D}{1\,\mathrm{km}} \right)^{-2.5}, \tag{10.2}$$

where the normalisation is from the *Pan-STARRS1* survey (Boe et al., 2019) and the power-law distribution is partly supported by observations and theory (Dohnanyi, 1969; Boe et al., 2019). On substituting $D = 2{,}000$ km (i.e., radius of 1,000 km) in (10.2), we arrive at $n_> \sim 8.4 \times 10^3$, compatible with the estimate of $\sim 10^4$ objects with radii ≥ 800 km (Mojzsis, 2021). Hence, there may be as many as $\sim 10^4$ objects with radii of $\gtrsim 1{,}000$ km in the Oort cloud. Not all of them will consist of substantial inventories of ice, and by extension, only a fraction of those worlds will harbour subsurface oceans. Nevertheless, so many 'medium-sized' candidates for icy worlds in our solar system hint that such objects might be common in the Milky Way.

It is natural to wonder whether extrasolar analogues of the Kuiper belt and the Oort cloud exist, and the answer is affirmative. About 20% of planetary systems orbiting F-, G-, and K-type stars have debris discs (Hughes et al., 2018), which are counterparts of the Kuiper belt and the asteroid belt. Likewise, exocomets are confirmed in multiple planetary systems, as reviewed in Strøm et al. (2020).

---

[1] https://lifeng.lamost.org/courses/astrotoday/CHAISSON/AT319/HTML/AT31902.HTM.

The exocomet size distribution $n_>$ in the β Pictoris planetary system has a power-law exponent of -2.6 ± 0.8 (Lecavelier des Etangs et al., 2022), nearly equal to the index of -2.5 in (10.2).

Aside from the class of icy bodies near stars, another category of icy worlds is conceivable – to wit, those not bound to any star(s). Such free-floating bodies could originate via ejection from their parent planetary systems. Dynamical processes such as planet–disc interactions, planetesimal–planet interactions, and planet–planet scattering are operational during planet formation (see Section 3.3), and these mechanisms are capable of ejecting objects of various sizes. One may attempt to gauge the frequency of free-floating worlds per star. This procedure is hampered by uncertainties arising from a paucity of empirical data. Nonetheless, based on observational constraints from gravitational microlensing (explained in Section 12.1.3), there might be $\lesssim 10^3$ Moon-sized free-floating worlds per star on average (Lingam et al., 2023b). However, it should be recognised that not all of these objects would actually have sufficient H_2O to support subsurface oceans.

In summary, there might be $\sim 10^3$ worlds per star (bound or free-floating) with the potential for subsurface oceans, although merely a subset of these worlds will harbour them. In comparison, the occurrence rate of rocky worlds in the HZ is ~ 0.1–1, and only a fraction of those would fulfil the criterion of hosting water on the surface. Hence, we can pose the question: could icy worlds actually be more common abodes for water (and possibly habitability) relative to rocky planets in the HZ (a category that includes Earth and Mars)? As per these estimates, even if a fairly small portion of icy objects support subsurface oceans, it is still feasible for the answer to be affirmative.

Hence, the majority of water-bearing, and thence potentially habitable, worlds in the Milky Way, and perhaps the Universe, might consist of subsurface oceans beneath icy envelopes, and may consequently not resemble Earth. This conclusion runs counter to the prevailing theme (e.g., in popular science or science fiction) of depicting worlds with life as loosely akin to Earth, namely: characterised by surface water and an atmosphere. If icy worlds are indeed the most prevalent class of habitable worlds (Mojzsis, 2021), we have a powerful rationale for studying them in their own right.

10.2   Icy worlds in the solar system

Icy worlds may be the most common type of habitable worlds, a compelling reason for investigating them. However, this statement is abstract since we did not specify any particular objects in the solar system with subsurface oceans. To empirically survey such worlds in the future, it is necessary to identify appropriate targets that are confirmed to harbour subsurface oceans: of this family, we shall highlight the trio of Enceladus, Europa, and Titan, on account of their astrobiological relevance, as well as their higher level of characterisation (Hendrix et al., 2021).

10.2.1   Enceladus

We begin by examining the basic characteristics of Enceladus, and briefly explain why this small moon of Saturn is considered in many quarters as the most promising abode for life in the solar system. Readers may consult the review by Cable et al. (2021) for further information, and Figures 10.1 and 10.2 to gain knowledge of the salient features of Enceladus.

At first glimpse, Enceladus is unlike the other icy worlds we shall encounter with respect to its size. It exhibits a mean radius of $R \approx 252$ km (around 4% that of the Earth) and a mean density

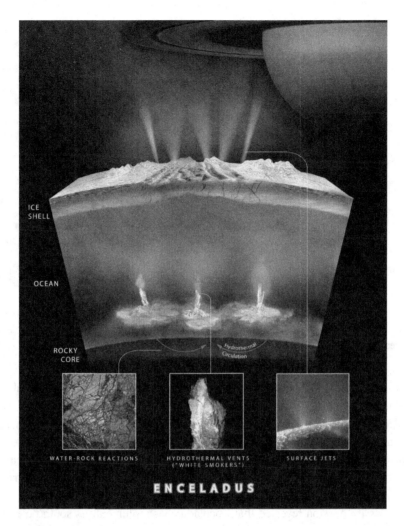

Figure 10.1 Schematic rendition of the structure of Enceladus. Water–rock interactions and hydrothermal vents at the bottom of the ocean floor are noteworthy due to their significance in astrobiology. (Credit: NASA; https://photojournal.jpl.nasa.gov/catalog/PIA21442)

of approximately 1.6 g/cm^3. As this value is not far removed from the density of water, it suggests that Enceladus is not dominated by rocky material, which typically has a density of $\sim$3 g/cm^3. The general consensus for the age of Enceladus is $\gtrsim$1 Gyr, although a few publications have advocated a younger age of $\sim$0.1 Gyr.

The small size of Enceladus is surprising given that it harbours a subsurface ocean. Smaller worlds possess thicker ice shells if all other factors are held fixed; we shall touch on this aspect in Section 10.3. Thus, if the only sources of heat were radioactive decay and primordial heat after formation, the chances of Enceladus hosting a long-lived subsurface ocean would be very slim. Tidal heating – which stems from differential gravitational forces (i.e., tidal forces from Section 8.3.4) – injects a substantial amount of energy into Enceladus, with a global heat production of $>2 \times 10^{10}$ W (Hand et al., 2020, Table 1), enabling the sustenance of its global ocean.

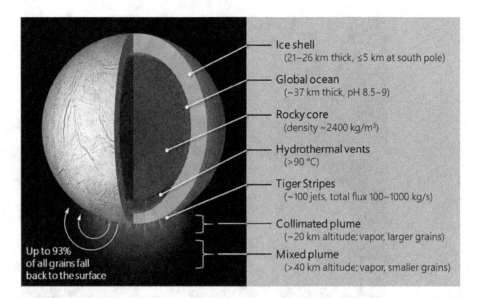

Figure 10.2 Structure and properties of the subsurface ocean of Enceladus based on *Cassini* data. In addition to the parameters furnished herein, it is predicted that the salinity is about more than two times times smaller compared to seawater on Earth (Kang et al., 2022). (Credit: Cable et al. 2021, Figure 2; CC-BY 4.0 license)

Most of what we have learnt about Enceladus is derived from the remarkable *Cassini–Huygens* mission, which surveyed the Saturn system for over a decade (2004–2017). The *Cassini* mission accorded two independent lines of evidence for a global subsurface water ocean. First, measurements of this moon's rotation revealed a wobble (*libration* to be precise) that was compatible with a global ocean separating the icy shell and rocky core. Second, gravitational field measurements indicated that Enceladus was not in hydrostatic equilibrium, and that its internal structure was best explained by a global ocean beneath the icy envelope.

The *Cassini* mission revealed that Enceladus has a plume at its south pole, and analysis of that material has yielded a wealth of information regarding the putative habitability of this world. For starters, multiple lines of evidence favour ongoing water–rock interactions and hydrothermal activity:

- Nanometre-sized silica (SiO_2) particles discovered by the Cosmic Dust Analyser (CDA) were interpreted to originate from hydrothermal reactions occurring at high temperatures of $>90°C$ (Hsu et al., 2015).
- The Ion and Neutral Mass Spectrometer (INMS) detected molecular hydrogen (H_2) in the plume, and detailed modelling by Waite et al. (2017) concluded that it could arise due to serpentinisation, which is associated with certain water–rock interactions in hydrothermal systems; this geological process is described in Section 7.1.3.
- Measurements of some other gases (e.g., methane) in the plume by INMS suggested that the relative composition of these compounds was broadly consistent with hydrothermal reactions generating them.

Note that, among other advantages, submarine hydrothermal systems on Enceladus would be valuable from the standpoint of facilitating the origin(s) of life, as sketched in Section 10.4. However, one caveat as per some models is that serpentinisation from hydrothermal activity on Enceladus may only last for ~0.1 Gyr, and might be approaching cessation (Daval et al., 2022).

Setting aside the water requirement for habitability, which is automatically met, two other crucial criteria are free energy sources and bioessential elements. Although we shall tackle these constraints in Section 10.5 for icy worlds, we will briefly comment upon them with regard to Enceladus. First, insofar as energy is concerned, the ingredients for the metabolic pathway of methanogenesis are readily available on Enceladus, as mentioned in Section 7.1.3. In fact, the prospects for the functioning of this pathway in Enceladus are potentially bright as per both laboratory experiments (Taubner et al., 2018) and numerical models (Affholder et al., 2021).

Turning our gaze towards bioessential elements, the CDA and INMS instruments have revealed both low- and high-mass compounds consisting of carbon, hydrogen, nitrogen, and oxygen, namely: four out of six bioessential elements. It is worth highlighting that a subset of ice particles were determined to contain large organic molecules with diverse structures (e.g., linear and cyclic), which may be fragments derived from (organic) compounds of greater mass and complexity. Moreover, certain nitrogen- and oxygen-bearing chemical species from the ice grains could serve as precursors of biomolecular building blocks such as amino acids (Khawaja et al., 2019).

In the case of sulfur, the claimed detection of hydrogen sulfide (H_2S) is ambiguous, but numerical models indicate that dissolved sulfides might be supplied to the ocean via hydrothermal activity. Next, sodium phosphates have been confirmed in the plume (Postberg et al., 2023), contradicting earlier modelling, which suggested that dissolved species comprising phosphorus would be scarce. This study concluded that phosphorus concentrations in Enceladus' ocean may be ~100 times higher relative to Earth. Aside from the basic bioessential elements (CHNOPS), some trace metals could be necessary for life, but their abundances are poorly constrained.

Hence, it might be reasonable to contend that Enceladus fulfils most, and plausibly all, of the vital conditions for habitability, thereby cementing its status as one of the prime targets for future missions. Perhaps because the *Cassini* mission ended in 2017, less than a decade ago, no immediate missions to Enceladus are on the horizon. With that said, several mission concepts have been developed and are under active study (see Question 10.2), motivated by the 2023–2032 *Planetary Science and Astrobiology Decadal Survey*.

10.2.2 Europa

Europa, one of the four Galilean moons of Jupiter, has a radius of 1,561 km, which is slightly less than Earth's Moon, whose radius is 1,737 km. Europa has a bulk density of ~3 g/cm^3, implying that it is likely to have a primarily rocky composition. It was suspected since the 1970s that Europa may harbour a subsurface ocean, but the first concrete line of evidence emerged from the *Galileo* mission in the 1990s.

One of the cornerstones of electromagnetism is Faraday's law of induction, which states that an electromotive force (EMF), denoted by $\mathcal{E}$, is induced because of a changing magnetic flux (Φ_B) as follows:

$$\mathcal{E} = -\frac{d\Phi_B}{dt}.$$ (10.3)

This EMF generates an induced electric current and magnetic field that is encapsulated by Lenz's law. In the case of Europa, the external magnetic field is provided by Jupiter, and the temporal variation is a consequence of the satellite's eccentric orbit. By measuring Europa's induced magnetic field (using a magnetometer), in accordance with the above description, an estimate of the satellite's conductivity can be obtained: the inferred value is best explained through the existence of a subsurface ocean.

Analysis of Europa's surface geological features (composed of 'chaos terrain', bands, and ridges) has suggested that they may have arisen through interactions with an underlying subsurface ocean. Furthermore, the detection of salts like sodium chloride (NaCl) correlates with certain surface features, and has been interpreted as originating from an interior source (Trumbo et al., 2019).[2] Last, there is spectroscopic and magnetic evidence for water vapour plume(s), but the source is not well understood (a subsurface ocean is one candidate), and it is not clear whether this plume is continuously maintained (Jia et al., 2018; Paganini et al., 2020).

Europa may have a global heat production in the vicinity of 10^{12} W (Hand et al., 2020, Table 1), with tidal heating and radioactive heating constituting the two major sources. Initially, the global thickness of the icy shell was modelled to be $\sim$10 km, even approaching $\sim$1 km in some regions. Current models support an estimate of $\sim$25 km for the ice layer (Howell, 2021), beneath which is situated a deep global subsurface ocean with a possible depth of $\sim$100 km. It is suspected that Europa's ocean floor permits water–rock interactions, albeit direct evidence to date is minimal.

Two other notable characteristics of Europa are relevant for our subsequent discussion. First, Europa is intensely bombarded by charged particles in the radiation belts of Jupiter, as illustrated in Figure 10.3. It is estimated, based on *Galileo* data, that Europa receives a particle energy flux

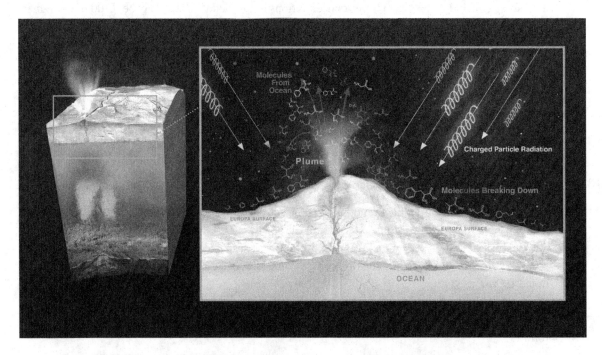

Figure 10.3 Schematic rendition of the structure of Europa. The prominent role played by radiation on the surface is depicted. (Credit: NASA; https://photojournal.jpl.nasa.gov/catalog/PIA22479)

---

[2] Europa's subsurface ocean is predicted to be salty, with a mean salinity of $\sim$1.5 times that of seawater on Earth (Hand and Chyba, 2007).

of ~ 0.13 W/m^2 (Cooper et al., 2001). This source drives radiolysis (i.e., dissociation of chemical species), and stimulates both the creation and destruction of many molecules (Carlson et al., 2009). As regards the latter, the production of oxidants such as O_2 (molecular oxygen), SO_4^{2-} (sulfate), and H_2O_2 (hydrogen peroxide) is feasible; the pathway for H_2O_2 is expressible as

$$2H_2O \rightarrow 2H_2O^+ + 2e^- \rightarrow 2H + 2OH \rightarrow H_2O_2 + H_2. \tag{10.4}$$

The second aspect is closely tied to the first, and involves the transport of the aforementioned oxidants generated via radiolysis across the icy shell and into the subsurface ocean. Europa is geologically active, and observations indicate that it has a young surface, with an average age of $t_{surf} \sim 40$–90 Myr. It has been proposed, therefore, that oxidant delivery might occur on timescales of $\lesssim t_{surf}$. One possible scenario invokes a version of plate tectonics. Despite tentative evidence for the recycling of material through such a mechanism (Kattenhorn and Prockter, 2014), Earth-like plate tectonics on Europa may be considered unlikely based on geophysical modelling.

The other two pathways that have been evaluated are drainage of brines (i.e., high-concentration salt solutions) and impact-generated melting. In the former, near-surface partial melting driven by some process (e.g., tidal heating) generates brines that incorporate surface oxidants and percolate through the icy shell. Detailed numerical modelling by Hesse et al. (2022) found that $\sim 85\%$ of surface oxidants could be delivered to the ocean on a short timescale of 2×10^4 yr. The latter can be enabled by moderate impacts that indent (but do not penetrate) the icy shell and trigger melting close to the surface, with these melts eventually draining into the ocean on rapid timescales of $\sim 10^3$–10^4 yr (Carnahan et al., 2022).

Two missions will investigate Europa in the coming decade. ESA's *Jupiter Icy Moons Explorer* (JUICE),[3] which departed Earth in 2023, is designed to chiefly survey Ganymede (and Callisto), but is planned to perform two flybys of Europa, which may yield valuable information regarding its internal structure and its plumes. The second mission, NASA's *Europa Clipper*,[4] launched in 2024, will significantly advance our knowledge of Europa (Howell and Pappalardo, 2020). Aside from mapping Europa's surface and interior, the spacecraft will be equipped with a dust analyser and mass spectrometer, both of which can detect organic and inorganic compounds.

10.2.3 Titan

Titan is one of the most fascinating objects in our solar system. This large moon of Saturn has a radius of 2,575 km ($0.4 R_{\oplus}$), making it slightly larger than Mercury. Moreover, it has a thick atmosphere with a surface pressure of ~ 1.5 bar, and hosts liquid bodies composed of methane and ethane on its surface; we shall return to this theme in Section 11.3.2.

Titan also possesses a subsurface water ocean, as illustrated in Figure 10.4. The *Cassini* mission implied its presence in a few different ways (Sotin et al., 2021). First, akin to Enceladus, measurements of Titan's fluctuating gravitational field, specifically its so-called degree-2 tidal Love number, suggested that a global subsurface ocean was a good fit to the data. Second, measurements of the electric field concluded that the detected signal(s) could arise from wave propagation in a cavity, with

---

[3] www.esa.int/Science_Exploration/Space_Science/Juice.

[4] https://europa.nasa.gov/.

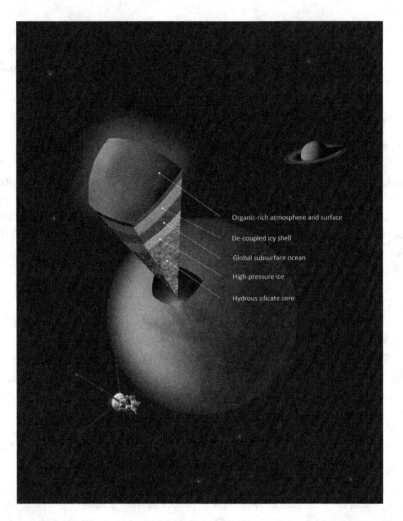

Organic-rich atmosphere and surface

De-coupled icy shell

Global subsurface ocean

High-pressure ice

Hydrous silicate core

Figure 10.4 Schematic rendition of the structure of Titan. This moon harbours liquid bodies on both the surface and subsurface; in the latter case, the water ocean is probably sandwiched between two ice layers. (Credit: NASA; https://photojournal.jpl.nasa.gov/catalog/PIA15607)

the lower conducting boundary being compatible with a subsurface ocean; this interpretation has, however, been challenged. Third, the obliquity (axial tilt) of Titan is higher than that of an entirely solid body, which lends credence to a subsurface ocean.

Titan's internal structure is not clearly understood, owing to which the thickness of its outer ice shell and the depth of the subsurface ocean are uncertain. Empirical constraints from *Cassini* data support a subsurface ocean at $\sim$50–200 km below the surface (MacKenzie et al., 2021b), which represents the thickness of the outer ice shell; the range is consistent with numerical models. The ocean thickness is perhaps $\sim$100 km (Vance et al., 2018, Table 8), albeit less than $\sim$1,000 km (Sotin et al., 2021, Table 2).

A unique attribute of Titan is that its subsurface ocean is believed to be bounded from both above and below by ice layers. The ice at the bottom of the ocean could be composed primarily of Ice VI, a high-pressure ice (Soderlund et al., 2020). In Figure 11.1, the regimes wherein high-pressure ices are manifested are discernible. Although these regimes depend on both pressure and temperature, a crude rule of thumb is that formation of such ices may be initiated at pressures greater than $P_c \approx 10^9$ Pa (1 GPa). Titan is expected to have a few 100 km of high-pressure ices beneath its ocean.

In other words, interactions between liquid water and the rocky core would seem impossible at first glimpse. This scenario is deemed problematic because water–rock interactions can provide a supply of reducing agents and nutrients, both of which are vital for habitability, and the associated submarine hydrothermal systems are promising candidates for facilitating abiogenesis. However, recent research has demonstrated pathways whereby such materials could be delivered from the core to the ocean, that is, across the high-pressure ice layer (Journaux et al., 2020). One mechanism relies on partial melting of the ice, and the transport of the generated water (and dissolved substances) through the porous ice via convection.

Titan was investigated by the *Cassini* mission, but there is much that we still do not know about its interior. The *Dragonfly* mission to Titan,[5] expected to launch in 2028, will shed light on not just its surface environments and habitability but also on its internal structure (e.g., subsurface ocean) by performing seismic measurements (Barnes et al., 2021).

10.2.4 Miscellaneous icy worlds

Aside from the trio of Enceladus, Europa, and Titan, several other objects in the solar system are either confirmed or suspected to host subsurface oceans, most of which are sketched next. Mimas, Saturn's tiny moon with a radius of about 200 km, reportedly possesses an ocean ∼30 km below its surface, as per measurements of its orbital motion (Lainey et al., 2024).

Ganymede: With a radius of 2,631 km, Ganymede is the largest moon in the solar system, and is unique in possessing an intrinsic magnetic field. Measurements of the time-varying magnetic field near Ganymede hinted at the existence of a subsurface ocean generating an induced magnetic field, along similar lines as Europa (see Section 10.2.2). In addition, observations by the *Hubble Space Telescope* found that Ganymede's auroral belts oscillate less than predicted, which is explainable by a conductive subsurface ocean that partly screens the effect of Jupiter's magnetic field on these belts.

Very little is known about the characteristics of the subsurface ocean, which is probably sandwiched between two ice layers, analogous to Titan. The thickness of the outer ice shell and the ocean depth may be ≲150 km and ≲500 km, respectively (Vance et al., 2018, Table 5). Despite a high-pressure ice layer at the bottom of the ocean, transport of substances derived from the underlying rock into the ocean is feasible, as noted in Section 10.2.3. The JUICE mission should be able to characterise Ganymede's subsurface H_2O reservoirs, and determine the extent of its subsurface ocean.

Callisto: Callisto, the third largest moon in the solar system, is virtually equal to the size of Mercury. Its interior is even less understood than Ganymede, and the existence of a subsurface ocean is only partly confirmed. The ocean was inferred by means of magnetic field measurements, paralleling the approach described in Section 10.2.2 for Europa, but this signal admits multiple interpretations (Soderlund et al., 2020).

[5] https://dragonfly.jhuapl.edu/.

In the same vein as Ganymede and Titan, Callisto may comprise high-pressure ice at the bottom of its ocean. Neither the thickness of the outer ice shell nor the depth of the ocean are constrained: values of $\sim$100 km and $\lesssim$140 km, respectively, have been proposed based on models (Vance et al., 2018, Tables 9 & 10). Although the JUICE mission emphasises Ganymede, it will carry out multiply flybys of Callisto, which should enable us to get a better handle on its internal structure and H_2O layers.

Apart from these objects, a number of candidates for icy worlds exist, but the evidence is equivocal or missing. A complete summary is furnished in Hendrix et al. (2021), from which a truncated list is provided here.

1. Neptune's moon, Triton ($R = 1{,}353$ km), has a young surface (<0.1 Gyr old) with distinctive morphology, and the *Voyager 2* spacecraft detected active geysers. These properties might be compatible with, but are not exclusive to, the presence of a subsurface ocean.
2. Pluto, once the ninth planet of the solar system, exhibits geological features consistent with a subsurface ocean. The Sputnik Planitia, an icy basin, has been argued to have undergone reorientation facilitated by such an ocean. Theoretical models have also shown that maintaining a subsurface ocean in Pluto on Gyr timescales is feasible.
3. Some moons of Uranus (e.g., Ariel, Titania, and Miranda) have young surfaces and display morphologies suggestive of geological activity (e.g., cryovolcanism) that might be indicative of enough energy to harbour subsurface oceans; theoretical models tentatively support oceans with thickness of $\lesssim$30–50 km (Castillo-Rogez et al., 2023).
4. Tethys and Dione, two moons of Saturn, were originally suspected to emit weak plumes (thus resembling Enceladus), but subsequent observations failed to corroborate this equivocal claim.
5. The *Dawn* mission explored the dwarf planet Ceres ($R = 470$ km), and discovered brines, hydrated minerals, and other markers of water (McCord et al., 2022). Hence, it is conceivable that Ceres harbours quite extensive, if not global, subsurface liquid water even in the current epoch.

10.3   Physical properties of icy worlds

Having illustrated some concrete examples of icy worlds in our solar system, we can now move on to examine the physical properties that characterise the generic members of this class of objects. We start off by introducing some notation employed throughout the chapter. The radius and mass of the subsurface icy world are denoted by R and M, respectively. The mean thickness of the (outer) ice shell is H_i, whereas the average ocean depth is H_o. If there are no high-pressure ices beneath, the total H_2O thickness (in both ice and water form) is $H_t = H_i + H_o$.

10.3.1   Thermal profile of the icy shell

For a generic icy world, it is instructive to perform a simplified derivation of the thermal profile in the icy shell. The transport of heat in this region can occur via either conduction or convection, with the former typically creating a colder thermal profile than the latter (Soderlund et al., 2020, Section 3.1). A dimensionless quantity, the Rayleigh number (see Section 4.1.2), regulates when convection is more efficient than conduction at heat transfer. We shall not delve into the specifics, which are complex and lie beyond the scope of the book. For the sake of constructing a simplified heuristic (albeit not accurate) model, we assume that conduction is the dominant mode of heat transfer in the icy shell (Journaux et al., 2020, Section 5.3).

For simplicity, we shall suppose that the curvature of the world is ignored, that is, we treat the ice shell as a slab, with z serving as the only relevant coordinate. This assumption performs poorly for small icy worlds, comparable in size to Enceladus, but is otherwise an acceptable approximation (Lingam and Loeb, 2019c). As per Fourier's law, we have

$$Q = -k(T)\frac{dT}{dz}, \tag{10.5}$$

where $T(z)$ is the temperature in the icy shell, Q is the heat flux (in units of W/m²), and $k(T) = C/T$ is the conductivity of ice; here, $C = 651$ W/m is constant. If the icy shell is not very thick, which is once again reasonable for larger worlds, we can hold the heat flux constant and equal to

$$Q \approx 7.5 \times 10^{-2}\,\text{W/m}^2\,\Gamma\left(\frac{R}{R_\oplus}\right)^{1.3}, \tag{10.6}$$

where the power law on the RHS arises from three factors. First, the heat flux is the heat flow divided by the surface area. Second, the radiogenic component of the heat flow is conventionally modelled as being proportional to the mass M, and the surface area is proportional to R^2. Third, we utilise the mass–radius relationship $M \propto R^{3.3}$ for H_2O-rich worlds with $R \lesssim R_\oplus$ (Sotin et al., 2007). The normalisation of 7.5×10^{-2} W/m² is adopted from the radiogenic heat flow of Earth (The Borexino Collaboration, 2020a).

The quantity Γ in the RHS of (10.6) admits multiple interpretations. When $\Gamma \approx 1$, this scenario would suggest that the heat is predominantly contributed by radioactivity. On the other hand, if $\Gamma > 1$, then other energy sources such as tidal heating could play a prominent role, or the icy world may comprise a higher inventory of radioactive elements per unit mass relative to Earth. Likewise, if $\Gamma < 1$, those icy worlds ought to contain a lower abundance of radioactive elements per unit mass than Earth. As all three options are viable, we leave Γ as a free parameter.

By solving the ODE in (10.5), we end up with the result

$$H_i \approx 8.7\,\text{km}\,\frac{1}{\Gamma}\left(\frac{R}{R_\oplus}\right)^{-1.3}\ln\left(\frac{T_m}{T_s}\right), \tag{10.7}$$

where T_s is the surface temperature of the icy world, and T_m is the melting point of ice. Note that the latter is not automatically equal to 273 K, because it depends on the pressure; for pure ice, we have $T_m \approx 250$–273 K. Furthermore, the melting point is lowered when certain impurities (e.g., salts or ammonia) are incorporated in the ice.

The surface temperature T_s is governed by the type of icy world under consideration. Objects around a star, unless they are very far from it, would receive the majority of their surface heat from the star, and the stellar flux would therefore govern T_s via the Stefan–Boltzmann law. In contrast, when the icy worlds are far removed from the star or are free-floating, the surface temperature is regulated by the internal heat. In the latter case, upon setting $Q = \sigma_{SB}T_s^4$, and invoking (10.6) to solve for T_s, we obtain

$$T_s \approx 33\,\text{K}\,\Gamma^{1/4}\left(\frac{R}{R_\oplus}\right)^{0.33}, \tag{10.8}$$

implying a fairly weak dependence on Γ and R for those class of worlds.

Even though (10.7) was obtained after several simplifying assumptions, it still yields useful quali-tative and quantitative information. For starters, we notice that H_i would increase on smaller icy worlds; Enceladus is an exception because of its anomalously high heat flow. Next, along expected lines, we find that H_i would decrease when the surface temperature is higher – as the ice would reach its melting point at a shallower depth – and vice versa. Finally, boosting the value of Γ, which is effectively equivalent to raising the internal heat, would cause thinner ice shells.

10.3.2   Constraints on subsurface oceans

A commonly encountered quantity in the literature is the H_2O (i.e., water in all phases) fraction, which represents the fraction of the total mass of the object existing as H_2O. We shall denote this quantity by f_w, such that the mass of H_2O would be $M_{H_2O} \equiv f_w M$ by definition.

A necessary, but not sufficient, condition for hosting a subsurface ocean is that M_{H_2O} must exceed the amount of mass present in the (outer) icy shell. If this criterion is not met, then all H_2O would manifest in the ice phase, thus suppressing the possibility of the subsurface ocean. We assume, for the sake of simplicity, that both water and ice have a density of $\rho_w \approx 1$ g/cm^3. Hence, the above constraint is expressible as

$$M_{H_2O} \gtrsim 4\pi \rho_w R^2 H_i, \tag{10.9}$$

where the RHS is obtained by multiplying the density with $4\pi R^2 H_i$, the volume of the icy shell. By invoking (10.7) and the mass-radius scaling $M \propto R^{3.3}$ from earlier, (10.9) is transformed into

$$f_w \gtrsim 7.4 \times 10^{-4} \frac{1}{\Gamma} \left(\frac{R}{R_\oplus} \right)^{-2.6} \ln \left(\frac{T_m}{T_s} \right). \tag{10.10}$$

As a point of comparison, we note that f_w for modern Earth's oceans is around 2×10^{-4}, which suggests that Earth's oceans would freeze over completely if our planet was ejected suddenly into space, consequently losing the heat supplied by the Sun. An interesting aspect of (10.10) is that f_w monotonically decreases with R, implying that smaller worlds would need higher H_2O fractions to permit subsurface oceans, if all other factors are held fixed. Even though this trend may seem incompatible with a subsurface ocean on Enceladus, this issue is counterbalanced by the substantial tidal heating of the moon, which boosts Γ and thereby lowers the bound on f_w.

Earlier, we had touched on the importance of water–rock reactions and hydrothermal systems, and we shall again revisit these themes in due course. Cutting off these features through the formation of high-pressure ices at the bottom of the ocean floor may be detrimental for habitability, although some pathways can perhaps ameliorate this issue, along the lines sketched in Section 10.2.3. We will, therefore, examine the condition(s) needed to permit water–rock interactions and rule out high-pressure ices.

Let us consider a generic icy world that does not harbour high-pressure ices, so that its internal structure comprises the icy shell and the subsurface ocean. If most H_2O in this world is locked up in these two layers (i.e., minimal H_2O occurs in the mantle/core), then by construction we end up with

$$f_w M \equiv M_{H_2O} \approx 4\pi \rho_w R^2 \left(H_i + H_o \right). \tag{10.11}$$

The first equality corresponds to the definition of M_{H_2O}, and the second follows from the assumption made in the prior paragraph, namely: there is essentially no H_2O outside the ice layer and the ocean.

Next, for the class of worlds we are tackling (i.e., those devoid of high-pressure ices), the pressure at the bottom of the ocean floor must be lower than the previously introduced pressure limit P_c (in Section 10.2.3) for the formation of high-pressure ices; otherwise, we would contradict our initial premise that high-pressure ices are absent. In quantitative terms, this constraint is given by

$$\rho_w g \left(H_i + H_o \right) \lesssim P_c, \tag{10.12}$$

where g is the acceleration due to gravity, which is held constant in the H_2O layers. The LHS is the pressure at the bottom of the ocean floor, since both ice and water are taken to have roughly the same density.

Now, after rearranging (10.11), we arrive at

$$\rho_w \left(H_i + H_o \right) \approx \frac{f_w M}{4\pi R^2}, \tag{10.13}$$

and, upon substituting this expression in (10.12), we end up with

$$f_w \lesssim 8.7 \times 10^{-3} \left(\frac{R}{R_\oplus} \right)^{-2.6} \left(\frac{P_c}{10^9 \, \text{Pa}} \right), \tag{10.14}$$

where we have used $g \propto M/R^2$ and the mass–radius relationship $M \propto R^{3.3}$. This bound on f_w is a monotonically decreasing function of R, indicating that smaller worlds would have a higher value of the upper bound.

On comparing (10.10) and (10.14), two key characteristics stand out. First, the former constitutes a lower bound for the H_2O fraction, while the latter represents an upper bound. Hence, worlds that lack high-pressure ices are approximately restricted to this range, which may become narrow in some instances. We point out, however, that (10.10) could apply to all icy worlds, even those with high-pressure ices, unlike (10.14) that was constructed explicitly for objects sans high-pressure ices. Second, we notice that the scaling with respect to R is identical for both of them.

We reiterate that (10.10) and (10.14) are inherently heuristic, because of the assumptions made while deriving H_i, the approximate nature of the mass–radius relationship, and the simplified choice of the pressure threshold P_c, among other limitations. Hence, caution is warranted when using them.

10.4 Origin(s) of life

Icy worlds are clearly distinct from Earth with regard to sites conducive to abiogenesis. Land-based settings on Earth such as hot springs, beaches, and lakes, to name a few, are ruled out, leaving submarine environments as the remaining contenders. Icy worlds with hydrothermal systems, where water–rock interactions are feasible, are promising because of their capacity to host hydrothermal vents and sediments, both of which are potential sites for the origin of life, as delineated in Section 5.5.1.

We have already summarised the theories for the origin(s) of life in Chapter 5, owing to which we shall not retread old ground. Instead, we shall highlight a few challenges and paths that are directly relevant to icy worlds, in the sense that their specific structure is explicitly involved.

10.4.1   Prebiotic chemistry: monomers

A popular, but by no means universal, paradigm postulates that life emerged from the self-organisation of appropriate building blocks. The first step involves the abiotic synthesis of monomers, which are subsequently polymerised and transformed into biomolecules. In principle, multiple energy sources could drive the synthesis of these monomers. We select a few sources, mirroring the order-of-magnitude analysis in Lingam and Loeb (2019c), and devote our attention to amino acids and proteins (their polymers), because the latter comprise the bulk of a cell's dry mass (Moran et al., 2010).

Objects around stars receive ultraviolet (UV) radiation from the star. Ultraviolet light is a promising avenue for powering prebiotic synthesis, but one crucial distinction needs to be spelt out. Unlike Earth, where UV radiation would pass through the atmosphere and reach various surface environments (e.g., ponds), icy worlds would have UV light incident on their icy surfaces. Hence, this route shares some loose similarities with Section 2.4.3 (see Table 2.4), although the latter was characterised by lower temperatures. The maximal yield of amino acids ($\mathcal{M}_{UV}$ in kg/yr) that might be generated by UV light, given a suitable mixture of ices, is

$$\mathcal{M}_{UV} \sim 10^{10} \, \text{kg/yr} \left(\frac{R}{R_\oplus} \right)^2 \left(\frac{a}{1 \, \text{AU}} \right)^{-2} \left(\frac{M_\star}{M_\odot} \right)^\zeta, \tag{10.15}$$

where a is the orbital radius, and ζ depends on the stellar mass $M_\star$ as follows: $\zeta \sim 1.2$ for $M_\star \lesssim M_\odot$ and $\zeta \sim 6.8$ for $M_\odot < M_\star \lesssim 2 M_\odot$.

The second candidate takes inspiration from Europa, and applies to moons around Jovian planets. As stated in Section 10.2.2, Europa's surface is subjected to constant bombardment by charged particles in the radiation belts of Jupiter. Saturn has a magnetic dipole moment that is ~ 30 times lower than Jupiter, and has weaker radiation belts as a consequence. Energetic particles have the advantage of facilitating prebiotic synthesis at relatively high efficiencies. A rough estimate for the amino acid yield ($\mathcal{M}_{GP}$) is

$$\mathcal{M}_{GP} \sim 1.9 \times 10^8 \, \text{kg/yr} \left(\frac{R}{R_\oplus} \right)^2 \left(\frac{a_m}{4.5 \times 10^{-3} \, \text{AU}} \right)^{-2}, \tag{10.16}$$

where a_m denotes the distance of the moon from a Jupiter-analogue.

These two paths are surface based. On icy worlds where submarine hydrothermal systems exist (addressed in Section 10.3.2), the high temperatures and pressures in those environments could favour the synthesis of organic compounds, including amino acids, on thermodynamic grounds (McCollom and Seewald, 2007). As these settings are on the ocean floor, their spatial extent is approximately proportional to R^2, that is, the area of the icy world. Hence, the amino acid yield would be likewise proportional to R^2, but we emphasise that the rates of synthesis are deeply sensitive to multiple parameters such as the rock composition and temperature. The following yield ($\mathcal{M}_{HV}$) should therefore be viewed only as a fiducial estimate.

$$\mathcal{M}_{HV} \sim 6.7 \times 10^8 \, \text{kg/yr} \left(\frac{R}{R_\oplus} \right)^2. \tag{10.17}$$

This list is not comprehensive. Other scenarios for prebiotic compounds invoke cosmic rays, exogenous delivery by comets and dust particles, and radioactivity. Most of these avenues, barring exceptions like hydrothermal vents, lead to the synthesis of prebiotic molecules on the surface, where they are susceptible to dissociation and degradation by high-energy sources like UV radiation, X-rays, and charged particles. Hence, it is not enough to just produce these compounds on the surface – they must be transported across the icy shell into the subsurface ocean.

A number of transport mechanisms are plausible for geologically active worlds. We have encountered some of them in Section 10.2.2, namely: tectonic activity, impact melting, and brine-mediated drainage. These processes can facilitate rapid turnover, but there are slower mechanisms at work as well. Impact gardening causes vertical mixing from bombardment by micrometeorites (with sizes of $\lesssim 1$ mm). The mixing rates (units of length per time) vary across worlds and over time, due to which identifying a characteristic value is difficult. It has been estimated that, on Europa, the gardening rate may attain a maximum of $\sim 1 \times 10^{-6}$ m/yr (Cooper et al., 2001).

10.4.2 Prebiotic chemistry: polymers

Once the formation of monomers has taken place, the next desirable step is polymerisation – for instance, the synthesis of peptides from amino acids. In certain environments like hydrothermal systems, such polymerisation reactions are thermodynamically favoured and could, in principle, occur spontaneously. An experimental study by Takahagi et al. (2019), resembling the hydrothermal systems expected on Enceladus based on *Cassini* data, showed that peptide formation is feasible. As hydrothermal systems were already elucidated in Section 5.5.1, we will highlight a different environment that can also aid polymerisation – ice. The desirable properties of ice as a setting for prebiotic chemistry have been extensively studied, as outlined below.

Before doing so, it is necessary to qualitatively understand the phenomenon of *eutectic freezing*. Suppose that some 'impurities' (in our case, prebiotic molecules) are present in water. As this mixture is cooled to the point where ice crystals start forming, these impurities are excluded from the crystals, and are thence concentrated within small pockets of liquid. The process of concentration is valuable for polymerisation because chemical reaction rates often depend on the abundances of the constituent species, indicating that they are sped up when these abundances are boosted. Hence, in a nutshell, eutectic freezing can promote polymerisation. On a closely related note, ice also enables synthesis of relevant monomers such as amino acids and nucleobases (e.g., Miyakawa et al., 2002). With regard to polymerisation and related topics, ice has played a central role in the following publications:

1. Eutectic freezing was harnessed by Monnard et al. (2003) to synthesise polymers of nucleotides, the building blocks of nucleic acids.
2. Attwater et al. (2013) empirically showed that an RNA enzyme (ribozyme) specially adapted to function in ice was able to synthesise RNA molecules of nearly the same length as itself.
3. By performing experiments reliant on freeze-thaw cycles, Mutschler et al. (2015) demonstrated that the assembly of ribozymes was rendered feasible from smaller RNA molecules.
4. Freeze-thaw cycles were empirically demonstrated to aid the activation and polymerisation of nucleotides (Zhang et al., 2022).

Aside from such experiments (only a few of which are reported here), theoretical models have demonstrated that, in principle, the polymerisation of glycine is thermodynamically favourable at temperatures lower than $T_p \approx 120$ K (Kimura and Kitadai, 2015). Hence, if the surface temperature is less than this value, it is possible to estimate the depth H_p up to which polymerisation could happen spontaneously, which is determined to be

$$H_p \approx 8.7\,\text{km}\,\frac{1}{\Gamma}\left(\frac{R}{R_\oplus}\right)^{-1.3}\ln\left(\frac{T_p}{T_s}\right) \tag{10.18}$$

after following the procedure outlined in Question 10.6.

10.5   Habitability

There are four key criteria required for instantaneous habitability (see the beginning of Chapter 7), of which one (i.e., water as solvent) is automatically satisfied on icy worlds we are studying.

Among the remaining trio, we shall not tackle the issue of suitable physicochemical conditions for life. The reasons are threefold: (1) with the exception of Enceladus, to an extent, we have sparse data about the salient properties of subsurface oceans in icy worlds; (2) only a few experiments have been conducted on the survival and growth of Earth-based organisms while fully mimicking such environments; and (3) as per our current understanding, physicochemical conditions may not pose substantial barriers to oceanic habitability of the well-known icy worlds. Many of these statements are also applicable when it comes to assessing the abundances of nutrients.

Thus, we are left with two signposts of habitability: energy sources and nutrients. We shall briefly delve into them in this section.

10.5.1   Energy sources

A consequence of having an ocean underneath the ice envelope is that the latter will effectively block off all radiation, including photosynthetically active radiation, from reaching the former. Hence, except perhaps for specialised environments (e.g., cracks and pores in the ice), phototrophy may be eliminated as a major contributor. This leaves us with chemotrophy, since we ignore exotic avenues like pathways powered by tidal or magnetic energy. As we have seen in Section 7.1.3, a rich variety of metabolisms fall under the umbrella of chemotrophy; we single out only a few examples here.

The first route is hydrogenotrophic methanogenesis, whose significance in astrobiology was underscored in the exposition below (7.10). To recapitulate, this pathway is ancient, has simple reactants, and is presumably feasible in astrobiological settings like Mars and Enceladus, among other advantages. The reaction for methanogenesis is often written as

$$4H_2 + CO_2 \to CH_4 + 2H_2O, \tag{10.19}$$

demonstrating that carbon dioxide and molecular hydrogen are the necessary reactants. Of this duo, CO_2 availability may not be an issue compared to that of H_2, making the latter a potential limiting factor. Molecular hydrogen can be produced via serpentinisation, which is possible when water–rock interactions occur; the simplified reaction is described in (7.11). Even if that option is ruled out, energy derived from radioactive decay could split water and generate H_2 and O_2. In Table 10.1, the H_2 production rates for Europa and Enceladus by means of these channels are furnished.

Table 10.1 Supply rates of reductants and oxidants to subsurface oceans in icy worlds

Icy world	Molecule	Avenue	Rate (mol/yr)	Reference
Enceladus	H_2	Serpentinisation	$1-5 \times 10^9$	Waite et al. (2017)
Enceladus	H_2	Water radiolysis	$0.2-3 \times 10^8$	Waite et al. (2017); Ray et al. (2021)
Europa	H_2	Serpentinisation and volcanism	$0.1-5 \times 10^{10}$	Vance et al. (2016)
Europa	H_2	Water radiolysis	$\lesssim 5 \times 10^9$	Chyba and Hand (2001)
Enceladus	O_2	Oxidant delivery from surface	$\sim 10^7$	Ray et al. (2021)
Enceladus	O_2	Water radiolysis	$\sim 10^7$	Ray et al. (2021)
Europa	O_2	Oxidant delivery from surface	$\sim 3 \times 10^8$ to $\sim 3 \times 10^{11}$	Vance et al. (2016)
Europa	O_2	Water radiolysis	$\lesssim 10^{10}$	Chyba and Hand (2001)

Note: Avenue (third column) refers to the pathway by which the relevant molecule (second column) is produced. Rate (fourth column) is expressed in units of mol/yr, and should be considered a representative value. The rates in the second, fourth, and fifth rows were calculated by dividing the total amount delivered by the time interval.

The second route also requires H_2 and involves sulfate reduction (to yield sulfide) of the particular form (Hand et al., 2009):

$$SO_4^{2-} + 4H_2 \rightarrow S^{2-} + 4H_2O. \tag{10.20}$$

On Europa, sulfur may be available endogenously through water–rock interactions and supplied by exogenous sources (e.g., Io's volcanism). Sulfate salts on Europa's surface are probably formed via radiolysis (Brown and Hand, 2013), and can be delivered to the ocean. Likewise, in the case of Enceladus, oxidants like O_2 and H_2O_2 could react with appropriate compounds and produce sulfates, as per numerical models (Ray et al., 2021). If sulfate abundance is not the bottleneck, then the quantity of H_2 might govern how much biomass is generated, much like methanogenesis earlier.

Analogous to H_2, we can similarly enquire whether there are chemotrophic pathways wherein O_2 (an electron acceptor) is a reactant. An inspection of Tables 7.1 and 7.2 reveals multiple candidates, of which the best known is aerobic respiration. Another noteworthy pathway is (aerobic) methanotrophy, which is encapsulated as (Cockell, 2020, Chapter 6):

$$CH_4 + 2O_2 \rightarrow CO_2 + 2H_2O. \tag{10.21}$$

On icy worlds with hydrothermal systems, the synthesis of hydrocarbons is thermodynamically viable through Fischer–Tropsch type reactions, as established by numerical, theoretical, and experimental studies (McCollom and Seewald, 2007); refer to (5.9) for details. Hence, at least in principle, methane availability ought not be an issue. When it comes to O_2, one prominent source is the splitting of water by energy derived from radioactive decay, which may operate on a wide variety of icy worlds.

The second route for O_2 supply involves bombardment of the icy surface by energetic charged particles to generate oxidants, which are then transported across the icy shell into the ocean. The oceanic delivery rate of O_2 is proportional to the area of the icy world, burial depth of the oxidants, and mean molar concentration of the oxidants in the ice; and is inversely proportional to the average delivery time. The possible transport mechanisms are delineated in Sections 10.2.2 and 10.4.1. Due to the multiple avenues and timescales at play, the delivery rates of O_2 exhibit significant variation, as seen from Table 10.1. As the O_2 supply may be higher near the water–ice interface at the top of the ocean, the prospects of detecting a biosphere in that region might be brighter (Russell et al., 2017).

Turning to aerobic respiration, this is undoubtedly an energy-rich pathway that can support complex macroscopic life forms, but it simultaneously demands fairly high dissolved O_2 levels. If the upper limits of O_2 delivery rates on Europa are assumed, the evolution of such organisms might be tenable. If we denote the O_2 supply rate by $\mathcal{N}_{O_2}$ and use the fact that a typical fish consumes ~ 100 mol O_2 yr^{-1} kg^{-1} (Greenberg, 2010), the upper bound on the mass of fish-like macrofauna (M_{fauna}) that could exist is

$$M_{\mathrm{fauna}} \sim 10^8\, \mathrm{kg} \left(\frac{\mathcal{N}_{O_2}}{10^{10}\, \mathrm{mol/yr}} \right), \tag{10.22}$$

but we caution that the existence of these macrofauna is speculative. Instead, it seems more likely that putative biospheres on icy worlds will be comfortably dominated by microbes.

Lastly, a number of publications have attempted to take energetic constraints (oxidant/reductant abundances) into account, or invoked parallels with Earth-based environments, to estimate the cell density of microbes in Enceladus, which is the best characterised of the icy worlds. Table 10.2 compiles the predictions in this area, and highlights the diversity of approaches and wide range of values. As a point of comparison, glacial ice in Antarctica's Lake Vostok harbours cell densities of ~ 100 cells/mL. In general, subglacial lakes such as Lake Vostok are extreme environments exhibiting some resemblance to the subsurface oceans of icy worlds, and are thus considered semi-analogues (Preston and Dartnell, 2014).

10.5.2 Nutrients

We preface our short exposition with the caveat that assessments of nutrients are uncertain due to the paucity of data. Furthermore, as stated at the beginning of Chapter 7, it is hard to draw general conclusions about nutrient abundances because they are sensitive to geological, chemical, and physical processes operational on a particular world.

The two well-known icy worlds in the solar system are Enceladus and Europa. As remarked in Section 10.2.1, all bioessential elements are potentially documented in the plume of Enceladus. Phosphorus, for example, is reportedly plentiful in the subsurface ocean as per numerical models and observations. In the case of Europa, compounds of carbon (e.g., carbon dioxide), nitrogen (e.g., ammonia), and sulfur (e.g., sulfates) are confirmed through surface observations, or are theoretically predicted to abound in the icy surface and/or subsurface ocean (Hand et al., 2009; Brown and Hand, 2013).

On Earth, marine biological productivity is chiefly limited by two nutrients – nitrogen and phosphorus – of which the latter is theorised to regulate the productivity on long timescales. It is worth recalling that phosphorus comprises an essential component of nucleic acids (sugar-phosphate backbone), cell membranes (phospholipids), and metabolism (ATP). A major reason underpinning P limitation is that minerals containing this element, such as apatites, are relatively insoluble in water. As the major supply of P to Earth's oceans is via continental weathering, at first glimpse, it would

Table 10.2 Predicted cell densities in various environments of Enceladus

Environment	Cell density (cells/mL)	Notes
Bulk ocean	6×10^{-6}–890	Estimated from bioavailable energy flux and bioenergetic requirements of methanogens
Bulk ocean	80–4250	Optimistic estimate when 100% of H_2 is consumed by methanogens
Bulk ocean	8.9×10^{-5}	Estimated by coupling bioenergetic model of methanogens to simulations of hydrothermal vent environment
Hydrothermal vent fluids	8.5×10^8	Estimated from hydrothermal vent data on Earth and methanogens' maintenance energy
Hydrothermal vent fluids	10^5	Estimated from hydrothermal vent data on Earth and available energy
Plume	8.5×10^7	Estimated from Earth's hydrothermal vent data, methanogens' maintenance energy, and dilution in plume
Plume	10^4–10^7	Estimated from hydrothermal vent data on Earth, available energy, plume dilution, and concentration in bubbles
Plume	2×10^6	Estimated by coupling bioenergetic model of methanogens to simulations of hydrothermal vent environment
Ocean floor	0.5–600	Estimated from bioavailable energy flux and microbial maintenance energy

Note: Last row is derived from Ray et al. (2021), penultimate row is taken from Affholder et al. (2022), the third row is found by dividing the result from Affholder et al. (2022) by the hypothesised ocean volume of Enceladus (2.7×10^{16} m^3), and the remaining rows are adopted from Cable et al. (2021).

appear as though icy worlds are hampered in this respect because they lack continents. This argument was initially invoked to suggest that P might constrain biological productivity on some icy worlds (Lingam and Loeb, 2018b).

However, recent research has uncovered various paths by which this putative hurdle could be overcome. Modelling by Hao et al. (2022) for Enceladus showed that an alkaline and (bi)carbonate-rich ocean can remove certain cations through the formation of carbonate minerals, thereby paving the way for the release of phosphate and its accumulation to high concentrations. Similarly, if the subsurface oceans are acidic – hypothesised by some scientists to hold true on Europa – the dissolution rates of P minerals may be enhanced, thus liberating greater quantities of P into the ocean from the seafloor. Note that this discussion is implicitly centred on icy worlds with water–rock interactions. If a high-pressure ice layer blocks these interactions, the situation is perhaps rendered very different.

Looking beyond P, metals such as iron, nickel, or molybdenum might function as limiting nutrients on icy worlds, but we are confronted by several attendant unknowns. First, the chemical composition of the ocean (e.g., whether oxidants or reductants are dominant) will determine the solubility of certain metals. Second, the sources and sinks will vary from one world to another, as they are regulated by factors like the presence/absence of hydrothermal systems and the composition of the rocky core. Third, among the metals, it is not clear which of them are truly indispensable for life.

10.6   Evolutionary trajectories

Forecasts based on physical and chemical principles may be regarded as reliable to some degree, because they are perceived as universal. However, envisioning the course of biological evolution is far more challenging, owing to the notion of 'chance' (contingent) events (Powell, 2020). Hence, attempting to trace the paths taken by biological evolution on icy worlds is a speculative enterprise. Notwithstanding this caveat, a few tentative inferences might be discernible from Earth's evolutionary record (see also Chapter 6).

In Section 10.5.1, we touched on the interesting possibility that complex macroscopic life, perhaps resembling aquatic animals on Earth, may be feasible on icy worlds from an energetic standpoint. Given that the habitats are predominantly oceanic, Earth's oceans offer us a proxy (of sorts) to carry out some thought experiments. These qualitative conclusions might also pertain to worlds with *surface* oceans, that is, the so-called water worlds or ocean planets, which are expected to be quite common in the Universe.

The diversification of animals may have commenced in the Ediacaran (635–539 Myr ago), after which one of the crucial developments, from an evolutionary standpoint, was the transition of vertebrates from sea to land ~400 Myr ago (Sections 6.1.3 and 6.2.2). Subsequently, the majority (11 out of 13) of high-performance and metabolically expensive innovations emerged first on land, as chronicled by Vermeij (2017), making the latter the locus of such breakthroughs. A few examples are provided here, where Δt is the gap between the appearance of the innovation on land and in the sea.

Animal endothermy ($\Delta t \approx 222$ Myr): Endothermy (i.e., capacity to generate body heat internally) has several advantages, such as occupying niches unsuitable for ectotherms (which lack this attribute), enhanced muscle power and activity, and reduced vulnerability to environmental fluctuations.

Vascular structure in plants ($\Delta t \approx 345$ Myr): The vascular system involves specialised tissues and fibres that permit efficient transport of nutrients and water across the plant body, and confer mechanical support.

Eusociality ($\Delta t \approx 116$ Myr): Eusocial animals (e.g., termites) are characterised by features like reproductive division of labour and cooperative care. Eusociality may grant benefits such as broader access to resources (food and territory) and improved protection against predation.

On average, high-performance evolutionary innovations appear to have evolved ~100 Myr earlier on land than in seas and oceans (Vermeij, 2017). This temporal gap might be linked to the fact that the density of air on the surface of Earth (which is where land-based animals dwell and travel) is $\sim 10^3$ times smaller than that of water, consequently diminishing the impediments to organismal activity, and therefore facilitating these innovations. For instance, the friction force is proportional to the density, indicating that $\sim 10^3$ times less energy would need to be expended by an organism to move on land compared to water, if all other factors are held equal.

Hence, although we started off by examining biological aspects, it is apparent that physical forces are still vital mediators. The physical properties of the ambient medium associated with land and oceans (i.e., air and water, respectively) may have, as per some hypotheses, even modulated profound cognitive divergences in animals dwelling in these habitats (MacIver and Finlay, 2022). Mammals and birds have brains that are ~ 10 times larger than fish of the same body size, which could potentially stem from the cognitive demands of enhanced decision-making, foresight, and planning. The latter trio, in turn, might be valuable, or perhaps necessary, since the visibility range on land is much higher than in water.

The last, and possibly most intriguing, question that springs to mind is: what is the likelihood of the evolution of 'high intelligence' and perhaps technology? We refer the reader to Lingam and Loeb (2021, Chapter 7.7) for a detailed exposition. On Earth, aquatic animals like cetaceans (whales, dolphins and porpoises) and octopuses possess high intelligence, which might bolster the prospects of its emergence in extraterrestrial oceans.[6] A short synopsis of select traits documented in aquatic animals is furnished next.

- Bottlenose dolphins have reportedly passed the mirror self-recognition test, which is conventionally argued to signify self-awareness.
- Cetaceans and octopuses both utilise 'tools'. Bottlenose dolphins in Shark Bay (Australia) employ sponges to protect their beaks while foraging for food on the seafloor. Veined octopuses carry coconut shells with them, so that they may be assembled together to form shelters.
- Social learning (i.e., learning from members of the same species) has been observed in both cetaceans and octopuses. Long-term tracking of humpback whale songs has revealed striking evidence for information transmission across individuals via social learning on large spatial scales.
- Humpback whale songs are famously intricate and quite lengthy (single songs are of order 10 minutes). These songs possess substantial information content, and are classic examples of complex animal communication.

Equipped with this knowledge, it is tempting to tentatively conclude that high intelligence may emerge in oceans, but caution is clearly warranted. The question of whether species with advanced technology can arise in such environments remains unresolved. A preliminary study by Lingam et al. (2023a) suggested that the likelihood is low, as technological activity is suppressed by the denser and more viscous medium of water (relative to air). If we are to make genuine progress on this front, the quest for technological signatures (the theme of Chapter 14) acquires added importance.

10.7 Problems

Question 10.1: As per (10.2), how many objects with diameters greater than that of Mars may occur in the Oort cloud? Likewise, estimate the number of objects with diameters greater than 1.5×10^4 km. From this result, do you find it likely that rocky worlds larger than Earth (super-Earths) could exist in the outer reaches of the solar system? The hypothetical Planet Nine (Batygin et al., 2019), if confirmed, might belong to this group.

---

[6] Note, however, that cetaceans evolved from land mammals, establishing that their heritage is not solely aquatic in nature.

Question 10.2: No missions to Enceladus are on the horizon (Section 10.2.1). However, multiple missions to Enceladus have been mooted, and some are under serious consideration. A recent in-depth proposal is the *Enceladus Orbilander* concept, which was articulated in MacKenzie et al. (2021a). After perusing this paper, briefly answer the following questions.

   a. What were the alternative mission concepts studied (list their key characteristics)? Why was the Orbilander chosen over its competitors?

   b. What are the core goals of the Orbilander, and what instrumentation would it necessitate to fulfil its objectives?

   c. Can you think of any extra features that you would include on the science payload of such a mission, if money is not a bottleneck?

Question 10.3: Assume that the density of ice and liquid water are nearly equal to ~ 1 g/cm^3. By harnessing this simplification and the data in Section 10.2.1 (that is, the thickness of the icy shell and the ocean), estimate the pressure at the ocean floor of Enceladus. Repeat this calculation for Europa by utilising Section 10.2.2. In both cases, is the pressure greater or smaller than P_c, that is, the heuristic threshold for high-pressure ice? Is your result consistent with expectations of the internal structure of these two moons?

Question 10.4: By starting with (10.5) and following the steps outlined in Section 10.3.1, solve the ODE and derive (10.7). Use the fact that $T = T_s$ at $z = 0$, and that the melting point is attained at depth H_i; make sure to handle the signs correctly. Next, tackle the following parts.

   a. By substituting $\Gamma = 1$, $T_m \approx 260$ K, and the parameters for Europa into (10.7), what icy shell thickness is obtained?

   b. To arrive at $H_i \approx 25$ km for Europa (consistent with observations and models), what value of Γ would be needed? Assume that $T_m \approx 260$ K, and the other parameters are set equal to that of Europa. Is this value consistent with significant tidal heating on Europa?

Question 10.5: In this question, employ part (b) of Question 10.4 for Europa.

   a. Estimate the lower bound of f_w – to permit the existence of a subsurface ocean – for the above array of inputs.

   b. Estimate the upper bound on f_w – to avoid formation of high-pressure ices – using the preceding choices in tandem with $P_c \approx 1$ GPa.

Question 10.6: Repeat the same steps in Question 10.4 to determine the polymerisation depth H_p given by (10.18). You will start with the ODE in (10.5), draw on the accompanying discussion, and integrate it to find H_p, which is the depth at which the polymerisation temperature T_p is attained. Note that $T = T_s$ at $z = 0$, and the signs must be properly accounted for.

Question 10.7: Assume that the volume of Europa's ocean is ~ 3 times that of Earth's oceans. As a crude yardstick, we suppose that a minimum O_2 concentration of $\sim 10^{-5}$ moles per kg is sufficient to support animal-like organisms (Mills et al., 2014). If O_2 supply from the surface occurs continuously over a timescale of ~ 1 Gyr, and O_2 sinks are ignored, what is the minimum O_2 delivery rate (units of mol/yr) necessary to sustain these lifeforms? How does this delivery rate compare with those in Table 10.1?

Question 10.8: Free-floating icy worlds may be detectable via their blackbody radiation. The observable quantity, the spectral flux density (denoted by S_{max}), was roughly estimated in Abbot and Switzer (2011, Section 3). Explain why the inverse-square law scaling with the distance of the object from Earth is manifested. Next, by substituting (10.8) into the expression in that publication, obtain the final scaling for S_{max} as a function of R.

Question 10.9: Simulations performed by Laughlin and Adams (2000) suggest that Earth could be expelled from the solar system due to an encounter with a passing star. By combining (10.7) and (10.8) with $\Gamma = 1.2$ and $R = R_\oplus$ for Earth, estimate H_i and compare this value with Earth's average ocean depth of ~ 3.7 km. Using this result, alongside carefully reading this paper, summarise the expected consequences of such an event for the biosphere, and highlight connections with the topics covered in this chapter.

Question 10.10: In Section 10.6, we delineated a few characteristics of high intelligence in aquatic animals on Earth. By citing peer-reviewed publications, summarise at least two other traits emblematic of high intelligence identified in aquatic animals. In your opinion, after consulting appropriate references, justify whether you deem it plausible that technological species analogous to humans might emerge in oceans.

11   'Exotic' Life

As the preceding chapters have clearly illustrated, 'life' is often implicitly treated as being synonymous with *life-as-we-know-it* (or LAWKI for short). The latter, in turn, is fundamentally built on liquid water (as the solvent) and carbon (as the backbone). It is natural to wonder, therefore, whether alternative forms of life reliant on different solvents and backbones are conceivable and plausible. The thrust of this chapter is to address the fascinating, yet poorly explored, realm of 'exotic' life. The rationale behind placing exotic in quotes is that the putative lifeforms under scrutiny are only exotic if viewed from our patently geocentric (Earth-centric) perspective; this crucial point should be borne in mind henceforth.

As one of the chief objectives of astrobiology is to seek out (signatures of) extraterrestrial life, it is along expected lines that the subject of exotic life and alternative biochemistries has an intricate and lengthy history. In fact, the early decades of astrobiology witnessed a number of books and publications devoted to intriguing conjectures about exotic life, most of which are now out-of-print or forgotten. The interested reader is invited to peruse Firsoff (1963) and Feinberg and Shapiro (1980) for historical glimpses of speculations concerning exotic life during that era.

During the same time period, other scientists like Sagan (1973b) and Wald (1974) advocated the opposing stance, to wit, that the basis of biochemistry is universal and that putative extraterrestrial life might also be based on carbon and/or liquid water. This outlook has been revived sporadically, and remains a genuine possibility to date (see, for example, Fry 2000; Pace 2001). The core reason for the lack of progress in modelling exotic life is not because of such arguments in favour of universality, but arguably more due to the intrinsic challenges associated with conceptualising the underlying biochemistries from an experimental or theoretical standpoint.

The twenty-first century has heralded a fair amount of much-needed progress on this front. The advances made were stimulated to an extent by two unusual worlds in our solar system, Titan and Venus, which are separately explored towards the end of this chapter. In-depth treatments of alternative biochemistries that justify their viability and present concrete scenarios for astrobiological targets are provided in Bains (2004), Benner et al. (2004), and Baross et al. (2007). However, as some of these publications presume chemistry and biology prerequisites beyond those of the book, we will instead mirror the exposition in Schulze-Makuch and Irwin (2018, Chapters 6 and 7), which can be consulted for additional details.

11.1   Water and its alternatives

In this section, we will take a close look at liquid water and its advantages as a solvent,[1] before delineating a few alternatives and their pros and cons.

---

[1] Instead of employing the term 'liquid water' every time, we resort to the shorthand 'water' and trust the readers to differentiate between them as per the context.

Before doing so, we may ask ourselves why a solvent is necessary in the first place. A key reason is summarised in Chapter 7: liquids offer an appealing intermediate degree of localisation, whereas gases offer too much and solids too little. There are many criteria that determine what constitutes a 'good' solvent, some of which are listed below and elaborated in Section 11.1.1.

1. If the liquid state of a substance has large thermal range, this factor enables it to exist (in this form) in diverse settings.
2. To facilitate the breaking or formation of certain chemical bonds (e.g., ionic or hydrogen bonds), a finite polarity is preferred. We will delve into this aspect further in Section 11.1.1.
3. A high specific heat capacity and enthalpy (latent heat) of fusion or vaporisation permit the liquid to act as an efficient thermal buffer, and reduce susceptibility to thermal fluctuations. This trend is manifested because it would be comparatively difficult to change the temperature of the liquid unless a high amount of heat is consumed or released.
4. A high cosmic abundance of the solvent suggests that it may be widespread and/or accumulate to yield extended liquid bodies (e.g., seas).

11.1.1 Water

There are manifold reasons why water represents an excellent solvent and mediator of (bio)chemical reactions. A subset of these are subtle and require a deeper dive into chemistry; consult Pohorille and Pratt (2012) and Ball (2017) for such analyses. Fortunately, however, several compelling benefits accorded by water are relatively easy to elucidate and comprehend.

First, water has a significant thermal range of 100 K (or equivalently 100°C) at the pressure of 1 bar (10^5 Pa); note that the pressure must be specified since the thermal range is sensitive to this parameter. This feature is intertwined with hydrogen bonding, which is a vital facet of water and shapes several of its remarkable properties. Intermolecular hydrogen bonds are held to be the primary cause of the elevated boiling point of water, as additional (thermal) energy is needed to break these bonds.[2] The chemistry of hydrogen bonds is quite intricate, and we recommend Atkins and De Paula (2006, Chapter 11.9) for a succinct description. However, for the purposes of our analysis, it may be visualised as follows.

A hydrogen bond is typically expressed as $A-H \cdots B$, where A and B are highly electronegative atoms, namely: they have a strong tendency to attract electron pairs; in the case of water, A and B correspond to oxygen (O). In the (over)simplified picture, the hydrogen bond arises because the partial positive charge associated with the proton from hydrogen (H) is attracted to the partial negative charge of B, and the ensuing Coulomb force comprises the basis of the hydrogen bond. It is worth reiterating that this explanation does not properly account for the complexity and subtlety of the interactions that mediate the hydrogen bond, but it suffices for this chapter.

The next aspect we wish to highlight is the *polar* nature of water: it has a finite (non-zero) electric dipole moment of 1.85 D (where 1 D is 3.34×10^{-30} C m), as outlined in Question 11.1. The polarity of water makes it feasible to initiate bonds with ions and polar molecules. The classic Coulomb force tells us that the force between two single charges (i.e., monopoles) scales as $1/r^2$, where r is the

---

[2] The bond dissociation energy (i.e., change in enthalpy) for the hydrogen bonds of water is around 20 kJ/mol, although the exact value depends on the temperature.

separation between the charges. Likewise, we can determine the force between a dipole and other charge configurations by adding powers of r to the denominator. In particular, the dipole–monopole and dipole–dipole forces are proportional to $1/r^3$ and $1/r^4$. These scenarios are respectively interpreted as interactions between water and ions, and water and polar substances. In contrast, if the solvent has a dipole moment of zero (viz., it is non-polar), such interactions may be effectively ruled out.

The prior paragraph illustrates why water would be capable of dissolving polar molecules (e.g., amino acids like cysteine and serine) and ions. The importance of the former is apparent since many biomolecules are polar (and can thus interact with water), but the significance of the latter should not be underestimated. Ions play key roles in biology: they are involved in intra- and inter-cellular communication, stabilisation of Watson–Crick–Franklin base pairs in DNA, and regulating osmostic pressure, to name a handful. Hence, the benefits from the polar nature of water are clear.

We now turn our attention to one more profound attribute of water: the hydrophobic effect, whose relevance for biology was articulated in a famous paper by Tanford (1978). Given that a rigorous treatment is impractical, we will pursue a loose qualitative summary. It is well-known that oil and water do not mix, a classic real-world example of this effect. In a nutshell, when hydrophobic substances (often non-polar molecules) are added to water, the ability of water to form hydrogen bonds is diminished in the vicinity of those molecules. As a result, the orientation of the hydrogen bonds formed in this region is restricted, and consequently attains a higher degree of order. This entropic ordering is equivalent to $\Delta S < 0$, which could cause $\Delta G > 0$ as seen from (7.3), thereby initiating the hydrophobic effect.

The dielectric constant of water is yet another crucial feature because of its high value of $\varepsilon_r \approx 80$. In electrostatics, it is straightforward to prove that the force and potential energy between a pair of charges (or charge configurations) scale inversely with the dielectric constant (Young and Freedman, 2018, Chapter 24). At high values of ε_r (as for water), the reduction of the interaction strength ensures that electrostatic forces do not overwhelmingly dominate, and that they are of moderate (and conceivably ideal) strength for facilitating regulation of cellular processes (Pohorille and Pratt, 2012).

Moving on, as mentioned in Section 11.1, a high specific heat capacity and/or latent heat of vaporisation are desirable. Water scores highly on this front, which is partly attributable to the effect of hydrogen bonding. The thermodynamic heat capacity C may be generally written as

$$C = \frac{\Delta Q}{\Delta T},\tag{11.1}$$

where ΔT is the change in temperature associated with the change in heat ΔQ. On rearranging this expression for fixed ΔQ, we notice that $\Delta T \propto 1/C$. In other words, liquids with high values of C cause lower shifts in temperature, indicating that they are reliable thermal buffers. Water belongs to this category compared to most other liquids.

The high surface tension – a measure of the energy supplied to raise the surface area (by unity) – of water is noteworthy on account of its connections with droplet formation. If a liquid droplet is suspended in a gas (e.g., water droplet in air), the pressure difference between the interior and exterior ΔP can be worked out using the Young–Laplace equation:

$$\Delta P = \frac{2\gamma_l}{R_D},\tag{11.2}$$

where γ_l is the surface tension of the droplet material, and R_D is the radius of the droplet. For fixed ΔP, it is evident that raising γ_l would imply a commensurate increase of R_D. Hence, broadly speaking, liquids with higher surface tension appear to be compatible with the formation of larger droplets. The ability to produce droplets has multiple uses, one of which we encountered in Section 5.5.3 pertaining to the origin(s) of life, because droplets are possible sites of abiogenesis, especially in terms of concentrating macromolecules and functioning as miniature prebiotic reactors.

The cosmic abundance of water is another astrophysical facet that is sometimes overlooked in discussions centred on water's chemical properties. Hydrogen is the most abundant element in the Cosmos, and oxygen is close behind in the third position; the only element in between is helium, which is quite inert. Hence, it is natural to suppose that water must be likewise abundant, and this assumption is borne out by observations. However, a subtle distinction must be made here between water and liquid water in this context. The majority of water in the Cosmos apparently exists as vapour or ice; for instance, even hot objects like the Sun and quasars (active supermassive black holes) host water vapour. Examining Figure 11.1, the phase diagram of water, reveals that the liquid state is feasible only across a fairly restricted range of temperature and pressure.

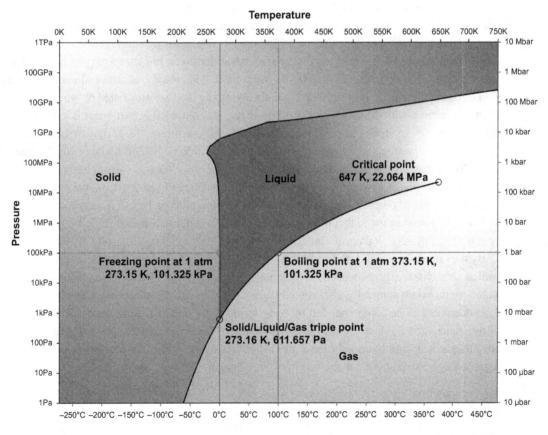

Figure 11.1 The phase diagram of water for the three major states of matter, with the pressure and temperature on the y-axis and x-axis, respectively. Beyond the critical point, we enter the regime of a supercritical fluid, which exhibits physicochemical properties resembling both gases and liquids. (Credit: https://commons.wikimedia.org/wiki/File:Phase_diagram_of_water_simplified.svg; CC-BY-SA 3.0 license)

Last, the anomalous expansion of water warrants highlighting. Introductory physics and chemistry textbooks often state that substances expand on heating and contract on cooling, and that these effects are directly linked to shifts in the kinetic energy of the atoms/molecules. As the mass stays constant, a decrease in volume at lower temperatures would translate to an enhanced density since these two quantities are inversely proportional to each other. However, water is a striking exception to this trend: at the pressure of 1 bar, it attains a maximum density of virtually 1 g/cm^3 at 4°C and *declines* very slightly thereafter (by ∼0.013%) on decreasing the temperature to 0°C. Of even greater significance is the fact that the density of ice (specifically ice I$_h$) at 0°C and 1 bar is distinctly lower at 0.917 g/cm^3. This anomalous behaviour is a notable manifestation of how hydrogen bonds modulate the properties of water.

The relevance of the preceding results to astrobiology can be gleaned by performing a simple thought experiment. If we have a liquid water body and progressively lower the temperature until it falls below 0°C, we would expect water ice to exist at the top (due to its lighter density), followed immediately by liquid water at 0°C, and the bottom layer would be composed of liquid water at 4°C. If the system is continuously cooled, the ice layer can grow in thickness, but the temperature of the bottom layer ought to stay roughly fixed at 4°C, until the liquid water has primarily frozen into ice.

Therefore, it would be possible for (a) organisms to survive in the deep waters, and (b) water–rock interactions at the floor to occur, provided that the pressure at the bottom is not high enough to induce the formation of high-pressure ices (see Figure 11.1). To put it differently, even in extremely cold environments, liquid water may survive underneath an ice layer. In fact, this is precisely the scenario that we covered in the previous chapter. If water behaved similarly as conventional liquids, the ice would have settled at the bottom, consequently stymieing the two pros listed above.

In conjunction with the aforementioned positives, it is remarkable that water is consumed or produced in 30–50% of all biochemical reactions, and that it constitutes 99.4% (by molar concentration) of all reactants, products, or intermediaries involved in *Escherichia coli* metabolism (Frenkel-Pinter et al., 2021). It would be a mistake, however, to presume that water as solvent has no downsides; we have merely emphasised water's assets hitherto. The tendency of water to dissolve polar molecules is ostensibly rendered problematic for certain biomolecular building blocks. Experiments by Levy and Miller (1998) demonstrated that the half-life of cytosine (refer to Section 5.1.2) in water is 1.7×10^4 yr at 0°C and merely 19 days at 100°C.

By the same token, water molecules are liable to cause hydrolysis of proteins and nucleic acids, as well as some vital reagents needed for metabolism. One other drawback is that reactions that generate water are thermodynamically disfavoured if they occur in aquatic solution(s). The physical basis of this statement can be comprehended by considering (7.5), and analysing what would happen if one of the anticipated products (water) is overabundant. Therefore, while water still boasts of enviable advantages, it is simultaneously beset by difficulties to the point that the *water problem* might be regarded as a crucial hurdle to the origin of life (Benner, 2014).

11.1.2   Alternatives to water

The prior discussion has highlighted a number of benefits inherent to water. We will now examine some of the commonly hypothesised alternatives to water, and determine how they stack up against it. We divide up the candidates into polar and non-polar solvents. The salient physical data are furnished in Table 11.1, which can be consulted when necessary.

Table 11.1 Physical characteristics of putative life-hosting solvents

Solvent	MP (in K)	BP (in K)	DM (in D)	ΔH_{vap} (in kJ/mol)	ε_r
H_2O	273	373	1.85	40.7	80.1
NH_3	195	240	1.47	23.3	16.6
H_2SO_4	283	610	$\sim$2.7	$\sim$56	$\sim$100
$HCONH_2$	275	483	$\sim$3.7	$\sim$62	$\sim$110
HCN	260	299	2.99	25.2	115
HF	190	293	1.83	30.3	83.6
CH_4	91	112	0	8.2	1.7
C_2H_6	90	185	0	14.7	$\sim$1.9
N_2	63	77	0	$\sim$5.6	$\sim$1.4

Note: 'MP' and 'BP' are melting point and boiling point at 1 bar; the liquidity range is BP−MP. The columns 'DM', ΔH_{vap}, and ε_r represent the solvent's electric dipole moment, enthalpy of vaporisation, and the dielectric constant.

11.1.2.1 Polar alternatives

Ammonia: Of all the potential solvents, none has perhaps received the same attention as ammonia (NH_3). The notion of ammonia serving as the solvent dates back to at least the 1950s, and was proposed by J. B. S. Haldane, one of the pioneers of modern evolutionary biology and origin(s)-of-life studies. Ammonia is particularly appealing because a number of biochemical reactions on Earth have analogues in ammonia. The synthesis of the peptide bond is essentially a dehydration reaction in which water is released (refer to Section 5.1.1). The same bond may be synthesised through the release of ammonia in principle, as identified by Haldane.

Aside from the above parallel with water, analogues of metabolism, cell membranes, and the phosphate group (PO_4^{3-}) in ammonia solutions have been advanced, albeit without much systematic research in astrobiology. For instance, in the realm of metabolism, molecules with C=O (carbonyl) bonds – which constitute the bedrock of Earth-based metabolic pathways (Baross et al., 2007) – may be replaced with C=N (imine) units instead, when NH_3 is chosen as the solvent. Computational modelling by Hamlin et al. (2017) suggests that ammonia can facilitate the formation of the correct Watson–Crick–Franklin base pairs, and could thus permit DNA replication.

Ammonia has the capacity to dissolve sodium and potassium ions; both these ions are widely employed in biology on Earth (e.g., for chemical equilibrium and cell signalling). The dynamic viscosity of ammonia is several times less than that of water, which would boost diffusion to an appreciable degree, as demonstrated by (7.17). Hence, reactants and waste products can be rapidly transported inside and outside the cell interior, respectively. The abundance of ammonia is another easily overlooked advantage. Ammonia–water solutions are believed to abound in the clouds of Jupiter, and a substantial fraction of the inventory of comets is composed of ammonia.

However, in a variety of respects, liquid ammonia seems underwhelming relative to water: (1) it is more susceptible to dissociation when exposed to ultraviolet (UV) radiation; (2) it has a narrower thermal range (at the same pressure); (3) it possesses a lower surface tension, which may pose hindrances to droplet formation, as indicated by the discussion below (11.2); and (4) it has smaller dipole moment and dielectric constant, which could diminish its ability to dissolve crucial polar molecules and ions.

Sulfuric acid: Perhaps the most conspicuous feature of sulfuric acid (H_2SO_4) is its incredible liquidity range. At pressure of 1 bar, sulfuric acid remains in liquid state between 10°C and 337°C, a range much higher than that of water. Furthermore, H_2SO_4 has a higher dipole moment and dielectric constant, making it clearly well-suited for dissolving a plethora of polar and ionic substances. It also has a higher enthalpy (latent heat) of fusion than water, which should enhance its capacity as a thermal buffer.

In metabolism, it is conceivable that the roles played by C=O (carbonyl) compounds could be replaced by molecules with C=C bonds in H_2SO_4. If we consider the question of cosmic abundance (an important criterion), sulfuric acid oceans might be common, perhaps even as abundant as liquid water oceans (Ballesteros et al., 2019). It is not necessary to restrict ourselves to pure H_2SO_4. As sulfuric acid mixes well with water, the possibility of water–sulfuric acid mixtures is now attracting serious assessment owing to the renewed interest in the clouds of Venus, which are thought to predominantly contain sulfuric acid, with much lower amounts of water. We shall revisit this intriguing environment in Section 11.3.1.

The extreme reactivity of sulfuric acid is often perceived as its overriding downside. An in-depth computational study by Bains et al. (2021) concluded that $\sim 70 \pm 5\,\%$ of all biochemical compounds have lifetimes of <1 s. This finding may appear to eliminate the prospects for life, but recall that this chapter is devoted to scenarios beyond LAWKI. In this vein, Bains et al. (2021) analysed the broader chemical space of synthetic molecules manufactured by humans, and determined that 70–85% of them have relatively stable lifetimes of $>1,000$ s. Moreover, this expanded space consisted of molecules with diverse functionalities (e.g., potential components of cell membranes) and varying levels of solubility. Finally, experimental results have recently established that amino acids and nucleobases (see Section 5.1.2) are fairly stable in sulfuric acid (Seager et al., 2023b), and that this solvent may permit the synthesis of a bevy of complex organic compounds by starting from simple molecules like formaldehyde (Spacek et al., 2024).

While these papers augur well for sulfuric acid as solvent, we emphasise that concrete paths to abiogenesis are still rather elusive at the moment.

Other solvents: Looking beyond ammonia and sulfuric acid, several other promising contenders might exist, but we will merely touch on them in passing here; additional details are furnished in the references cited previously.

1. Formamide ($HCONH_2$) is endowed with an appealing combination of high thermal range, dipole moment, and is regularly utilised as a feedstock molecule in the prebiotic synthesis of biomolecules (e.g., nucleic acids). The foremost drawback, however, is that sizeable bodies of formamide may be unlikely to form in most settings.
2. On the one hand, hydrogen cyanide (HCN) is quite abundant on Earth and elsewhere, and its significance as a feedstock molecule in prebiotic chemistry for assembling the building blocks of major biomolecules is rising every year (Ruiz-Bermejo et al., 2021). On the other hand, its thermal range (at 1 bar), enthalpy of vaporisation, and surface tension are all perceptibly lower than that of water.
3. Hydrofluoric acid (HF) has the key metabolic benefit that fluorination yields more energy compared to oxidation, suggesting that the former could substitute for the latter. HF also evinces

solvent properties similar to, or at times better than, those of water. However, fluorine has a very low cosmic abundance in comparison to oxygen, raising questions about whether it can give rise to extended liquid bodies.

11.1.2.2 Non-polar alternatives

We have dealt hitherto with polar solvents, but we will now relax the assumption that the solvent must have a finite dipole moment. Such non-polar solvents are of immediate relevance to Titan in our solar system (addressed in Section 11.3.2), and similar worlds elsewhere. Non-polar solvents, despite their conceivable ubiquity, are poorly explored in astrobiology; systematic empirical and theoretical analyses are needed.

As per the standard textbook narrative, a non-polar solvent lacks a dipole moment and cannot participate in dipole–monopole and dipole–dipole interactions. At first glimpse, it may be supposed that non-polar solvents are not well-suited for facilitating reactions with organic compounds (many of which are polar). However, there is adequate evidence from synthetic organic chemistry that this statement is not valid; with that being said, most of the experiments and modelling did not attempt to explicitly simulate astrobiological environments. An immediate advantage of non-polar solvents is that the obstacle of degradation of biomolecular building blocks (e.g., nucleobases) is perhaps not as pronounced as it would be for water.

A couple of widely investigated non-polar solvents are methane (CH_4) and ethane (C_2H_6). Not only are these two solvents confirmed in the lakes and seas of Titan – with liquid CH_4 ostensibly exhibiting higher average abundance (Hörst, 2017) – but they also represent quintessential simple hydrocarbons that are prevalent in the Universe. Furthermore, liquid methane and ethane possess substantially lower melting and boiling points than water, implying that objects located in regions farther away from the star could support these solvents on their surfaces. This aspect is quantified in Question 11.4 and Ballesteros et al. (2019); the latter proposed that oceans of ethane may be as much as ∼10 times more common than liquid water oceans.

Liquid nitrogen (N_2) has a boiling temperature of 77 K at 1 bar, making it viable as a solvent for objects at large distances from the star. We caution, however, that this option remains scarcely explored from an experimental standpoint. A genuinely unusual choice of solvent would be a supercritical fluid (loosely intermediate between gas and liquid), which is implicit in Figure 11.1. For example, supercritical carbon dioxide (CO_2) displays both high diffusion rates and solubility of various substances. Moreover, at certain habitats on the ocean floor, mixtures of water and supercritical CO_2 harbour cell densities of ∼10^7 cells/mL (Budisa and Schulze-Makuch, 2014).

11.2   Carbon and its alternatives

A number of alternatives to liquid water have been identified, but the situation is rather different for carbon, where the choice is limited (chiefly silicon). In fact, if one had to hazard a guess, carbon-based biochemistry might be the favoured outcome, and perhaps even universal to a degree. We caution that our exposition in this section, as well as this chapter overall, sacrifices some nuances for the sake of brevity.

11.2.1 Carbon

It is instructive to begin our journey with summarising the pros of carbon prior to contemplating alternatives. For starters, it is self-evident that the scaffolding of LAWKI is carbon, given that it constitutes the backbone of lipids, sugars, nucleic acids, and proteins.

Carbon readily enters into bonds with other elements, thus enabling the synthesis of compounds with a wide array of functional groups (i.e., structural units with specific chemical properties), including –OH (alcohol), –CHO (aldehyde), –COO–R (ester), and –COOH (carboxylic acid). Many ensuing carbon compounds are stable over a broad thermal range, allowing numerous environments to host carbon-based (bio)chemistry. The stability of carbon molecules is partly attributable to the high carbon–carbon bond strength (dissociation energy of $\sim$368 kJ/mol) and the high activation energies associated with the appropriate chemical reactions; raising the latter lowers the rates, as is apparent from inspecting (7.16).

Carbon compounds are documented to assemble into myriad structures, both linear and cyclic. Carbon has the capacity to form double and triple bonds, which alter the chemical and physical properties of the resultant molecules. For instance, as we discussed in Section 7.2.1, psychrophiles take advantage of the enhanced flexibility of lipids with double bonds to resist cold temperatures. The ability of carbon to generate multiple configurations has proven amenable to the simultaneous existence of lipids (primarily linear molecules) and proteins (with manifestly 3D structures).

An oft-overlooked, yet pivotal, point is that the solvent and the scaffolding ought not be viewed purely in isolation because their synergistic interactions are crucial. In this respect, carbon forms strong bonds with both hydrogen and oxygen (comprising water), with bond dissociation energies of $\sim$413 kJ/mol and $\sim$360 kJ/mol, respectively. Moreover, as underscored earlier, a plethora of organic carbon compounds are soluble in water, further emphasising the deep connections between carbon and water. Carbon is the fourth most abundant element in the Universe, a datum in its favour that ought not be underestimated, although it must be prefaced with the caveat that silicon is presumably more prevalent on terrestrial worlds.

Until now, we have singled out carbon's relationship with biomolecular building blocks. However, we must also recognise its significance in Earth-based metabolisms. Carbon can cycle between its most oxidised and reduced versions, respectively embodied by CO_2 and CH_4. Note that these molecules are employed as reactants by methanogens and methanotrophs, both of which are well-known chemotrophs outlined in Section 7.1.3. Since both these compounds are gases in Earth-like settings, that makes it easier for their uptake or disposal by organisms.

Last, we highlight that extraterrestrial life may entail carbon as the scaffolding, and yet not use the same building blocks. For example, such life could potentially consist of different amino acids, nucleobases, and so on. The Murchison meteorite alone contains $\sim$100 amino acids, much more than the 20–22 proteinogenic amino acids in Earth biology. Likewise, non-canonical nucleobases like xanthine have been recovered from this meteorite. As per state-of-the-art computational analyses (e.g., Cleaves et al., 2019), there are sufficient grounds to hypothesise that a plethora of non-standard biomolecular building blocks might be utilised elsewhere. The putative life forms that would emerge are not truly exotic (i.e., non-carbon-based) in a sense, but can nevertheless diverge substantially from those on Earth.

11.2.2 Alternatives to carbon

We will single out silicon, with other candidates relegated to the closing paragraphs, since the former has attracted more scrutiny.

Silicon has a rich and lengthy history of being considered an alternative to carbon because it belongs to the same group (group IV) of the periodic table. In fact, the *Star Trek* episode entitled 'The Devil in the Dark' (1967) presciently envisioned a silicon-based organism. At the outset, we remark that extraterrestrial life need not 'choose' between carbon and silicon. The field of organosilicon chemistry revolves around the C–Si bond, which is endowed with nearly the same bond strength as the C–C bond; the dissociation energies are ~360 kJ/mol and ~368 kJ/mol, respectively.

As silicon is larger than carbon, to wit, composed of more 'shells' of electrons, this feature engenders notable divergences. A few major differences between the properties of silicon and carbon are as follows.

1. The valence electrons (i.e., in the outermost shell) in carbon are closer to the nucleus, and are therefore often capable of forming stronger bonds due to the attractive force exerted by the positively charged nucleus; the converse holds true when we consider silicon.
2. Silicon's increased size is responsible for the larger bond angles (i.e., angle between two bonds with a common atom) associated with it.
3. The preceding trait is reported to usually impede the formation of double and triple bonds, as well as analogues of the so-called aromatic hydrocarbons that are distinguished by chemical structures resembling benzene (which crudely has alternating single and double bonds).
4. The Si–O bond has a very high bond dissociation energy of ~452 kJ/mol, indicating that silicon has a strong tendency to become oxidised.

Given that life on Earth essentially comprises carbon polymers, we will briefly venture into the world of silicon polymers.

Silanes are analogues of hydrocarbons, endowed with the general chemical formula Si_aH_b. The Si–H bond dissociation energy is ~393 kJ/mol, which is distinctly lower than the C–H bond dissociation energy of ~413 kJ/mol and the Si–O bond dissociation energy. From these data, we gather that the Si–H bond is comparatively weak, and that the presence of oxygen can rapidly oxidise silanes. Silanes are typically not stable under Earth-like conditions, but their stability may be enhanced at lower temperatures and/or higher pressures. In this case, the solvent would have to be conducive to silane-based biochemistry, and thus possess appropriate physical and chemical characteristics. Alternatively, replacing hydrogen with organic functional groups could boost the thermal and chemical stability of the resultant molecules.

Despite these stumbling blocks, silanes have desirable chemical properties, as reviewed in Benner et al. (2004). They can consist of as many as ≥26 Si–Si bonds, albeit their stability generally declines with increasing length. Some silanes are chiral, a vital attribute that was summarised in Section 5.1.1. Depending on which functional groups are attached to the Si–Si backbone, they may be soluble in either polar or non-polar solvents. Certain silanes are loosely similar to lipids, and evince the capacity to assemble in water and produce compartment-like structures. Last, electron delocalisation (observed in aromatic compounds) and electronic excitation (a cornerstone of photosynthesis; see Section 7.1.4) are realisable in silanes.

Owing to the aforementioned drawbacks of silanes, some environments are likely more favourable to their existence (Schulze-Makuch and Irwin, 2018). First, the settings must be ideally devoid of water and molecular oxygen, as the O can readily react with Si and disrupt the structure of silanes. Second, as mentioned earlier, low temperatures and high pressures could aid in stabilising silanes. Third, the presence of solvents compatible with silane chemistry under these specific physicochemical conditions is necessary, as remarked previously; liquid methane might be one feasible candidate. Last, environments depleted in carbon are perhaps preferable, as otherwise C may hinder Si by virtue of its greater flexibility. Intriguingly, Titan appears to fulfil most of these criteria, as documented in Section 11.3.2.

Instead of silanes, it has been theorised that *polysiloxanes* (better known as *silicones*) could serve as the basis of silicon-based chemistry. Silicones are made up of (organic) functional groups attached to a $Si-O$ backbone. The sturdy $Si-C$ bond and the even stronger $Si-O$ bond jointly confer high thermal and chemical stability on silicones. Furthermore, silicone polymers have reduced susceptibility to oxidation, hydrolysis (they are markedly hydrophobic), and UV photolysis; they also happen to be effective dielectrics. To avoid being outcompeted by purely carbon-based chemistry, a temperature of 200–400°C may be beneficial as organics are often unstable in this range. However, candidate solvents (barring sulfuric acid) rarely occur in the liquid state at such temperatures (for pressure of 1 bar).

Irrespective of whether silicon biochemistry is tenable, it must be appreciated that silicon has played a profound role in biology on Earth. Silica (SiO_2) augments rigidity in both plants and animals. Diatoms are microalgae whose shells are made of silica, thereby requiring the uptake of the latter; diatoms are important because they might constitute ~10–20% of the total carbon fixation (biosynthesis of organics from inorganic carbon) on Earth. Silicon compounds can take part in bacterial and fungal metabolisms as per some studies. Last, certain minerals composed of Si (and other elements) could have acted as catalysts in the origin of life.

It is appropriate at this juncture, before commenting on alternatives to silicon, to assess the overall plausibility of silicon-based chemistry. After a comprehensive review and computational analysis, Petkowski et al. (2020) concluded that lifeforms centred on silicon chemistry are implausible in most solvents, although sulfuric acid may fare better than water in this context. Hence, it is potentially safe to presume that carbon comfortably remains our best bet according to our current knowledge of (bio)chemistry.

If we consider elements aside from C and Si, their prospects for being the bedrock of alternative biochemistries are dimmer. Of the contenders, boron is perhaps more promising compared to the rest. As this element is from group III of the periodic table, it contains one less valence electron than either carbon or silicon (from group IV). It displays a high affinity for ammonia and the formation of $B-N$ bonds, which indicates that ammonia may represent a viable solvent for boron biochemistry. On account of the higher reactivity of boron compounds – such as *borazine* ($B_3N_3H_6$), informally dubbed 'inorganic benzene' – the lower temperatures at which ammonia is liquid could moderate the reaction rates to acceptable levels. Thus, the possibility of boron-based biochemistry cannot be ruled out, but a major obstacle is the meagre cosmic abundance of boron, as well as the expected scarcity of boron in the crust of terrestrial planets. Boron seems, on the whole, an unlikely replacement for carbon.

Other elements contemplated in lieu of carbon are nitrogen, phosphorus, sulfur, and germanium. However, all of them apparently have noteworthy drawbacks elucidated in Schulze-Makuch and Irwin

(2018, Chapter 6). For instance, germanium is from group IV (the same as C and Si), but is hampered by its metallic nature, very low abundance, and large size.

11.3 Exotic life in the solar system?

Until now, we have restricted ourselves to generalities that may be applicable in diverse settings. However, in order to empirically probe the likelihood of exotic life in sufficient detail, it is imperative to dig into actual targets. Fortunately, there are a minimum of two worlds in our solar system that fall under this category, which we tackle hereafter.

11.3.1 The clouds of Venus

Venus, the nearest neighbour to Earth and often whimsically called Earth's twin or sister planet, has a mean surface temperature of $\sim464°C$. Hence, when we envisage Venus, a hellish world comes to mind, and this conception is quite accurate inasmuch as the surface is concerned. However, this simplistic picture is giving way to a more nuanced, albeit speculative, perspective of Venusian – also (seldom) known as 'Cytherean' – habitability.

In sharp contrast to the surface, the cloud decks of Venus harbour physical conditions partially akin to those on Earth (life has been detected in the latter). More precisely, at the altitude of ~47–57 km, the temperatures span $\sim0°C$ to $\sim100°C$, and the pressures are in the range of 0.8–1.2 bar (Limaye et al., 2021, Table 1). Even before these parameters were fully understood, Carl Sagan (1934–1996) and Harold Morowitz presciently conjectured that the clouds of Venus might harbour an aerial biosphere (Morowitz and Sagan, 1967). This proposal was intermittently revived in the subsequent decades.

A puzzling feature of the Venusian cloud layers is the sharp rise in reflectance from ~0.2 at ~300 nm to ~0.8 at ~500 nm. Abiotic models have explained the strong UV absorption via organics or inorganic compounds like sulfur dioxide, iron chloride, and iron sulfate. By synthesising observational data with constraints on energy and nutrient sources, Limaye et al. (2018) hypothesised that biology might contribute to this spectral signature, and that microbes may dwell inside the protective cocoons of cloud droplets/aerosols; this work built on earlier suggestions of this kind. Potential candidates include photosynthetic pigments and UV screening biomolecules.

Greaves et al. (2021) (see also Greaves et al., 2022) reported the detection of the gas *phosphine* (PH_3) at an abundance of ~1–10 ppb in/above the cloud layer at ~50–60 km altitudes. This paper stirred intense debate because phosphine on Earth is produced chiefly by biology, which raised the possibility of biogenic origin for Venusian phosphine. The majority of independent analyses have failed to identify unambiguous signatures of phosphine (e.g., Lincowski et al., 2021; Villanueva et al., 2021; Cordiner et al., 2022). Ultimately, settling the riddle of whether phosphine is prevalent in the cloud deck is feasible through *in situ* measurements and samples.

At this stage, it is vital to avoid conflating the two chief unknowns of Venusian aerial life and phosphine. Even if the latter is absent, it does not rule out the former, and vice versa. Motivated by the aforementioned works, a variety of publications have begun to rigorously reassess the habitability of Venusian clouds; as the results are numerous, we refer the reader to Limaye et al. (2021) and Bains et al. (2024). In brief, ongoing research has posited that putative microbes in the cloud layer have access to sufficient photon fluxes for oxygenic photosynthesis; feedstock molecules for chemotropic pathways like nitrate reduction and sulfur oxidation; and soluble phosphorus concentrations $\gtrsim 10$ times higher than nutrient thresholds on Earth.

However, since a multitude of unknowns exist, by no means can we definitively say that the Venusian cloud decks are habitable. In fact, this environment faces profound challenges, summarised in Seager et al. (2021) and Schulze-Makuch (2021). The first is the extreme acidity – perhaps many orders of magnitude higher than the most acidic habitats on Earth – arising from the canonical expectation that the dominant component of the clouds is sulfuric acid. On the other hand, the presence of ammonium or hydroxide salts might be able to mitigate this issue and reduce the pH to $\lesssim 1$ (Mogul et al., 2021; Rimmer et al., 2021), which is close to the bounds associated with extremophiles on Earth, as revealed by Section 7.2.4.

The second problem is the paucity of water in the clouds of Venus. As these settings are presumed to consist primarily of sulfuric acid, with sparse amounts of water, Hallsworth et al. (2021) estimated a water activity of $a_w \sim 0.004$, around 150 times smaller than the minimum value tolerated by life on Earth (refer to Section 7.2.3); this obstacle may be overcome in theory by pumping water into the cell interior (Bains et al., 2024). Moreover, sulfuric acid abundance is decreased by impurities like salts, and the ensuing water activity inside droplets might reach the lower bound for a_w (Mogul et al., 2021). Alternatively, if truly exotic forms of life based on sulfuric acid are viable, involving biomolecules not found on Earth (see Sections 11.1.2 and 11.2.2), the obstacle posed by water activity is rendered moot.

A conceivable life cycle for microbes in the Venusian clouds was advanced by Seager et al. (2021), and consists of five steps shown in Figure 11.2. At the start, desiccated microbial spores are hosted in the lower haze, which is stable to atmospheric convection and functions as a depot. Second, some of these spores are sporadically transported upward, where they act as cloud condensation nuclei. Third, after encapsulation within a liquid droplet, conditions eventually become habitable and the spores germinate. Fourth, microbes metabolise and grow in the droplets on timescale of months to years, thereby driving the expansion and coagulation of droplets. Last, once the droplet gets too large, it sinks downward and enters less hospitable conditions, which leads to deactivation (i.e., retreat to the spore phase).

To round off, sophisticated 3D climate models are stimulating us to revisit our earlier understanding of the evolution of Venus' climate over time (Way and Del Genio, 2020; Gillmann et al., 2022). A subset of these models support the claim that Venus was possibly habitable until as recently as ~ 0.7 Ga, and that this planet may have therefore retained clement conditions for >3 Gyr of its history. However, significant uncertainties regarding the initial conditions on Venus abound, which can drastically alter these predictions. If these projections are borne out by future evidence, we might gain further insights about the likelihood of the putative aerial biosphere. Perhaps the hypothetical microbes dwelt on the surface and could have gradually migrated upward over time as the climate warmed, or maybe they were deposited into the atmosphere from a different world.

Our knowledge of Venus' mysterious past should improve substantially with upcoming missions like *DAVINCI* (Deep Atmosphere Venus Investigation of Noble Gases, Chemistry, and Imaging)[3] and *VERITAS* (Venus Emissivity, Radio Science, InSAR, Topography, and Spectroscopy).[4]

---

[3] https://ssed.gsfc.nasa.gov/davinci/.

[4] www.jpl.nasa.gov/missions/veritas.

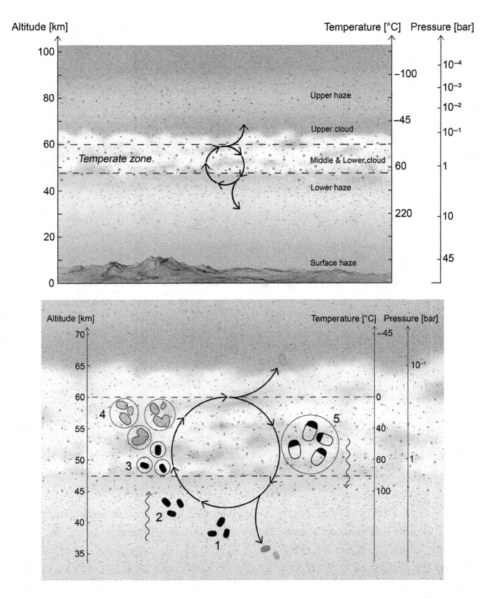

Figure 11.2 A postulated life cycle for putative microbes in the Venusian cloud decks. The five stages shown herein are described towards the end of Section 11.3.1. (Credit: Seager et al., 2021, Figure 1; CC-BY-NC 4.0 license)

11.3.2 The lakes and seas of Titan

We have already commented on the surface and internal structure of Titan, to an extent, when discussing its subsurface ocean in Section 10.2.3. Among the moons of our solar system, Titan is unique in possessing a thick atmosphere with a surface pressure of $\sim$1.5 bar, that is, somewhat higher than that of Earth. Titan's atmosphere is dominated by molecular nitrogen (N_2) and methane, with molar abundances of $>90\%$ and $\sim5\%$, respectively.

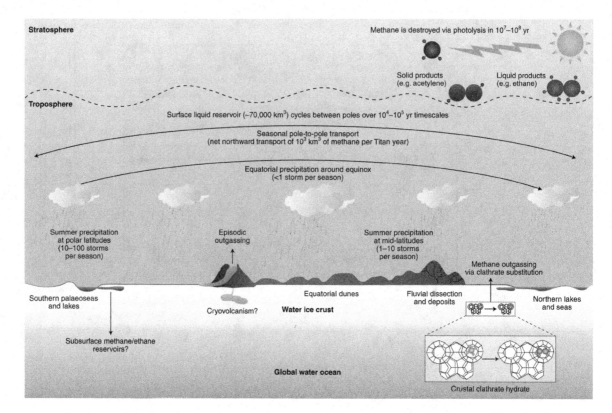

Figure 11.3 Methane-based hydrological cycle on Titan, with northward movement of $\sim 5 \times 10^{14}$ kg of CH$_4$ per year. Multiple processes comprise the cycle, each with their attendant timescale(s). (Credit: Hayes et al., 2018, Figure 4; reproduced with permission from Springer Nature)

The surface of Titan is markedly cold, given that it has a mean temperature of $\sim$94 K (Hörst, 2017). Despite this major divergence from Earth's surface temperature of 288 K, Titan evinces some striking features in common with Earth. It is probably the only place in our solar system, barring Earth, which hosts stable (extended) surficial liquid bodies – the larger ones are dubbed *maria* (seas) and the smaller ones *lācus* (lakes). The majority of these lakes and seas ($\sim$97% by area) are situated in the north polar region. The liquid bodies span $\sim$1% of Titan's surface area, which is comparable to the fraction of Earth's land area covered by water bodies (Hayes, 2016).

Akin to Earth, Titan displays an active hydrological cycle with the exchange of materials between atmospheric, surface, and subsurface reservoirs. The dominant compound is methane, owing to which Titan's hydrological cycle is methane-centric, as opposed to water on Earth. A schematic of Titan's intricate hydrological cycle is depicted in Figure 11.3. On the surface, organic molecules form 'molecular minerals' such as *co-crystals* (Maynard-Casely et al., 2018), analogous to minerals (viz., solid substances with definite chemical compositions and ordered structures) on Earth. Co-crystals are crystalline structures with more than one component that are bound together by non-covalent interactions like hydrogen bonds.

We have barely scratched the surface of Titan's potential for habitability; recent reviews are provided in Hörst (2017), Nixon et al. (2018), and Barnes et al. (2021). We have listed four central criteria for habitability in Chapter 7, among which the nutrient abundances are harder to ascertain, because they are sensitive to the specific biochemistry and physiology of the putative organisms. Nutrients on the surface of Titan might be limited in terms of diversity (McKay, 2016), although this topic is far from settled. This leaves us with the solvent, energy sources, and physicochemical conditions. Of this trio, let us scrutinise the last criterion: certain parameters like the surface temperature are well outside the biospace of LAWKI, which is why life on Titan's surface ostensibly falls under the category of exotic life.

Prior to tackling the issue of (pre)biotic chemistry in methane as the solvent, a comment on energy sources is in order. Photosynthesis is viable in principle insofar as the photon fluxes at the surface are concerned, because they comfortably surpass the threshold in Section 7.2.5. Next, if we contemplate chemotrophy, McKay and Smith (2005) evaluated various pathways for methanogenesis on Titan. The authors concluded that the following reaction has $\Delta G = -334$ kJ/mol, and thus represents a credible energy source for putative metabolism(s):

$$C_2H_2 + 3H_2 \rightarrow 2CH_4. \tag{11.3}$$

It is known that acetylene (C_2H_2) and molecular hydrogen (H_2), the two reactants, are available in Titan's environments. Recall that methane production explains the nomenclature of methanogenesis.

We will now delve into studies that have sought to assess the plausibility of methane as a solvent for biochemistry in Titan-like settings. This moon sustains rich organic chemistry: aerosols (small particles of sizes $\lesssim 1$ μm) may be vital for synthesising, concentrating, and transporting organic molecules to the surface, which are presumably formed after input from UV light, cosmic rays, electrical discharges, and other energy sources (Cable et al., 2012). At this stage, we highlight an important distinction that sets Titan apart from Earth. The atmospheric and surface environments of Titan are dominated by reduced compounds, and contain a limited inventory of oxygen-containing species (Raulin et al., 2012; Nixon et al., 2018).

Oxygen is an essential ingredient of biomolecules on Earth, such as proteins, nucleic acids, and sugars. Moreover, as mentioned in Section 11.1.2, the carbonyl bond is central to metabolism. There are three avenues to offset the paucity of oxygen. First, the meagre oxygen-bearing molecules in the atmosphere – with the exception of carbon monoxide, which is relatively abundant ($\sim$50 ppm) – might still suffice for the synthesis of prebiotic compounds. Second, liquid water (or water-ammonia mixtures) may become transiently available on the surface, thanks to cryovolcanism or impact-generated pools (Hörst, 2017), and hydrolysis of the aforementioned aerosols can produce biomolecular building blocks (Neish et al., 2010).

Third, biomolecules sans oxygen might exhibit similar functionality as their Earth-based analogues if pathways for their synthesis are feasible. Computational modelling by Lv et al. (2017) suggests that the equivalents of nucleobases, amino acids, and sugars (among others) may be generated from a simple mixture of hydrocarbons, HCN, and nitriles (i.e., compounds with the $-C\equiv N$ group), all of which are documented on Titan. The building blocks so formed can potentially assemble further, and yield analogues of biomolecules like proteins and nucleic acids. The underlying premise is that imines (C=N) act as substitutes for carbonyls (C=O),[5] and that this replacement is possible in liquid methane or ethane. While these results are intriguing, they are either theoretical or conjectural for now.

---

[5] Polyiminies could form on the surface of Titan, where they have the capacity to serve as catalysts and/or drive photochemistry.

In Section 5.1.4, we encountered a fundamental property of cell membranes, namely: they are composed of lipid bilayers. Compartmentalisation may be a core attribute of life, as stated in Section 5.1. The constituent molecules are *amphiphilic*, as they are equipped with a hydrophilic (i.e., polar water-loving) head that faces towards the water and a hydrophobic (non-polar) tail that is directed away from the water; we have touched on the hydrophobic effect in Section 11.1.1. It is worth remembering that this configuration is manifested in water, which is a polar solvent (refer to Section 11.1.1). What if we seek to understand a non-polar solvent instead?

Naïvely, it seems reasonable to invert the polarity of the membranes (made up of the component molecules) such that the hydrophobic end faces outward and the hydrophilic end faces inward (e.g., Norman, 2011). This elegant solution was investigated by Stevenson et al. (2015) from a theoretical standpoint, wherein the assembly of cell membranes (dubbed *azotosomes*) in liquid methane was modelled. These azotosomes consist of small molecules with a polar nitrogen-containing head and a hydrocarbon tail group. Stevenson et al. (2015) focused on acrylonitrile (C_2H_3CN) due to the expectation that the ensuing azotosomes would be robust. In the numerical simulations, compartments of diameter ~ 10 nm were obtained.

Thus, at first glimpse, the prospects for cell membrane formation appeared bright. Furthermore, acrylonitrile was shortly identified in Titan's atmosphere, thereby boosting the credibility of azotosomes. In addition, prior theoretical and experimental analyses had explored formation of polarity-inverted membranes in non-polar solvents, and did not cast major doubts on this process. However, subsequent sophisticated quantum-mechanical calculations by Sandström and Rahm (2020) indicate that the assembly of azotosomes is thermodynamically unfavourable in Titan-like conditions, which may conceivably diminish/negate the earlier findings.

Pivoting from lipids (comprising cell membranes) to nucleic acids, it is necessary to determine analogues of nucleic acids that could function in non-polar solvents, given that this class of biomolecules are devoted to information storage and transmission. Some basic prerequisites must be fulfilled by plausible candidates: (1) they must be constructed from fairly simple monomers (as otherwise the assembly of these monomers itself may be challenging); and (2) they must be soluble in the non-polar solvent (like DNA/RNA in water) to enable both cleavage and formation of bonds.

McLendon et al. (2015) hypothesised that *polyethers*, that is, molecules with repeating R−O−R' (ether) linkages, fulfil the preceding two criteria. The solubility of polyethers was analysed in liquid propane, a non-polar solvent that possesses a substantial liquidity range of 85–231 K at the pressure of 1 bar. It was found that polyethers are quite soluble in propane up to 200 K, but the solubility declined strongly thereafter until 170 K, at which point it became immeasurably low. Therefore, the second requirement from the previous paragraph is clearly violated, and that too at a laboratory temperature much higher than the surface temperature of Titan.

All in all, Titan remains a compelling astrobiological target, especially on account of its exciting potential to host exotic life in the liquid bodies on its surface; if such life is discovered, our knowledge of the roads to abiogenesis may well be revolutionised. Yet, as we have illustrated hitherto, there are a plethora of key unresolved mysteries pertaining to whether the functional analogues of biomolecules on Earth are tenable in non-polar solvents like liquid methane. A handful of numerical models broadly support or incline towards this perspective, but specific theoretical, computational, and experimental analyses are unfortunately lacking for the most part.

To round off, the cryogenic temperature of Titan's surface could drastically suppress the rates of (bio)chemical reactions by many orders of magnitude even if these reactions are thermodynamically favourable, consequently necessitating the action of powerful enzymes (see Question 11.5). Some of the aforementioned unknowns and uncertainties might, however, be unravelled by the *Dragonfly* mission,[6] scheduled to launch in 2028, which will shed light on Titan's prebiotic chemistry, habitability, and biosignatures.

11.4 Final remarks: comprehending exotic life

Are carbon and water cornerstones of all life, and not just of LAWKI? This conundrum motivated the entirety of this chapter. As we have witnessed so far, persuasive grounds may exist for believing that water and carbon are well-suited for biochemistry (Baross et al., 2007), possibly even uniquely so. Yet, we must avoid conflating 'persuasive' with 'definitive', given that a great deal remains unexplored or unsettled. Therefore, while we can continue searching for LAWKI with *some* degree of confidence that such life might be preferred in comparison to alternatives such as silicon biochemistry, it is vital to keep investigating other potential avenues at the same time.

This parallel programme entails four extensive pathways. First, we could perform experimental studies that attempt to precisely mimic the physical and chemical conditions in environments of astrobiological interest. However, this endeavour is challenging, and we may opt instead to simulate the outcomes with state-of-the-art numerical models (e.g., density functional theory), which represents the second avenue. Third, we can seek out exotic life on Earth itself in extreme habitats, that is, search for a shadow biosphere (Davies et al., 2009); refer to Question 11.8. Last, nature has been kind to us, so to speak, in furnishing a minimum of two worlds in our solar system that might harbour alternative biochemistries. Hence, sending life-detection and/or habitability assessment missions to Venus and Titan arguably merits high priority, and opportunely such missions are actually on the horizon.

If any of this quartet, especially the last one, yield positive outcomes, humanity would have invaluable data concerning non-canonical routes to the genesis of living systems. We might then be able to gain a deeper understanding of the prevalence of myriad types of life in the Universe.

11.5 Problems

Question 11.1: The shape of the water molecule may be envisioned as a 'V'. Each of the two segments is $\sim$96 pm (i.e., the O$-$H bond length). The bond angle between the two O$-$H segments is $104.5°$. If the magnitude of the charge on the oxygen and hydrogen atoms is e, using the above information, estimate the dipole moment of water (in units of D). Is this value compatible with 1.85 D mentioned in Section 11.1.1? If not, write about what hidden assumptions are erroneous in this calculation.

Question 11.2: Section 11.1.1 delineated multiple advantages associated with liquid water as the solvent, whereas a few other liquids were introduced in Section 11.1.2. All these solvents have

---

[6] https://dragonfly.jhuapl.edu/.

at least one notable drawback with respect to water. Amongst these candidates, which one do you think has the fewest drawbacks and thus represents a genuine alternative to water? Draw on peer-reviewed data and/or models beyond those encountered in Section 11.1 for bolstering your argument.

Question 11.3: Not all possible alternatives to water were covered in Section 11.1. By carrying out a literature search for peer-reviewed sources, identify two other solvents, and describe their pros and cons with respect to water.

Question 11.4: A world by the name of Solaris orbits a star with radius $R_\star$ and temperature $T_\star$.[7] The Bond albedo of this world is A_b and the orbital radius is a. The blackbody temperature of Solaris (T_{eq}) is

$$T_{eq} = T_\star \sqrt{\frac{R_\star}{2a}} (1 - A_b)^{1/4}. \tag{11.4}$$

Familiarise yourself with the derivation of this formula from Section 8.1.1.

a. Let us suppose that the surface temperature of Solaris is equal to T_{eq}, implying a minimal green-house effect. We desire oceans of a given solvent to exist on the surface of Solaris. The lower bound (melting point) and the upper bound (boiling point) of that solvent are T_{min} and T_{max}. Calculate the corresponding orbital radii, which are denoted by a_{max} and a_{min}.

b. From these values, determine the width of the annular region, which possesses the definition $\Delta a = a_{max} - a_{min}$.

c. Choosing stellar parameters equal to that of the Sun and $A_b = 0.3$, estimate a_{min}, a_{max}, and Δa for (i) water, (ii) ammonia, and (iii) methane.

Question 11.5: In (11.3), we outlined how methanogenesis could operate on Titan through the reduction of acetylene. Without knowing the specifics of the hypothetical methanogens, it is not feasible to estimate the activation energy E_a, which governs the reaction rates as seen from (7.16). The cyanobacterium *Anabaena variabilis* performs the reduction of acetylene to ethylene (C_2H_4), with an activation energy of 166 kJ/mol at temperatures <21°C (Jensen and Cox, 1983). Likewise, for the sake of argument, let us assume that $E_a \approx 166$ kJ/mol for methanogens on Titan.

a. If the temperature diminishes from 288 K (mean surface temperature of Earth) to 94 K (mean surface temperature of Titan), what is the factor by which the reaction rate would decline for the above value of E_a?

b. In simplified terms, a catalyst can lower the activation energy. By what factor would E_a need to reduce for this pathway if the reaction rate on Titan is to be the same as that on Earth? If you compare this value with the magnitude of the decrease in E_a stimulated by enzymes on Earth, is this outcome feasible? Survey and cite published data on enzyme action.

Question 11.6: A handful of forthcoming missions to Venus and Titan were referenced earlier. In view of what you have learnt in this chapter, look up the anticipated instruments comprising the (scientific) payload. Report on which instruments are explicitly designed to gauge the habitability of these worlds, and how they would ideally address this matter. Setting aside these missions, are there any missing governmental or private missions?

---

[7] The novel *Solaris* (1961) by Stanisław Lem (1921–2006) is regarded by many as one of the most thought-provoking works in the science-fiction genre.

Question 11.7: Two targets for potential exotic life in our solar system were elucidated: Venus and Titan. On the basis of this chapter, and by perusing the properties of major objects in our solar system (i.e., with sizes of $\gtrsim 1,000$ km), justify whether any other candidates for harbouring exotic life were omitted. If you identify such candidates, explain your reasoning, and speculate about what biochemistries might be tenable.

Question 11.8: The goal is to perform a thought experiment relating to the notion of Earth hosting a shadow biosphere with exotic life.

    a. At what depth below the surface of Earth is the temperature of 200°C attained, for a geothermal gradient of ~ 25 K/km in continental regions?

    b. Estimate the pressure at this location if the continental crust has an average density of ~ 2.7 g/cm^3. At this pressure and the above temperature, could water survive in liquid state based on Figure 11.1?

    c. If liquid water does indeed exist, the temperatures would be rather high for carbon-based biomolecules, as indicated in Section 7.2.1. Drawing on the current chapter, discuss whether any other elements may serve as the scaffolding for alternative biochemistry in these conditions.

12   Detecting and Characterising Exoplanets

As highlighted in Section 1.2, the compelling notion of the plurality of worlds dates back to thousands of years. However, over the span of this vast period, we lacked empirical confirmation or refutation of this idea. The status quo underwent a dramatic change in the late twentieth century, thanks to impressive advances in science and technology.

The first exoplanets were detected a few decades ago, as the twentieth century drew to a close. An exoplanet, also called an extrasolar planet, is essentially a planet beyond our solar system. Despite the relatively recent discovery of exoplanets, this field has evolved into one of the fastest growing fields in astrophysics, with the number of confirmed exoplanets having exceeded 5,700 by virtue of telescopes like *Kepler* (2009–2018), *TESS* (Transiting Exoplanet Survey Satellite), *CHEOPS* (CHaracterising ExOPlanets Satellite), and *JWST* (James Webb Space Telescope),[1] the last three of which are operational. Future large-scale surveys such as *ARIEL* (Atmospheric Remote-sensing Infrared Exoplanet Large-survey), *PLATO* (PLAnetary Transits and Oscillations of stars), and *Roman* (Nancy Grace Roman Space Telescope) will expand our catalogue of detected and characterised exoplanets.

Recounting the early history of exoplanet detection (and speculation) is beyond the scope of this book, owing to which the comprehensive encyclopaedia by Perryman (2018) may be consulted instead. It has been suggested that the first indirect signature of an exoplanet is over a century old, stemming from the discovery of heavier elements – ostensibly derived from planets – in the atmosphere of a white dwarf, Van Maanen 2 (Zuckerman, 2015). A few early notable breakthroughs in exoplanets are delineated next; this list is not meant to be comprehensive.

1. A companion to the star Gamma Cephei A (13.8 pc from Earth) was detected by Campbell et al. (1988), but its existence was confirmed only in 2003. This exoplanet, named Gamma Cephei Ab, apparently has a mass many times higher than Jupiter.
2. A sub-stellar companion of the star HD114762 (38.6 pc from Earth) was reported by Latham et al. (1989). It was suggested that this object might be an exoplanet, but later data ruled out this possibility.
3. The pulsar (a type of neutron star) PSR B1257+12, at ∼710 pc from Earth, possesses a planetary system with at least three planets, of which one of them has a mass merely about twice that of the Moon. This planetary system was discovered in 1992 by Wolszczan and Frail (1992).

Therefore, if we consider the exoplanet scene in 1994, there were no *confirmed* (i.e., unambiguously detected) exoplanets around *main-sequence* stars.

However, this state of affairs was completely overturned with the momentous discovery of 51 Pegasi b by Michel Mayor (1942–present) and Didier Queloz (1966–present) in 1995 (Mayor and

---

[1] The name of this telescope has garnered substantial controversy for multiple reasons, and may well undergo change(s) in the future.

Queloz, 1995). The 2019 Nobel Prize in Physics was awarded to these two astronomers *'for the discovery of an exoplanet orbiting a solar-type star'*.[2] This planet orbits the Sun-like star 51 Pegasi, which has a mass of approximately $1.1\,M_\odot$ and is 15.5 pc from Earth. 51 Pegasi b has a mass around half that of Jupiter, and the planet's most striking feature is its semimajor axis of $a \approx 0.05$ AU. Due to the close proximity of 51 Pegasi b to the star, it is believed to possess a temperature of $>1,000$ K, making it the archetype of a prominent class of planets known as *Hot Jupiters*. We will not delve into Hot Jupiters, as their astrobiological potential is virtually nil for life-as-we-know-it; a modern review of this topic can be found in Fortney et al. (2021).

Ever since the 1990s, the exoplanet revolution has proceeded apace, and an exhaustive overview of this subject up to the year 2018 is furnished in Perryman (2018). In addition, readers are strongly encouraged to familiarise themselves with the NASA Exoplanet Archive (Akeson et al., 2013),[3] as well as the Extrasolar Planets Encyclopaedia (Schneider et al., 2011),[4] for a wealth of up-to-date resources and data; some of the problem sets (especially Question 12.1) require access to these resources. As there are over 5,700 known exoplanets, only a couple of exoplanetary systems are singled out here for their relevance to astrobiology.

Proxima Centauri is an unremarkable low-mass star (an M-dwarf) with mass $M_\star \approx 0.12\,M_\odot$ and blackbody temperature $T_\star \approx 3,000$ K. What makes it unique, however, is the fact that this star is the closest neighbour of the Sun at a distance of 1.3 pc. Hence, if this star were to harbour any planets, they would correspond to the exoplanets (around other stars) nearest to Earth. Anglada-Escudé et al. (2016) reported the detection of an exoplanet termed Proxima Centauri b (Proxima b) with minimum mass of $1.07\,M_\oplus$ and semimajor axis of $a \approx 0.05$ AU. Aside from Proxima b, it is believed that at least two other planets may orbit Proxima Centauri. However, Proxima b is of particular interest because of its rocky composition and its location in the habitable zone (HZ), a vital concept tackled in Section 8.1.

TRAPPIST-1 is an old ultracool dwarf star with mass $M_\star \approx 0.09\,M_\odot$, temperature $T_\star \approx 2,566$ K, and age $\sim$7.6 Gyr, at distance of 12.5 pc from Earth. In 2016–2017, scientists identified a minimum of seven planets around this star (Gillon et al., 2017). Two rare features of this planetary system stand out. First and foremost, all seven planets are roughly Earth-sized, with radii ranging from 76% to 113% that of the Earth (Agol et al., 2021, Table 6); their composition is expected to be broadly rocky, although the exact breakdown of their internal structure is unknown. Second, three of the planets lie within the canonical habitable zone, and another two might support habitable conditions on their surfaces in certain scenarios.

The time is now ripe to understand how terrestrial exoplanets (in the HZ) are detected and characterised, which is addressed in this chapter.

12.1 Methods for detecting exoplanets

The avenues for detecting exoplanets have not altered substantially since their inception – in-depth technical reviews are provided in Seager (2010), Seager and Deming (2010), Perryman (2018), and

---

[2] www.nobelprize.org/prizes/physics/2019/summary/.

[3] https://exoplanetarchive.ipac.caltech.edu/.

[4] http://exoplanet.eu/.

Hatzes (2020). We will chiefly focus on the *radial velocity* (RV) and *transit* methods, with brief overviews of *direct imaging, gravitational microlensing,* and *astrometry.*

The rationale behind our choice is straightforward. This quintet of observational techniques are responsible for the vast majority of exoplanets discovered so far, as well as those detected in the future. Hence, we skip alternative techniques such as the *transit timing variation* and *transit duration variation* methods, respectively based on perturbations in the timing and duration of transits caused by other planets (Agol and Fabrycky, 2018).

12.1.1   Radial velocity

For the sake of simplicity, we shall work with perfectly circular orbits hereafter as it makes the algebra more tractable.

If we consider an object executing circular motion, its total speed v stays constant and only the direction changes. Let us envision an observer (at rest) who is monitoring this object. For now, we assume that the normal to the orbital plane is perpendicular to the observer's line of sight, implying that the orbit is being viewed edge-on. The velocity along the line of sight, which is denoted by v_r (i.e., the radial velocity perceived by the observer), will vary along the orbit, and attains a maximum and minimum of $+v$ and $-v$, respectively; at two points in the orbit, note that v_r becomes zero.

Next, consider the general scenario wherein the normal to the orbital plane is inclined at an angle i with respect to the line of sight, as depicted in Figure 12.1. The velocity at any point in the orbit is tangential to the orbital plane, implying that the radial component of the velocity should be multiplied by a factor of $\cos(90° - i) = \sin i$. In this case, the maximal value of v_r (which we label $v_{r,max}$) is no longer equal to v, but is accordingly generalised to

$$v_{r,max} = v \sin i. \tag{12.1}$$

If we set $i = 90°$, we recover the result from the preceding paragraph. At this stage, a couple of clarifications are warranted.

1. The object of interest to us, which is characterised by circular motion, is the star. The centre of the circular orbit is the centre-of-mass (COM), which we shall revisit shortly.
2. In actuality, the COM will have a net velocity, as measured by the observer, and this relative velocity must be subtracted beforehand to ascertain v. In our subsequent discussion, we will omit this extra step, that is, we presume that such a subtraction procedure is already implemented.

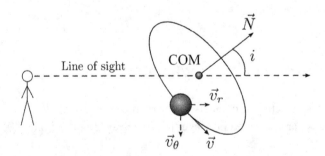

Figure 12.1 The motion of the star about the centre-of-mass on an inclined orbit, as seen by the observer; the inclination of the orbit is i. From geometric considerations, the maximum radial velocity discerned by the observer is $v \sin i$. (Credit: Frode Kristian Hansen; reproduced with permission)

Measurement(s) of v_r (and thence $v_{r,max}$) allow estimation of v when i is known. Fortunately, there is such a method to determine v_r: the Doppler effect. A succinct derivation of the (relativistic) Doppler effect for electromagnetic waves is furnished in Young and Freedman (2018, Chapter 37). If λ and λ_0 represent the photon wavelengths as measured in the rest frame of the observer and source (i.e., the star), respectively, they are related via

$$\lambda = \lambda_0 \gamma \left(1 + \frac{v_r}{c} \right), \tag{12.2}$$

where $\gamma = \left(1 - v^2/c^2 \right)^{-1/2}$ is the Lorentz factor from special relativity. Note that v is distinct from v_r, as previously mentioned. Upon taking the limit of $v/c \ll 1$ and rearranging (12.2), we end up with

$$\frac{\lambda - \lambda_0}{\lambda_0} = \frac{v_r}{c}, \tag{12.3}$$

indicating that the observed wavelengths are longer (redshift) when the source is moving away from the observer, and blueshift (observation of shorter wavelengths) occurs in the opposite scenario. Hence, if we could precisely measure the magnitude of the shift (in either direction), it is feasible to calculate $v_{r,max}$ and $v_{r,min}$, where the latter signifies the minimum value of the radial velocity. For a purely circular orbit, these two quantities are not independent because $v_{r,min} = -v_{r,max}$.

In our subsequent analysis, we will suppose that the radial velocity of the star is computed through the Doppler effect. The next step is to determine how this quantity may be used to infer the mass of a planet around this star. Before describing this procedure, we adopt the notation that the subscripts '$\star$' and 'p' refer to stellar and planetary variables, respectively. We must also be cognisant of a crucial point that is often glossed over for the sake of convenience – strictly speaking, planets do *not* orbit stars. On the contrary, the star and the planet orbit the COM (of this two-body system).[5] We will henceforth employ the symbol $\mathcal{T}$ for the orbital period.

When dealing with circular motion, it is easy to calculate the planetary and stellar orbital speeds, which are

$$v_\star = \frac{2\pi a_\star}{\mathcal{T}} \quad \text{and} \quad v_p = \frac{2\pi a_p}{\mathcal{T}}, \tag{12.4}$$

where $a_\star$ and a_p are the orbital radii of the star and the planet, respectively. This formula stems from the fact that the object must complete the total length of its orbit (viz., the circumference of a circle with given orbital radius) in a single orbital period. Next, we avail ourselves of the freedom that we can specify the origin in a coordinate frame. In our setup, the COM is chosen to be the origin, which leads us to

$$\mathbf{r}_{\text{COM}} = \frac{M_\star \mathbf{r}_\star + M_p \mathbf{r}_p}{M_\star + M_p} \quad \Rightarrow \quad M_\star \mathbf{r}_\star + M_p \mathbf{r}_p = 0, \tag{12.5}$$

where $\mathbf{r}_{\text{COM}}$, $\mathbf{r}_\star$, and $\mathbf{r}_p$ are the position vectors of the COM, star, and planet. The second equality is a consequence of selecting the COM as the origin. A couple of corollaries follow from (12.5), of which the first is

$$M_\star \mathbf{r}_\star = -M_p \mathbf{r}_p \quad \Rightarrow \quad M_\star |\mathbf{r}_\star| = M_p |\mathbf{r}_p| \quad \Rightarrow \quad M_\star a_\star = M_p a_p, \tag{12.6}$$

---

[5] Yet, we will continue to harness the description of planets orbiting stars as a heuristic, since it is an easy-to-visualise and fairly accurate picture.

and the last equality is obtained from invoking the definitions $|\mathbf{r}_\star| = a_\star$ and $|\mathbf{r}_p| = a_p$ for circular orbits. The second equation derivable from (12.5) differentiates this expression with respect to time, thereby yielding

$$M_\star \mathbf{v}_\star = -M_p \mathbf{v}_p \quad \Rightarrow \quad M_\star |\mathbf{v}_\star| = M_p |\mathbf{v}_p| \quad \Rightarrow \quad M_\star v_\star = M_p v_p. \tag{12.7}$$

Simplifying (12.6) to separate out masses and orbital radii, we have

$$\frac{M_p}{M_\star} = \frac{a_\star}{a_p}, \tag{12.8}$$

from which we notice that $a_\star$ approaches zero in the formal mathematical limit of $M_p/M_\star \to 0$. In most instances, the planet's mass is negligible compared to the star, therefore demonstrating why it is commonplace to view the planet as orbiting the star (equivalent to $a_\star \to 0$) instead of the COM in actuality. However, it is essential for now to avoid taking the aforementioned limit, as otherwise further progress is hindered.

By merging (12.1), (12.4), and (12.8), we obtain the relationship between the stellar and planetary radial velocities:

$$v_{pr,max} = v_{\star r,max} \frac{M_\star}{M_p}, \tag{12.9}$$

wherein $v_{pr,max}$ and $v_{\star r,max}$ are the maximum planetary and stellar radial velocities. The semimajor axis of the planet (effectively the star-planet distance in circular orbits) can be found using $a = a_\star + a_p$. Although this result is not derived here, it is intuitive when the star and the planet are diametrically opposite to the COM, and have orbital radii of $a_\star$ and a_p. By plugging (12.4) into this definition of a, we find

$$a = \frac{T}{2\pi} (v_p + v_\star). \tag{12.10}$$

Let us take stock of where we stand. In (12.10), the stellar velocity $v_\star$, up to the factor of $\sin i$, is determined from the Doppler method. Once $v_\star$ is estimated, we can express v_p in terms of $v_\star$, $M_\star$, and M_p, as seen from (12.7). For now, the orbital period may be regarded as a known quantity, which is addressed shortly. Thus, if we were to rewrite the LHS of (12.10) in terms of M_p and other quantities, we would have an algebraic equation that could be solved for M_p. It is possible to achieve this goal by invoking Kepler's Third Law of planetary motion, which states that

$$M_p + M_\star = \frac{4\pi^2 a^3}{T^2 G}. \tag{12.11}$$

It is easy to verify that, when all other factors are held fixed, the more familiar form of the law emerges – to wit, the scaling $a^3 \propto T^2$.

To solve for the planetary mass M_p, we first calculate a as a function of the other variables and constants by rearranging (12.11). Next, we utilise (12.10) in tandem with (12.1), (12.7), and (12.9) to solve for a in terms of the desired quantities, as indicated in the paragraph below (12.10). Lastly, we equate these two expressions for a and isolate M_p, which leads us to

$$M_p \sin i = \frac{v_{\star r,max} M_\star^{2/3} T^{1/3}}{(2\pi G)^{1/3}}, \tag{12.12}$$

where the ordering $M_p/M_\star \ll 1$ was employed to arrive at this equation. However, we caution that the result is valid only for circular orbits. If we consider elliptical orbits with eccentricity $e \neq 0$, the derivation is more complex. The analogue of (12.12) for this case is expressible as

$$M_p \sin i = \frac{\sqrt{1-e^2}\,\mathcal{K}\,(M_\star + M_p)^{2/3}\,\mathcal{T}^{1/3}}{(2\pi G)^{1/3}}, \qquad (12.13)$$

where we have introduced the radial velocity semi-amplitude defined as $\mathcal{K} = (v_{\star r,max} - v_{\star r,min})/2$. It is possible to show that (12.13) reduces to (12.12) in the appropriate limit (refer to Question 12.2).

Let us inspect the RHS of (12.12) carefully and interpret the various variables; we work with this equation due to its relative simplicity.

1. The first quantity in the numerator, the maximal radial velocity $v_{\star r,max}$, is estimated through the Doppler method.
2. The second term, which is the stellar mass $M_\star$, is treated as being known beforehand. In reality, accurate inferences of the stellar mass are not always feasible, with some exceptions such as stars in stellar binaries.
3. The third quantity, namely the period $\mathcal{T}$, can be ascertained from the Doppler method. As already mentioned, the wavelengths are redshifted and blueshifted because of the star's orbital motion about the COM. If the time period of the oscillations associated with these wavelengths is found, it would correspond to $\mathcal{T}$.

From these points, it is apparent that we are in a position to compute $M_p \sin i$, which is the LHS of (12.12) and (12.13). However, it is apparent that we run into a degeneracy issue because we cannot determine M_p in isolation unless and until the orbital inclination i is known. Therefore, the radial velocity method permits us to infer the *lower bound* on the planet mass, which follows from the fact that $\sin i \leq 1$.

Some of the most famous exoplanets referenced at the beginning of this chapter were discovered by means of the radial velocity method. For instance, 51 Pegasi b was detected through this avenue, as were many of the early exoplanets; in fact, the majority of exoplanets unearthed up to 2010 were discerned through this technique.[6] Jumping ahead to recent discoveries, Proxima b is a noteworthy example. Only the minimum mass of $1.07\,M_\oplus$ has been obtained for this planet because, as explained above, the radial velocity method yields the lower bound.

Another crucial aspect of the radial velocity method is worth underscoring. On solving (12.13) for the radial velocity semi-amplitude $\mathcal{K}$ and taking the reasonable limit $M_p/M_\star \ll 1$, we find

$$\mathcal{K} \propto M_p \mathcal{T}^{-1/3} M_\star^{-2/3}. \qquad (12.14)$$

A large value of $\mathcal{K}$ is desirable since that would enhance the magnitude of the Doppler effect, as seen from (12.3), making the shift in wavelength easier to detect. Hence, the radial velocity method is well-suited for planetary systems wherein $\mathcal{K}$ is high. The above formula reveals that this condition is fulfilled for massive planets on close-in orbits near low-mass stars. In light of this deduction, it is clear why the radial velocity method has historically favoured the detection of Hot Jupiters, which are simultaneously massive and close to their host stars.

---

[6] https://exoplanetarchive.ipac.caltech.edu/exoplanetplots/.

Before tackling the next technique, a brief historical digression is warranted. In a classic paper from 1952, Otto Struve (1897–1963) suggested that planets with the mass of Jupiter and orbital periods of $\sim$1 day (equivalent to $a \sim 0.02$ AU) might be detectable around Sun-like stars – potentially by telescopes of that time – by applying the radial velocity method (Struve, 1952, pg. 200).[7] However, as the notion of giant planets existing at such close distances was deemed implausible (or impossible), comprehensive searches were not initiated until a few decades later. If such surveys had been conducted much earlier, it is quite conceivable that the landscape of exoplanetary science would be markedly different than what it is nowadays.

12.1.2   Transits

We have learnt in the preceding section that the radial velocity method permits us to estimate $M_p \sin i$. However, a small subset of planets are characterised by $i \approx 90°$, which enables us to gauge M_p quite accurately because the degeneracy induced by $\sin i$ goes away. If we examine the attendant geometry in Figure 12.2, it is straightforward to verify that the orbit of the planet is being viewed edge-on for

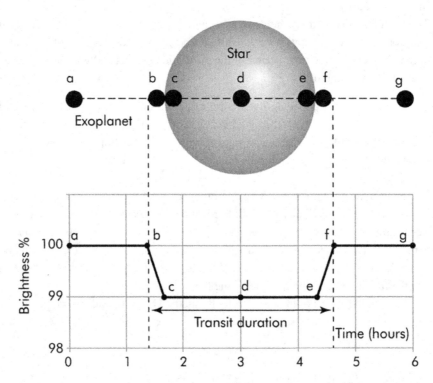

Figure 12.2 Stages of a planet transiting the host star are sketched, with accompanying changes in the light curve. Note that the transit duration can be envisioned as the timescale of the entire transit, and that the planet has travelled a distance equal to its diameter between points *b* and *c*. (Credit: IOP Publishing; https://spark.iop.org/detecting-exoplanets; reproduced with permission)

[7] An even earlier article touching on detecting exoplanets via the radial velocity method was published by David Belorizky (1901–1982) in 1938 (Briot and Schneider, 2018).

$i = 90°$, that is, the normal of the orbital plane is perpendicular to the line of sight. In this scenario, the planet will *transit*, to wit, it will pass by in front of the star from the observer's viewpoint and block out a small fraction of the starlight. Aside from the immediate advantage of directly calculating M_p (and not the minimum planetary mass), transiting exoplanets are endowed with other benefits, as elaborated hereafter.

The first question that comes to mind is: what is the probability that a particular planet transits a star? This quantity is known as the transit probability (P_{transit}), which is defined to be

$$P_{\text{transit}} \approx \frac{R_\star}{a}, \tag{12.15}$$

where $R_\star$ is the stellar radius and a is the semimajor axis, or equivalently the orbital radius for circular orbits. Even though (12.15) has a simple functional form, the expression is not derived from first principles herein because the procedure is somewhat intricate for the general case: a compact derivation of this formula is furnished in Catling and Kasting (2017, pg. 434). On inspecting P_{transit}, we notice that it is enhanced at low values of a, for fixed $R_\star$. Hence, akin to the radial velocity method, the transit method favours (in probabilistic terms) the detection of close-in planets.

Next, let us turn our attention to the heart of the transit method. As the planet travels (along its path) in front of the star, there is a slight dip in the observed brightness by virtue of the planet blocking the starlight. If the intensity of light prior to the transit is I_0 and falls to I_1 when the planet is completely transiting the star (i.e., entirely 'inside' the stellar disc), we may define the transit depth δ as

$$\delta = \frac{I_0 - I_1}{I_0}. \tag{12.16}$$

Given that both I_0 and I_1 can be computed through observations, it is evident that δ is an empirically determinable quantity. We will now express δ in terms of planetary and stellar parameters. To leading order, the intensity is proportional to the area of the stellar disc (as perceived by the observer) emitting radiation, which yields

$$I_0 \propto \pi R_\star^2 \quad \text{and} \quad I_1 \propto \pi \left(R_\star^2 - R_p^2 \right), \tag{12.17}$$

where R_p is the planet's radius. The second relation follows from recognising that a fraction of the area of the stellar disc is blocked by the planetary disc, whose area is πR_p^2. On substituting (12.17) into (12.16), we arrive at

$$\delta \approx \left(\frac{R_p}{R_\star} \right)^2. \tag{12.18}$$

Given that the LHS is measurable, if the stellar radius is known, it is feasible to ascertain the radius of the planet. On scrutinising this equation, it is clear that the transit depth is boosted for larger planets, indicating that this technique works well for detecting massive planets.

Alternative avenues for estimating R_p by means of the transit method are documented. As the planet commences its transit (moving from left to right), the right side of the planetary disc will graze the (left) edge of the star followed by the left side grazing at a later time. These two events correspond to the start of the transit and the attainment of maximum transit depth, respectively, signifying that the time interval between them can be measured via observations (denoted by Δt_p), as depicted in Figure 12.2. Next, the relative velocity of the planet with respect to the star is $v_p + v_\star$ because these velocities are calibrated relative to the COM. In the case of circular motion, as the speed stays constant, this leads us to

$$2R_p = (v_p + v_\star)\, \Delta t_p. \qquad (12.19)$$

In terms of the underlying physics, this equation informs us that the planet has travelled a distance of $2R_p$ (the diameter) over the time interval of Δt_p while moving at the above relative velocity.

As our goal is to calculate the LHS, we must make sure all quantities on the RHS are obtainable through observations. We have already noted that Δt_p falls under this category. Next, let us turn our attention to v_p, which is expressible in terms of $v_\star$, M_p, and $M_\star$ from (12.7). Of this trio, M_p is deducible if radial velocity measurements are available, because $\sin i \approx 1$ (see Section 12.1.1). Access to radial velocity data implies that $v_\star$ can be determined, likewise, from (12.3) and (12.1). Lastly, $M_\star$ is treated as a known constant. Hence, by invoking the radial velocity method, R_p could be inferred from (12.19), although this procedure is rather convoluted.

Before moving on to other methods, we comment that the transit method has an equally rich, if not even richer, history than the radial velocity method, which is summarised in Briot and Schneider (2018). The first allusions to employing the transit method are currently traceable all the way back to the nineteenth century by Dionysius Lardner (1793–1859), and more precise/quantitative proposals were put forward by David Belorizky in 1938 and Otto Struve in 1952; we encountered both of these individuals before in connection with the genesis of the radial velocity method.

12.1.3 Gravitational microlensing

The physics underpinning gravitational microlensing is heavily reliant on Albert Einstein's general theory of relativity, which goes far beyond the scope of this book. Hence, out of necessity, the treatment will be qualitative, and only the final salient formulae are presented, paralleling the discussion and notation in Armitage (2020).

The core principle is that massive objects can 'bend' the light received from distant sources, thereby acting as a lens. In the lensing setup of interest to us, the source and the lens are both specified to be stars and are perfectly aligned with the observer, that is, the observer, lens, and source all lie on the same axis; this configuration is illustrated in Figure 12.3. It is essential to appreciate that many more photons emitted by the source would be detected by the observer than usual (i.e., sans the lens) because the lens star behaves analogous to a convex lens, thereby bending the light towards the observer.

The deflection angle α of photons caused by the lens (a star of mass $M_\star$ in our case) was derived by Einstein in 1915, and has the form

$$\alpha = \frac{4GM_\star}{bc^2}, \qquad (12.20)$$

where b is the impact parameter, and crudely encapsulates the perpendicular distance between the photon's path and the lens star. In the case of perfect alignment depicted in Figure 12.3, by virtue of the lensing effect, the light emitted by the source seemingly arises from an angle θ_E above the axis. However, due to the symmetry of the setup, it also appears to come from θ_E below the axis, from the left and right of it (at the angle θ_E), and so forth. In total, therefore, instead of a point source, we

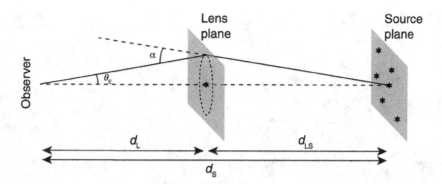

Figure 12.3 In this lensing setup, the observer receives light from the source star that is bent towards them by the lens star; the corresponding deflection is denoted by α. When perfect alignment of the observer, lens, and source occurs, the point source is viewed as an Einstein ring with angular radius θ_E. (Credit: Armitage, 2020, Figure 1.10; reproduced with permission from Cambridge University Press)

perceive a ring of light circumscribing the source, which is known as the *Einstein ring*. The angular radius θ_E of the Einstein ring can be estimated from

$$\theta_E = \sqrt{\frac{4GM_\star d_{LS}}{c^2\, d_L d_S}}, \tag{12.21}$$

where d_S and d_L are the source–observer and lens–observer distances, respectively, and $d_{LS} = d_S - d_L$ is the source–lens distance.

If the source star is situated in the bulge of the Milky Way and the lens of $M_\star \approx M_\odot$ in the Milky Way disc, we end up with $\theta_E \sim 1$ mas,[8] which cannot be readily resolved. Fortunately, analogous to a conventional lens, there is an increase in detected brightness of the source caused by the gravitational lens. The underlying reason has to do with the larger effective area associated with the source on account of the lensing, as implied by the prior paragraphs. Hence, by monitoring the source star, it is feasible to infer properties of the lens star from the light curve of the former.

Now, let us perform a thought experiment. Suppose that the lens star has a planet located around it, whose mass is M_p. We have already remarked that the lens star magnifies the source, which is detectable via tracking the latter's light curve. Now, the addition of the planet would cause a smaller magnification of the source, in turn functioning as a perturbation; this should show up as a small 'spike' in the light curve. Moreover, this spike will not occur at the peak of the light curve, since its manifestation is based on geometric considerations of the lens star, planet, and the source star. A schematic representation of microlensing is included in Figure 12.4.

The typical timescale of the lensing event t_E is approximately determined by dividing the radius of the Einstein ring R_E by the (relative) transverse velocity v_{tL} of the lens across the sky, as seen by the observer. Note that $R_E \approx d_L \theta_E$, which follows from Figure 12.4 and $\tan x \approx x$ for small x. With this information, we find that t_E is expressible as

$$t_E \sim 70\,\text{days}\, \sqrt{\frac{M_\star}{M_\odot}}\left(\frac{d_S}{8\,\text{kpc}}\right)^{1/2}\left(\frac{v_{tL}}{200\,\text{km/s}}\right)^{-1}\sqrt{(1-\beta_{LS})\,\beta_{LS}}, \tag{12.22}$$

---

[8] Note that 1 mas (milliarcsecond) is a unit of angle, often employed in astronomy, that is approximately equal to 4.85×10^{-9} radians.

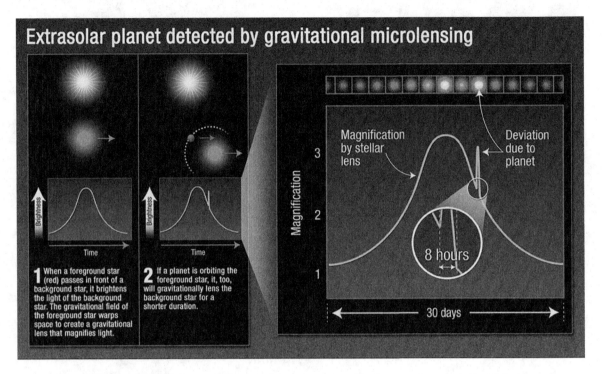

Figure 12.4 In panel #1, the magnification of the source caused by lens star (sans any planets) is shown. In contrast, panels #2 and #3 demonstrate how the inclusion of a planet modifies the magnification, which is materialised as a spike in the light curve. (Credit: NASA, ESA, and K. Sahu (STScI); https://exoplanets.nasa.gov/resources/53/extrasolar-planet-detected-by-gravitational-microlensing/)

where we have introduced the notation $\beta_{LS} = d_L/d_S$. The various parameters on the RHS are normalised by their characteristic values; for example, 8 kpc is the distance to stars in the bulge of the Milky Way, which serve as the sources. If we set $\beta_{LS} = 0.5$, the lensing time is roughly one month for Sun-like lens stars in the disc.

Equation (12.22) is applicable when the source and the lens are both stars. What about the scenario of microlensing, namely, when a planet is added to the mix? For carrying out order-of-magnitude calculations, the microlensing event timescale is taken to be $t_{ML} \sim \sqrt{M_p/M_\star}\, t_E$. It is possible to obtain this formula in a heuristic fashion. Instead of a star (with mass $M_\star$) serving as the lens, imagine that a planet (with mass M_p) does so. We would accordingly need to replace $M_\star$ with M_p in (12.22), consequently leading us to the aforementioned expression for t_{ML}.

Although we have commented on the timescale for the microlensing event, we have not addressed the probability of its occurrence. It turns out that the geometric configuration necessary for inducing the perturbation of the light curve has a likelihood that is proportional to $\sqrt{M_p/M_\star}$. Thus, for both the microlensing timescale and the probability, we see that the detection of larger planets is favoured. However, given that the dependence is rather weak – these quantities are proportional to $\sqrt{M_p}$ – the odds of finding relatively smaller planets are not significantly lowered.

Gravitational microlensing is capable of detecting planets that are quite far from the star. If the orbit of the planet is close to the radius of the Einstein ring, the source images comprising the ring are amenable to additional magnification by the planet. For source and lens stars in the Milky Way bulge and disc, respectively, the Einstein ring radius is a few AU. Therefore, unlike the radial velocity or transit methods, gravitational microlensing may be deployed to search for comparatively distant (i.e., long-period) planets, corroborating the earlier statement.

12.1.4 Astrometry

The fundamental premise of the radial velocity method in Section 12.1.1 was that both the star and the planet orbit their COM. In other words, the position of the star is not fixed, and it will appear to wobble back-and-forth from the observer's perspective. If this wobble is detectable, it would represent an indirect method for discovering exoplanets. Astrometry entails measuring the positions and velocities of objects, rendering it a prospective avenue for detecting exoplanets.

Let us quantify the magnitude of the wobble. The orbital radius of the star is $a_\star$, which is rewritten as $a_\star \approx a\,(M_p/M_\star)$ after using (12.6) and $a = a_\star + a_p \approx a_p$; the second relation in the latter formula is strictly valid when $M_p/M_\star$ is negligible. The angular displacement $\theta_\star$ is thus given by

$$\theta_\star \approx \frac{a_\star}{d_\star} \approx \frac{M_p}{M_\star}\frac{a}{d_\star}, \tag{12.23}$$

where $d_\star$ is the distance of the star (or planetary system) from Earth. The first equality is obtained by invoking $\tan x \approx \sin x \approx x$ for small x. In this formula, there is no factor of $\sin i$, implying that the inherent degeneracy of the radial velocity method is avoided. Another noteworthy aspect of astrometry is that this method is suitable for detecting massive planets on wide orbits around low-mass stars that are close to Earth. On normalising the various parameters by their typical values, we end up with

$$\theta_\star = 5 \times 10^{-4}\,\text{arcsec}\left(\frac{M_p}{M_J}\right)\left(\frac{M_\star}{M_\odot}\right)^{-1}\left(\frac{a}{5\,\text{AU}}\right)\left(\frac{d}{10\,\text{pc}}\right)^{-1}. \tag{12.24}$$

Although this angular displacement is undoubtedly minuscule, current and future astrometry missions have the requisite sensitivity. The *Gaia* mission, for instance, can reach astrometric precision of $\sim 10^{-5}$ arcsec,[9] although this bound is only true for a subset of all stars surveyed by this mission.

12.1.5 Direct imaging

The reader may have noticed that, until now, all methods of detecting exoplanets were indirect. For instance, in the radial velocity and astrometry methods, we relied on identifying the planet via its effect on the star's orbit (about the COM). Likewise, if we contemplate the transit method, the planet is discovered through the dip in the stellar light curve that it causes.

It is natural to wonder, therefore, as to whether the planet could be detected directly (viz., by means of direct imaging). However, there are two crucial issues that spring to the forefront. The first is the difficulty of spatially resolving the planet and the star, to wit, perceiving the former as a distinct source of light. The angular distance of the exoplanet θ_p at distance $d_\star$ from Earth is expressible as

$$\theta_p \approx \frac{a}{d_\star} \approx 100\,\text{mas}\left(\frac{a}{1\,\text{AU}}\right)\left(\frac{d_\star}{10\,\text{pc}}\right)^{-1}. \tag{12.25}$$

---

[9] www.cosmos.esa.int/web/gaia/science-performance#astrometric%20performance.

However, in order to resolve the planet–star system, we require θ_p to be larger than the diffraction limit θ_R. The diffraction limit represents a more-or-less fundamental bound on the resolution that is achievable by an idealised imaging system (e.g., telescope), and is known to be

$$\theta_R \approx 1.22 \frac{\lambda}{D} \approx 252 \,\text{mas} \left(\frac{\lambda}{1\,\mu\text{m}} \right) \left(\frac{D}{1\,\text{m}} \right)^{-1}, \tag{12.26}$$

where λ is the photon wavelength, and D is the aperture (diameter) of the telescope. On comparing (12.25) and (12.26), the challenges are obvious.

We are compelled to either undertake observations at shorter wavelengths or utilise bigger telescopes to reduce θ_R (a preferred objective). However, the latter is economically costly, may pose technological and social impediments, and cannot be readily pursued. What about the notion of searching for exoplanets at shorter wavelengths? At this stage, we run into the obstacle imposed by Wien's displacement law, which roughly states that the peak of the blackbody spectrum will be at a wavelength λ_{max}, given by

$$\lambda_{\text{max}} \approx \frac{2{,}898\,\mu\text{m K}}{T}, \tag{12.27}$$

where T is the temperature of the blackbody. If we consider a star like the Sun, with blackbody temperature of $T_\odot \approx 5{,}780$ K, we obtain $\lambda_{\text{max}} \approx 0.5\,\mu$m. In contrast, for a planet akin to Earth, with blackbody temperature $T_{\text{eq},\oplus} \approx 255$ K, we find $\lambda_{\text{max}} \approx 11.4\,\mu$m. Therefore, as we move towards shorter wavelengths, the flux from the star will preferentially dominate, making it increasingly harder to discern the planet. Cooler worlds face a related constraint – the bulk of their emission is at longer wavelengths, and probing this regime would elevate θ_R, which is undesirable.

Alternatively, if we leave θ_R untouched, then we must boost θ_p. For a fixed value of $d_\star$, the only path for achieving this goal is by increasing a. However, raising the semimajor axis initiates another drawback that is further explored in Section 12.2.4. We shall show that the fraction of starlight intercepted by the planet falls off as a^{-2}, making it harder to detect the planet if a is boosted. Furthermore, if the planet is located far away, then its blackbody temperature could decline commensurately, and this aspect may become problematic, as indicated in the preceding paragraph.

We move on to the next section, but we reiterate that the prospects for direct imaging are revisited in Section 12.2.4.

12.2   The characterisation of exoplanets

Hitherto, our focus has been on *detecting* exoplanets, and not so much on inferring their physical and chemical properties. While the discovery of nearly any exoplanet is good news from the perspective of (exo)planetary science, the situation with respect to astrobiology is more nuanced. First, merely detecting an exoplanet tells us (almost) nothing about its potential habitability. Second, even if we were to characterise exoplanets, certain classes are deeply unlikely to support life-as-we-know-it (e.g., Hot Jupiters).

In this section, we open with a description of how the mean density of an exoplanet can be computed, and what it could tell us about that world. However, we devote the bulk of our analysis to various methods of analysing photons linked to the planet. In particular, we will elucidate the benefits of spectroscopy.[10] We can divide spectroscopy into two primary categories: absorption spectroscopy and emission spectroscopy. In the former, absorption lines/bands are produced when photons of specific wavelengths are absorbed, and cause a transition from one quantum mechanical state to another. The latter may be interpreted as the converse of the former. Absorption lines are typically generated if the material is cool (and absorbs incoming hotter radiation), while the opposite is often valid for emission lines.

There are three general avenues whereby electromagnetic radiation can interact, or be associated, with an exoplanet:

1. Starlight passes *through* the planet's atmosphere (transmitted light).
2. The planet is a blackbody with temperature T_{eq}, and consequently emits thermal radiation (thermal emission).
3. Starlight is reflected from the atmosphere and/or the surface of the terrestrial planet (reflected light).

In our subsequent discussion, we will closely follow the approach in the review by Fujii et al. (2018), which may be consulted for additional details.

12.2.1 Mean density of exoplanets

If we have transiting exoplanets for which radial velocity data are available, we are in a position to calculate both the planet's mass and radius, as outlined in Sections 12.1.1 and 12.1.2. Once these two quantities are known, the mean density ρ_p is readily determined via

$$\rho_p \left(\frac{4\pi}{3} R_p^3 \right) = M_p \quad \Rightarrow \quad \rho_p = \frac{3M_p}{4\pi R_p^3}. \tag{12.28}$$

If we consider the planets in our solar system, there is a clear divide in the mean densities of the terrestrial planets and the giant planets. While there is certainly some internal variation, the gas giants and ice giants (i.e., giant planets) broadly exhibit densities of $\rho_G \sim 1$ g/cm^3, whereas the terrestrial planets possess densities of $\rho_T \sim 5$ g/cm^3.

Thus, if we were to detect an exoplanet with $\rho_p \approx \rho_T$, it would suggest that the planet may be rocky, although other compositions cannot be ruled out. In contrast, if the exoplanet is found to have $\rho_p \approx \rho_G$, this density is emblematic of a substantial volatile inventory (thick atmospheres and/or ices and liquids). Such worlds are less Earth-like and might pose impediments to habitability (for life-as-we-know-it), but it must be acknowledged that our understanding of these planets is rudimentary.

Clearly, the density in isolation is not enough to construct an in-depth picture of the planet's overall composition. We run into issues with degeneracy, as illustrated by Question 12.5. Hence, we move onward to the information garnered from spectroscopy and photometry.[11]

12.2.2 Transmitted light

As the title implies, transmitted light refers to light that passes through an exoplanet's atmosphere, undergoes scattering or absorption, and is detected by an observer on Earth. If we consider the

---

[10] Many excellent monographs cover this subject, of which one of them is Svanberg (2001).

[11] Unlike in spectroscopy, where we must resolve the wavelengths of the detected photons, in photometry we merely monitor the total flux in some particular wavelength band.

geometry, it is evident that this setup requires the planet to pass in front of the star, that is, to transit it. Spectroscopy of transmitted light is known as transmission spectroscopy, and the resolved light is the transmission spectrum.

For starters, it is necessary to comprehend the important concept of the atmospheric scale height $\mathcal{H}_a$, which is defined to be

$$\mathcal{H}_a = \frac{k_B T_a}{\mu_a g} = 7.6\,\text{km} \left(\frac{T_a}{250\,\text{K}}\right) \left(\frac{R_p}{R_\oplus}\right)^2 \left(\frac{M_p}{M_\oplus}\right)^{-1} \left(\frac{\mu_a}{28\,m_p}\right)^{-1}, \tag{12.29}$$

where T_a is the atmospheric temperature, μ_a is the mean molecular mass for this atmosphere, and $g = GM_p/R_p^2$ is the acceleration due to gravity. The scale height may be envisioned as the height to which the particle (with mass μ_a) is raised such that its gravitational potential energy becomes equal to the thermal energy of the atmosphere.

Now, let us quantify the strength of the signal arising from transmitted light. When light from the host star traverses the atmosphere, it interacts with the atoms and molecules of the latter. If the chemical species absorb strongly in some wavelengths, the atmosphere would appear opaque, that is, light would not be transmitted through it; the converse applies when the absorption is weak. In simpler terms, the atmosphere will manifest at certain wavelengths and will appear to 'vanish' at others. The difference in the two transit depths yields the signal strength S:

$$S \approx \frac{\pi (R_p + N_H \mathcal{H}_a)^2 - \pi R_p^2}{\pi R_\star^2} \approx \frac{2 N_H \mathcal{H}_a R_p}{R_\star^2}$$

$$\approx 8 \times 10^{-7} \left(\frac{N_H}{4}\right) \left(\frac{T_a}{250\,\text{K}}\right) \left(\frac{R_p}{R_\oplus}\right)^3 \left(\frac{M_p}{M_\oplus}\right)^{-1} \left(\frac{\mu_a}{28\,m_p}\right)^{-1} \left(\frac{R_\star}{R_\odot}\right)^{-2}$$

$$\approx 8 \times 10^{-7} \left(\frac{N_H}{4}\right) \left(\frac{T_a}{250\,\text{K}}\right) \left(\frac{\rho_p}{\rho_\oplus}\right)^{-1} \left(\frac{\mu_a}{28\,m_p}\right)^{-1} \left(\frac{R_\star}{R_\odot}\right)^{-2}, \tag{12.30}$$

where N_H is a dimensionless parameter that encapsulates the effective thickness of the atmosphere (equal to $N_H \mathcal{H}_a$), and $\rho_\oplus \approx 5.5\,\text{g/cm}^3$ represents the average density of Earth. In the top line, the first term in the numerator of the RHS embodies the generalised version of the transit depth (12.18) with the atmosphere added, while the second term is merely the transit depth (12.18) sans an atmosphere. The RHS in the second line follows after substituting $\mathcal{H}_a$ from (12.29); the last line is obtained after invoking (12.28).

There are a couple of valuable broad inferences that can be drawn from (12.30), which are elucidated next.

1. If all other parameters are held fixed, then we see that $S \propto R_\star^{-2}$, indicating that analysing transmitted light is easier for planets around low-mass stars. This scaling shows why, at least in the near future, transmission spectroscopy programmes will focus on M-dwarf exoplanets. Recall that M-dwarfs are the smallest and coolest main-sequence stars. They are also the most plentiful, comprising $\sim 75\%$ of all stars in the Milky Way.

2. The signal strength is boosted when the atmosphere is hot, when it consists of lighter chemical species (e.g., H_2 and He), and when the mean density is low. All these factors make it easy to characterise Hot Jupiters, which explains why this class of planets have already been widely investigated. The same factors stymie the characterisation of rocky planets in the habitable zone – the orthodox targets of interest for astrobiology.

As mentioned at the start of Section 12.2, absorption bands are produced when atoms and molecules absorb photons of specific wavelengths, thereby decreasing the spectral flux. Given that absorption features of a particular chemical species are unique (at high enough spectral resolution), we can search for their 'fingerprints'. The wavelength range probed is obviously dependent on the design of the telescope, and is dictated by multiple constraints. The mid-infrared ($\sim$5–25 μm) is fairly well-suited for transmission spectroscopy because the stellar flux remains quite high, and many molecules evince strong absorption features. The Mid-Infrared Instrument on board *JWST* approximately spans this wavelength range.

The transmission spectra of exoplanets are not only dependent on their own (atmospheric) composition but also the spectral type of the host star, as illustrated in Figure 12.5. Some molecules of (in)direct astrobiological relevance – a handful of which are tackled afterwards in Section 13.2 – and their prominent spectral features are delineated next.

1. 2.7, 4.3, and 15 μm for carbon dioxide (CO_2)
2. 0.94, 1.13, 1.9, and 6 μm for water (H_2O)
3. 0.69, 0.76 (O_2–A band), and 1.27 μm for molecular oxygen (O_2)
4. 0.4–0.7 (Chappuis absorption bands), 3.3, 4.7, and 9.65 μm for ozone (O_3)
5. 2.3, 3.3, and 7.7 μm for methane (CH_4)
6. 2.9 μm, 4.5, and 17 μm for nitrous oxide (N_2O)

12.2.3 Thermal emission

The planet absorbs some of the light from its host star, which is subsequently irradiated away as heat in the infrared. In Section 12.1.5, we touched on how planets at Earth-like blackbody temperatures preferentially emit radiation at wavelengths of $\sim$10 μm according to Wien's law.

In principle, it is possible to detect (and characterise) exoplanets by their thermal emission. This avenue is easier to pursue for transiting exoplanets. Consider the radiation received from the planetary system just before the planet passes *behind* the star, as viewed by the observer. At this moment, the radiation would consist of both the stellar and planetary components. Next, when the planet is eclipsed by the star, only the stellar contribution is measured, because the radiation from the planet is blocked by the star. Therefore, by taking the difference, the radiation associated with the planet could be disentangled from its stellar counterpart.

The contrast ratio $\mathcal{C}_{TE}$ embodies the signal strength of the planet's thermal emission, and constitutes the analogue of (12.30). To calculate this quantity, we utilise the fact that the spectral radiance of a blackbody at temperature T is prescribed by Planck's law:

$$B_\lambda(\lambda;\, T) = \frac{2hc^2}{\lambda^5} \left[\exp\left(\frac{hc}{\lambda k_B T} \right) - 1 \right]^{-1}, \tag{12.31}$$

where h is the Planck constant. Since the spectral radiance is the power per unit area per unit solid angle per unit wavelength, multiplying it with the cross-sectional area of the blackbody implies that the ensuing product is proportional to the power emitted by the blackbody. Hence, calculating the ratio of this power for the planet and the star yields the contrast ratio $\mathcal{C}_{TE}$:

$$\mathcal{C}_{TE} \approx \frac{\pi R_p^2 \, B_\lambda(\lambda;\, T_{\text{eq}})}{\pi R_\star^2 \, B_\lambda(\lambda;\, T_\star)} \approx \left(\frac{R_p}{R_\star} \right)^2 \frac{B_\lambda(\lambda;\, T_{\text{eq}})}{B_\lambda(\lambda;\, T_\star)}. \tag{12.32}$$

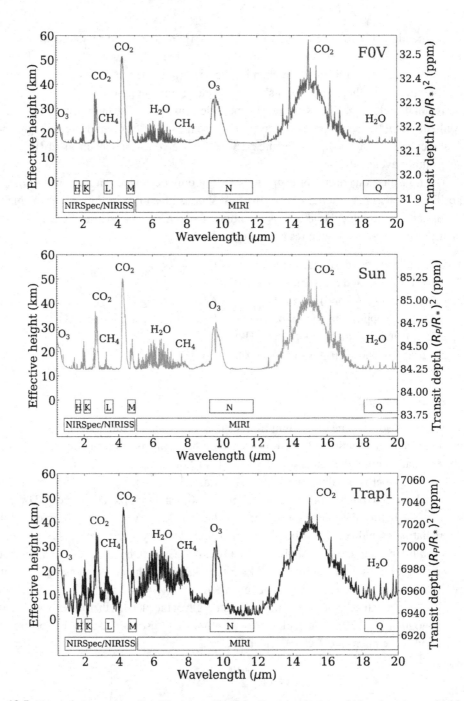

Figure 12.5 Transmission spectra (0.4–20 μm) predicted for Earth-analogues orbiting three stars with blackbody temperatures of 7,000 K (F0), 5,780 K (Sun), and ∼2,560 K (TRAPPIST-1). Note that the transmission spectrum on the y-axis is expressed either in terms of the effective atmospheric thickness (left) or the expected transit depth (right); the latter increases significantly for low-mass stars. (Credit: Kaltenegger and Lin, 2021, Figure 3; ©AAS, reproduced with permission)

Although it is not directly evident, the detection of thermal emission is favourable for larger planets at hotter temperatures orbiting smaller stars. As before, we find that this method is conducive to studying planets around M-dwarfs, because the latter have smaller radii than the Sun. We have accordingly normalised (12.32) for the typical values of an M-dwarf on the smaller end of the mass range (such as TRAPPIST-1).

$$
\mathcal{C}_{TE} \approx 5.4 \times 10^{-5} \left(\frac{R_p}{R_\oplus} \right)^2 \left(\frac{B_\lambda(\lambda; T_{eq})}{B_\lambda(10\,\mu m;\ 300\,K)} \right)
$$
$$
\times \left(\frac{R_\star}{0.1\,R_\odot} \right)^{-2} \left(\frac{B_\lambda(\lambda; T_\star)}{B_\lambda(10\,\mu m;\ 2500\,K)} \right)^{-1}.
\tag{12.33}
$$

The prospects for detecting exoplanets via their thermal emission are maximised in the mid-infrared. Certain molecules of significance in astrobiology exhibit the following distinctive spectral features in this region, which may show up in thermal emission spectra.

1. 8.9, 9.65, and 14 μm for ozone (O_3)
2. 8.5 and 17 μm for nitrous oxide (N_2O)
3. 7.7 μm for methane (CH_4)
4. 15 μm for carbon dioxide (CO_2)

A representative thermal emission spectrum for a particular composition of TRAPPIST-1e (namely an ocean planet) is shown in Figure 12.6, where some of the aforementioned features are manifested.

A valuable advantage must be mentioned in connection with thermal emission. It is suspected that most planets around M-dwarfs are tidally locked into a state of synchronous rotation (refer to Section 8.3.4) wherein their rotation period is equal to the orbital period, akin to the dynamics of

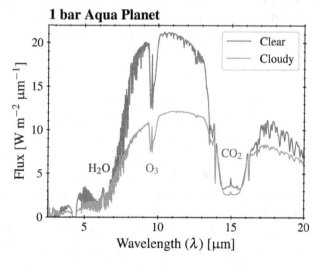

Figure 12.6 Thermal emission spectra predicted for TRAPPIST-1e if the entire surface of this planet was covered by an ocean (i.e., an ocean planet), with other properties equivalent/identical to those of Earth (e.g., 1-bar atmosphere). Top and bottom curves depict clear (cloud-free) and cloudy atmospheres. Thermal emission spectral flux (y-axis) is plotted as a function of the wavelength. (Credit: Lincowski et al., 2018, Figure 13; ©AAS, reproduced with permission)

Earth's Moon. In consequence, one side of the planet would perpetually face the star (dayside) and the other would be directed away from it (nightside). As a result, we may naïvely expect the equilibrium dayside temperature $T_{eq,D}$ to be much hotter than the corresponding nightside temperature $T_{eq,N}$. However, climate models have demonstrated that the presence of a sufficiently thick atmosphere (and/or oceans) can efficiently redistribute heat.

In the case of near-zero or tenuous atmosphere, we would expect a substantial difference between $T_{eq,D}$ and $T_{eq,N}$. In turn, as revealed by (12.32), the contrast ratio C_{TE} for the dayside should deviate significantly compared to the nightside, to wit, the former must be much higher than the latter. On the other hand, this difference in the contrast ratio ought to be less pronounced for planets that possess thick atmospheres (and/or oceans). Hence, if we were to monitor the time-varying thermal emission received at Earth over the orbital period of the planet (dubbed a *thermal phase curve*),[12] we may be able to discern whether the exoplanet has an atmosphere.

The technique of thermal phase curves has already proven its worth. For example, Kreidberg et al. (2019) inferred that the presumably rocky exoplanet LHS 3844b (with $R_p \approx 1.3\,R_{\oplus}$) – which has a short orbital period of 11 hours and dayside temperature of $>1{,}000$ K – lacks a thick atmosphere, and might be essentially airless. Thermal phase curves could likewise be harnessed to seek out signatures of atmospheric heat distribution on well-known terrestrial exoplanets such as Proxima b by deploying *JWST*.

In closing, we note that other techniques reliant on thermal emission to characterise exoplanets exist. For instance, in the presence of an atmosphere, the dayside temperature $T_{eq,D}$ should be reduced with respect to the dayside temperature of a bare rock (i.e., an airless world), because of the redistribution of some heat to the nightside. As per this prediction, the dayside thermal emission flux would be perceptibly lower than an equivalent bare rock. Hence, by conducting observations of the dayside alone, it seems possible in theory to detect signatures of atmospheric heat redistribution (Koll et al., 2019). Measurements of thermal emission by *JWST* in 2023 are consistent with the absence of a thick atmosphere on TRAPPIST-1b and TRAPPIST-1c (Greene et al., 2023; Zieba et al., 2023), although some degree of ambiguity and latitude of interpretation remains.

12.2.4 Reflected light

We now turn our attention to reflected light detected from an exoplanet. Let us denote the power (per unit wavelength) emitted by the star as L_λ. For a circular orbit, the stellar flux at the planet's location is $L_\lambda/(4\pi a^2)$ because the orbital radius is a; the denominator represents the area of a sphere with radius a. Radiation emitted by the star perceives the planet as a disc with radius R_p, implying that the power intercepted by the planet is $\pi R_p^2 \times \left(L_\lambda/(4\pi a^2)\right)$. Of this intercepted power, the planet reflects a fraction A_p into space, where A_p is the planet's albedo (reflectivity).

Therefore, combining these factors, the reflected power (per unit wavelength) from the planet is $A_p \times \pi R_p^2 \times \left(L_\lambda/(4\pi a^2)\right)$. The contrast ratio of this power to the total power emitted by the star is denoted by C_R (serving as a measure of the signal strength), which is found to be

$$C_R = \frac{1}{4}\left(\frac{R_p}{a}\right)^2 A_p \approx 1.4 \times 10^{-10} \left(\frac{R_p}{R_{\oplus}}\right)^2 \left(\frac{a}{1\,\text{AU}}\right)^{-2} \left(\frac{A_p}{0.3}\right). \qquad (12.34)$$

---

[12] This variation would arise because the fractions of the dayside and nightside (each with their own temperature) visible from Earth would change over the course of the orbit.

It is immediately apparent on inspecting this equation that C_R is very low; in fact, it is orders of magnitude smaller than (12.30), its analogue for transmitted light. This simple calculation highlights the key stumbling block for direct imaging that we hinted at in Section 12.1.5. Before proceeding ahead, we note that a slightly different expression for C_R is obtained after a more realistic treatment (Fujii et al., 2018, pg. 755):

$$C_R \approx 10^{-10} \left(\frac{R_p}{R_\oplus} \right)^2 \left(\frac{a}{1\,\text{AU}} \right)^{-2} \left(\frac{A_g}{0.3} \right), \tag{12.35}$$

where A_g is the so-called *geometric albedo*, which will not be defined here because we do not utilise this variable hereafter. In Section 12.4, the above formula can be utilised in place of (12.34).

On inspecting (12.34) or (12.35), we see that the contrast ratio is boosted for larger planets. Moreover, the prospects for direct imaging are brighter (pun intended) if the planet is close to the star. If we consider exoplanets like Proxima b and the TRAPPIST-1 planets, the contrast ratio is boosted by about three orders of magnitude relative to the Earth–Sun system. At first glimpse, it would seem that such exoplanets are well-suited for direct imaging. However, this positive is counterbalanced by an accompanying drawback. When a is decreased, the angular separation between the star and the planet is proportionally diminished, as demonstrated by (12.25). We are confronted with the problem of resolving the planet, which typically necessitates larger telescopes in accordance with (12.26) and the associated discussion.

Returning to (12.34) or (12.35), it is clear that we must block out the starlight effectively to discern the reflected light (and spectra) of exoplanets. Two common instruments for achieving this goal are *coronographs* and *starshades*. We will not delve into the mechanisms underlying these devices, as doing so would take us far afield of our focus on the astrobiological facets; the reader may consult Crill and Siegler (2017) for a technical review. It suffices to mention that coronographs are attachments to telescopes, while starshades are positioned externally, that is, at distances of tens of thousands of kilometres from the telescope. Of the duo, coronographs are the more well-established instruments, with designs dating back to the early days of solar physics nearly a century ago.

The reflectance spectra for Earth along with select worlds within and beyond the solar system are plotted in Figure 12.7. The distinctive spectral features of Earth stand out, of which some arise from molecules indicative of habitability or life itself, as we shall discover when we venture into Section 13.2. In addition to revealing fingerprints of atmospheric gases, reflectance spectra could furnish valuable information about the composition of the planet's surface (e.g., presence of oceans), the surface pressure and temperature, rotation rate, and clouds. Atmospheric spectral features that are identifiable in reflected light include the following:

1. 0.63, 0.69, 0.76, and 1.27 µm for molecular oxygen (O_2)
2. Chappuis absorption bands (0.4–0.7 µm) for ozone (O_3)
3. 0.82, 1.13, and 1.4 µm for water (H_2O)
4. 1.2, 1.6, and 2.0 µm for carbon dioxide (CO_2)
5. 0.78, 1.0, and 1.66 µm for methane (CH_4)

It is worth reiterating that some surface materials may also show up in reflectance spectra. One of the most prominent among them, labelled in the topmost panel of Figure 12.7, is the aptly named vegetation red edge at $\sim$0.7 µm, which we elucidate subsequently in Section 13.2.2.

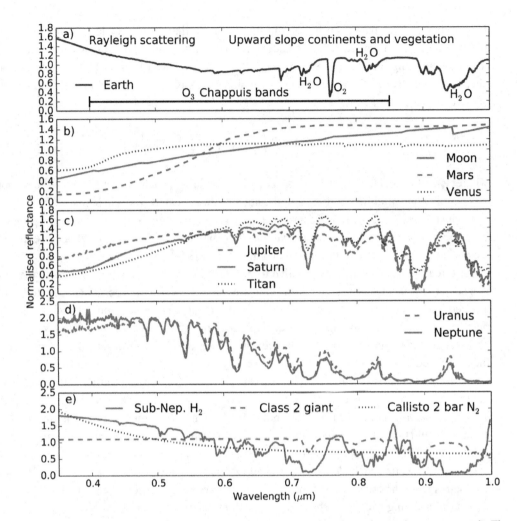

Figure 12.7 Reflectance spectra for solar system objects and hypothetical exoplanets (lowermost panel). The (scaled) reflectance is plotted as a function of the wavelength. The x-axis is truncated at 1 μm since the reflected light is dominated by optical and near-infrared photons for Sun-like stars. Note the distinct spectrum of Earth compared to other worlds, and the appearance of spectral features due to both atmospheric and surface components. (Credit: Krissansen-Totton et al., 2016, Figure 1; ©AAS, reproduced with permission)

12.3   Final remarks: future of exoplanet characterisation

We have provided a synopsis of the myriad avenues through which exoplanets can be detected and characterised by current and forthcoming telescopes. Our emphasis was firmly on the physical fundamentals underpinning exoplanet observations, given that these principles are likely to remain invariant even as the exoplanet revolution continues to play out in real time.

Looking ahead, the future of exoplanet characterisation is undoubtedly rosy. The majority of exoplanets whose atmospheres were analysed hitherto do not meet the conventional standards of habitability (e.g., Hot Jupiters), but this state of affairs is rapidly evolving. *JWST* has launched, and is already garnering valuable data on rocky exoplanets in the habitable zone such as TRAPPIST-1e.

Under optimal circumstances, it might detect molecules of astrobiological interest in the atmospheres of habitable worlds, although there are manifold attendant uncertainties (e.g., clouds). For instance, a tentative detection of dimethyl sulfide – a possible biological marker – in the atmosphere of the sub-Neptune K2-18b ($R_p \approx 2.6\,R_\oplus$) has been recently proposed (Madhusudhan et al., 2023), but this finding is equivocal for now.

The ground-based *Extremely Large Telescopes* (ELTs), with apertures of approximately 30 ± 10 m, should become operational in the second half of the 2020s, and complement the observations by space-based telescopes like *JWST* and *ARIEL*. If we peer further beyond into the future, ambitious proposals have been mooted: a $50-100$ m telescope on the Moon (Schneider et al., 2021) and the *Nautilus Array* comprising a fleet of 35 light-weight telescopes of 8.5 m in space (Apai et al., 2019) come to mind.

Lastly, we have highlighted a handful of molecules (e.g., O_2 and O_3) on account of their ostensible astrobiological importance in Section 12.2. However, even though we outlined the spectral features attributable to these molecules and how they may be detected, we did not actually elaborate on their significance as potential signposts of life. Our omission is deliberate since we shall revisit this matter in Section 13.2, where we enumerate putative biosignatures and clarify why certain molecules are perceived as signatures of life-as-we-know-it. As we search for such biosignatures in the future, we should ideally strive to keep alternative biosignatures produced by exotic life in mind, especially if any puzzling anomalies are discovered.

12.4   Problems

Question 12.1: Nowadays, it is commonplace to employ online databases in astronomy such as the NASA Exoplanet Archive (e.g., Akeson et al., 2013). Access this database and familiarise yourself with its usage.

  a. Go to 'Tools' and access the 'Confirmed Planets Plotting Tool'. Generate a histogram for the number of exoplanets detected as a function of the radius. Comment on the structure of the histogram – for example, is it monotonic or bimodal or neither? You can also attempt to justify why this trend is manifested by digging into peer-reviewed literature.

  b. Go to 'Data' and then pull up the 'Direct Imaging Table'. At the top of this table, there is an entry called 'Plot Table'. Plot all the planets discovered via this technique as a function of their temperature (y-axis) and mass (x-axis). Identify the most common type of planets detected, and then explain why this trend is observed by harnessing Section 12.1.

  c. In 'Tools', look up the 'Pre-generated Plots' and select the plot of 'Mass-Period distribution'. For the radial velocity and transit methods, what are the most common type of planets discovered so far? Are the data compatible with what we have covered in this chapter?

Question 12.2: This problem is centred on the radial velocity method.

  a. The procedure for deriving (12.12) is sketched in the paragraph preceding it. However, since the precise details are missing, fill out the missing steps, and show that (12.12) is indeed the final result.

  b. At first glimpse, (12.13) and (12.12) look quite different, albeit with some common factors. What are the two assumptions that are necessary to simplify the former and obtain the latter? After identifying these postulates, show that (12.13) does transform into (12.12).

Question 12.3: You have discovered that a new planet orbits a star of mass $M_\star = 0.18 M_\odot$ and radius $R_\star = 0.21\, R_\odot$. By conducting transit measurements, you infer that the planet has an orbital period $T = 25$ days.

a. You ascertain that the transit depth for this planet is $\delta \approx 2.7 \times 10^{-3}$. From this information, determine the radius of the planet in units of $R_\oplus$.

b. Next, you utilise the radial velocity method, and deduce that the maximal stellar radial velocity is $v_{\star r, max} \approx 20$ m/s. From the available data, compute the mass of the planet (units of $M_\oplus$) for a circular orbit.

c. From parts **a** and **b**, estimate the mean density of the planet. Is this value reasonable or not? Make sure to justify your answer.

d. From Kepler's Third Law, what is the semimajor axis a of the planet?

Question 12.4: Let us suppose that we have a stellar lens with $M_\star = 0.8\, M_\odot$ at a distance of $d_L = 4$ kpc from Earth, while the source star is situated in the Milky Way bulge at $d_S = 8$ kpc. The lens is moving at transverse velocity of $v_{tL} = 200$ km/s with respect to Earth.

a. What is the angular radius and physical radius of the Einstein ring?

b. Prove that the timescale of the lensing event generally attains a maximum for $\beta_{LS} = 1/2$. Verify that this condition is fulfilled for the system under consideration. Calculate the lensing timescale using the given data.

c. The lens star harbours a planet with $M_p = 0.5\, M_\oplus$. What is the typical timescale predicted for the microlensing event?

Question 12.5: For simplicity, consider a planet with merely two layers: the inner and outer layers have densities ρ_1 and ρ_2.[13] The planet's mass is M_p, of which the mass fraction of the inner layer is f_1. By means of radial velocity and transit measurements, the planet's mean density is inferred to be ρ_p.

a. What is the fraction of the planetary mass associated with the outer layer, when expressed in terms of these parameters?

b. Calculate the fraction f_1 as a function of ρ_1, ρ_2, and ρ_p. In place of this analytical treatment, easy-to-handle numerical tools such as the Exoplanet Composition Interpolator are also useful.[14]

c. Akin to Earth, we will roughly adopt $\rho_1 \approx 13$ g/cm^3, $\rho_2 \approx 4.5$ g/cm^3, and $\rho_p \approx 5.5$ g/cm^3. Estimate f_1 for this world, and compare it with the value for Earth; perform a literature search to find the latter.

d. Let us envision an icy world where the upper layer consists of water ice and/or liquid water – both of which exhibit $\rho_2 \approx 1$ g/cm^3 – and the inner layer is composed of rocks with $\rho_1 \approx 3.5$ g/cm^3. If the mean density is $\rho_p \approx 3$ g/cm^3, what would be the value of f_1?

e. Lastly, for a planet with three layers (labelled as #1, #2, and #3), their densities and mass fractions are ρ_i and f_i, where $i = \{1, 2, 3\}$. Demonstrate quantitatively that it is impossible to determine all the f_i's if only the mean density ρ_p is known; this degeneracy was mentioned in Section 12.2.1.

Question 12.6: By starting with (12.31), work out the following problems.

a. Obtain Wien's law (12.27) by taking the derivative of (12.31) with respect to λ, and setting it equal to zero. Either graphically or mathematically, verify that the value of λ thus computed is a maximum.

---

[13] Even though the Earth has multiple layers, we may crudely partition it into the mantle and the core (ignoring further divisions like the inner and outer core).

[14] https://tools.emac.gsfc.nasa.gov/ECI/.

b. By combining equations (12.32) and (12.31), and then taking the Rayleigh–Jeans limit (i.e., the long-wavelength limit of $hc/(\lambda k_B T) \ll 1$), simplify (12.32) to determine how the contrast ratio C_{TE} scales with the blackbody temperature of the star and the planet.

Question 12.7: In Sections 12.1.5 and 12.2.4, we elucidated the pros and cons of high-contrast direct imaging. Suppose that the target exoplanet is 10 pc from Earth, and that we intend to detect it at the wavelength of 0.5 μm.

   a. Estimate the minimum size of the telescope aperture required to image the planet when (i) $a = 0.05$ AU, (ii) $a = 0.3$ AU, and (iii) $a = 1.5$ AU.

   b. What is the desired contrast ratio for these three cases? You may assume that the planet has $R_p = R_\oplus$, and a constant geometric albedo of 0.3.

   c. Among the above trio of contrast ratios, which ones are realisable by current technology? You will need to conduct a literature review of contrast ratios for state-of-the-art instruments.

   d. The Extremely Large Telescope, scheduled to begin operations in 2028, will have a diameter of 39.3 m. What is the smallest value of the semimajor axis a that is resolvable at $\lambda = 0.5$ μm, for a planet 10 pc from Earth?

Question 12.8: The chief goal is to acquire a feel for the magnitudes of the various signal strengths for rocky exoplanets.

   a. In (12.32), assume that $R_p = R_\oplus$ and $T_{eq} = 300$ K. Calculate C_{TE} for the following four cases: (i) $\lambda = 2$ μm, $R_\star = 0.15 R_\odot$, and $T_\star = 3,000$ K; (ii) $\lambda = 8$ μm, $R_\star = 0.15 R_\odot$, and $T_\star = 3,000$ K; (iii) $\lambda = 2$ μm, $R_\star = R_\odot$, and $T_\star = 5,780$ K; and (iv) $\lambda = 8$ μm, $R_\star = R_\odot$, and $T_\star = 5,780$ K. Are there any general trends that are discernible from this quartet?

   b. Consider an exoplanet with $R_p = 1.5 R_\oplus$ orbiting a star of mass $M_\star = 0.1 M_\odot$, radius $R_\star = 0.15 R_\odot$, temperature $T_\star = 3,000$ K, with an orbital radius of 4.5×10^{-2} AU. Calculate the values of S (transmission), C_{TE} (thermal emission), and C_R (reflection). All other parameters in the relevant formulae should be held fixed at their prescribed fiducial values.

Question 12.9: In Section 12.3, we alluded to a couple of futuristic proposals for telescopes. Aside from this duo, several other mission concepts have been proposed. Carry out a literature search of peer-reviewed publications to identify one of them. Briefly describe the specifications, objectives, and advantages of this propounded telescope. Write about what major drawbacks exist, if any, when trying to make it a reality.

Question 12.10: In Section 12.1, only the common techniques of detecting exoplanets were addressed, that is, some avenues were excluded. After a careful survey of the peer-reviewed literature, select one method omitted in this chapter. Outline who came up with this technique, how this process is meant to discover exoplanets, whether any worlds have been detected accordingly (if so, name a few), and its future prospects.

Question 12.11: We did not pose a vital question: are any moons around exoplanets (*exomoons*) unambiguously confirmed? Dig into the peer-reviewed literature, and summarise two of the promising candidates unearthed so far, their physical properties, and what methods were employed for detecting them. Are any characteristics of these putative exomoons surprising?

Question 12.12: We have taken the definition of an exoplanet to be self-evident, but it has some latent subtleties. Read the review by Lecavelier des Etangs and Lissauer (2022) on this topic, sketch the arguments presented, and then discuss whether you subscribe to them (along with your reasons).

Part V

Detecting Life

13   Biosignatures

In the preceding chapters, we have tackled multiple facets of habitability and encountered a plethora of ostensibly habitable worlds. These topics immensely contribute towards resolving one of the fundamental questions in astrobiology, '*Are we alone?*'; and yet, they do not truly answer this question empirically. The canonical method of settling this matter is to discover markers of extraterrestrial life, either alive or extinct.

One straightforward example of such a marker would be, say, stumbling across an extraterrestrial 'bug' scuttling around in a habitable environment. In actuality, however, it is plausible that any *biosignatures* we unearth in future missions will be (much) more equivocal. In the 2008 NASA Astrobiology Roadmap, the concept of a biosignature, which has latent subtleties, was defined by Des Marais et al. (2008, pg. 729) as follows.

A biosignature is an object, substance and/or pattern whose origin specifically requires a biological agent. The usefulness of a biosignature is determined, not only by the probability of life creating it, but also by the improbability of non-biological processes producing it.

The first sentence conveys the notion that a genuine biosignature should be generated by a suitable biological agent. The second sentence warrants further examination (see Gillen et al., 2023 for a careful exposition).

All biosignatures evince a finite degree of ambiguity. This uncertainty can manifest through a couple of broad channels. First, the signal linked with a given biosignature could arise from non-biological (abiotic) mechanisms, thereby rendering it a *false positive*. Second, a particular biosignature may leave no tangible trace(s) or fall below the threshold of detection, despite the existence of the organism(s) responsible for that biosignature; this scenario is a *false negative*. To offer a classic example, molecular oxygen (O_2) is susceptible to both false negatives and false positives (Meadows et al., 2018). A robust biosignature would be one that minimises the complicating effects induced by false positives and negatives, among other desirable factors.

However, the above duo are not the only impediments. The ambient environment can cause contamination, degradation, and dilution of biosignatures, to name a handful of detriments. To mitigate this drawback, as well as address the possibility that the biosignature candidate could have an abiotic origin, a thorough understanding of the processes operational in the environment is necessary, that is, *contextual information* is crucial for determining whether a given signal is a genuine biosignature (Barge et al., 2022). Aside from the significance of contextual information, detecting more than one type of putative biosignature at the same spatiotemporal location may boost the likelihood of existence of extraterrestrial life.

Last but not least, biosignatures are based on the implicit postulate that they reflect some innate features of living systems. Molecular oxygen, for instance, is a metabolic product of oxygenic photosynthesis, as explained in Section 7.1.4. Hence, the emphasis on O_2 as a biosignature is not just a premise involving one specific metabolism but also invokes a deeper assumption that metabolism is fundamental to life. However, since the universal attributes of life are still subject to debate (Cleland, 2019), it is natural to suppose that a portion of this ambivalence is carried over to biosignatures. Therefore, the search for biosignatures is intertwined with the profound question of deciphering the universal characteristics of life.

It is apparent, on the whole, that a robust biosignature must ideally satisfy a number of conditions, which are collectively illustrated in Figure 13.1. For example, it must exhibit a low risk of false positives and false negatives, as already mentioned. If a biosignature is characterised by high reliability and survivability, it would partially help meet such criteria. A strong biosignature should also epitomise the maxim: '*Life is the hypothesis of last resort*' (Sagan et al., 1993, pg. 716). In other words, a signal may justify the tag of a robust biosignature if conceivable abiotic mechanisms are ruled out at high confidence, thereupon leaving only life as a known viable explanation.

Having briefly clarified the role(s) of ambiguity and approaches for grappling with it, we introduce the two overarching categories of biosignatures, each with their own strengths and weaknesses.

1. In situ biosignatures (ISBs): This class of biosignatures are unearthed through on-site explorations of habitable environments; breaking with convention, we place biosignatures from sample return under this umbrella. In situ biosignatures require more-or-less direct access to appropriate materials. A major benefit of ISBs is that a wider battery of tests and experiments can be conducted, consequently raising the credibility of life detection. The downside is that not many astrobiological targets exist, namely: some worlds in our solar system, and perhaps a few interstellar objects.
2. Remote-sensing biosignatures (RSBs): Detecting this group of biosignatures chiefly entails performing spectroscopic and photometric observations of targets. The immediate advantage is that a vast number of worlds (mostly exoplanets) fall under the scope of this avenue, but the weakness is that the quantity and quality of relevant information (viz., the signal and context) would be relatively limited due to the absence of on-site data.

In this chapter, we restrict ourselves to a modest subset of ISBs and RSBs, and outline the attendant false positives/negatives where feasible.

13.1   In situ biosignatures

We refer to Neveu et al. (2018), Chan et al. (2019), and Cavalazzi and Westall (2019) for detailed overviews of ISBs. Unlike remote sensing, which is primarily based on spectroscopy (briefly summarised in Section 12.2), a diverse array of techniques are employed for studying ISBs. Given that knowledge of analytical chemistry is not presumed, we shall skip the actual methods; the latter are reviewed in Seaton et al. (2021) and Abrahamsson and Kanik (2022). In situ biosignatures can be further categorised as substances, objects, and patterns, as illustrated in Figure 13.2 along with the accompanying methods for identifying ISBs. However, because some ISBs could straddle different classes, we do not adopt this scheme.

13.1.1   Isotopic fractionation

Metabolism, as witnessed in Sections 1.1.3 and 5.3.1, is one of the vital characteristics of life. Hence, markers ensuing from metabolism are expected to comprise reliable biosignatures, albeit provided that they are not easily mimicked or distorted by abiotic processes.

CRITERIA — **EXPLANATION** — **EXAMPLES / NOTES**

INSTRUMENTAL CRITERIA

Sensitive
WHY? Avoid false negatives | TEST Positive controls

Signal for feature of life selectively quantified above instrumental limit of quantitation, within response time and dynamic range

Encompasses all quantitative measures of instrumental performance (Armbruster & Pry 2008)

Contamination-free
WHY? Avoid false positives | TEST #1 Neg. controls | TEST #2 Blanks

Signal for feature of life not selectively detected (below instrumental LoD) in abiotic samples

Distinguish indigenous signals from contamination signals arising from:

'Below LoD' implies signal indistinguishable from noise given instrument sensitivity and stability (Armbruster & Pry 2008)

- Hardware
- Other samples (cross-contamination)

Repeatable
WHY? Avoid fluke | TEST ≥Triplicate

$N \geq 3$ measurements is the typical burden of proof in microbiology and chemistry, also depends on other factors (Table 2)

N measurements per sample, for as many samples as needed to capture the heterogeneity of the setting

Detectable
WHY? Avoid false negatives | TEST Positive controls

Physical, chemical, or geological conditions in the sample's current environment do not prevent the measurement from being made

- Reaction of organics with oxidant upon heating
- Ionization suppression by salts
- Some antibody methods in "sticky" briny liquid

CONTEXTUAL CRITERIA

Survivable
WHY? Avoid false negatives | TEST $T_{residence} > N*T_{degradation}$

Physical, chemical, or geological conditions in the suite of environments encountered by the sample between its synthesis and its measurement have not destroyed targeted signs of life

- Photo-destruction of biosignature gases
- Radiolysis of organics
- Racemization

Test from Hoehler (2017); $N \sim 1$ to a few units

Reliable
WHY? Avoid false positives | TEST Measure [bio]≠[abio]

Propensity to be produced by life and distinguished from abiotic backgrounds from any of the environments encountered by the sample between its synthesis and its measurement

$$\frac{[bio]}{[bio]+[abio]} > \text{instrument precision}$$

Compatible
WHY? Consistent with what we know of life | TEST Within bounds of NRC (2007)

The feature must not be excessively different from what is known of life on Earth (specificity vs. genericity). Limits can be pushed within bounds.

- Organic molecules, carbon-based (this would include e.g., proposed "arseno-DNA" (Wolfe-Simon et al., 2011)).
- Temperature not too high. Prions survive combustion at 600°C (Brown et al. 2004).

Last-resort hypothesis
WHY? Precludes abiotic origin | TEST Decision rules (Table 3)

The measurement, either alone or taken together with sufficient complementary measurements, precludes an abiotic origin with sufficiently little ambiguity.

Not just an appeal to authority (Sagan et al., 1993), as this criterion has been informally adopted by the astrobiology community as a standard.

Figure 13.1 A select list of criteria that must be fulfilled by a robust biosignature, such that the reported detection of extraterrestrial life is supported with high confidence. (Credit: Neveu et al., 2018, Table 1; CC-BY-NC 4.0 license)

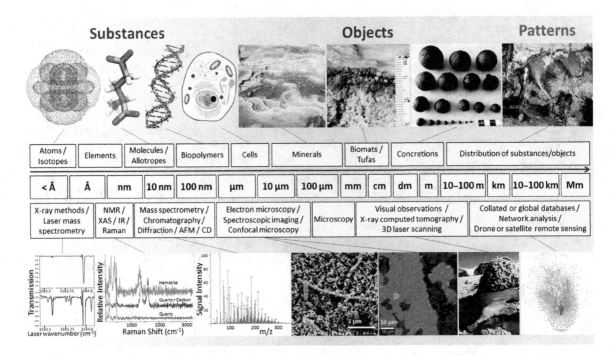

Figure 13.2 A subset of in situ biosignatures on Earth, the groups into which they can be classified, and the techniques needed for detecting them. These biosignatures span many orders of magnitude in length, as biology itself operates on multiple scales. (Credit: Chan et al., 2019, Figure 2; CC BY-NC 4.0 license)

Recall that isotopes are multiple versions of a chemical element with the same number of protons, but varying numbers of neutrons. Molecules that harbour heavier isotopes are broadly known to exhibit lower (vibrational) energy levels, higher stability, and slower reaction rates than their counterparts with lighter isotopes (Cavalazzi and Westall, 2019, pg. 65). In turn, these features would promote changes in the isotope ratio(s) because some products (of chemical reactions) will become enriched or depleted in certain isotopes. This phenomenon is termed *isotopic fractionation*, which we briefly covered in Section 3.1.7 pertaining to protoplanetary discs.

As a crude rule of thumb, lighter isotopes preferentially interact with biology on account of their faster reaction rates. For instance, carbon fixation pathways and the synthesised biomass are distinguished by increased assimilation of the lighter ^{12}C with respect to the heavier ^{13}C. Therefore, distinctly negative values of $\delta^{13}C$ are anticipated for biological samples (with some caveats below), where $\delta^{13}C$ is defined as

$$\delta^{13}C = \left(\frac{(^{13}C/^{12}C)_{sample}}{(^{13}C/^{12}C)_{standard}} - 1 \right) \times 1{,}000\,\%_0, \qquad (13.1)$$

with $\%_0$ representing the symbol for parts per thousand. The numerator $(^{13}C/^{12}C)_{sample}$ is the ratio of the atomic abundances of ^{13}C and ^{12}C in the given sample, whereas the denominator $(^{13}C/^{12}C)_{standard}$ is the equivalent isotope ratio for a predetermined standard.

On Earth, $\delta^{13}C$ exhibits considerable variation, ranging from approximately $0\,\%_0$ for seawater and carbonates (both of which are abiotic) to roughly between $-10\,\%_0$ and $-30\,\%_0$ for organic compounds synthesised via photosynthesis, and attaining even lower values for methane generated through biology (Cockell, 2020, Chapter 12). A typical range of $\delta^{13}C = -25 \pm 5\,‰$ is documented for kerogen, namely: insoluble organic matter largely derived from the remains of dead organisms. Therefore, if we encounter samples with $\delta^{13}C \lesssim -20\,‰$, they would have a seemingly biological origin.

However, this optimistic picture is complicated by geological processes, along with the ever-present spectre of contamination by later biological activity, underscoring the importance for contextual information (Barge et al., 2022). Metamorphism, wherein rocks are subjected to intense temperatures and pressures, is proven to modify $\delta^{13}C$; in addition, such an environment would be patently inhospitable for life, consequently diminishing claims of biogenicity. Likewise, the mantle is reported to host materials with clearly negative values of $\delta^{13}C$, which could mix with an abiotic sample and thence make it appear biological in origin. For such reasons, the ^{13}C to ^{12}C ratio is not viewed as an unambiguous biosignature in isolation.

Graphite (a carbon mineral) extracted from a $\sim$4.1 Ga zircon in Jack Hills, Western Australia, has $\delta^{13}C = -24 \pm 5\,‰$ (Bell et al., 2015), albeit it remains unclear whether this carbon is biogenic. Along the same lines, graphite from the $\sim$3.7 Ga Isua supracrustal belt in West Greenland was interpreted as isotopic evidence for life due to its low value of $\delta^{13}C \approx -20\,‰$ (Mojzsis et al., 1996; Rosing, 1999), but this claim has been contested. One alternative explanation posits that the graphite was formed abiotically by means of high-temperature reactions involving *siderites* ($FeCO_3$).

Last, other isotope ratios beside carbon may serve as biosignatures. Sulfate-reducing bacteria (see Section 7.1.3) metabolically produce sulfides that are depleted in ^{34}S, the heavier isotope of sulfur. Denitrification, whereby nitrates are converted to N_2, is known to consume smaller quantities of the heavier ^{15}N, thus causing a relative enrichment of nitrates with ^{15}N.

13.1.2 Microfossils

It is virtually taken for granted that life must display a cellular structure (consult Section 5.1), with a well-defined boundary (viz., the cell membrane). If one subscribes to the notion that this trait is universal to extraterrestrial life, then fossilised (or living) cells would constitute credible biosignatures. The chemistry by which organisms get fossilised is intricate (e.g., electrostatic and hydrogen bonds), and is therefore not elucidated. Some examples of widely accepted microfossils are depicted in Figure 13.3.

Evaluating the authenticity of putative microfossils is challenging, to say the least. It is not only imperative to take the environmental context into account (e.g., verifying that the structures were not embedded in rocks at unrealistic temperatures for life) but also to rule out false positives stemming from abiotic phenomena. Several criteria have been advanced for bolstering the biogenicity of microfossils; some of them are summarised here, based on Javaux (2019) and Cockell (2020, Chapter 12.7).

1. Sizes of these structures must be consistent with the minimal length scales predicted for microbes using biophysical frameworks.
2. Boundaries should be made of carbonaceous (carbon-containing) material resembling cell walls, with hollow structures internally.
3. Evidence of cell division or cell colonies (i.e., multiple structures with similar morphology at the same location).

Biosedimentary structures

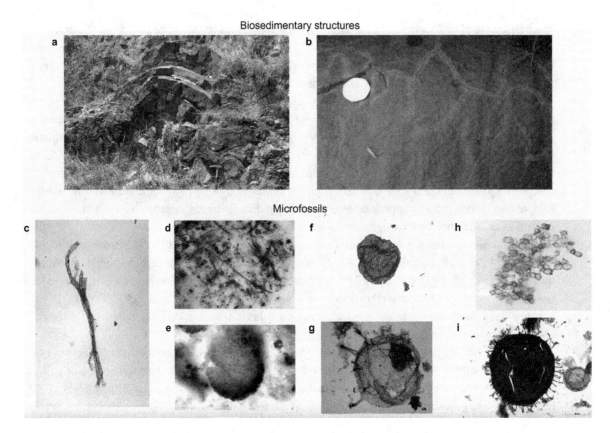

Microfossils

Figure 13.3 Stromatolites are depicted in panel (a), microbially induced sedimentary structures in panel (b), and confirmed (viz., less ancient) microfossils in panels (c) to (i). (Credit: Javaux, 2019, Figure 2; reproduced with permission from Springer Nature)

4. Identification of co-localised organic matter (typically kerogen) and isotopic fractionation compatible with life.
5. Structures should occur in sedimentary rocks, and not in settings that were subjected to environmental extremes inhospitable for life.

It is hard to simultaneously fulfil all these conditions, which are themselves characterised by mixed interpretations. Alternatively, the morphological features of potential microfossils can be quantified (e.g., aspect ratio), and statistical trends arising from the entire sampled *population* may help in discerning biogenicity (Rouillard et al., 2021).

As for false positives, intricate structures mimicking microfossils have been synthesised in the laboratory in diverse settings (McMahon and Cosmidis, 2022), implying that even complex morphologies are not necessarily indicative of life. Plant-like structures termed *chemical gardens* are produced when certain salts (with transition metals) dissolve in silicate solutions; the resultant tube-like shapes can also resemble bacteria and fungi. The aptly named *biomorphs* are inorganic objects often comprising silica (SiO_2) and alkali metal carbonates. Biomorphs exhibit a remarkable variety of shapes such as leaf-like sheets, helical filaments, and flower-like arrangements.

It is not surprising, therefore, that Earth's oldest purported microfossils are controversial. Dodd et al. (2017) proposed that tubules from the 3.77–4.28 Ga Nuvvuagittuq belt in Quebec, Canada, were biogenic, but other candidates (e.g., chemical gardens) cannot be ruled out (McMahon and Cosmidis, 2022). The 3.4–3.5 Ga structures from the Pilbara Craton region in Western Australia were originally perceived to be microfossils, but several publications have subsequently favoured an abiotic origin on the grounds of morphology, environmental context, and contamination by fluid percolation (Javaux, 2019); some of these criticisms also apply to the ∼3.4 Ga Kromberg Formation in South Africa. The ∼3.43 Ga Strelley Pool Formation in Western Australia is considered one of the oldest convincing examples of authentic microfossils for multiple reasons (Cavalazzi and Westall, 2019, Chapter 7.2), although even this claim is not universally accepted.

In the same vein, the alleged microfossils recovered from the famous Martian meteorite ALH84001 – which we encountered in Section 9.4.2 – are now thought to be unreliable markers of life on Mars; instead, they can be explained solely by appealing to non-living mechanisms.

13.1.3   Microbialites

In a nutshell, microbialites are deposits of organic matter and sediments (i.e., organosedimentary deposits) that are (in)directly produced by microbial communities. There are four major classes of microbialites: stromatolites (layer-like), dendrolites (tree-like), thrombolites (clot-like), and leiolites (structureless). Of this quartet, we shall focus on stromatolites hereafter.

Stromatolites are '*laminated rocks produced by microbial precipitation and/or trapping of minerals*' (Javaux, 2019), and span a truly impressive range of length scales (reaching kilometres in size). Stromatolites are formed via complex interactions of microbial mats with their environments (consisting of water), and one such example is shown in panel (a) of Figure 13.3. It is apparent that stromatolites are akin to microfossils in the sense that their morphology plays a crucial role. Hence, some of the ambiguities that surround microfossils are likewise applicable to stromatolites.

In evaluating the biogenicity of stromatolites, the proposed criteria are intrinsically scale-dependent. At the macroscale, the overall shape and structure are important – the regular spacing of conical shapes associated with certain stromatolites is conceivably not easy to reproduce by abiotic processes (McMahon and Cosmidis, 2022). At the intermediate scale (i.e., mesoscale), lamination composed of thinner dark layers interspersed with thicker light layers (involving sedimentary strata) is held to support a biological origin (Chan et al., 2019, pg. 1091). At the microscale, where microscopes are required, the positioning of sediment grains and the sharpness of the boundaries demarcating the light and dark layers constitute useful metrics.

In spite of these diagnostics, it is still difficult to conclusively identify stromatolites because of environmental effects; abiotic mechanisms can imitate some key characteristics of stromatolites. Triangular structures from the ∼3.7 Ga Isua supracrustal belt (Greenland) were interpreted by Nutman et al. (2016) as early stromatolites, whereas Allwood et al. (2018) have asserted that these features arose from abiotic deformations of these rocks at a later juncture. While this debate remains ongoing, the presence of two potential biosignatures in the Isua supracrustal belt (see Section 13.1.2) might elevate their likelihood of comprising genuine signposts of early life in this region. Travelling ahead in time, the ∼3.43 Ga Strelley Pool Formation (Australia) is widely thought to consist of authentic stromatolites.

13.1.4 Homochirality and isomeric preferences

In Section 5.1.1, we introduced enantiomers and highlighted that biology on Earth is characterised by (near-)homochirality, that is, virtually all biological amino acids are levorotatory (L-amino acids) and sugars are dextrorotatory (D-sugars), implying that one enantiomer overwhelmingly dominates over the other. In quantitative terms, the enantiomeric excess (ee) is

$$ee = |X_D - X_L| \times 100\,\%, \tag{13.2}$$

where X_D and X_L are the mole fractions of the dextrorotatory and levorotatory enantiomers, respectively; note that $X_D + X_L = 1$. Given that life on our planet exhibits near-homochirality, we would thus expect ee close to 100% for such amino acids and sugars. Hence, it has been conjectured since the 1950s – arguably extending back to Louis Pasteur (1822–1895) in the nineteenth century – that homochirality is a universal attribute of living systems, and that high ee would therefore represent a robust biosignature (Wald, 1957).

For a long time, high ee was presumed to be one of the most compelling biosignatures, but research in meteorites and synthetic chemistry is stimulating a re-evaluation of this stance. In the latter domain, experiments have established that even a slight ee may be amplified substantially by non-biological autocatalytic reactions. In tandem, analyses of meteorites have shown that some of them possess an ee up to 60% in the absence of biology (Glavin et al., 2019); intriguingly, meteorites also display an excess of L-amino acids (akin to Earth's biology). On a related note, our understanding of false negatives has improved – for instance, the decay of dead organisms could serve to decrease ee considerably.

Notwithstanding these caveats, there are still firm grounds to believe that high ee is a good biosignature. In particular, if we were to find a sample with high ee, but *opposite* chirality – in the sense that, say, we discover most amino acids are dextrorotatory – this feature might not only confirm the existence of extraterrestrial life but also strongly suggest that this life emerged independently of Earth (i.e., there was no common origin mediated by panspermia). Even if the chirality is the same as that observed on Earth, as long as the ee approaches 100%, the latter (viz., near-homochirality) may be adequate to claim that the sample is of biological origin.

A variety of techniques permit the detection of enantiomers and ee in principle. It is possible to employ a polarimeter – which measures the so-called optical activity of chiral substances – to eventually compute X_L, X_D, and ee. If the ee is low, one could instead utilise chiral selectors that bind to the enantiomers, and yield compounds that may be readily differentiated based on their physical properties. This differentiation is experimentally realisable by means of either mass spectrometry or gas chromatography; these methods are summarised in Seaton et al. (2021).

Enantiomeric excess is merely one among many types of isomeric preference; recall that isomers possess varying spatial arrangements of atoms, but share an identical chemical formula. We will now sketch another significant isomeric preference linked to life on Earth. If we inspect the ~20 proteinogenic amino acids in biology (refer to Section 5.1.1), a remarkable feature is that they almost invariably occur as α-amino acids, in which the amino acid group and carboxyl group are each attached to the α-carbon. In contrast, abiotic amino acids do not ostensibly evince such a distinct preference, because they also exist as β-, γ-, and δ-amino acids, to list a few.

It is tempting to conclude that discovering prominent isomeric preferences of the above kind would function as an effective biosignature. However, a couple of general counterpoints are worth heeding:

(a) abiotic mechanisms could induce some degree of isomeric preference; and (b) the feasibility of life entailing alternative combinations of isomer mixtures cannot be excluded.

13.1.5   Biological molecules and their abundances

As documented in Section 5.1, the four major categories of biomolecules are lipids, carbohydrates, proteins, and nucleic acids. The latter trio can be roughly envisioned as polymers of sugars, amino acids, and nucleotides, respectively. We have already witnessed in Sections 2.4.3, 5.4.1, and 10.4.1 that these monomers could be produced through abiotic channels, thereby downplaying their credibility as biosignatures. Proceeding further, polymers of moderate length can be synthesised abiotically – especially in non-equilibrium settings (Ianeselli et al., 2023) – but abiotic synthesis of complex biomolecules endowed with homochirality and extensive chain length seems improbable, suggesting that they may be reliable biosignatures.

It is feasible to quantify our statement as follows. If the probability of elongating a polymer with k units into $k+1$ units is $\mathcal{P}$, then the probability P_N of synthesising an N-mer (with N monomers) in this fashion is

$$P_N \approx \underbrace{\mathcal{P} \cdot \mathcal{P} \dots \mathcal{P}}_{N-1 \text{ times}} \approx \mathcal{P}^{N-1} \approx \exp\left[(N-1)\ln\mathcal{P}\right], \tag{13.3}$$

revealing that P_N declines exponentially with N; this is because the exponential function has a negative argument, due to $\ln\mathcal{P} < 0$. Needless to say, this picture is extremely simplified, but it suffices for our purpose.

If we subscribe to this scenario, where complex polymers analogous to biomolecules on Earth are viable biosignatures, the question arises: what would these polymers look like? To tackle this question, let us focus on the counterparts of nucleic acids since they are the information-carrying molecules. Computational chemistry models have illustrated that nucleic acids may be composed of non-canonical 'alphabets' (i.e., monomers), and that the collection of functional nucleic acid analogues could easily run into the millions (Cleaves et al., 2019). Hence, detecting 'exotic' analogues clearly divergent from RNA and DNA may not only corroborate the existence of life but also support abiogenesis independent of Earth.

However, the previous paragraph glosses over an important question: what constitutes a true nucleic acid analogue? Nucleic acids are not merely repositories of (genetic) information, but they also enable Darwinian evolution to operate by allowing for both mutations (i.e., 'errors') and their transmission via replication. As underscored in Section 1.1.3, Darwinian evolution is enshrined in the NASA definition of life, and is considered a powerful biosignature (Neveu et al., 2018). Therefore, we would ideally want these analogues to exhibit the same capacity for Darwinian evolution.

Building on earlier publications, Benner (2017) proposed that the preceding features may be fulfilled by polymers with repeating charges (in their backbone), the latter of which perform multiple roles by acting as chemical buffers, promoting template-based replication, modulating strand interactions, and so forth. If this hypothesis is correct, it is plausible that extraterrestrial nucleic acid analogues might be composed of repeating charges. In that event, electrostatic interactions could be harnessed to accumulate these polymers on an oppositely charged surface, allowing us to detect and characterise them by methods like mass spectrometry and fluorescent tagging.

Hitherto, we have centred our attention on individual biomolecules. Changing tack, by drawing inspiration from (13.3), the abiotic synthesis of longer polymers might not be favourable on thermodynamic and/or kinetic grounds in sundry locations, whereas living systems would strive to find workarounds if the resultant molecules possess the requisite functionality. Hence, it is plausible that even *distributions* of the abundances of biological building blocks, such as amino acids, could be manifestly dissimilar in biological and abiotic samples (McKay, 2004; Dorn et al., 2011).

In this context, measuring the concentrations of amino acids relative to glycine (the simplest of the lot) may represent a promising biosignature candidate (Davila and McKay, 2014), with the major caveat that we do not have a clear picture of the patterns associated with extraterrestrial life. By the same token, the biological synthesis of fatty acids (viz., vital components of cell membranes) often involves either two- or five-carbon units, whereas the abiotic production of linear carbon molecules is typically achieved through the addition of one-carbon units (Georgiou and Deamer, 2014). In consequence, unusual length distributions of fatty acids (e.g., dominance of compounds with even number of C atoms) are conceivably an effective biosignature, albeit with the aforementioned caveat in mind.

13.1.6   Agnostic biosignatures

To reiterate, biosignatures are closely intertwined with our conception(s) of life. The potential ISBs tackled so far are directly motivated by biology on Earth. To mitigate this drawback in recent times, intriguing research has sprung up in *agnostic* biosignatures (Conrad and Nealson, 2001), which seek general properties of life that are not specific to Earth.

One of the fundamental postulates in the so-called *assembly theory* is that living systems are far more 'complex' than non-living ones (Sharma et al., 2023). We may formulate a heuristic and crude justification as follows. Suppose that we intend to build an object from an initial set of building blocks: we will presume that there are M unique blocks, and that the object is made up of N blocks.[1] The probability $P_{M,N}$ of assembling one specific object, with all the desired functionality, purely at random is

$$P_{M,N} = \frac{1}{\underbrace{M \cdot M \ldots M}_{N \text{ times}}} = M^{-N} = \exp\left(-N \ln M\right), \tag{13.4}$$

which declines exponentially with N. Hence, this simple calculation demonstrates that the formation of complex objects (with large values of N) through random processes might have negligible probability. In contrast, living systems may ostensibly evince a certain goal-directed nature (i.e., some loose version of purpose), which is patently non-random and could therefore drive the assembly of such objects at higher probability by means of targeted synthesis.

Molecular assembly number (MA), or *assembly index*, encapsulates the minimal aggregate of steps necessary to construct a given object from a set of building blocks (Sharma et al., 2023). Reportedly, biological samples from Earth exhibit MA $\gtrsim$ 15, whereas abiotic materials fall below this threshold

---

[1] The building blocks can be molecules with different chemical formulae or even isomers (which are endowed with the same chemical formula).

(Marshall et al., 2021). Hence, it was argued that MA can help discern life from non-life. Tandem mass spectrometry could be capable of gauging MA because of a claimed linear relationship between the quantity of peaks in the mass spectra (i.e., plot of intensity versus mass-to-charge ratio) and MA (Marshall et al., 2021). Once a sample is analysed, MA may be inferred from the number of peaks, which might thence shed light on the biogenicity of the sample. Note, however, that the novelty, efficacy, and methodological soundness of assembly theory (e.g., MA) have all been critiqued.

Likewise, Guttenberg et al. (2021) analysed the mass spectra of non-biological compounds, living organisms, and remains of dead lifeforms. This preliminary study concluded that the mass spectra reveal qualitative and quantitative divergences among this trio of categories, lending credence to the premise that mass spectrometry might help us separate biogenic and abiotic materials. A significant practical benefit is that mass spectrometry for undertaking agnostic life detection, along the lines explained here, ought to be straightforward to implement in future missions (Chou et al., 2021).

The rationale underpinning the interesting agnostic biosignature advanced by Johnson et al. (2018) is that aptamers (short strands of nucleic acids) bind to a rich variety of inorganic and organic (including biological) compounds. Patterns in the distribution of bound aptamers may enable distinguishing between biological or abiotic samples; for instance, aptamers are anticipated to bind more readily with complex biogenic materials than simple abiotic ones. It might even be viable to tell apart Earth-based life from 'exotic' forms of life by employing this approach or variants thereof.

Lastly, cutting-edge machine learning could identify biosignatures based on statistical analysis of surficial patterns and morphologies (Warren-Rhodes et al., 2023). Machine learning, allied to experimental methods like gas chromatography and mass spectrometry, may also help extract distinct patterns in molecular distributions that demarcate biological samples from abiotic ones (Cleaves et al., 2023). While the reliability and false positives/negatives of agnostic biosignatures warrant further research, this paradigm holds immense promise in quests for both life-as-we-know-it and exotic life.

13.1.7 Miscellaneous biosignatures and synthesis

The list of putative ISBs we have encountered until now is far from exhaustive, owing to which a handful of other ISBs are elucidated next.

1. Motility: Many organisms have the capacity for spontaneous motion (motility), rendering it a convincing biosignature. An advantage of searching for motile lifeforms is that lower spatial resolution is required compared to their sessile (i.e., non-motile) counterparts (Nadeau et al., 2016).
2. Minerals: The abundance distribution of minerals may vary for environments that have/had life and those without it. A few minerals on Earth are exclusively associated with biological processes, such as oxammite and abelsonite, which are respectively derived from guano and chlorophyll.
3. Metabolic products: This potential ISB has a lengthy history, as it comprised the basis of the *Viking* life-detection experiments on Mars (reviewed in Section 9.4.2). The basic premise is that metabolic reactions generate heat and chemical products, both of which could be detected. However, supplying the appropriate reactants for initiating metabolism is challenging when the pathway is unknown a priori. Moreover, abiotic reactions can mimic metabolism, thereby acting as false positives.

4. <u>Co-location of oxidants and reductants:</u> Redox (reduction–oxidation) reactions lie at the heart of metabolism. Unearthing evidence of co-located oxidants and reductants has been hypothesised to constitute a signature of metabolic activity; this school of thought is traceable to the 1960s (Lovelock, 1965). However, the simultaneous operation of abiotic mechanisms may cause similar effects, on account of which knowledge of the environmental context becomes crucial.

5. <u>Growth and reproduction:</u> Both of these traits are often considered universal to life, even if they do not suffice to define life by themselves. If a sample harbours living organisms, by providing suitable energy sources and nutrients, it is theoretically feasible to witness growth and reproduction. Continued measurements of the output of metabolic products may also permit corroboration of these attributes.

In rounding off our exposition of ISBs, we emphasise that life-detection missions should search for as many potential ISBs as possible, especially those that complement each other and boost the credibility of life detection. In practice, the ISBs that can be detected are dependent on the onboard scientific instruments, which are regulated in turn by the mission design (e.g., flyby versus sample return). The specifics of the mission design are shaped by the habitable world that will be surveyed, since each world hosts its own unique environments (e.g., the cloud decks of Venus), and the accompanying challenges of exploring them would vary accordingly.

Hence, a life-detection mission to Mars is anticipated to carry a different suite of instruments than one to Enceladus. The rationale is that the former offers a plethora of surface environments to investigate, whereas the primary means of exploring the latter involves flying through the plume and sampling it; another drawback is the greater distance of Enceladus, which would translate to an increased fuel budget for the spacecraft. In Table 13.1, we provide a brief summary of what ISBs might be detected in the plume of Enceladus, along with the desired instrumentation.

Finally, imagine that we have discovered some potential ISBs during a life-detection mission. How confidently can we claim that we have found life? We examine this key question in Section 13.3.

13.2   Remote-sensing biosignatures

In defining RSBs at the start of the chapter, we highlighted that they are discernible via spectroscopy and/or photometry (using space- and ground-based telescopes), and that they typically focus on exoplanets. Reviews of exoplanetary biosignatures are furnished in Schwieterman et al. (2018), Meadows et al. (2020), and Lingam and Loeb (2021, Chapter 6).

We adopt the classification scheme and organisation of Schwieterman et al. (2018), which is a state-of-the-art overview of this subject. The three major groups of RSBs are summarised in Figure 13.4, while the spectral features of gaseous and surface RSBs are depicted in Figure 13.5. Before discussing them, we recommend that the readers peruse Section 12.2 to refresh their memory of how atmospheric and surface components are identifiable.

13.2.1   Gaseous biosignatures

An ideal gaseous biosignature should fulfil some criteria: (1) it should be produced abundantly, such that it can accumulate to relatively high levels and become detectable; (2) it should have a low likelihood of being generated by abiotic processes (i.e., low probability of false positives); and (3) it

Table 13.1 In situ life-detection mission to sample plume of Enceladus

ISB	Technique	Plume material
Amino acid abundance and isomeric distribution	MS + GC/LC	$\sim 10^{-4}$ kg
Enantiomeric excess in chiral amino acids	MS + GC/LC	$\sim 5 \times 10^{-4}$ kg
Distribution of chain length of long-chain organics	MS + GC/LC	$\sim 10^{-4}$ kg
Polymers with repeating charges in backbone	Nanopore sequencing	$\gtrsim 1$ kg
Biological polymers (e.g., proteins, nucleic acids)	Immunoassay and MS	$\gtrsim 10^{-3}$–10^{-2} kg
Cell-like morphologies	Atomic force/optical/digital holographic microscopy	$\gtrsim 10^{-3}$ kg
Molecular assembly number (agnostic biosignature)	MS + GC/LC	$\sim 10^{-4}$ kg
Isotopic fractionation	MS + GC	$\sim 10^{-4}$ kg
Co-located oxidants and reductants	MS, spectroscopy X-ray crystallography	$\lesssim 10^{-3}$ kg

Note: MS, GC, and LC refer to mass spectrometry, gas chromatography, and liquid chromatography. The last column shows the mass of plume material needed for ISB detection. All data and experiments are based on Neveu et al. (2020).

must exhibit strong and distinctive spectral features that do not overlap much with other gases. A comprehensive list of potential gaseous biosignatures was compiled by Sara Seager (1971–present) and collaborators (Seager et al., 2016); we will confine ourselves to a handful of examples here. We preface this discussion with the caveats that the majority of publications have investigated atmospheric compositions similar to early/modern Earth, and that the effects of space weather phenomena are not well understood.

Molecular oxygen (O_2):   Molecular oxygen as a gaseous biosignature was proposed in 1930 by Sir James Jeans (see Section 1.2). As explained in Section 7.1.4, this gas is a metabolic product of oxygenic photosynthesis:

$$CO_2 + 2H_2O \xrightarrow[\text{light}]{\text{pigments}} CH_2O + H_2O + O_2. \tag{13.5}$$

Notable spectral features of O_2 occur at approximately 0.69 μm, 0.76 μm (O_2–A band), and 6.4 μm; of this trio, the 0.76 μm band is well-suited for detection by means of measuring reflected light. A thorough overview of O_2 as a gaseous biosignature is presented in Meadows et al. (2018).

It is necessary to enquire whether O_2 has false positives, and the answer is affirmative. X-rays and extreme ultraviolet (jointly XUV) radiation, in the wavelength range of ~ 0.1–90 nm, can drive

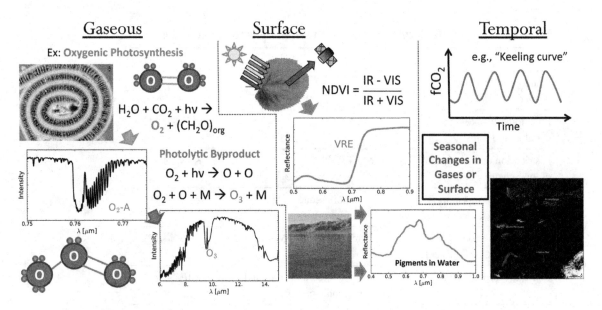

Figure 13.4 Remote-sensing biosignatures detectable by spectroscopy: (1) gaseous biosignatures are gases produced by biological (usually metabolic) activity; (2) surface biosignatures arise from biological materials on surfaces; and (3) temporal biosignatures are manifestations of temporal variations in the biosphere. (Credit: Schwieterman et al. 2018, Figure 1; CC BY-NC 4.0 license)

the splitting (photolysis) of molecules such as water and carbon dioxide, thus releasing O_2. This effect could be pronounced for exoplanets orbiting M-dwarfs, where high XUV fluxes are predicted during the pre-main-sequence phase, perhaps engendering O_2-dominated atmospheres with pressures of $\mathcal{O}(100)$ bars. As regards false negatives, Earth had minimal O_2 for ~50% of its history, and certain exoplanets (e.g., around M-dwarfs) may never accumulate biogenic O_2 to detectable levels even if oxygenic photosynthesis exists.

Fortunately, it seems possible to differentiate between abiotic and biogenic molecular oxygen. As the former may be characterised by massive O_2 atmospheres, the formation of the O_2–O_2 dimer is feasible, which would result in strong absorption bands at 1.06 μm and 1.27 μm. Hence, the detection of these features could help rule out a biological origin for O_2.

Ozone (O_3):   The formation of ozone is conditioned by the availability of O_2, because of its production from the latter as follows:

$$3O_2 \longleftrightarrow 2O_3, \tag{13.6}$$

which is mediated by the flux of UV radiation at wavelengths <240 nm. Hence, as the appropriate UV flux is raised, the production of O_3 is enhanced. Therefore, even if the O_2 inventory is held fixed, the spectral type of the star may dictate the abundance of O_3. Loosely speaking, hotter (higher-mass) stars are likely to support elevated O_3 levels owing to the increased UV flux, while the opposite trend is anticipated for biosignature gases (e.g., methane) that are (in)directly destroyed by photolysis (Rugheimer et al., 2013). Aside from UV flux, Section 8.2.4 conveyed that ozone levels are mediated by stellar energetic particles (SEPs), which trigger its depletion.

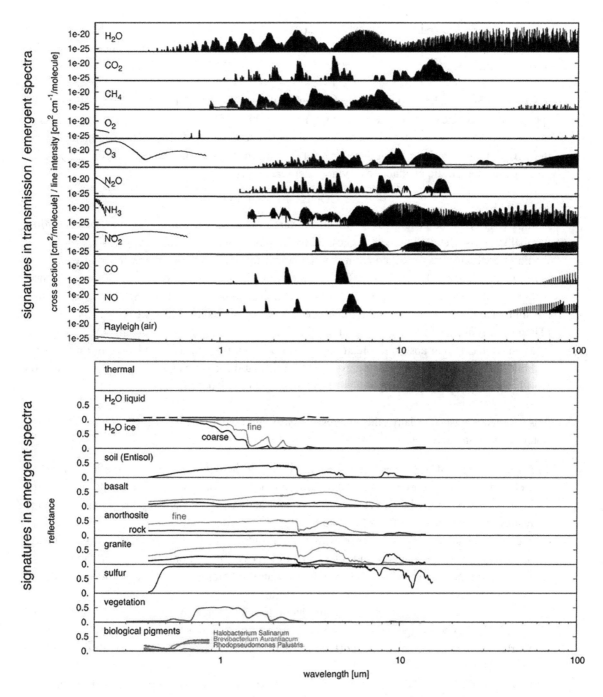

Figure 13.5 Spectral features of common gaseous and surface (remote-sensing) biosignatures. Transmission spectra were described in Section 12.2.2, whereas emergent spectra encompasses thermal emission (Section 12.2.3) and reflected (Section 12.2.4) light. (Credit: Fujii et al., 2018, Figure 2; CC BY-NC 4.0 license)

O_3 has strong spectral features such as the 0.2–0.3 μm Hartley–Huggins bands, the 0.4–0.7 μm Chappuis bands, and the 9.65 μm band. As O_3 production is linked to O_2, it may inherit some false positives/negatives of O_2.

Methane (CH_4):   The predominant source of methane on Earth is microbial metabolism; to be specific, the biological flux of methane is higher than abiotic fluxes by at least one to two orders of magnitude (Thompson et al., 2022). We have already encountered hydrogenotrophic methanogens in (7.10), which contribute about one-third of biogenic emissions, whereas the remaining two-third arises from acetoclastic methanogens, whose net reaction is

$$CH_3COOH \rightarrow CH_4 + CO_2. \tag{13.7}$$

Abiotic CH_4 could be generated by volcanism, large impacts, and water–rock reactions at hydrothermal systems. Detailed models suggest that CH_4 molar ratios of $\gtrsim 10^{-3}$ are plausibly biological for Earth-analogues; as remarked earlier, hotter stars can suppress the methane abundance and vice versa. In addition, the detection of CH_4 in conjunction with gases such as O_2 and CO_2 may support its biological provenance, on the grounds of *chemical disequilibrium* (refer to Section 13.2.4), while the simultaneous presence of carbon monoxide (CO) is presumed to constitute an *anti-biosignature* (Thompson et al., 2022), that is, a signature disfavouring biogenic origin (of methane).

Methane has strong absorption features at 1.65 μm, 2.4 μm, 3.3 μm, and 7–8 μm in the near- and mid-infrared. One point to recognise, however, is that these bands may not be easy to detect on worlds with low levels of atmospheric CH_4, namely, akin to modern Earth.

Miscellaneous gases: Many other interesting gases are potential RSBs, of which we sketch some noteworthy contenders here.

1. Nitrous oxide (N_2O): N_2O is chiefly produced during the denitrification metabolic pathway, as well as human agricultural production. Atmospheric levels of N_2O are sensitive to not only the input fluxes but also the stellar temperature and activity. This gas exhibits strong absorption features at 2.9 μm, 4.5 μm, 7.8 μm, and 8.5 μm.
2. Methylated halogens: These gases (e.g., methyl chloride and methyl bromide) arise from pathways involving plants, fungi, algae, and organic matter decay. Simulations indicate that their concentrations may be higher in M-dwarf exoplanetary atmospheres. The major absorption features of these gases occur at 3.3 μm, 9.9 μm, and 13.7 μm.
3. Ammonia (NH_3): On Earth, ammonia production is dominated by biogenic and anthropogenic sources. It has been suggested that ammonia may accumulate to detectable levels on 'Cold Haber Worlds', thanks to a runaway effect (Ranjan et al., 2022). A widely investigated absorption band of ammonia corresponds to 1.45–1.55 μm.
4. Phosphine (PH_3): We have already touched on phosphine in Section 11.3.1, when discussing the clouds of Venus. Despite phosphine being a trace gas on Earth, on some anoxic worlds it might reach atmospheric concentrations up to ∼100 ppm. This gas has strong absorption bands at 2.7–3.6 μm, 4.0–4.8 μm, and 7.8–11.5 μm (Sousa-Silva et al., 2020).

For most biosignature gases, false positives are not rigorously characterised, on account of the tendency to invoke Earth as a proxy. Besides Earth-like planets, Hycean worlds from Section 8.1.4 might host gaseous biosignatures. These worlds are now popular targets of observing programs, which are hinting at intriguing gases perceived as RSBs (Madhusudhan et al., 2023).

13.2.2 Surface biosignatures

When light from the star penetrates to the surface, it will be absorbed and reflected to varying degrees depending on the wavelength and surface composition. Hence, through monitoring the reflected light via direct imaging (outlined in Section 12.2.4), it is theoretically possible to discern surface biosignatures; some of the notable candidates are shown in Figure 13.5.

Photosynthesis: We have extensively reviewed photosynthesis in Section 7.1.4, due to which we shall not tackle it again. It is worth recalling that the simplified version of the net photosynthetic reaction is

$$CO_2 + 2H_2X \xrightarrow[\text{light}]{\text{pigments}} CH_2O + H_2O + 2X, \tag{13.8}$$

where H_2X is the reducing agent, and X is the product. In oxygenic photosynthesis, water is the reducing agent, while in anoxygenic photosynthesis, reductants like H_2, H_2S, and Fe^{2+} are employed. As water is plentiful on Earth, maybe it is not surprising that eventually life adopted this molecule as the electron donor, and achieved high biological productivity.

As elucidated in Section 7.1.4, the primary photopigments are the chlorophylls and bacteriochlorophylls, the latter of which exist in anoxygenic phototrophs. These pigments essentially consist of tetrapyrroles: molecules with four heterocyclic rings. In principle, photopigments may evolve on exoplanets with clearly different structures (Lingam et al., 2021), but empirical evidence is lacking hitherto. The absorption maxima of chlorophylls is sensitive to the pigment type and the environmental conditions; with this caveat in mind, the last peak is manifested at $\sim700 \pm 50$ nm (Schwieterman et al., 2018, Table 1). In contrast, the bacteriochlorophylls can absorb strongly at longer wavelengths, up to $\sim1{,}040$ nm for bacteriochlorophyll b.

The absorption of the prevalent chlorophylls declines shortly after 700 nm, thence translating to a steep increase in the reflectance beyond this wavelength; this characteristic is evident upon inspecting Figure 13.4. Hence, this 'spectral edge' at red wavelengths may be a tenable biosignature of vegetation. In fact, this notion of the *vegetation red edge* (VRE) was proposed over a century ago by Vladimir Artsikhovski (see Section 1.2), and it has witnessed a revival in the twenty-first century. A couple of limitations to bear in mind are the following: (a) the spectral edge might be shifted to longer/shorter wavelengths based on the stellar spectral type and the planet's atmospheric composition (Kiang et al., 2007b; Lingam et al., 2021); and (b) minerals display similar features, like cinnabar at ~600 nm.

The VRE is a challenging feature to detect, as revealed by Figure 12.7, where this edge is not easily identifiable at ~700 nm. It is necessary for the vegetation-analogue to not only span a sizeable fraction of the surface but also for this region to be relatively free of clouds; the cloud-free fraction of vegetation may need to exceed 20%. Recent simulations have shown that a signal-to-noise ratio of 10 (i.e., high, yet achievable) can enable us to discern the VRE on exoplanets in tandem with constraining the location of the edge to within 70 nm (Gomez Barrientos et al., 2023). Apart from vegetation, other oxygenic photoautotrophs might produce distinctive and detectable spectral edges (O'Malley-James and Kaltenegger, 2019b).

Anoxygenic photosynthesisers exhibit their own unique spectral features governed by their photopigments, which might be likewise detectable. However, such an outcome would require these organisms to cover a substantial portion of the planet's surface, perhaps as microbial mats.

Rhodopsins and other pigments: In Section 7.1.4, we stated that many microbes possess rhodopsins that fulfil manifold roles (e.g., ion pumps). For instance, bacteriorhodopsin is found in purple-coloured *Haloarchaea* with an absorption peak at $\sim$568 nm. A 'Purple Earth' dominated by equivalent microbes might therefore be detectable. Looking beyond phototrophy, biological pigments execute myriad functions (e.g., signalling and UV protection), and some of their reflection spectra are illustrated at the end of Figure 13.5. We caution, however, that the false positives for the pigments mentioned in this context are poorly understood.

Homochirality: It may seem odd at first glimpse to revisit homochirality since we tackled it in Section 13.1.4, but it has been studied independently as an RSB. The expectation is that surface-based chiral biomolecules (which includes pigments like chlorophylls) would interact with light, and engender signals discernible via polarisation spectroscopy.

Over 50 years ago, it was experimentally proven that circular polarisation signatures are present in scattered light from photosynthetic organisms; these signatures are considered a consequence of homochirality. The normalised amplitude of circular polarisation might reach up to $\sim$1%, thereby making it feasible for next-generation ground- and space-based telescopes to detect this signal (Sparks et al., 2021). Likewise, linear polarisation spectroscopy may reveal signatures of photosynthetic organisms at higher amplitudes (thus improving the prospects for detection), but at the potential cost of enhanced risk of false positives. Preliminary detection of the VRE has been achieved via linear polarisation spectroscopy of Earthshine (Sterzik et al., 2019; Gordon et al., 2023), and further research is warranted for exoplanets.

Biofluorescence: Biological materials on the surface can be detected by their emission, and not just through their absorption properties. Fluorescence entails absorption of photons at higher energies, and emission at longer wavelengths; both abiotic and biotic substances evince this trait. Chlorophyll *a* fluorescence exhibits maxima at $\sim$685 and $\sim$740 nm, but constitutes a small fraction of the reflected light at these wavelengths ($\sim$1% of spectral flux). However, on planets bombarded by repeated stellar flares, fluorescence might evolve for UV protection, in which case its signature(s) may be rendered more prominent (O'Malley-James and Kaltenegger, 2019a).

13.2.3 Temporal biosignatures

The biosphere is intrinsically dynamic. Hence, on inhabited worlds, one would expect some detectable signals (biogenic or otherwise) to display temporal variations that are modulated, at least in part, by biological activity. In essence, this is the underlying rationale for temporal biosignatures.

Let us contemplate the atmospheric CO_2 abundance. In the summer, the growth of vegetation should be enhanced, leading to an increased drawdown of CO_2, because this gas is utilised for carbon

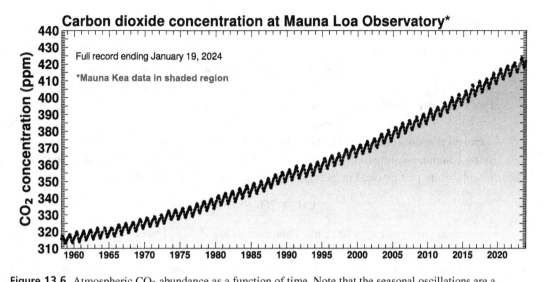

Figure 13.6 Atmospheric CO_2 abundance as a function of time. Note that the seasonal oscillations are a biosignature, whereas the steady increase is a techosignature. (Credit: https://keelingcurve.ucsd.edu/; CC BY-NC 4.0 license)

fixation. In the winter, the opposite trend is anticipated, namely: CO_2 consumption declines as vegetation decays. The net result is that atmospheric CO_2 should decline in summer and increase in winter, giving rise to seasonal oscillations. This behaviour is indeed observed in Figure 13.6, but we also notice a long-term increase in CO_2 content that can be linked to the burning of fossil fuels (anthropogenic climate change). Hence, it seems appropriate to label this curve as a 'bio-technosignature' since one of its components (oscillation) is a biosignature, while the other (rapid growth) is currently a technosignature.

It is apparent from (13.5) that CO_2 and biogenic O_2 are anti-correlated, implying that O_2 seasonal oscillations also ought to be anti-correlated with fluctuations in CO_2. Since ozone is derived from O_2, as evidenced by (13.6), it may parallel the O_2 oscillations. The aforementioned temporal variations in the atmospheric abundances of CO_2, O_2, and O_3 (and methane) are discernible in principle, but the amplitude of these fluctuations is $\lesssim 10\%$ (Mettler et al., 2023). This aspect poses not only an issue for detection but also runs the danger of being distorted by unforeseen false positives.

Transitioning to surface RSBs, as outlined above, vegetation should grow and decay in summer and winter, respectively. Hence, the strength of the VRE signal can exhibit oscillations, indicative of biological processes, which may be detectable either by means of regular spectroscopy or combining it with polarimetry (i.e., polarisation spectroscopy). Microbial mats could also experience temporal fluctuations, which might translate to detectable variations. In a similar vein, transient biological phenomena (e.g., large-scale algae-like blooms) are temporal biosignatures in theory, but the ability to discern such a signal in practice is doubtful.

Last, analogous to Section 13.1.6, formulating agnostic RSBs may be possible. Bartlett et al. (2022) demonstrated that temporal variations in reflected or emitted light could enable the computation of statistical measures of complexity. It was argued that planets with higher complexity are likely to signify biospheres, but this interesting hypothesis is not yet verified.

13.2.4   Methods for assessing plausibility

We have seen that the majority of gaseous RSBs are susceptible to ambiguity. Hence, if we were to detect some set of these biosignatures and other compounds, it is necessary to develop methods for gauging whether this set is truly an indicator of life. We will summarise two of them here; a detailed exposition is furnished in Lingam and Loeb (2021, Chapter 6.7).

Chemical disequilibrium: The notion of employing chemical disequilibrium as a guide for identifying genuine biosignatures has been around since the 1960s (e.g., Lovelock, 1965). The classic disequilibrium pair invoked in astrobiology is CH_4 and O_2, since they would combine to yield

$$CH_4 + 2O_2 \rightarrow CO_2 + 2H_2O, \tag{13.9}$$

on a transient timescale of ~ 10 years (Thompson et al., 2022). Hence, if methane is not continuously replenished, it would be rapidly depleted in O_2-rich atmospheres. The fact that biology on Earth contributes the vast majority of atmospheric CH_4 (refer to Section 13.2.1) is what permits this gas to coexist with O_2 in Earth's atmosphere. In other words, this documented chemical disequilibrium is directly traceable to life.

Krissansen-Totton et al. (2018) conducted a comprehensive analysis of chemical disequilibria in Earth's history attributable to biological activity. In the Proterozoic (2500–539 Ma) and Phanerozoic (539–0 Ma), the predominant source of biogenic disequilibrium was determined to be the coexistence of N_2, O_2, and liquid water, because these molecules would otherwise produce nitric acid, and therefore get depleted as follows:

$$5O_2 + 2N_2 + 2H_2O \rightarrow 4H^+ + 4NO_3^-, \tag{13.10}$$

implying that detecting this trio may corroborate the presence of life. In the Archean, the coexistence of CH_4, CO_2, N_2, and liquid water is disfavoured in the absence of life, because these molecules would react to yield

$$5CO_2 + 4N_2 + 3CH_4 + 14H_2O \rightarrow 8NH_4^+ + 8HCO_3^-, \tag{13.11}$$

suggesting that detecting this quartet could support the presence of biology. This framework can be applied to other worlds, but it may lose accuracy if unknown chemical, geological, or biological processes are at play.

Biomass estimation: This approach was elucidated by Seager et al. (2013) to assess biogenicity of certain gases; for example, gases that are products of metabolic reactions entailing chemical (redox) energy (tackled in Section 7.1.3). In the first step, the gas concentration that results in a detectable spectral feature is computed, given suitable planetary and stellar inputs. Next, an atmospheric photochemistry model is utilised to infer the source flux needed to achieve this gas abundance. In the final step, a thermodynamic model is harnessed, along with the calculated source flux, to estimate the required biomass for generating the spectral feature.

Once the biomass has been estimated, it is necessary to ask whether this biomass is plausible. If the answer is yes, this lends credence to the hypothesis that the gas is an actual RSB. On the other hand, if the biomass is deemed unrealistic, it lowers the possibility of biological origin. We caution, however, that this framework may suffer from uncertainties in the input parameters, as well as in the choice of the thermodynamic model itself.

13.3   Rubrics for life detection

During our exposition, we have delved into myriad sources of ambiguity surrounding ISBs and RSBs. In light of this challenge, any detection of putative biosignature(s) must be carefully evaluated using appropriate rubrics. These rubrics are not exact 'laws', owing to the known and unknown sources of ambiguity, and should rather be viewed as heuristics broadly agreed upon by the astrobiological community (albeit not universally).

The first route involves Bayesian analysis, which builds on the famous *Bayes' theorem*. The detected potential biosignatures constitute the evidence E, and the hypothesis that this evidence stems from biological activity is labelled H. As per Bayes' theorem, the probability $P(H|E)$ (which we wish to determine) of the hypothesis H given the evidence E is expressed as

$$P(H|E) = \frac{P(E|H)P(H)}{P(E)}, \qquad (13.12)$$

where $P(H)$ is the probability that the hypothesis is true (called the *prior*), $P(E)$ is the probability of the evidence E, and $P(E|H)$ is the probability of detecting the evidence E given that the hypothesis H is true (known as the *likelihood*). The magnitude of $P(H|E)$ sets the confidence level of the existence of extraterrestrial life. This framework is quantitative, and offers a systematic means of investigating presumed biosignatures, due to which it is widely advocated (Catling et al., 2018; Walker et al., 2018; Lingam et al., 2023c). However, one drawback is that the prior(s) and the odds of false positives are weakly constrained for many candidates, thereby complicating the estimation of $P(H|E)$. Bayesian approaches have facilitated analysis of several facets of ISBs and RSBs (e.g., Balbi and Grimaldi, 2020).

Moving to a qualitative picture, the aptly named 'Ladder of Life Detection' was proposed by Neveu et al. (2018). Chiefly ISBs and their false positives were assessed, with an emphasis on identifying either individual or groups of robust biosignatures, along the lines outlined at the start of this chapter. A collection of decision rules was consequently developed, wherein some sets of potential ISBs would strongly corroborate, if not essentially confirm, the existence of extraterrestrial life; and other rules would not fulfil this purpose. The ISBs comprising the Ladder of Life Detection are depicted in Figure 13.7. A subset of the decision rules is provided next, with the full description and rationale delineated in Neveu et al. (2018, Table 3).

- Rule A: Detection of Darwinian evolution (if practically possible) alone is enough to exclude an abiotic origin.
- Rule B: Detection of complex evolved biological pigments by means of their spectral features is purportedly sufficient to rule out an abiotic origin.
- Rule D: Detection of cell-like morphologies and structural preferences (see Section 13.1.4) is adequate to exclude an abiotic origin.
- Rule F: Detection of metabolic products in response to substrate addition (refer to Section 13.1.7) in combination with structural preferences is satisfactory to rule out an abiotic origin.
- Rule I: Detection of structural preferences without any other supporting evidence is *not* sufficient to exclude an abiotic origin.

The Ladder of Life Detection

Left-margin groupings: **LIFE** (Darwinian evolution → Molecules & Structures Conferring Function); **SUSPICIOUS BIOMATERIALS** (Potential biomolecule components → Biofabrics).

Column group headers: **INSTRUMENTAL CRITERIA** (Quantifiable, Contamination-free, Repeatable); **CONTEXTUAL CRITERIA** (Detectable, Survivable, Reliable, Compatible, Last resort).

RUNG (Roughly: subjectively ordered by (top to bottom); 1. decreasing strength of evidence for life; 2. increasing ease of measurement)	FEATURE (Listed in no specific order within a given rung)	MEASUREMENT TARGET	LIKELIHOOD (that the feature would be a biosignature, given the criteria to the right)	Quantifiable (Detectability)	Contamination-free (Likelihood of false positive)	Repeatable	Detectable (Detectability)	Survivable (Likelihood of false negative)	Reliable (Ambiguity of feature)	Compatible (Specificity to Earth life)	Last resort (Ambiguity of interpretation)
Darwinian evolution	Changes in inheritable traits in response to selective pressures	Not practical under mission constraints • In situ • Sample return	No	-	-	Number of replicates depends on: • Characteristics of the instrument • Heterogeneity of the sample • Likelihood of systematic errors • Required values of the relevant statistical parameters • Value and cost of information.	-	N/A (extant)	-	-	-
Growth & Reproduction	Concurrent life stages or identifiable reproductive form, motility	Cell-like(?) structures in multiple stages • In situ • Sample return	Low	Hard	Low		Med (don't identify stages, timing off, sample size low)	High?	Ambiguous. What is a cell? What morphological differences exist?	Earth	Med / High
Growth & Reproduction	Major element or isotope fractionations indicative of metabolism	Deviation from abiotic fractionation controlled by thermodynamic equilibrium and/or kinetics • Remote sensing • In situ • Sample return	Low / Med	Easy	High		Medium	High	Hinges on understanding of context	Earth?	Low
Metabolism	Response to substrate addition	Waste output (compound, heat) • In situ • Sample return	Low / Med	Easy	Low		High	N/A (extant)	Hinges on understanding of context	Earth	Medium
Metabolism	Co-located reductant and oxidant	Deviation from abiotic distribution controlled by thermodynamic equilibrium and/or kinetics • Remote sensing • In situ • Sample return	Med / High	Med (linked to specificity of instrument)	Low / Med		Med / High	High	Need reactions, large inventory of chemistries	Generic	Low / Med
Molecules & Structures Conferring Function	Polymers that support information storage and transfer for terran life (DNA, RNA)	Polymer with repeating charge • In situ • Sample return	Low	Hard (instrument specifically must be high); RNA hard to measure on Earth	DNA: high; RNA: low (reactive)		Low (technology limited, only terran); RNA highly reactive	Low (hydrolysis in water)	Reliable	Earth	Negligible
Molecules & Structures Conferring Function	Structural preferences in organic molecules (non-random and enhancing function)	Enantiomeric excess > 20% in multiple amino acid types • In situ • Sample return	High	Need a lot of material and overprinting must be discernible	Low		Low	Low (hydrolysis in water, diagenesis)	How much preference needed to detect?	Generic	Low
Molecules & Structures Conferring Function	Pigments as evidence of non-random chemistries (e.g. specific preferences)	Spectral feature and/or color, otherwise see "Structural preferences" • Remote sensing • In situ • Sample return	Low / Med	Easy (fluorescence)	Low		Med / High	Medium	Mixed sample both processes present	Generic	Low
Potential biomolecule components	Organics not found abiotically (e.g. hopanes, ATP, histidine)	Presence • In situ • Sample return	Medium	How much excess necessary?	Low		Low	Low (diagenesis)	How to define pigment as we don't know it?	Earth (can one abstract?)	Very low
Potential biomolecule components	Complex organics (e.g. nucleic acid oligomers, peptides, PAH)	Presence • Remote sensing (PAH) • In situ • Sample return	High	Easy if enough material	Low		High	High	Low	Earth?	High
Potential biomolecule components	Monomeric units of biopolymers (nucleobases, amino acids, lipids for compartmentalization)	Presence • Remote sensing • In situ • Sample return	Med / High	Easy if enough material	High		High	Med (diagenesis)	Abiotic production known	Generic	Med / High
Potential metabolic byproducts	Distribution of metals e.g. V in oil or Fe, Ni, Mo/W, Co, S, Se, P	Presence • In situ • Sample return	Medium	Limit of detection, need a lot of material	Low		High	High	Abiotic pathways known	Generic	Medium
Potential metabolic byproducts	Patterns of complexity (organics)	Deviation from equilibrium (P/Poisson distribution of pathway complexity) <0.01?) or abiotic kinetic distribution • In situ • Sample return	High	Background issue, material limited	Low		High	Medium?	Background known; Limited documentation of abiotic vs. biotic differences	Generic	Medium
Biofabrics	Textures	Biologically mediated morphologies, preferably with co-located composition • In situ • Sample return	Medium	Medium	Low		Medium	High?	Highly ambiguous	Earth	High
Habitability	Liquid water, building blocks, energy source, gradients								Redox, temperature, pH, energy, disequilibria		

Figure 13.7 Components of the Ladder of Life Detection. First column: biosignature candidate; second column: observable characteristic(s); third column: avenues for detecting this potential biosignature; fourth column: likelihood of the candidate representing a detectable and plausible biosignature; sixth and seventh columns: predictions of how the potential biosignature stacks up against the criteria in Figure 13.1. (Credit: Neveu et al., 2018, Table 2; CC BY-NC 4.0 license)

Table 13.2 Confidence of Life Detection (CoLD) scale

Level	Measurement indicators	Qualitative assessment
Level 1	Signal ensuing from biological activity detected	Biogenic signal detection
Level 2	Possibility of contamination effectively ruled out	Biogenic signal detection & Environmental relevance inferred
Level 3	Demonstration of biogenic signal in host environment	Environmental relevance inferred & Abiotic false-positives discrimination
Level 4	Known abiotic sources proven unlikely in that environment	Abiotic false-positives discrimination & Independent biosignature established
Level 5	Additional independent signals from biology detected	Independent biosignature established & Alternative hypotheses ruled out
Level 6	Future observations for ruling out alternatives proposed	Alternative hypotheses ruled out & Confirmation of biology
Level 7	Follow-up observations confirm predicted biological activity	Confirmation of biology

Note: The progressively increasing CoLD scale of Green et al. (2021) for gauging the viability of evidence potentially suggestive of extraterrestrial life. 'Qualitative assessment' of these levels often spans two ranges, owing to which both are presented.

These decision rules are preliminary, of which some involve a certain degree of subjective interpretation; therefore, their validity can be debated. Another caveat is that these rules permit only two outcomes, instead of a continuum.

In a similar vein, the 'Confidence of Life Detection' (CoLD) scale was proposed by Green et al. (2021), which is shown in Table 13.2. As one travels downward from the top, the conditions become more stringent, and the associated confidence level (of detecting life) is enhanced commensurately. However, the boundaries demarcating the various stages will not always be sharp in reality. Hence, it is not straightforward to gauge the level, and they might not follow a sequential order. For this reason, the CoLD scale should be envisioned as a first step towards initiating a dialogue.

13.4 Problems

Question 13.1: The mole fraction of an *L*-amino acid in a sample is 0.3. Calculate the mole fraction of the corresponding *D*-amino acid, and the enantiomeric excess *ee* of the sample. Would this value of *ee* constitute a robust biosignature? Support your answer by elucidating two reasons.

Question 13.2: Suppose that a small protein is composed of $N = 15$ amino acids. What is the probability of this protein assembling together randomly, assuming that we have a set of $M = 20$ amino acids? Outline whether this protein can be regarded as a viable biosignature.

Question 13.3: By consulting and citing peer-reviewed literature, summarise at least two putative in situ biosignatures not addressed in this chapter. Discuss why they are considered indicators of life, the associated false positives, and what experimental techniques can extract them.

Question 13.4: The *ExoMars* mission may launch sometime in the 2020s.[2] Delineate the planned onboard scientific instrumentation, and thereby determine which in situ biosignatures may be detectable. If all these signatures are indeed identified, would they be adequate to claim the discovery of life on Mars? Justify your reasoning by harnessing a suitable rubric.

Question 13.5: Repeat the exact tasks specified in Question 13.3, but for the category of remote-sensing biosignatures instead.

Question 13.6: We introduced the notion of a 'bio-technosignature'. Describe at least one more such potential signature. Explain your reasoning and cite peer-reviewed sources; for instance, if you propose a gas, you may want to look up the fluxes arising from biological and anthropogenic activity.

Question 13.7: The plausibility of gaseous biosignatures may be assessed by quantifying the topology of the atmospheric chemical reaction network (Schwieterman et al., 2018). Conduct a literature review of the appropriate references, and summarise the possible pros and cons of this approach.

Question 13.8: The *Viking* mission is conventionally deemed unsuccessful in unearthing life on Mars, whereas future space- and ground-based telescopes may possess the capacity to validate life on exoplanets. Clarify this difference by utilising the 'Ladder of Life Detection' from Section 13.3; in particular, consult Neveu et al. (2018, Tables 2–4) and accompanying text.

Question 13.9: Imagine that a life-detection mission to the clouds of Venus discovers reliable evidence of isotopic fractionation (e.g., depletion of ^{13}C) and microscopic structures that display cell-like morphologies. As per the 'Ladder of Life Detection' from Section 13.3, are these two criteria sufficient to declare that we have found life on Venus? Justify your answer in depth.

---

[2] www.esa.int/Science_Exploration/Human_and_Robotic_Exploration/Exploration/ExoMars.

14   Technosignatures

Thus far, we have explored one of the main questions of astrobiology ('*Are we alone?*') by tackling putative extraterrestrial life in general, from microbial life forms to complex multicellular organisms. However, there is a subset of life that, for lack of a precise definition, is (erroneously) labelled as '*intelligent life*'. Defining intelligence is deeply challenging and contested, and we shall not attempt an exposition of this topic here. We will, rather, adopt the pragmatic strategy adopted ever since the earliest empirical attempts to *search for extraterrestrial (technological) intelligence* (SETI for short), namely, to seek signs of extraterrestrial technological activity.

This is a relatively well-posed problem that is a logical extension of the search for biosignatures in the previous chapter. Hence, in recent times, SETI has been reframed as a search for *technosignatures*, broadly defined by SETI pioneer Jill Tarter (1944–present) as follows (Tarter, 2007, pg. 20):

[E]vidence of some technology that modifies its environment in ways that are detectable …[to] infer the existence, at least at some time, of intelligent technologists.

This definition encompasses a variety of possible sources and signals, each requiring distinct observational strategies; the most important of them are elucidated later in Section 14.4. Of course, the search for technosignatures inherits many of the methodological problems encountered in Chapter 13 with respect to biosignatures: most notably, non-zero ambiguity and the risk of incurring false positives. On the other hand, technosignatures might have inherent advantages – in principle, they may be more abundant, more easily detectable, and less ambiguous than biosignatures (Wright et al., 2022).

Therefore, it could be argued that the search for technosignatures is not only appealing from a human perspective (and for the cultural and societal consequences that discovering extraterrestrial intelligence may engender) but also because it could complement the search for biosignatures and enhance the probability of successfully detecting extraterrestrial life in general.

14.1   The Drake equation

No treatment of SETI can omit a discussion of the renowned *Drake equation* (Drake, 1965). This mathematical expression was devised by Frank Drake (1930–2022) in 1961, during a pivotal meeting on SETI convened at the *Green Bank Observatory*. In order to guide the discussion, Drake drew up a list of the myriad factors that contribute to the probability that technologically capable life has developed elsewhere in our Galaxy. Combining these variables yields an estimate of the expected number N_T of planets that may harbour remotely detectable technosignatures:[1]

$$N_T = R_\star f_p \, n_e \, f_l f_i f_c \, L, \tag{14.1}$$

where $R_\star$ is the star formation rate (number of stars per year) in the Milky Way, f_p is the fraction of stars hosting planetary systems, n_e is the number of habitable planets per planetary system, f_l is the fraction of habitable planets where life develops, f_i is the fraction of inhabited planets where intelligent species appear, f_c is the fraction of planets with intelligent species that produce detectable signatures of technological activity, and L is the duration (in years) over which such signals manifest themselves.

It is illuminating to rewrite the Drake equation as

$$N_T = \Gamma L, \tag{14.2}$$

with $\Gamma \equiv R_\star f_p n_e f_l f_i f_c$. This highlights the fact that the equation is a specific example of a remarkable result of queueing theory, the so-called *Little's Law* (Little, 1961), which states that the average number of items in a stationary system (e.g., customers waiting to be served) is equal to the average arrival rate of the items multiplied by the average time that an item spends in the system. Clearly, in our case, the system is the Galaxy (or, more broadly, a volume of space around our location), Γ is the appearance rate of technosignatures (measured in yr^{-1}), and L is their average longevity.

Despite its simplicity (or perhaps on account of that), the Drake equation has spawned extensive discussions and a vast corpus of literature in the decades after its formulation. We refer the reader to Vakoch and Dowd (2015) for a detailed scientific and historical analysis of the various factors, and our state of knowledge concerning them. We shall not attempt to explicitly derive quantitative estimates for N_T – an exercise that, though potentially instructive, is marred by large uncertainties; the reader is nevertheless invited to experiment with (14.1) (see Question 14.1). It is straightforward to realise that, with appropriate choices of the parameters, N_T can easily be made to span the entire spectrum from extreme optimism (with N_T only a few orders of magnitude smaller than the number of stars in our Galaxy, $\sim 10^{11}$) and extreme pessimism (with $N_T \to 1$, implying that no species barring humans are technologically active in the Galaxy).

The variables appearing in the Drake equation may also be regrouped into clusters pertaining to astrophysics, biology, and sociology, respectively. In other words, we can rewrite (14.1) as

$$N_T = f_{ast} \, f_{bio} \, f_{soc}, \tag{14.3}$$

with $f_{ast} = R_\star f_p n_e$, $f_{bio} = f_l f_i$, and $f_{soc} = f_c L$. This sorting emphasises how challenging it is to obtain reliable estimates of N_T, since one has to invoke knowledge derived from disparate fields, which becomes progressively ambiguous moving from one parameter to the next. It might be contended that f_{ast} is now empirically determined to fall in the range $f_{ast} \sim 0.1\text{--}1 \, \mathrm{yr}^{-1}$, since observations over the last three decades have provided reliable demographics of temperate terrestrial planets in the Milky Way. On the other hand, the biological variable is essentially unconstrained, as we currently have evidence of merely one planet hosting life (the Earth), which leads to precarious (to say the

---

[1] In Drake's original formulation, N_T was the number of technological 'civilisations' that could communicate (intentionally or not) via radio signals over interstellar distances.

least) inference of the frequency of life in the Galaxy. Finally, the sociological factor can only be speculated upon.

Another interesting aspect to ponder is the possible spatial distribution of technosignatures in the Galaxy. In particular, we may wonder about the average separation between worlds (canonically taken to be planets) hosting technosignatures, if their total number is N_T. This can be estimated by modelling the Milky Way as a thin disc with radius $R_G \approx 15$ kpc and scale height $H_G \approx 0.35$ kpc, so that its volume is

$$V_G = \pi R_G^2 H_G. \tag{14.4}$$

If there are N_T planets with technological life, separated by an average distance d_T, they occupy a total volume $\sim N_T \times 4\pi d_T^3/3$. By equating the latter expression to (14.4), and solving for d_T, we obtain:

$$d_T \approx 180 \, \text{pc} \left(\frac{N_T}{10^4} \right)^{-1/3}. \tag{14.5}$$

It is to be expected that detecting any given technosignature would become progressively difficult at larger distances, owing to the weakening of signals, and because the number of potential targets that ought to be surveyed for putative signals grows as d_T^3. Hence, we see that the above expression places a constraint on the minimal value of N_T needed for searches to be viable. We will return to this theme in Section 14.4.

14.1.1 Shortcomings

It is perhaps an attestation of the profound influence that the Drake equation has exerted on the SETI community that, over the years, a host of critiques have been raised. Far from downplaying its utility, such examinations serve to highlight some intrinsic limitations, and underscore the importance of acquiring a better understanding of the subtleties at play.

One of the most crucial among them, although not often explicitly stated, is that the Drake equation deals with average quantities. Therefore, the various factors appearing in this equation should each be interpreted as the mean value of a random variable, with its attendant variance. By harnessing a statistical generalisation of the Drake equation, Mieli et al. (2023) suggested that the average value of N_T is $\sim$2,000, but we reiterate and caution that many unknowns are operational.

While it can be shown that (14.2) holds independently of the probability distributions of the variables involved, it does assume that the underlying statistical process is stationary, that is, unchanging in time, which is invalid. Not only has the rate of star formation changed over cosmic history but also processes thought to influence planetary habitability on the Galactic scale (consult Section 8.4) have evolved in time. For example, the radiation environment was harsher in the past – due to higher rates of supernovae and gamma ray bursts – and the availability of heavier elements (needed for forming terrestrial planets and biomolecules) was diminished compared to the present. Therefore, the appearance rate of technosignatures Γ is not constant in time. Accounting for this temporal variation may lead to significantly different estimates of N_T compared to those ensuing from the postulate of stationarity (Cirkovic, 2004; Balbi, 2018).

Likewise, the spatial distribution of planets suitable for the emergence of complex life may not be uniform, due to the existence of a Galactic habitable zone (which we introduced in Section 8.4), so that estimates of N_T could fluctuate considerably depending on the volume under consideration. All other factors held constant, searching for technosignatures in regions with higher density of stars, such as the Galactic bulge, might elevate the prospects for detection relative to sparser environments. Additional complexities that warrant scrutiny are as follows: (1) the stellar type may regulate the frequency of habitable planets (e.g., low-mass stars could be disadvantaged, as remarked in Section 8.2); (2) planets in the habitable zones of stars need not be the only sites of abiogenesis and intelligence (e.g., life might appear on icy moons, as outlined in Chapter 10); and (3) technosignatures, in principle, can originate from artefacts not necessarily bound to planets or stellar systems (e.g., free-floating worlds).

An even more extreme scenario of spatial and temporal dependence is to presume that life is not confined to the planet(s) where it originally appeared, but might spread across the Galaxy, either by some natural process (e.g., lithopanspermia) or because technological species develop interstellar travel. The Drake equation can thus be modified, with the introduction of an additional *settling factor*, $S_T \geq 1$, which accounts for the propagation of intelligent life in the Galaxy (Walters et al., 1980).

14.1.2 Detectability

The spatiotemporal distribution of technosignatures is relevant in one more respect, which can be encapsulated as follows.

Consider a single technosignature, located at distance d_T from Earth, that commenced being detectable at t_T years before present (where the current epoch is $t = 0$), and remained this way for an interval L_T. Since any signal leaving from d_T can, at most, travel at the speed of light c, we will be able to observe this technosignature from Earth today, if and only if

$$t_T - L_T \leq d_T/c \leq t_T, \tag{14.6}$$

as depicted in Figure 14.1. This causal constraint has interesting implications, since it dictates that a technosignature is detectable only if t_T and L_T do not differ more than d_T/c, which is $\lesssim 10^5$ years for locations within our Galaxy. To put it another way, because the time it takes electromagnetic signals to cross the entirety of our Galaxy is $\sim 10^5$ years, any technosignature that ceased to be 'active' more than $\sim 10^5$ years ago is not currently observable (given that its signals have left the Milky Way).

While the causal constraint of (14.6) has to hold for any single technosignature detectable today, it may be applied to a distribution of technosignatures with random t_T, d_T, and L_T. In this case, the condition (14.6) acts as a filter that selects only a subset N_{det} of the N_{tot} technosignatures that ever existed over the history of the Galaxy. The fraction $f_{det} = N_{det}/N_{tot}$ of technosignatures that can be detected (i.e., they are in causal contact with us) depends on the details of their spatiotemporal distribution (Grimaldi, 2017; Balbi, 2018). For example, if most technosignatures manifested recently, a large fraction could be in causal contact with us; conversely, if the epoch of appearance of technosignatures peaked around, say, a few billion years ago, then we are very unlikely to receive signals from any of them, unless their technosignatures remained perceptible for $\sim 10^9$ years.

To quantify this further, suppose that there was no preferred time for the emergence of intelligent life (and technosignatures) during the history of the Galaxy, so that t_T is a uniformly distributed random variable. Under this assumption, the number of detectable technosignatures may be simply obtained from the Drake equation written in the form (14.2), where the rate of appearance of

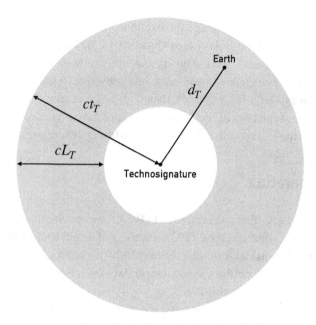

Figure 14.1 Schematic illustration of the causal constraint set by (14.6). The annular region represents the region spanned by any signal coming from the technosignature source located at the centre.

technosignatures is constant and given by $\Gamma = N_{tot}/T_G$, where $T_G \approx 10^{10}$ years is the age of the Galaxy. By inserting this rate and the average longevity $L = \langle L_T \rangle$ in (14.2), we end up with

$$N_{det} = N_{tot} \frac{\langle L_T \rangle}{T_G} \qquad (14.7)$$

so that the detectable fraction is $f_{det} = \langle L_T \rangle / T_G$. Unless the average longevity is of the same order of magnitude of T_G, this quantity will be very small. Therefore, a large number of technosignatures over the history of the Galaxy can translate to just a few detectable today.

This calculation reinforces a crucial point recognised since the beginning of SETI, that is, the chances of finding evidence of extraterrestrial intelligence are sensitive to L. In fact, optimistic interpretations of the Drake equation almost invariably demand $L \gtrsim 10^6$ years. Earlier studies viewed L as the average lifespan of technological agents, and had to argue in favour of long-lived 'super-civilisations' to strengthen the case for SETI (Shklovskii and Sagan, 1966). Today, L is more broadly understood as the duration of technosignatures. In theory, as they could outlive the civilisation (and even the species) that produced them, the plausibility of attaining high values of L is somewhat enhanced. A recent analysis of the role of technosignature longevity in SETI is provided in Balbi and Ćirković (2021).

Finally, if we ignore the requirement of detectability, and focus instead on the total number of technosignatures that ever existed during T_G, we can write $N_{tot} = \Gamma T_G$, which is equivalent to dropping L from the Drake equation and multiplying all the other factors by the total age of the Galaxy. If we make the extra simplifying assumption that the rate of star formation remained approximately constant during T_G, and adopt the previously mentioned fiducial interval of $f_{ast} = 0.1 - 1$ yr^{-1}, we obtain

$$N_{tot} = 10^9 - 10^{10} f_{bio} f_c. \qquad (14.8)$$

This expression may be used to gauge the likelihood that ours is the only technological species that evolved in the history of the Galaxy (see Question 14.3). Taken at face value, it is not implausible that $N_{tot} \to 1$, since f_{bio} and f_c are unknown, and either or both of them could be small. On the other hand, if we invoke a *principle of mediocrity*, and subscribe to the naïve stance that what transpired on Earth is the norm (refer to Chapter 6), then $f_{bio}f_c \sim 1$ can seem plausible, and we are left with the conclusion that a vast number of technological species have preceded us in the past billions of years. However, we are then faced with a different conundrum: *'Where are they?'*

14.2   The Fermi paradox

In the summer of 1950, Enrico Fermi (1901–1954)[2] was having lunch with his colleagues Edward Teller (1908–2003), Herbert York (1921–2009), and Emil Konopinski (1911–1990) at the Los Alamos National Laboratory cafeteria. Prompted by a casual conversation pertaining to the probability of extraterrestrial intelligence and interstellar travel, Fermi is reported to have abruptly asked: *'But where is everybody?'* (Jones, 1985).

This simple question constitutes the essence of what was subsequently christened the *Fermi paradox*: in short, the purported conflict between the supposed abundance of extraterrestrial intelligences and the absence of any obvious proof of their existence. Although it was later extended to include the lack of evidence for any technosignature, which is often termed *The Great Silence*, the Fermi paradox originally arose from the observation that Earth has, ostensibly, not been visited by alien space probes.

The first quantitative formulation of the argument was delineated in Hart (1975), along with some potential solutions, and is summed up as follows. The radius of the Galaxy is $R_G \sim 15$ kpc, while the speed attainable by even basic space probes (i.e., similar to humanity's spacecraft) is $v_s \sim 10^{-4}c$. One such probe may thus traverse the entire Galaxy in a timescale

$$t_s = 2R_G/v_s \sim 10^9 \text{yr}, \tag{14.9}$$

which is an order of magnitude smaller than the age of the Galaxy, $T_G \approx 10^{10}$ yr. Therefore, in the past billion years, any automated probe launched by a putative ancient technological (extraterrestrial) species would have theoretically had plenty of time to reach even the farthest stars.

This quantitative estimate can be refined by adding further details. Detailed numerical simulations by Carroll-Nellenback et al. (2019) found that the entire Galaxy might be populated by one space-faring technological species on a timescale much smaller than T_G. Tipler (1980) proposed that advanced technological species might build self-replicating probes, drawing on pioneering work by John von Neumann (1903–1957), and achieve speeds up to $v_s \sim 0.1\,c$ via advanced propulsion systems. This would considerably shorten the timescale needed for visiting every star in the Galaxy, thereby exacerbating the Fermi paradox. Hence, this stimulated the *Hart–Tipler argument*: no technological species capable of interstellar travel has ever existed, thus negating the Fermi paradox.

Countless solutions to the Fermi paradox have emerged in the past decades, too many to list. We will rather present a broad classification of such solutions, and provide a few illustrative examples.

---

[2] Fermi was awarded the 1938 Nobel Prize in Physics for his research in nuclear physics.

The interested reader may peruse Webb (2015) and Cirkovic (2018) for thorough treatments. Paralleling the scheme of Webb (2015), we categorise most possible answers to the Fermi paradox in at least one of following three classes:

- *They are here*: extraterrestrial technological agents have reached the Earth (or the solar system), but we have not found evidence of them yet.
- *They exist, yet are not here*: technological species exist elsewhere, but their technology never reached Earth (or the solar system). Moreover, they have produced detectable technosignatures, although we have yet to unearth evidence of these technosignatures.
- *They do not exist*: intelligent species capable of sending probes to Earth (or to the solar system) and/or producing detectable technosignatures never appeared in our Galaxy (or perhaps even in the entire Universe).

Now, we shall delve into each of this trio of possibilities in turn.

14.2.1 They are here

Admittedly, we might be tempted to dismiss this solution as too far-fetched or glaringly absurd. However, the datum that no evidence is documented for visitation of the Earth and its neighbourhood by extraterrestrial probes does still leave room, however little, for the proposition that extraterrestrial artefacts are present in the solar system, yet have been undetected.

Of course, addressing this scenario engenders serious methodological problems, because it is well-known that scientific investigations cannot always conclusively prove that some postulated entity does not exist, as exemplified by the aphorism: '*Absence of evidence is not evidence of absence*'. Nevertheless, a failure to prove a hypothesis false does not automatically imply the converse, namely, that it is true. This is sometimes illustrated by a vivid analogy devised by the mathematician and philosopher Bertrand Russell (1872–1970), who also received the 1950 Nobel Prize in Literature: we cannot disprove that a fine porcelain teapot, too small to be detected by our astronomical instruments, is present somewhere in space, but this fact does not necessarily imply that we must seriously contemplate such a conjecture.

Keeping these caveats in mind, we review one quantitative analysis of this class of solutions. Haqq-Misra and Kopparapu (2012) analysed the likelihood of *non-terrestrial artefacts* (NTAs) in the solar system by adopting a Bayesian formalism (encountered in Section 13.3). They assumed that a certain volume V is scanned up to a finite spatial resolution $\mathcal{R}$ (resulting in a sample volume V_R), and that no evidence of NTAs is found. Next, they evaluated the likelihood $\mathcal{L}$ that NTAs are indeed absent in V. Crucially, the outcome of the analysis depends on the probability assigned a priori (i.e., before conducting the search) to the absence of NTAs in V, measured relative to their probability of existence; we label such prior odds by Θ.

Figure 14.2 depicts the results ensuing from the analysis. In general, when $\Theta \gg 1$ (i.e., if we assign very high prior odds to the hypothesis that NTAs are absent in V), we arrive at $\mathcal{L} \to 1$ regardless of the completeness of the search, namely: for any value of the ratio V_R/V. In other words, we will be inclined to rule out the presence of NTAs in V with near-certainty. On the other hand, if $\Theta \ll 1$ (viz., if we ascribe very high prior odds to the hypothesis that NTAs exist in V), then we have $\mathcal{L} \to 0$, that is, no amount of negative evidence will convince us that V is devoid of NTAs.

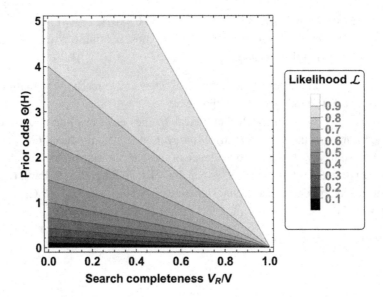

Figure 14.2 The likelihood $\mathcal{L}$ that NTAs are absent within a volume V as a function of the prior odds Θ of their absence and the search completeness ratio V_R/V, based on Haqq-Misra and Kopparapu (2012).

A more interesting outcome is obtained in the case of equal prior odds, $\Theta = 1$, that is, when we have no a priori preference for the existence versus absence of NTAs in V. In this case, the likelihood depends on the typical size of the artefacts compared to $\mathcal{R}$. The rationale here is that NTAs smaller than the search resolution will remain hidden. Clearly, for the solar system as a whole, we may specify $V_R/V \ll 1$, such that $\mathcal{L} \sim 0.5$ for $\Theta \approx 1$. The bottom line is that, for an agnostic observer, there are plenty of ways whereby NTAs might exist in the solar system and yet be undetected.

This stance immediately begs the question: where could they be hiding? A number of proposals have sprung up in response to this issue. If NTAs reached the Earth in the remote past, some evidence might have theoretically survived the geological activity of our planet, which can however completely reprocess its surface on timescales of $\sim 10^8$–10^9 yr. Schmidt and Frank (2019) analysed this scenario, and described potential technological markers such as radioactive waste produced by nuclear fission, large-scale mining, artificial materials, and pollutants (e.g., plastics) buried in sediments.

Davies and Wagner (2013) suggested that the Moon would represent a better target for NTAs searches, both because geological activity is minimal and its surface features can be mapped with very high resolution. Some other speculations include the locations where a probe could be maintained in a stable position with respect to the Earth–Moon or the Earth–Sun systems (i.e., the *Lagrange points*) (Freitas and Valdes, 1985), or the asteroid belt between Mars and Jupiter (Papagiannis, 1978). Although none of these possibilities seem likely (and may even be deemed fanciful), they cannot be rigorously ruled out given the current state of knowledge, and therefore, they might offer a solution to the Fermi paradox.

In the past few years, the notion of NTAs visiting our cosmic neighbourhood was (re)ignited by the discovery of the first interstellar object passing through the solar system (Meech et al., 2017). This object, called 'Oumuamua, exhibited certain unexpected characteristics, which stimulated multiple hypotheses for its provenance ranging from an object composed of nitrogen/hydrogen/water ice to a fractal aggregate (dust 'bunny'). Besides natural explanations, the hypothesis that it is a thin artificial extraterrestrial artefact was also broached (Loeb, 2022). The scientific consensus patently favours a natural origin (see 'Oumuamua ISSI Team, 2019). Not all of its anomalies are wholly resolved, however, and unconventional hypotheses might conceivably help the search for more robust models and data.

Getting back to settling the Fermi paradox via undetected NTAs, an additional question persists: why should extraterrestrial probes be so elusive? A species that made the effort to intentionally send their spacecraft to other stellar systems might be anticipated to employ them to signal its presence and attempt direct contact. The premise that a technologically sophisticated species could build autonomous interstellar space probes with the explicit purpose of communication was hypothesised by Bracewell (1960); therefore, devices of this kind are occasionally dubbed *Bracewell probes*.

A Bracewell probe might be as simple as an electromagnetic beacon, but it may also contain and transmit more elaborate information. If equipped with artificial intelligence, it could even evince the capacity to initiate and sustain a dialogue of sorts with other technological species, overcoming the long delays that are implicit in interstellar communication through electromagnetic signals. Since Bracewell probes would, by definition, actively attempt to manifest their existence, a lack of evidence of such devices in a volume surrounding the Earth unambiguously signifies their absence (at least within the limits of modern observational capabilities).

We are then left with either the possibility that no effort was made on behalf of the builders of NTAs to manifest themselves or that they intentionally designed their probes to go undetected. A 'sociological' explanation as to why an advanced species would send probes across interstellar distances in 'stealth mode' is then needed. One such explanation is the so-called *zoo hypothesis* (Ball, 1973), according to which advanced technological species would deliberately choose to not interfere with the evolution of life on other planets, and content themselves with passively collecting information for scientific purposes. Readers might appreciate the striking similarity with the *Prime Directive* from the TV series and movies of the *Star Trek* franchise.

In a similar vein, the *interdict hypothesis* assumes that the Earth was inspected and recognised as inhabited early in its history, after which it was promptly sealed off from contact and settlement to preserve its isolation (Fogg, 1987). In a more pessimistic rendition of these viewpoints, intelligent species capable of interstellar travel may view other similar species as potential threats, and could therefore decide to remain silent to guarantee their survival. This perspective has gained popularity as the *dark forest hypothesis* in the semi-eponymous science fiction novel by Liu Cixin (1963–present).

The issue with all such 'contact avoidance' hypotheses is that they require that *every* technologically advanced species, without exceptions, functions in a similar fashion and decides to stay invisible to us. Such pronounced uniformity in behaviour appears hard to justify, and would require additional complications and constraints to make it feasible (e.g., a *Galactic Club* demanding common protocols and actions from its members).

14.2.2 They exist, yet are not here

This class of solutions postulates that no NTAs have ever reached the solar system, although we could be in causal contact with technosignatures produced in other stellar systems that we have failed to detect hitherto. Since we will devote a separate discussion to the search for remote technosignatures in Section 14.4, we will only review some proposed explanations for why no interstellar probe may have ever visited the solar system.

One straightforward answer is that interstellar travel is innately (almost) impossible. This stance does not appear plausible. Space travel is certainly very demanding and risky when the transport of living organisms is involved, and sending them to other stars at non-relativistic speeds would require incredible planning, multi-generational 'crews', and solutions to manifold daunting technological problems. On the other hand, automated spacecraft do not necessarily face insurmountable hurdles. Humanity has already approached the threshold of sending probes to interstellar space, although at the slow speeds allowed by chemical propulsion. As remarked earlier, even at such speeds reaching nearby stellar systems in a period much shorter than the age of the Galaxy or the lifetime of main sequence stars is tenable.

Nonetheless, one caveat should be made more explicit at this juncture. Several calculations of the minimal time it would take for a single technological species to visit the entire Galaxy invoke the use of self-replicating probes. Upon reaching a target, each probe must be able to automatically construct a functioning copy of itself by assembling raw material collected locally. Therefore, the total number of active probes will grow exponentially, permitting the exploration of all planetary systems in the Galaxy in an estimated time of $\lesssim 10^7$ years (see Question 14.5). However, this estimate is not guaranteed to be valid, because it makes the rather strong assumption that the underlying mechanism is indeed realisable.

Even though the design of self-replicating probes seems achievable with current technology (Borgue and Hein, 2021), objections to the above scenario can be raised. For example, there may be a non-zero probability that the replication halts at some stage during the process for various reasons (e.g., malfunctioning, lack of energy/resources, deliberate choice). This would result in the formation of isolated regions that are never visited (e.g., Landis, 1998). On the opposite side, Sagan and Newman (1983) proposed that if self-replication were perfectly efficient, it could consume the entire mass of the Galaxy in $\sim 10^6$ years (see Question 14.6), which might be a detriment for its instantiation by advanced technological species.

If the use of exponentially self-replicating probes is discounted, the duration for the exploration of the entire Galaxy by a single advanced species would certainly increase. However, earlier papers that obtained upper bounds as high as $\sim 10^{10}$ yr are contradicted by state-of-the-art models (Carroll-Nellenback et al., 2019). Hence, the Fermi paradox arguably remains unresolved if one strictly resorts to this class of explanations.

14.2.3   They do not exist

Finally, we should seriously admit the scenario wherein technologically intelligent life does not exist beyond Earth. This notion is usually dismissed by appealing to some variant of Copernicanism, namely: the assumption that humans do not occupy a privileged or special position in the Cosmos. However, it is not inherently anthropocentric to acknowledge that the number of technological species that ever emerged in the Galaxy (or in the Universe) might be minimal, just as there is no chauvinism associated with the fact that only a fraction of all planets exhibit 'Earth-like' attributes, or that only one species among the billions that evolved on Earth discovered quantum mechanics and invented the radio telescope.

As revealed by the Drake equation, the number of technological species existing in the Galaxy hinges on multiple probabilistic factors that could prove to be minuscule (individually or in combination), making the outcome $N_T \rightarrow 1$ a definite possibility even if the population of habitable worlds in our Galaxy is extremely large. A host of bottlenecks may be encountered on the long and winding road that goes from the formation of a habitable world to the advent of intelligence and technology. Thus, only a handful of 'successful' outcomes might arise from numerous independent trials, resulting in what Robin Hanson (1959–present) has called the *Great Filter*.[3] We shall not explore this topic in greater depth, as we believe that the sundry potential bottlenecks would have become quite apparent to readers who have perused the preceding chapters of this book.

The bottom line is that an uncritical appeal to the 'mediocrity' of what transpired on Earth to draw generic conclusions about the prevalence of technological species in the Universe might be regarded as fallacious (see Chapter 6), and must be generally avoided where suitable (Balbi and Lingam, 2023). We may very well be alone in the Cosmos, and this would certainly constitute a viable (albeit perhaps disappointing) solution to the Fermi paradox.

Ultimately, a scientifically sound response to the question *'Where is everybody?'* can only emerge through sustained empirical investigation. The subsequent sections of this chapter will address this topic.

14.3   Interlude: the Kardashev scale

Before we tackle methods for discerning signs of extraterrestrial intelligence, let us take a momentary detour to explore a related topic, namely: how to quantitatively formulate expectations regarding the anticipated characteristics of putative technological species.

The best-known framework in this context is the so-called *Kardashev scale*, proposed by Nikolai Kardashev (1932–2019) in 1964 (Kardashev, 1964). This scale categorises technological species as per their power consumption and utilisation capacities. It consists of three primary types (Gray, 2020):

- *Type I (planetary)*: A Type I species harnesses and controls nearly all the available energy resources on its home planet, with power consumption of $\sim 10^{16}$ W. This species can ostensibly manage its planetary environment effectively, employing renewable energy sources, and potentially hosting a global network of communication and transportation.
- *Type II (stellar)*: A Type II species has progressed to the point where it can mobilise and manipulate the whole energy output of its host star, with power consumption of $\sim 10^{26}$ W. This level of energy mastery could enable extensive space travel and settlement.
- *Type III (galactic)*: A Type III species can harness energy on a Galactic scale, with power consumption of $\sim 10^{36}$ W. This energy budget may conceivably facilitate intergalactic travel.

By contemplating where a hypothetical extraterrestrial species falls on the Kardashev scale, we might better anticipate the types of technosignatures it may produce, and the methods by which

---

[3] https://mason.gmu.edu/~rhanson/greatfilter.html.

we could detect them. It is possible to construct intermediate (i.e., fractional) positions within this scheme by adopting a logarithmic formula (Gray, 2020):

$$K = \frac{1}{10} \log_{10} \left(\frac{P}{10^6 \, \text{W}} \right),$$ (14.10)

where P is the power available to the technological species and K is the corresponding level on the Kardashev scale. On specifying $P \approx 2 \times 10^{13}$ W,[4] humanity would be classified as a $K = 0.73$ species.

The Kardashev scale offers the advantage of being grounded in a well-defined physical quantity (i.e., power) closely tied to technological activity. However, it is vital to appreciate that it is not the sole method for categorising extraterrestrial intelligences. Over time, the scale has undergone numerous adjustments and refinements to address various intricacies and nuances, as reviewed in Cirkovic (2018). These adaptations include considerations such as a species' capacity to manage and govern information, its extent of space settlement, and its proficiency in micro-scale environmental manipulation. Furthermore, since the detectability of technosignatures may be linked to their duration, a complementary classification system based on longevity could be created, as discussed in Balbi and Ćirković (2021).

14.4   Searching for technosignatures

The modern search for technosignatures is conventionally traced to the seminal publication by Cocconi and Morrison (1959). The authors conjectured that extraterrestrial intelligences might choose to broadcast radio interstellar communications near the emission line of neutral hydrogen at wavelength $\lambda = 21$ cm (viz., a radio frequency of $\nu = 1.42$ GHz), and that we could detect them by means of radio observations. The paper motivated the first actual search, the whimsically named *Project Ozma*, conducted by Frank Drake in 1960. For about 200 hours, the *Green Bank Observatory* in West Virginia scanned two nearby Sun-like stars (Tau Ceti and Epsilon Eridani) for signs of artificial radio emissions (Drake, 1961). While the observations yielded null results, they heralded the empirical dawn of SETI.

These historical circumstances serve to explain why the search for technosignatures has long been the province of radio astronomy. However, as we have already implied, extraterrestrial technological activity can manifest itself in myriad ways. In recent times, this standpoint has been increasingly appreciated by the scientific community, and the scope of investigations has grown commensurately. A summary of ongoing and planned technosignature surveys is furnished in Socas-Navarro et al. (2021).

Therefore, we start this section by reviewing the expanding range of technosignatures that astronomers are now pursuing in SETI. This emerging perspective acknowledges that technologically advanced species may employ a variety of technologies, from radio emissions and laser signals to megastructures and other yet-to-be-conceived methods of communication or energy capture. By casting a wider net during exploration, we can increase the likelihood of detecting subtle or unconventional signs of intelligent alien life.

---

[4] See https://ourworldindata.org/energy-production-consumption.

14.4.1 Types of technosignatures

14.4.1.1 Radio signals

As we saw, radio signals seem a realistic outcome of technological activity from sufficiently advanced species. Humans have broadcast such signals for the past century (mostly unintentionally, although a few attempts at direct communication do exist), covering a volume that encompasses some $\sim 10^4$ stars. We might, therefore, assume that similar signals could be produced by extraterrestrial technological species, either as a by-product of their internal communication systems or because of deliberate attempts to reach out to other intelligent life (e.g., by deploying powerful beacons).

A general criterion to distinguish an artificial signal from radio emissions produced by astrophysical processes is not easy. It was conventionally held that artificial transmissions would be confined to an extremely narrow electromagnetic bandwidth $\Delta \nu$ characterised by $\Delta \nu / \nu \lesssim 10^{-12}$, as opposed to natural phenomena that tend to spread over a larger frequency interval, typically with $\Delta \nu / \nu \gtrsim 10^{-6}$ (Tarter, 2001). However, this is not a universal principle, since broadband artificial transmissions are also conceivable, and might be preferable in some cases (Siemion et al., 2015).

Another crucial variable to contemplate is the frequency of signals. For intentional communication, some choices appear more compelling than others. Aside from the previously mentioned 1.42 GHz line of neutral hydrogen (thoroughly investigated by radio astronomers), it was proposed that the frequency of the hydroxyl radical (OH) emission (1.66 GHz) would be an appealing choice. As H and OH arise from the dissociation of the water molecule, the frequency interval encompassing the two lines has been aptly termed the *water hole*.[5] This range of frequencies has the additional advantage of falling in the microwave window where the atmospheres of Earth-analogue planets are potentially transparent, and relatively free from contamination by other astrophysical processes (see Figure 14.3).

Finally, we list some reference values for the strength of radio signals emitted on modern Earth, often expressed as *equivalent isotropically radiated power* (EIRP): this quantity ranges from $\sim 10^5$ W for TV and radio stations to $\sim 10^{13}$ W for the most powerful planetary radars. Next, the minimal flux detectable by a radio telescope F_{min} is given by (Siemion et al., 2015):

$$F_{min} = \text{SNR} \times \text{SEFD} \times \sqrt{\frac{\Delta \nu}{\tau_{int}}}, \qquad (14.11)$$

where SEFD (the *system equivalent flux density*) is a measure of the receiver sensitivity, SNR denotes the signal-to-noise ratio, $\Delta \nu$ is the bandwidth of the receiver and τ_{int} is the integration time. The maximum distance that a given telescope can search for radio technosignatures is $d_{max} = \sqrt{L_E / 4\pi F_{min}}$, where L_E is the EIRP of the emitter.

14.4.1.2 Optical signals

While seeking optical signals may seem disadvantageous due to the contamination from starlight, highly collimated laser (or maser) beams, intentionally aimed at Earth, offer a promising avenue for signalling across interstellar distances, which was recognised since the dawn of modern SETI

---

[5] This is also a nod to the notion that, just as animals congregate around water holes in desert environments, technological species may 'meet' each other around these frequencies.

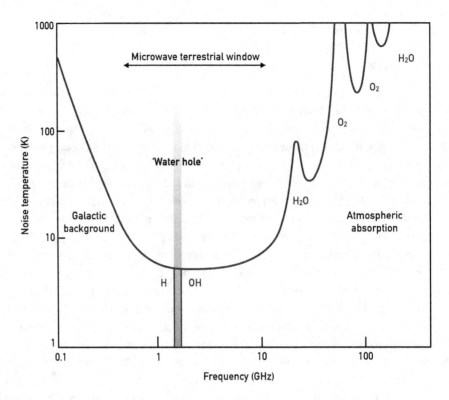

Figure 14.3 Noise temperature (i.e., measure of noise magnitude) as a function of frequency. The quiet terrestrial radio window in the microwave, and the position of the famous 'water hole' amid the neutral hydrogen (H) and hydroxyl radical (OH) spectral emission lines, are shown.

(Schwartz and Townes, 1961). In fact, the achievable SNR in the visible range might exceed that of radio signals, as it loosely increases with frequency. Furthermore, because lasers are nearly monochromatic, a pulse with $\sim 10^{-9}$ s duration can outshine the optical light emitted by a Sun-like star by several orders of magnitude, even using technology from two decades ago (Howard et al., 2004). Pulsed emissions have the additional benefit of becoming distributed over a broad frequency range (in contrast to radio narrowband signals), and therefore do not always require selecting a restricted frequency channel.

Another scenario for detecting optical pulses involves the deployment of swift propulsion systems known as laser-driven light sails for undertaking interplanetary and interstellar travel. In this hypothetical scenario, intense laser beams would be focused on large sails attached to spacecraft, propelling them through space at significant velocities through the action of radiation pressure (consult Section 15.2.2). The dynamics of the powerful lasers and the sail trajectory may potentially create optical signatures in the form of pulses or flashes, with distinct diffraction patterns that might be detectable even at kpc distances (Guillochon and Loeb, 2015).

14.4.1.3 Infrared (thermal) emission

A generic feature of technological activity, directly linked to intensive energy utilisation, is the production of *waste heat*, namely: the unavoidable emission of degraded (high-entropy) energy. This may be sought in the form of excess (i.e., anomalous) infrared energy radiated away by astrophysical sources.

As an illustrative example, we may consider an advanced technological species that intercepts a fraction α of total starlight for energy usage (e.g., via space-deployed solar panels) and radiates away a fraction γ as waste heat. Neglecting other forms of energy production and loss, conservation of energy indicates that $\alpha \approx \gamma$. If $T_\star$ is the stellar temperature and T_w is the temperature of waste heat, the predicted increase of the infrared flux ΔF_w, relative to the stellar flux $F_\star$, was estimated to be (Wright et al., 2014):

$$\frac{\Delta F_w}{F_\star} \approx \gamma \left(\frac{T_\star}{T_w}\right)^3 - \alpha. \tag{14.12}$$

On substituting $T_\star \approx 6,000$ K and $T_w \approx 300$ K, the expected boost in flux would be $\sim 10^3$ even for small values of $\gamma \sim 0.1$. Note that the observed flux in this regime will only decrease linearly as a function of α.

The extreme case wherein all the starlight is blocked, equivalent to $\alpha \approx \gamma \approx 1$, would entail the construction of a structure completely surrounding the star. This edifice was first delineated in the scientific literature by Freeman Dyson (Dyson, 1960), and is therefore usually termed a *Dyson sphere*.[6] In this case, the star ought not be visible to astronomical observations, and substantial anomalous infrared emission should be detected. The assembly of a full Dyson sphere, or comparable structures, could allow a technological species to transition to Type II on the Kardashev scale.

14.4.1.4 Transiting structures

The idea that advanced technological species might harness stellar energy by building large structures in space has further observational consequences. As implied in the preceding exposition, such gigantic artefacts, usually called *megastructures*, would block part/all of the radiation from the star, and may thus be detectable as attenuation features in the light curve, using the same photometric techniques for detecting transiting exoplanets (refer to Section 12.1.2). Since transit events are commonly occurring natural phenomena, convincing criteria are needed to accurately distinguish the dimming caused by artificial structures. Some guiding principles are outlined in Arnold (2005) and Wright et al. (2016), and include non-spherical and non-circular shapes; sizes, masses, and orbits markedly different from those of planets; non-gravitational accelerations; and so on.

A remarkable case of an anomalous transit garnered intense scrutiny in 2016, when it was determined that the light curve of the star KIC 8462852 observed by the *Kepler* telescope displayed highly unusual features, with large ($\sim 20\%$) and irregular dips uncommonly encountered in exoplanetary studies (Boyajian et al., 2016). Subsequent analysis essentially negated the possibility that the dips are produced by artificial structures, and instead found them to be consistent with attenuation by dust, although no single explanation has earned universal consensus. The example of KIC 8462852 demonstrates that current data already have the capability to empirically test the hypothesis of gigantic artificial structures in exoplanetary systems, thereby perhaps warranting the attention of the astrobiological community.

---

[6] This concept was actually elucidated earlier in the visionary science fiction novel *Star Maker* (1937) by Olaf Stapledon (1886–1950).

A similar example of artificial constructs that may leave an imprint in stellar light curves is a dense belt of geostationary or geosynchronous satellites orbiting an exoplanet, investigated by Socas-Navarro (2018). Such a structure, dubbed the *Clarke exobelt*,[7] might be produced even with moderately advanced technologies, and is predicted to be discernible by existing telescopes on Earth under favourable conditions.

14.4.1.5 Atmospheric gases

Human industrial activity is documented to modify the Earth's atmospheric chemical composition via the release of gaseous pollutants. Akin to gaseous biosignatures, the detection of specific molecules in exoplanetary atmospheres through spectroscopy (Section 12.2) may then serve as a smoking gun of extraterrestrial technology; we sketch a few examples.

Chlorofluorocarbons (CFCs), which were predominantly employed as refrigerants in the twentieth century, comprise a conspicuous group of terrestrial pollutants. Once released in the atmosphere, these molecules can persist for long timescales; some could last up to $\sim 10^5$ yr. As the majority of industrial CFCs are not generated by abiotic or biological (but non-technological) processes in sizeable amounts,[8] they have been rated as robust technosignatures ever since the 1980s. The *JWST* may be able to identify strong absorption features produced by CFCs (in infrared wavelengths) at concentrations comparable to present-day Earth in nearby exoplanetary atmospheres, given optimal conditions (Lin et al., 2014; Haqq-Misra et al., 2022a).

Another anthropogenic class of compounds suggested as atmospheric technosignatures is nitrogen oxides (NO_x). Although they are produced biologically by bacteria in the soil through nitrification or denitrification metabolic pathways, NO_x gases are likewise released as a consequence of combustion processes, such as vehicle emissions and power plants run on fossil fuels. Kopparapu et al. (2021) estimated that NO_2 (at current Earth concentrations) on an Earth-like planet around a Sun-like star at 10 pc could be detected by future infrared telescopes, albeit with very long integration times.

Nitrogen-containing species such as NH_3 and N_2O are outcomes of modern agriculture owing to the intensive use of fertiliser synthesised via the Haber–Bosch industrial process. The simultaneous detection of NH_3 and N_2O in an atmosphere that also consists of H_2O, O_2, and CO_2 may thus be a technosignature of extraterrestrial agriculture (Haqq-Misra et al., 2022b).

14.4.1.6 Surface constituents

Technological activity might sculpt a planetary surface, such as altering its albedo, reshaping natural features, or emitting anomalous radiation.

For example, large artificial structures, of sizes comparable to continents, might be built by advanced technological species to mitigate the effects of global warming (e.g., by reflecting a large fraction of stellar light into space), or to produce energy (e.g., by absorbing starlight through photovoltaic panels) (Berdyugina and Kuhn, 2019). The latter would modify the reflectance properties and

---

[7] This term was named after the famous science-fiction author Sir Arthur C. Clarke (1917–2008), who popularised the concept of the geostationary orbit in the 1940s.

[8] Fully fluorinated non-carbon gases like NF_3 and SF_6 might rank even more highly in this respect (Seager et al., 2023a), although their detection necessitates high concentrations of these molecules relative to that of Earth's atmosphere.

spectral features of the planetary surface, conceivably producing a distinct 'artificial edge' that could be likened to the vegetation red edge outlined in Section 13.2.2 (Lingam and Loeb, 2017a).

Another conjectured surface technosignature is the potential existence of *heat islands*, where the waste heat produced by technological activity clustered in specific geographical areas would be dumped into the surrounding environment (Kuhn and Berdyugina, 2015). This generated heat would manifest as excess infrared emission, which is clearly discernible in satellite images of large cities on Earth. However, detecting such islands on other worlds may call for future telescopes with extremely large apertures.

Strong artificial lights on exoplanets can also represent a detectable technosignature, which might be potentially identifiable through their spectral signatures (e.g., characteristics of LEDs or other illumination devices) (Loeb and Turner, 2012). Detailed modelling by Beatty (2022) of high-pressure sodium lamps concluded that city lights are detectable on Earth-like planets around neighbouring Sun-like stars only if the urbanisation fraction is $\gtrsim 10\%$, whereas the corresponding value on Earth is merely $\sim 0.05\%$.

14.4.1.7 Non-electromagnetic signals

Lastly, it is essential to acknowledge the prospect that extraterrestrial technologies could reveal themselves through non-electromagnetic signals (as opposed to prior technosignature candidates), such as neutrino beams or gravitational waves (Paprotny, 1977; Hippke, 2018). While admittedly speculative, we cannot dismiss this possibility outright, since it broadens the parameter space of potential technosignatures deserving of attention.

Gravitational waves, in particular, offer an intriguing avenue worthy of exploration. They are endowed with certain advantages over electromagnetic waves, as they are not scattered by interstellar matter and their signal strength decays as $1/r$ rather than $1/r^2$. Thus, they might be a compelling communication channel for technologically advanced species. On the other hand, the intentional production of a gravitational signal that could be detected at interstellar distances by current terrestrial antennas would entail the manipulation of immense masses (i.e., exceeding that of a terrestrial planet) arranged in symmetrical fashion and moving in coordination at very high speeds; or situating such a mass in close proximity to the Milky Way's supermassive black hole (Abramowicz et al., 2020). Therefore, it may be presumed that only highly sophisticated species (around Type II or higher on the Kardashev scale) would be able to produce these technosignatures.

Interferometric antennas on Earth achieved the necessary sensitivity to reliably detect gravitational wave signals arising from natural phenomena, such as black hole mergers or neutron star collisions, as recently as 2015. Consequently, we are witnessing the nascent stages of gravitational wave astronomy. The potential applications of this field to the search for technosignatures might blossom in the distant future.

14.4.2 Current limits

While it is accurate to state that no credible evidence has emerged so far regarding any markers of technological activity outside of Earth, it is equally important to acknowledge the considerable hurdles associated with converting these null findings into rigorous and quantitative empirical constraints on technosignatures. This complexity arises from several factors, including the vast and diverse parameter space that requires exploration, the paucity of concrete predictions concerning the nature of signals to seek out, the natural sources that might mimic artificial ones, and so forth.

An updated list of technosignature searches is maintained by the SETI Institute, and can be accessed from the *Technosearch* website.[9] By far, the most extensively studied category of technosignatures to date is artificial radio transmissions. Attempts to gauge the volume of multidimensional parameter space scanned by existing radio technosignature surveys yielded tiny fractions of $\sim 10^{-18}$ (Wright et al., 2018), although this value has increased by about a couple of orders of magnitude thereafter. These minuscule numbers underscore the fact that our cumulative efforts in SETI are still minimal, and the upper limits we have set on detecting technosignatures are weak. It is not surprising that, as per the statistical (Bayesian) model formulated by Grimaldi (2023), a waiting time of 60–1,800 yr may hence be necessary to detect artificial electromagnetic transmissions at 50% probability.

The largest and most comprehensive (radio) technosignature survey ever undertaken is the ongoing privately funded *Breakthrough Listen* project,[10] which was formally commenced in 2016 (Isaacson et al., 2017). Its central objectives are to investigate one million nearby stars, the Milky Way's centre and plane, a hundred neighbouring galaxies, and a plethora of astrophysical objects (e.g., asteroids, exoplanets) within a decade.

Figure 14.4 summarises some of the prominent radio technosignatures surveys conducted hitherto (Gajjar et al. 2021, and references therein). The comparison is expressed in terms of one potential figure-of-merit, among many existing candidates (see Wright et al. 2018 for other examples), namely: the transmitter rate defined as

$$\mathcal{T}_r \propto \frac{L_{E,min}}{N_{stars}(\Delta\nu/\nu)}, \tag{14.13}$$

where $L_{E,min}$ is the weakest EIRP detectable by a given instrument, $\Delta\nu/\nu$ is the fractional bandwidth of the receiver centred on the frequency ν, and N_{stars} is the total number of stars observed by the survey. The figure-of-merit $\mathcal{T}_r$ consequently represents a balance between sensitivity and the extent of the sky covered, because it assigns comparable significance to the pursuit of either a potent emitter within a vast ensemble of stars or a fainter signal emanating from a limited sample of stars.

Various searches for artificial optical signals have been conducted in the last two decades, returning no detection. For example, Howard et al. (2004) surveyed 1.3×10^4 stars for 2,400 hours, looking for pulses of $\sim$5 ns duration in the visible band, with sensitivity of $\sim$100 photons/m^2 per pulse. Howard et al. (2007) also performed an all-sky search in the Northern hemisphere for pulsed optical signals using the same methodology employed in the targeted search. Other notable optical technosignatures surveys include the observations of more than 5,000 stars performed at the *Keck Observatory* (Tellis and Marcy, 2017), the *NIROSETI* survey at the *Lick Observatory* that monitored 1,280 astronomical objects for light pulses shorter than 50 ns (Maire et al., 2019), and the ongoing *Panoramic SETI* survey (Maire et al., 2020).

Finally, anomalous infrared emissions produced by Dysonian megastructures were sought using the wide-field infrared surveys of the *IRAS*, *WISE*, and *Spitzer* satellites (e.g., Carrigan, 2009; Wright et al., 2014), finding no compelling candidates. The recent search for partial Dyson spheres in our

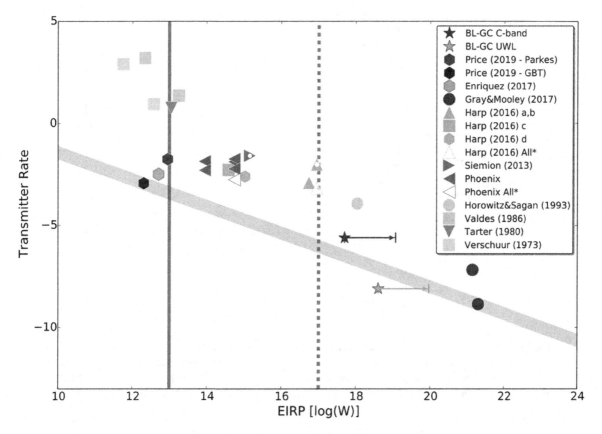

Figure 14.4 A comparison of the transmitter rate (see text) versus EIRP for some prominent radio technosignature surveys. The thick grey line signifies the slope of the transmitter rate as a function of the expected EIRP. The dotted line displays the total energy budget of a Kardashev Type I species, while the solid vertical line shows the EIRP output of the *Arecibo* planetary radar. (Credit: Gajjar et al. 2021, Figure 13; ©AAS, reproduced with permission)

Galaxy by Suazo et al. (2022, 2024) concluded that the fraction of stars that harbour 100–700 K Dyson spheres at 10–90% completion is $\lesssim 10^{-6}$.

On a related note, zero unusual transiting phenomena caused by Earth-sized or larger artificial occulters occurring within a span of hours have been identified in the $\sim 2 \times 10^5$ stars scrutinised by the *Kepler* telescope.

14.4.3 Future prospects

We conclude this chapter by delineating a handful of future prospects in the burgeoning field of technosignature searches. We parallel the recent review by Socas-Navarro et al. (2021), and adopt their useful classification of technosignatures in terms of the so-called *ichnoscale*, denoted by the quantity ι. The ichnoscale functions as the 'footprint' of a given technosignature,[11] to wit, its relative 'scale' (e.g., size) in units of the same technosignature manifested by present-day Earth technology.

---

[11] The ancient Greek root *ichnos* ($\iota\chi\nu o\varsigma$) means footprint.

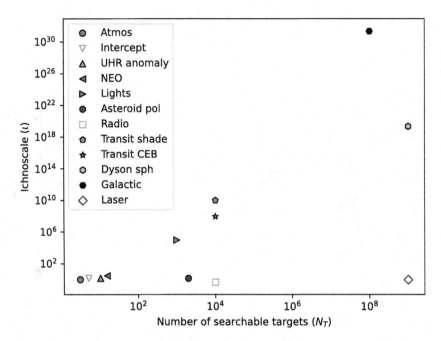

Figure 14.5 Ichnoscale ι (i.e., the relative scale of a given technosignature in units of present-day Earth technology) versus the number of searchable targets for several putative technosignatures discussed in the text. Filled symbols represent continuous observables, while empty symbols refer to discontinuous ones. (Credit: Hector Socas-Navarro, adapted from Socas-Navarro et al. 2021, Figure 1; reproduced with permission from Elsevier)

Given our current observational capabilities, different types of technosignatures will become detectable only above certain values of ι. In addition, for each technosignature, an expected number of potential targets can be surveyed. It must also be kept in mind that not all technosignatures are continuously visible, as some may conceivably operate only for a short time compared to the duration of a typical observation; an obvious example of discontinuous technosignatures is targeted artificial radio transmissions. Ideally, the optimal type of technosignature worth seeking would be abundant, continuous, and have a small ι, among other desirable qualities (Sheikh, 2020). Figure 14.5 shows a comparison of select technosignatures introduced in the preceding pages, based on the classification just described.

We outline a few major astronomical observations, either operational or in the planning stages, that offer valuable avenues for technosignature research:

- Atmospheric technosignatures may be detectable on transiting terrestrial planets in the habitable zones of stars at $\sim$10 pc in a few hundred hours of observation time with envisioned next-generation space telescopes.
- Pulsed laser technosignatures are being pursued by the *Panoramic SETI* (*PANOSETI*) instrument (Maire et al., 2020), which can search most of the available sky for the entire night, and is capable of rejecting false positives via coincidence detection. *PANOSETI* is predicted to have the capacity to discover laser pulses with $\iota \sim 1$ across the whole Galaxy.

- Future infrared space telescopes, endowed with high sensitivity to point sources and full-sky capabilities (i.e., an update of the IRAS mission), might be able to detect Dyson spheres at distances up to $\sim$1 kpc.
- Besides the ongoing *Breakthrough Listen* project, the search for radio technosignatures also comprises part of the 'Cradle of Life' programme of the future *Square Kilometre Array* radio telescope (Siemion et al., 2015). In the long-term future, the far side of the Moon is of great interest for radio astronomy because it is protected from contamination by human transmissions. A permanent lunar radio facility would be very relevant for general astronomical purposes, and would be of considerable significance in the search of radio technosignatures (Michaud et al., 2021).

In summary, the continuous advancement of technology (software and instrumentation), coupled with a deepening comprehension of the diverse forms by which intelligent extraterrestrial life might reveal its existence, is unveiling a multitude of fresh opportunities for the formulation and exploration of technosignatures. Thus, humanity is now facing an unprecedented and thrilling prospect: the chance, for the first time in history, to systematically and rigorously discern its authentic position amidst the Cosmos.

14.5   Problems

Question 14.1: On the basis of what you have gleaned from the previous chapters of this book, offer your own predictions for N_T by utilising (14.1). For each factor in the Drake equation, give a concise justification of your estimate and, if possible, provide both an optimistic and a pessimistic guess, so that your final answer will be a range of values rather than a single number. Comment on the implications of your results.

Question 14.2: Use the range of estimates for N_T that you have derived in Question 14.1, and input them in (14.5) to calculate the potential average distance between systems hosting technosignatures.

Question 14.3: Use (14.8) to rewrite the cumulative factor $f_t \equiv f_{bia}f_c$ as a function of N_{tot}. Based on the fact that N_{tot} is at least 1 (since we know that at least one technological species exists in the Galaxy), what can be said about f_t? Justify whether you find this constraint plausible in light of the contents of previous chapters.

Question 14.4: The average distance between stars in the Milky Way disc is $\sim$1 pc. Estimate the typical timescale τ_s required to cover this distance at the speed $v_s \sim 10^4$ m/s attainable by chemical rockets. Next, estimate the upper and lower timescales for reaching the closest stellar system that might host a detectable technosignature, as computed in Question 14.2.

Question 14.5: Assume that a perfect self-replicating probe can travel at the speed adopted in Question 14.4. Upon reaching the closest stellar system, the probe produces a new copy of itself in a timescale that is negligible compared to the travel time, after which both probes travel to another system, and so on. How many generations are required to produce a total number of probes equal to the number of stars in the Galaxy? What may be the corresponding total time needed to explore all the stars?

Question 14.6: Suppose that a single self-replicating probe possesses a mass of 10^6 kg; this is a conservative choice, because their mass might be much lower. Drawing on the results of Question 14.5, how long will it take, since the launch of the first probe, to consume the entire mass of the Galaxy, which is loosely estimated to be $\sim 3 \times 10^{42}$ kg?

Question 14.7: Assume that the minimal flux discernible by a radio telescope is approximately $F_{min} \sim 10^{-27}$ W m^{-2} (this could be achieved in about 10 minutes of observations from the future *Square Kilometre Array* telescope). What would be the maximum distance at which we can detect a radio beacon with an EIRP of $\sim 10^{13}$ W, namely, comparable to the most powerful planetary radar on Earth nowadays? Repeat the same calculation for a high-power radio station with an EIRP of $\sim 10^5$ W.

Question 14.8: Start from the present-day average world power consumption of around 2×10^{13} W, and assume that the energy consumption grows approximately 2% each year (i.e., in an exponential fashion). How long would it take for humanity to reach Type I on the Kardashev scale?

Question 14.9: To answer this question, consult Table 1 of Socas-Navarro et al. (2021). Use the results obtained in Question 14.2 to comment on the implications for technosignature searches as follows. Which type of technosignature, if any, would be improbable to detect, given the estimated closest distance from us? Which one(s) would be the most promising?

Question 14.10: By surveying the peer-reviewed literature, identify and describe one alternative to the Kardashev scale. Justify whether you find this alternative more or less compelling than the Kardashev scale.

Question 14.11: After a literature search, outline two potential technosignatures that were not covered in Section 14.4.1, and discuss their detectability by current or future telescopes; if applicable, mention what rung of the Kardashev scale would be occupied by this putative technological species.

Question 14.12: By adopting a bird's-eye view, motivate two advantages and disadvantages of biosignatures (see Chapter 13) with respect to technosignatures. If we were to hypothetically assign an annual budget of $1 billion for seeking extraterrestrial life, how would you partition this expenditure between biosignatures and technosignatures? Explain your reasoning.

Question 14.13: In each of the three categories of proposed solutions to the Fermi paradox (refer to Section 14.2), briefly summarise one *specific* candidate and comment on its plausibility, with suitable peer-reviewed sources.

Part VI

Futures

15   The Future of Humankind

As we approach the conclusion of this book, it is time to ponder the final fundamental question of astrobiology, namely: *'Where are we going?'*. There is scarcely any need to emphasise that striving to gauge the future of life and humanity on Earth and beyond is a highly uncertain endeavour. The outcome depends on a multitude of diverse variables, including astrophysical events, environmental changes, technological advancements, and human actions. Furthermore, all these factors influence each other in complex nonlinear ways, thereby rendering any attempt to anticipate their evolution similar to planning a trek in an ever-changing uncharted landscape.

Nevertheless, contemplating our future trajectory against the vast backdrop of the Cosmos has a direct bearing on how we approach and address some of the most pressing challenges and opportunities facing humanity. At the very least, confronting these paths prompts us to engage in rigorous scientific inquiry, as well as in crucial societal and philosophical reflection.

15.1   The future of humanity on Earth

We will first delineate the long-term future of the Earth, and then the bevy of existential risks that may impede humankind at some stage.

15.1.1   Future evolution of the Earth

Nothing that has a beginning may last forever, and our planet and its denizens are no exceptions. In this book, we have already elucidated the mechanisms that gave rise to atoms, molecules, planets, and life. We cannot discuss in similar detail the expected demise of the various structures that constitute the current material content of the Cosmos, and the distant fate of the Universe itself. The interested reader should consult the classic review by Adams and Laughlin (1997), which is still valid in many respects.

For our purposes, though, it is useful to acquire at least a modest understanding of the key processes that will sculpt the future trajectory of the Earth, and their accompanying timescales. The overall lifespan of our planet is obviously intertwined with the fate of the Sun. Up-to-date theoretical models of stellar evolution can be used to infer the long-term fate of the Sun and Earth (Schröder and Smith, 2008). Our star will exit the main sequence and enter the red giant phase ~ 5.5 Gyr from now (see Section 2.3.3), gradually growing cooler, while its luminosity and radius will increase.

At an age of about 12 Gyr, the Sun may attain its maximum luminosity of $\sim 2730 L_\odot$, with a radius of 1.2 AU. By that point, the Sun would be shedding mass at a rate of $\sim 2.5 \times 10^{-7} M_\odot$ per year, translating to a cumulative loss of around one-third of its mass. At a cursory glance, we might surmise that the diminished gravitational pull would permit the Earth's orbit to gradually expand outward. However, when the concomitant loss of angular momentum due to tidal interaction and dynamical friction is properly taken into account, the numerical result is that the Earth orbit will never exceed approximately 1 AU, so that it will most likely be engulfed by the outer shell of the expanding Sun $\sim$7.5 Gyr from now. Assuming that no intervening processes have shifted Earth to a safer distance, this event would herald the ultimate physical obliteration of our planet.

However, Earth would have become uninhabitable long before this epoch. The Sun is witnessing its luminosity increase at a rate of 1% every $\sim$110 Myr at present, as revealed by (8.11). This trend will cause the habitable zone (HZ) to move outward: adopting the criteria prescribed in Section 8.1.4, it may be demonstrated that the inner edge of the HZ should cross 1 AU in roughly 1 Gyr from now (Question 15.2). At this juncture, Earth will transition to a moist greenhouse regime, with the onset of rapid ocean evaporation. State-of-the-art climate models also indicate that the moist greenhouse effect could emerge $\lesssim$2 Gyr in the future (Wolf and Toon, 2015).

Subsequently, the UV dissociation of atmospheric water vapour will trigger hydrogen escape into space, ultimately driving the loss of all water on the planet. When the average surface temperature reaches $\sim$420 K, no surface life may survive (peruse Section 7.2.1): this event is predicted to happen $\sim$1.2–1.85 Gyr from now (refer to Figure 15.1). In theory, unicellular, anaerobic extremophiles might discover refuges in subsurface habitats, caves, and high-altitude regions, potentially allowing a residual biosphere to endure up to $\sim$2.8 Gyr in the future (O'Malley-James et al., 2013); this duration is a plausible upper temporal limit for Earth as a habitable world.

The decline of the biosphere, however, would commence prior to Earth's departure from the HZ. Rising temperatures could cause an increase in the rate of silicate weathering, and a consequent decrease of CO_2 levels in the atmosphere (see Section 8.3.2). Eventually, low CO_2 concentrations

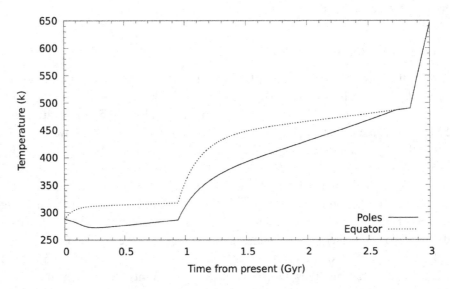

Figure 15.1 Average surface temperature of Earth computed as a function of time. Dotted and solid lines signify the temperatures in the equatorial and polar regions, respectively. Moist greenhouse regime might ensue $\sim$1 Gyr from now. (Credit: O'Malley-James et al. 2013, Figure 5; reproduced with permission from Cambridge University Press)

will make photosynthesis impossible for vascular plants. Once all plants die off, atmospheric oxygen will steeply drop and disappear in millions of years. The depletion of oxygen, coupled with the diminishing availability of primary food sources, will trigger the demise of all animals. This extinction will perhaps start with large vertebrates, progress towards smaller ones, and ultimately affect invertebrates. At this stage, $\sim$100 Myr after the demise of plants, equivalent to about $\sim$1 Gyr from now (Ozaki and Reinhard, 2021), only microbial life may inhabit the Earth (O'Malley-James et al., 2013).

15.1.2 Existential risks

The scenario outlined above is apparently quite robust, since it relies on well-established astrophysical and planetary phenomena. However, given that it lies so far in the distant future, it has minimal relevance to the fate of our species. From the historical record of life on Earth, the estimated average lifespan of species ranges from about 5 to 10 Myr. In particular, invertebrate species tend to survive for approximately 10 Myr, whereas mammals typically persist for around 1 Myr (Lawton and May, 1995, Chapter 1). Therefore, unless we prolong the lifetime of our species by means of (controversial) technological interventions like genetic engineering or hybridisation with artificial life, it is foreseeable that humans may face extinction due to natural causes long before our planet is rendered uninhabitable for animals.

In principle, certain catastrophic events could also engender the 'premature' demise of our species, perhaps in tandem with a mass extinction analogous to those that unfolded in the Phanerozoic eon, which are described in Section 6.4. One of the most tangible risks, in this regard, is the potential hazard of asteroid or comet impacts (e.g., Chapman and Morrison, 1994). A global (extinction-level) catastrophe might require the collision of a rocky object of at least 1–2 km in size: the expected frequency of such an event, as per the documented crater record, is less than one strike per Myr (consult Figure 15.2). Smaller impacts are unlikely to entirely eradicate the human species from the planet, but they happen more frequently, and may still result in substantial disruptions, to the extent of triggering widespread societal collapse. Thus, there is a vital need for diligent monitoring and preventive measures to address such scenarios.

In a similar vein, we have elucidated in Section 6.4 that most of the 'Big Five' mass extinctions seem to be strongly correlated with large igneous provinces (LIPs), that is, massive volcanic events. Although LIPs are partly stochastic in nature, an analysis of large-scale volcanism in the past 260 Myr suggests they might occur at intervals of $\mathcal{O}(10)$ Myr on average (Rampino et al., 2019). However, it is worth highlighting that this timescale is longer than the characteristic lifetime of mammals, as mentioned previously. On a cognate note, a single *supervolcano* eruption would produce $\gtrsim 10^3$ km^3 of magma, and its environmental perturbations may drive a substantial decline of human population and/or induce societal collapse.

In this category of 'natural' risks are also high-energy astrophysical phenomena such as supernovae, gamma ray bursts, and active supermassive black holes, all of which were delineated in Section 8.4; of this trio, supernovae and gamma ray bursts have been tentatively linked to mass extinction events (see Section 6.4), although their effects and timing remain disputed. The cumulative risk of a global catastrophe ensuing from many, if not all, of the aforementioned events is ostensibly $<10^{-5}$ per year (Ord, 2020, pg. 190). Hence, it might be concluded that the probability of natural disasters posing global extinction-level risks in the next century is minimal.

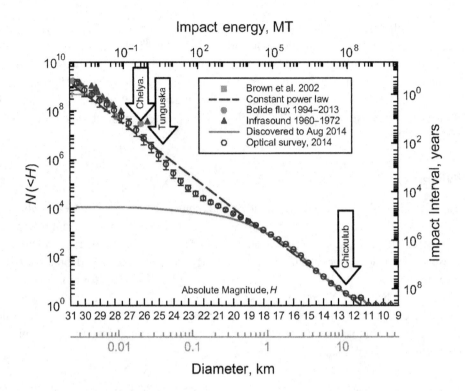

Figure 15.2 Interval between impacts by asteroids over a given size, and the corresponding energy released. (Credit: Harris and D'Abramo 2015, Figure 5; reproduced with permission from Elsevier)

The majority of existential threats that our species is poised to encounter in the immediate future are self-inflicted, and tied to the breakneck advancement of technology. Some prominent candidates are as follows (e.g., Bostrom, 2002):

- Nuclear annihilation: The existence of a vast arsenal of nuclear weapons comprises a palpable threat, whether through intentional use in a worldwide conflict or involving accidental mishaps.
- Deliberate or accidental misuse of nanotechnology: Self-replicating cellular-scale robots offer conceivable benefits in the realms of medical or environmental science, but could prove to be a menace to the entire biosphere if they get out of control, or are deployed with malicious intentions.
- Rogue artificial intelligences (AIs): Extremely advanced artificial intelligence (i.e., *superintelligence*) might either misinterpret or pursue too zealously its original goals, turning its enhanced capabilities against humanity.
- Pandemics: As the recent COVID-19 pandemic has shown, the emergence of new pathogens and their rapid diffusion into the world population poses a serious hazard to our globalised society. In addition to natural mechanisms such as *zoonosis* (i.e., transmission from animals to humans), new diseases can now be created (intentionally or accidentally) via genetic manipulation, thereby boosting the risk of a catastrophic situation.
- Climate change and environmental degradation: Human activity has been unequivocally demonstrated to elevate the global average temperature of Earth, and to exert a plethora of adverse impacts on the environment: biodiversity loss, ocean acidification, land degradation and deforestation,

atmospheric pollution, and more (Lewis and Maslin, 2018). In the most adverse scenario, such effects may escalate and exceed tipping points, potentially culminating in ecosystem and civilisation collapse.

The preceding factors are endowed with varying degrees of plausibility and inherent risk: thorough treatments of this subject are provided in Bostrom and Ćirković (2008), Rees (2018), and Ord (2020).

15.2 The future of humanity beyond Earth

With the dawn of the *Space Age* in the second half of the twentieth century (i.e., 1957 to the current day), humanity acquired the capability of circumventing Earth's gravity, and eventually reaching other celestial objects. This striking development opened up the possibilities of envisioning a future for our species beyond our planet. A quote often attributed to a pioneer of rocket science and astronautics, Konstantin Tsiolkovsky (1857–1935), encapsulates the notion that humanity may seek to broaden its cosmic horizons: '*Earth is the cradle of humanity, but one cannot live in a cradle forever*'.

This sentiment is in resonance with the statements and musings of contemporary space entrepreneurs, science fiction authors, and influential public figures, who have proposed that humanity might aspire to extend its presence to outer space and extraterrestrial locales. This expansion has been pitched (at times) not solely for scientific exploration but also for pursuing economic endeavours or bolstering the likelihood of long-term survival of our species. We highlight, however, that the opposite stance has likewise attracted many adherents, who contend that humankind should refrain from activities of this kind until the multifaceted suite of problems prevalent on Earth are resolved, lest we repeat the mistakes of the past.

With the knowledge we have acquired thus far in this book, we can undertake a succinct analysis of the prospects for space exploration.

15.2.1 Solar system settlements

Within the solar system, only two celestial bodies are arguably realistic for human settlements in the medium term: the Moon and Mars. Remarkably, the Moon stands as the sole extraterrestrial object to have been visited by humans. Twelve astronauts from NASA's *Apollo* missions surveyed the lunar surface between 1969 and 1972, amassing a total of approximately 80 hours spent in extra-vehicular activities.[1] A return to the Moon with crewed missions is currently scheduled within the next few years as part of the *Artemis* mission.[2] The aim of this programme is to renew human presence on the Moon, with the overarching goal of building an orbital space station (the *Lunar Gateway*) and creating a permanent lunar base.

A stable outpost on the Moon is often perceived as a stepping stone for the more ambitious goal of sending the first humans to Mars. This aspiration has indeed captivated the imaginations of many since the inception of the Space Age, likely driven by Mars' relative proximity to Earth and the fact that it is arguably the only world in the solar system that bears crude resemblance to our home

[1] www.nasa.gov/missions/a-few-things-artemis-will-teach-us-about-living-and-working-on-the-moon/.
[2] www.nasa.gov/specials/artemis/.

planet. However, as discussed in Chapter 9, this similarity is quite limited, particularly when it comes to habitability. Hence, the prospects for a long-term settlement on Mars are significantly constrained.

A historical overview of the plans formulated for human exploration of Mars during the latter half of the twentieth century can be found in Portree (2001). Over the past two decades, the notion of crewed missions to Mars has gained additional momentum, driven in part by the growing private sector involvement in space exploration. Notably, the *SpaceX* company has explicitly articulated its desire to establish a permanent human presence on Mars in the coming decades (Musk, 2017). It should be recognised, however, that setting up a human base on Mars would engender major challenges (Szocik, 2019). Some of the most prominent hurdles are as follows:

- Radiation: The Martian surface is exposed to intense ultraviolet (UV) radiation, solar energetic particles, and cosmic rays, as it lacks the protective shielding of a substantial atmosphere and a magnetosphere. Even short round-trip journeys to Mars, using tried-and-tested propulsion methods and shielding, are predicted to subject astronauts to a radiation dose of 0.66 ± 0.12 sievert (Sv) (Zeitlin et al., 2013).[3] This dose is undesirably close to the 'safe' threshold of 0.6 Sv prescribed for astronauts by NASA. The health risks for both travellers and potential settlers, if not properly mitigated, could consequently be serious, since the hazards encompass a bevy of medical conditions like acute radiation syndrome, nervous system damage, and an elevated risk of cancer (Durante and Cucinotta, 2011).

- Low surface gravity: On the Martian surface, humans would experience reduced gravity (38% that of Earth), which may thus induce multiple physiological alterations – including bone atrophy, muscle loss, visual impairments, and cardiovascular system weakening – in the absence of counteracting strategies (e.g., Kanas and Manzey, 2008; Patel et al., 2020).

- Life support: To maintain a permanent human base on Mars, the construction of artificial habitats simulating Earth's environment will be essential. These habitats would not only have to provide shelter from the hostile external environment (e.g., temperature, pressure, and radiation) but also be equipped to supply and recycle the requisite air, water, temperature regulation, energy, and food. The necessary resources should either be regularly shipped from Earth or produced on-site. Both options translate to formidable logistical and technological challenges.

- Isolation: A crewed vehicle will typically take six to nine months to reach Mars within practical flight plans, plausibly introducing psychological drawbacks to the endeavour (Patel et al., 2020). Moreover, the substantial Earth–Mars distance results in a one-way communication delay of anywhere between around 4 and 24 minutes. In actuality, Martian settlers would have to operate with limited support from Earth, and would probably strive for self-sufficiency, particularly for long-term sustainability.

The list above, albeit not exhaustive, is an overview of the obstacles faced by permanent Martian habitations. In reality, even just sending a crewed mission to Mars and returning them safely to Earth remains an immensely daunting task with the resources available nowadays.

Yet, on the longer term, it is not beyond the realm of possibility to envision humanity establishing a presence on Mars. This process could commence with small outposts, and eventually progress towards the creation of large-scale artificial habitats. Theoretical studies have also explored the

---

[3] One sievert embodies the biological damage caused when 1 J of ionising radiation is absorbed by 1 kg of human tissue

potential for extensive environmental engineering – known as *terraforming* – to transform the Martian climate and make the planet habitable (e.g., Fogg, 1998). As an initial phase, terraforming may entail injecting substantial quantities of CO_2 into the atmosphere, thereby intensifying the greenhouse effect and elevating the surface temperatures (Sagan, 1973a; McKay et al., 1991).

However, even under optimistic circumstances and assuming tremendous economic and technological capacity, this endeavour may warrant as much as $\sim 10^5$ yr overall (McKay and Marinova, 2001). In reality, as per state-of-the-art data on the total inventory of accessible Martian carbon dioxide, it appears that globally terraforming Mars is purportedly not achievable by dint of CO_2 alone, even in theory, with modern technology (Jakosky and Edwards, 2018). On the other hand, the construction of small-scale habitats with localised greenhouse effects (e.g., maintained by silica gel) may be feasible in principle (Wordsworth et al., 2019).

It is also important to acknowledge that the justifiable scenario of extant Martian indigenous (microbial) life introduces an additional layer of complexity to the presumption of human visits and potential settlements on Mars. Valid concerns about preserving evidence of independent abiogenesis, as well as the diverse ethical issues pertaining to possible interference (deliberate or otherwise) with biological evolution on another planet, should be carefully considered when assessing and deciding the modality of human exploration of Mars (e.g., Rummel et al., 2012; McKay, 2018).

If self-sustaining artificial habitats required for Mars settlement do become viable in the future, the deployment of similar life support systems might aid in establishing space settlements that are not tethered to a planetary surface. Such concepts were explored in the early twentieth century by scientists like Tsiolkovsky and J. D. Bernal (1901–1971), and subsequently by Gerard K. O'Neill (1927–1992) (O'Neill, 1977). Enormous space stations, ostensibly capable of accommodating many thousands of inhabitants, could create artificial gravity on their inner surfaces by means of controlled rotation (thereupon inducing centrifugal force). Myriad shapes were proposed and deemed suitable, encompassing cylindrical, toroidal, and spherical designs (Chen et al., 2021b). These space habitats may be strategically positioned at appropriate locations, such as the *Lagrange points* of the Earth–Moon system (where the net force on objects would be close to zero).

Needless to say, the implementation of such theoretical studies arguably lies beyond the practical domain of current and near-future technologies. It would necessitate, among other bottlenecks, the organisation of an extensive space transport infrastructure, and access to substantial reserves of raw materials, potentially through methods like asteroid mining. The efficacy and ethics of these enterprises remain debated and controversial.

15.2.2 Interstellar travel

Having conducted a brief examination of the prospects for human space exploration (and perhaps expansion) within the boundaries of the solar system, we can now turn our gaze to an even loftier goal: the notion of journeying to planets orbiting distant stars (Crawford, 2018). In our discussion of *interstellar travel*, note that most propulsion systems described could also be harnessed for deep space missions to the outer solar system.

When contemplating this daunting challenge of travelling to another star, the primary hurdle is the vast distances, often of order tens of light-years, that separate Earth from neighbouring planetary systems. On account of the fundamental limitation imposed by the expectation that faster-than-light travel is ruled out – notwithstanding exotic concepts such as *warp drives* (e.g., Davis, 2013) – this constraint would result in lengthy trips, even under optimal settings of voyaging at a fraction of the speed of light.

15.2.2.1 Rockets

To acquire a quantitative understanding of the difficulties associated with bridging such immense distances, we will turn to a famous relationship commonly referred to as the *Tsiolkovsky equation* or the *rocket equation*. This equation is derived directly from the principle of momentum conservation, and encapsulates the maximum change in velocity (Δv) that a rocket could achieve by expelling a portion of its mass at an *exhaust velocity* v_{ex}.

$$\Delta v \approx v_{ex} \ln \left(\frac{M_i}{M_f} \right), \tag{15.1}$$

where M_i and M_f are the initial and final mass, roughly equivalent to the fuel and payload masses, respectively. For our purposes, a 'rocket' is any spacecraft that carries its fuel onboard and propels itself by expelling this fuel at v_{ex}. Furthermore, we restrict ourselves to non-relativistic velocities, that is, speeds much lower than the speed of light. A fully relativistic version of the rocket equation was derived in the mid-twentieth century.

It is apparent that the rocket equation imposes stringent constraints on the maximum velocity attainable by a spacecraft, with only two parameters available for manipulation: the fuel mass and the exhaust speed. Chemical propulsion utilising liquid propellants can reach exhaust speeds of a few km/s, with the well-known combination of liquid oxygen and liquid hydrogen fuel having $v_{ex} \approx 4.4$ km/s. More reactive fuel combinations do achieve slightly higher v_{ex}, but they are comparatively prone to the risks of explosions. Due to the exponential dependence of M_i on $\Delta v / v_{ex}$, as seen from (15.1), it is obvious that substantial amounts of propellant must be employed to merely attain the escape velocity of 11.2 km/s from Earth.

As of now, humanity has predominantly relied on chemical propulsion for space exploration. This includes the quintet of probes – *Pioneer 10* and *Pioneer 11*, *Voyager 1* and *Voyager 2*, and *New Horizons* – currently travelling at speeds surpassing the velocity necessary to exit the solar system and journey into interstellar space. The typical velocities of these interstellar spacecraft are of order ~ 10 km/s, with the fastest among them, *Voyager 1*, currently moving away from the Sun at a relative speed of 16.9 km/s. At this speed, it would take about 75,000 years to reach the star closest to the Sun: Proxima Centauri. This calculation gives a tangible idea of the immense breakthroughs imperative to make interstellar travel a realistic prospect.

An alternative major form of propulsion that has been tested and proven in actual spaceflights entails the use of plasma (i.e., electrically charged particles), which is ejected using electromagnetic fields. The key advantage of this system is the capacity to achieve v_{ex} on the order of several tens of km/s, thereby requiring less propellant to reach the same final velocity of chemical rockets. This propulsion is practical in the vacuum of space, and therefore, traditional chemical thrusters are still needed to exit the Earth's atmosphere. Moreover, most plasma propulsion systems produce low thrust, implying that attaining the desired Δv can take a long time.

On expanding our perspective to incorporate admittedly futuristic rockets, nuclear propulsion is often perceived as a promising candidate. In theory, controlled nuclear fusion could be harnessed to generate substantial amounts of energy to power rockets. Such a concept was explored in the *Project Daedalus* and *Project Icarus* studies (Bond and Martin, 1986; Long et al., 2011). While this propulsion system may be capable of traversing the distance to nearby stars in a span of decades, and the hazard of radioactive contamination is diminished, its implementation is stunted by the fact that humanity has hitherto not produced sustained energy in a controlled fashion by means of nuclear fusion, despite decades of intensive research.

With current technology, one technically viable nuclear power concept is the so-called *nuclear pulse propulsion*, which is reliant on nuclear fission. In essence, this scheme involves deploying successive nuclear detonations to propel the spacecraft. This option was analysed in the *Project Orion* study (Dyson, 1968). A spacecraft powered by nuclear pulse propulsion could theoretically be accelerated to a few per cent of the speed of light, enabling it to reach the nearest stars in roughly a century. In practice, however, it would demand a staggering number of $\sim 10^5$ nuclear explosions (each releasing an energy of $\sim 4 \times 10^{15}$ J); the litany of profound risks and ethical concerns associated with this project are of greater importance.

A more far-fetched option, which nonetheless cannot be ruled out solely on the basis of physical principles, is the utilisation of matter–antimatter annihilation as a propulsion system (e.g., Forward, 1982; Semyonov, 2014). At first glimpse, *antimatter propulsion* is the most efficient power production mechanism imaginable, as it can fully convert matter into energy through the famous mass–energy formula, $E = mc^2$, where E and m are the energy and mass, respectively. Yet, in practice, its efficiency is not perfect.

Even so, speeds of about $\sim 50\%$ the speed of light may be actualised using this method (Westmoreland, 2010). However, matter–antimatter rockets would require ~ 10 kg of antimatter (along with 4 tons of liquid hydrogen) to accelerate a $\sim 10^3$ kg probe to 10% the speed of light (Forward, 1982); in sharp contrast, with present-day technology, producing just one gram of antimatter incurs economic costs exceeding one trillion dollars. Furthermore, the storage and manipulation of antimatter is deeply challenging, since it instantly annihilates into energy upon contact with matter. Therefore, it seems highly implausible that this propulsion system could be realised in the foreseeable future.

15.2.2.2 Sails and world ships

If we set aside the conventional paradigm of 'rocket' architectures, and turn towards the opposite stance of dispensing with the onboard propellant, alternative pathways are conceivable. A couple of propulsion systems worth highlighting are *magnetic sails* (Zubrin and Andrews, 1991) and *electric sails* (Janhunen, 2004), which are based on momentum exchange with the solar/stellar wind by virtue of deflecting the latter via magnetic and electric forces, respectively. However, their peak velocities are typically restricted to approximately the solar/stellar wind speed of ~ 500 km/s.

Perhaps the most noteworthy example in the propellant-free category is *light sail propulsion*, in which a spacecraft is propelled by radiation pressure (i.e., by the momentum carried by photons); the process resembles the commonplace phenomenon of sailboats pushed by winds. The first scientific publications on light sails date from over a century ago, and this intriguing concept has experienced a revival of sorts (Gong and Macdonald, 2019). From a practical standpoint, the *IKAROS* probe launched by JAXA in 2010 utilised solar radiation pressure to push a polymer sail measuring 14 m by 14 m and 7.5 μm in thickness to a velocity of ~ 400 m/s.

Let us suppose that we consider photons with linear momentum p_γ and energy E_γ; note that the energy–momentum relation for photons is $p_\gamma = E_\gamma/c$ (Young and Freedman, 2018, Chapter 37.8). Now, imagine that these photons strike the surface of a light sail and are perfectly reflected with momentum $-p_\gamma$. Since the change in photon momentum is $-2p_\gamma$, as per the conservation of momentum, the light sail receives an impetus of $2p_\gamma$. The force F exerted on the light sail is given by

$$F = \frac{d(2p_\gamma)}{dt} = \frac{2}{c}\frac{dE_\gamma}{dt} = \frac{2P_\gamma}{c},$$ (15.2)

where the second equality follows from the above energy–momentum relation, and the last equality from the fact that the rate of change of energy is the power P_γ emitted by the photon source. The power supplied by this source is not necessarily constant with distance (see Question 15.6). If the final velocity of the light sail (of mass M_s) is v_f, the total energy expenditure E_t to attain this speed is loosely twice the change in kinetic energy:

$$E_t \sim M_s v_f^2.$$ (15.3)

Note, however, that decelerating the spacecraft to rest demands another E_t.

In theory, powerful lasers could be harnessed to accelerate a light sail to a sizeable fraction of the speed of light, rendering this propulsion method particularly attractive for interstellar travel (Forward, 1984; Lubin, 2016). This setup was under active investigation by the *Breakthrough Starshot initiative*,[4] which sought to assess the feasibility of employing laser-powered light sails to propel miniaturised gram-mass probes towards Proxima Centauri, with an estimated travel time of approximately two decades. While undeniably appealing for uncrewed space exploration (if the technical obstacles are ironed out), this propulsion system would necessitate extremely high energy resources for accelerating substantial payload masses to fast speeds (refer to Question 15.7). Consequently, it has limited relevance for transporting humanity to extrasolar destinations in the anticipated future.

On contemplating the limitations encountered so far, it is probably realistic to conclude that, even in the most optimistic scenario(s), the duration of a hypothetical crewed mission to other planetary systems may amount to several centuries. One conceivable approach, from a purely theoretical perspective, for conveying a viable human population to an interstellar destination would involve the deployment of a *generation starship* or *world ship* (Bond and Martin, 1984; Hein et al., 2012). Such a vessel would be an enormous, self-contained, and self-sustaining interstellar vehicle designed to support multiple generations of humans throughout the extensive interval of time required to reach an ostensibly habitable exoplanet.

Besides the obvious technical challenges and perils inherent in this ambitious enterprise, a plethora of ethical and societal issues will have to be carefully weighed. A one-way trip to an insufficiently characterised target would arguably qualify as a last-resort option. Furthermore, the choice to undertake such a journey will affect not only the initial travellers but also their descendants. Last, the initiation of this long-term project would need to be strategically timed, as unforeseen technological advancements could make it obsolete shortly after its commencement (Kennedy, 2006).

15.3 Denouement

The preceding discussion, although limited in scope, may hopefully constitute a realistic appraisal of the actual prospects for a human future beyond Earth. Notwithstanding the exuberant visions of numerous space enthusiasts, the Martian, lunar, and orbital space habitats do not represent viable

---

[4] https://breakthroughinitiatives.org/initiative/3.

alternatives to our home planet for mass human habitation in the decades to come. As of now, interstellar travel largely belongs to the realm of science fiction rather than to actual tangible engineering programmes.

Of course, this seemingly pessimistic outlook does not imply that humanity is forever destined to be confined to Earth. Space exploration will continue, and plausibly expand, yielding greater insights into the hurdles that must be surmounted to extend our presence past the terrestrial orbit. Meanwhile, astronomical observations of our Universe and the growing comprehension of extraterrestrial environments may enhance our understanding of our own planet, and of the myriad factors that have rendered it suitable for sustaining a rich biosphere over an extended period. This knowledge might aid us in charting our future path, and possibly devising solutions for many of the existential threats that could arise at some stage.

As humanity has become the primary force behind planetary change, it must confront unprecedented risks and trials. At the same time, it has also unlocked the potential to manage and shape that change, at least to a certain degree. By illuminating the profound relation of life on Earth to the whole of our Cosmos, astrobiology can play a vital role in increasing awareness of the significance of our existence on this planet and within the Universe. We do not yet know whether extraterrestrial life is very common or exceptionally rare. But we do know that, in manifold ways, our species has been granted a singular opportunity. It is both a grave burden and a precious gift.

There is no readily available backup plan for humanity elsewhere. Maybe, one day, we will extend our reach to the stars. Perhaps we will become part of a tapestry of life spanning the entire galaxy, or we might be the ones responsible for weaving together its inception. Until then, it is our duty to keep the flame burning on this beautiful, fragile, solitary planet.

15.4   Problems

Question 15.1: From the conservation of angular momentum (for a circular orbit) and the assumption that the Sun will lose one-third of its current mass after exiting the main sequence, calculate Earth's new orbital distance.

Question 15.2: Draw on what you learned in Section 8.1.4 and the expected evolution of the solar luminosity to compute the locations of the inner and outer edges of the solar system habitable zone 1 Gyr from now.

Question 15.3: Perform a survey of the scientific literature pertaining to assessments of existential risks faced by humanity. List the three that you deem most relevant, and provide a detailed exposition of their inherent plausibility, as well as some suggestions for their mitigation.

Question 15.4: Derive a mathematical relation for the apparent gravitational acceleration experienced at the inner walls of a generic rotating space station as a consequence of the (fictitious) centrifugal force. Express your result in terms of the surface gravity of Earth, $g_\oplus = 9.8$ m s^{-2}. What are the variables that could be acted upon to create an artificial gravity of $g_\oplus$? What would you consider realistic choices of such variables?

Question 15.5: In this problem, apply the rocket equation (15.1) to analyse the acceleration of a single proton, using chemical propulsion with $v_{ex} \approx 4.4$ km/s, to reach $\Delta v = 10^{-3} c$. Assuming that all the propellant is expelled, so that the final mass M_f is simply the proton mass (1.67×10^{-27} kg), what is the required M_i to achieve this Δv? How does this compare against the estimated mass of ordinary matter in the observable Universe ($\sim 1.5 \times 10^{53}$ kg)?

Question 15.6: In the case of a light sail powered by stellar radiation pressure (i.e., dubbed a solar sail), prove that $P_\gamma = L_\star A_s/(4\pi r^2)$ in (15.2), where $L_\star$ is the stellar luminosity, r is the sail distance from the star, and A_s is the area of the sail (whose mass is M_s). Next, describe why the area density of the solar sail, defined as $\sigma_s = M_s/A_s$, constitutes the crucial parameter for determining its acceleration and terminal velocity.

Question 15.7: Imagine that we wish to employ laser-driven light sails to accelerate a modest-sized spacecraft with $M_s = 100$ tons to a final velocity of $v_f = 0.1\,c$. What is the total energy E_t necessary to accomplish this objective? By looking up the modern annual energy consumption of humankind, estimate the number of years that humanity would have to stockpile energy in order to attain this value of E_t. Is this timescale feasible in the twenty-first century? Justify your reasoning with appropriate references.

Question 15.8: By conducting a thorough literature review, motivate and discuss the prominent ethical and societal problems that are associated with the practical implementation of a world ship. Utilising an AI art generator of your choice, proceed to design, execute, and explain images of how such a starship, and the daily lives of its inhabitants, might look like.

Question 15.9: How may diverse human societies respond to the discovery of extraterrestrial intelligence(s), if this epochal event were to happen? Draw on the peer-reviewed literature to corroborate your standpoint.

References

Abbot, D. S., and Switzer, E. R. 2011. The Steppenwolf: A Proposal for a Habitable Planet in Interstellar Space. *Astrophys. J. Lett.*, **735**(2), L27.

Abrahamsson, Victor, and Kanik, Isik. 2022. In situ Organic Biosignature Detection Techniques for Space Applications. *Front. Astron. Space Sci.*, **9**, 360.

Abramowicz, Marek, Bejger, Michał, Gourgoulhon, Éric, and Straub, Odele. 2020. A Galactic Centre Gravitational-Wave Messenger. *Sci. Rep.*, **10**, 7054.

Acuna, M. H., Connerney, J. E. P., Wasilewski, P. et al. 1998. Magnetic Field and Plasma Observations at Mars: Initial Results of the Mars Global Surveyor Mission. *Science*, **279**, 1676–1680.

Adamala, Katarzyna, and Szostak, Jack W. 2013. Competition between Model Protocells Driven by an Encapsulated Catalyst. *Nat. Chem.*, **5**(6), 495–501.

Adams, Fred C. 2019. The Degree of Fine-tuning in Our Universe - and Others. *Phys. Rep.*, **807**, 1–111.

Adams, Fred C., and Grohs, Evan. 2017. Stellar Helium Burning in Other Universes: A Solution to the Triple Alpha Fine-Tuning Problem. *Astropart. Phys.*, **87**, 40–54.

Adams, Fred C., and Laughlin, Gregory. 1997. A Dying Universe: The Long-Term Fate and Evolution of Astrophysical Objects. *Rev. Mod. Phys.*, **69**(2), 337–372.

Adams, Fred C., Lada, Charles J., and Shu, Frank H. 1987. Spectral Evolution of Young Stellar Objects. *Astrophys. J.*, **312**, 788–806.

Adamski, Paul, Eleveld, Marcel, Sood, Ankush et al. 2020. From Self-Replication to Replicator Systems En Route to De Novo Life. *Nat. Rev. Chem.*, **4**(8), 386–403.

Affholder, Antonin, Guyot, François, Sauterey, Boris, Ferrière, Régis, and Mazevet, Stéphane. 2021. Bayesian Analysis of Enceladus's Plume Data to Assess Methanogenesis. *Nat. Astron.*, **5**, 805–814.

Affholder, Antonin, Guyot, François, Sauterey, Boris, Ferrière, Régis, and Mazevet, Stéphane. 2022. Putative Methanogenic Biosphere in Enceladus's Deep Ocean: Biomass, Productivity, and Implications for Detection. *Planet. Sci. J.*, **3**(12), 270.

Agol, Eric, and Fabrycky, Daniel C. 2018. Transit-Timing and Duration Variations for the Discovery and Characterization of Exoplanets. Pages 797–816 of: Deeg, Hans J., and Belmonte, Juan Antonio (eds.), *Handbook of Exoplanets*. Cham: Springer.

Agol, Eric, Dorn, Caroline, Grimm, Simon L. et al. 2021. Refining the Transit-Timing and Photometric Analysis of TRAPPIST-1: Masses, Radii, Densities, Dynamics, and Ephemerides. *Planet. Sci. J.*, **2**(1), 1.

Airapetian, V. S., Barnes, R., Cohen, O. et al. 2020. Impact of Space Weather on Climate and Habitability of Terrestrial-Type Exoplanets. *Int. J. Astrobiol.*, **19**(2), 136–194.

Akeson, R. L., Chen, X., Ciardi, D. et al. 2013. The NASA Exoplanet Archive: Data and Tools for Exoplanet Research. *Publ. Astron. Soc. Pac.*, **125**(930), 989.

Alexander, Richard. 2022. *Formation of Planetary System*. https://rdalexander.github.io/planets_2022.html. Accessed: 23-03-2023.

Algeo, Thomas J., and Shen, Jun. 2023. Theory and Classification of Mass Extinction Causation. *Natl. Sci. Rev.*, **11**(1), nwad237.

Allwood, Abigail C., Rosing, Minik T., Flannery, David T., Hurowitz, Joel A., and Heirwegh, Christopher M. 2018. Reassessing Evidence of Life in 3,700-million-year-old Rocks of Greenland. *Nature*, **563**(7730), 241–244.

Alvarez, Luis W., Alvarez, Walter, Asaro, Frank, and Michel, Helen V. 1980. Extraterrestrial

Cause for the Cretaceous-Tertiary Extinction. *Science*, **208**(4448), 1095–1108.

Ambrifi, A., Balbi, A., Lingam, M., Tombesi, F., and Perlman, E. 2022. The Impact of AGN Outflows on the Surface Habitability of Terrestrial Planets in the Milky Way. *Mon. Not. R. Astron. Soc.*, **512**(1), 505–516.

Amend, J. P., and Shock, E. L. 1998. Energetics of Amino Acid Synthesis in Hydrothermal Ecosystems. *Science*, **281**, 1659.

Ameta, Sandeep, Matsubara, Yoshiya J., Chakraborty, Nayan, Krishna, Sandeep, and Thutupalli, Shashi. 2021. Self-Reproduction and Darwinian Evolution in Autocatalytic Chemical Reaction Systems. *Life*, **11**(4), 308.

Andrews, Sean M., Huang, Jane, Pérez, Laura M. et al. 2018. The Disk Substructures at High Angular Resolution Project (DSHARP). I. Motivation, Sample, Calibration, and Overview. *Astrophys. J. Lett.*, **869**(2), L41.

Andrews-Hanna, Jeffrey C., Zuber, Maria T., and Banerdt, W. Bruce. 2008. The Borealis Basin and the Origin of the Martian Crustal Dichotomy. *Nature*, **453**(7199), 1212–1215.

Anglada-Escudé, Guillem, Amado, Pedro J., Barnes, John et al. 2016. A Terrestrial Planet Candidate in a Temperate Orbit around Proxima Centauri. *Nature*, **536**(7617), 437–440.

Ansari, Arif H. 2023. Detection of Organic Matter on Mars, Results from Various Mars Missions, Challenges, and Future Strategy: A Review. *Front. Astron. Space Sci.*, **10**, 30.

Apai, Dániel, Milster, Tom D., Kim, Dae Wook et al. 2019. A Thousand Earths: A Very Large Aperture, Ultralight Space Telescope Array for Atmospheric Biosignature Surveys. *Astron. J.*, **158**(2), 83.

Arcones, Almudena, and Thielemann, Friedrich-Karl. 2023. Origin of the Elements. *Astron. Astrophys. Rev.*, **31**(1), 1.

Argiroffi, C., Reale, F., Drake, J. J. et al. 2019. A Stellar Flare-Coronal Mass Ejection Event Revealed by X-ray Plasma Motions. *Nat. Astron.*, **3**, 742–748.

Aristotle. 1907. *De Anima: With Translation, Introduction and Notes*. Cambridge: Cambridge University Press.

Armitage, P. J., and Rice, W. K. M. 2008. Planetary Migration. Pages 66–83 of: Livio,

Mario, Sahu, Kailash, and Valenti, JeffE (eds.), *A Decade of Extrasolar Planets around Normal Stars Proceedings of the Space Telescope Science Institute Symposium*. Space Telescope Science Institute Symposium Series. Cambridge: Cambridge University Press.

Armitage, Philip J. 2007. Lecture Notes on the Formation and Early Evolution of Planetary Systems. *arXiv e-prints*, Jan., astro–ph/ 0701485.

Armitage, Philip J. 2020. *Astrophysics of Planet Formation*. 2nd ed. Cambridge: Cambridge University Press.

Armstrong, John C., Leovy, Conway B., and Quinn, Thomas. 2004. A 1 Gyr Climate Model for Mars: New Orbital Statistics and the Importance of Seasonally Resolved Polar Processes. *Icarus*, **171**(2), 255–271.

Arnold, Luc F. A. 2005. Transit Light-Curve Signatures of Artificial Objects. *Astrophys. J.*, **627**(1), 534–539.

Arnscheidt, Constantin W., and Rothman, Daniel H. 2022. Presence or Absence of Stabilizing Earth System Feedbacks on Different Time Scales. *Sci. Adv.*, **8**(46), eadc9241.

Atkins, Peter, and De Paula, Julio. 2006. *Physical Chemistry for the Life Sciences*. Oxford: Oxford University Press.

Atri, Dimitra, Hariharan, B., and Grießmeier, Jean-Mathias. 2013. Galactic Cosmic Ray-Induced Radiation Dose on Terrestrial Exoplanets. *Astrobiology*, **13**(10), 910–919.

Attwater, James, Wochner, Aniela, and Holliger, Philipp. 2013. In-ice Evolution of RNA Polymerase Ribozyme Activity. *Nat. Chem.*, **5**(12), 1011–1018.

Aubrey, A. D., Cleaves, H. J., and Bada, Jeffrey L. 2009. The Role of Submarine Hydrothermal Systems in the Synthesis of Amino Acids. *Orig. Life Evol. Biosph.*, **39**(2), 91–108.

Bailey, C. 1957. The Extant Writings of Epicurus. Pages 3–68 of: Oates, W. J. (ed.), *The Stoic and Epicurean Philosophers*. New York: The Modern Library.

Bains, William. 2004. Many Chemistries Could Be Used to Build Living Systems. *Astrobiology*, **4**(2), 137–167.

Bains, William, Petkowski, Janusz Jurand, Zhan, Zhuchang, and Seager, Sara. 2021. Evaluating

Alternatives to Water as Solvents for Life: The Example of Sulfuric Acid. *Life*, **11**(5), 400.

Bains, William, Petkowski, Janusz J., and Seager, Sara. 2024. Venus' Atmospheric Chemistry and Cloud Characteristics Are Compatible with Venusian Life. *Astrobiology* **24**(4), 371–385.

Balbi, Amedeo. 2018. The Impact of the Temporal Distribution of Communicating Civilizations on Their Detectability. *Astrobiology*, **18**(1), 54–58.

Balbi, Amedeo, and Ćirković, Milan M. 2021. Longevity Is the Key Factor in the Search for Technosignatures. *Astron. J.*, **161**(5), 222.

Balbi, Amedeo, and Frank, Adam. 2024. The Oxygen Bottleneck for Technospheres. *Nat. Astron.*, **8**, 39–43.

Balbi, Amedeo, and Grimaldi, Claudio. 2020. Quantifying the Information Impact of Future Searches for Exoplanetary Biosignatures. *Proc. Natl. Acad. Sci.*, **117**(35), 21031–21036.

Balbi, Amedeo, and Lingam, Manasvi. 2023. Beyond Mediocrity: How Common Is Life? *Mon. Not. R. Astron. Soc.*, **522**(2), 3117–3123.

Balbi, Amedeo, and Tombesi, Francesco. 2017. The Habitability of the Milky Way during the Active Phase of Its Central Supermassive Black Hole. *Sci. Rep.*, **7**, 16626.

Balbus, Steven A. 2003. Enhanced Angular Momentum Transport in Accretion Disks. *Annu. Rev. Astron. Astrophys.*, **41**(Jan.), 555–597.

Ball, John A. 1973. The Zoo Hypothesis. *Icarus*, **19**(3), 347–349.

Ball, Philip. 2017. Water Is an Active Matrix of Life for Cell and Molecular Biology. *Proc. Natl. Acad. Sci.*, **114**(51), 13327–13335.

Ball, Philip, and Hallsworth, John E. 2015. Water Structure and Chaotropicity: Their Uses, Abuses and Biological Implications. *Phys. Chem. Chem. Phys.*, **17**(13), 8297–8305.

Ballesteros, F. J., Fernandez-Soto, A., and Martínez, V. J. 2019. Diving into Exoplanets: Are Water Seas the Most Common? *Astrobiology*, **19**(5), 642–654.

Bambach, Richard K. 2006. Phanerozoic Biodiversity Mass Extinctions. *Annu. Rev. Earth Planet. Sci.*, **34**, 127–155.

Bar-Nun, A., Bar-Nun, N., Bauer, S. H., and Sagan, Carl. 1970. Shock Synthesis of Amino Acids in Simulated Primitive Environments. *Science*, **168**(3930), 470–473.

Bar-On, Y. M., Phillips, R., and Milo, R. 2018. The Biomass Distribution on Earth. *Proc. Natl. Acad. Sci.*, **115**(25), 6506–6511.

Barge, Laura M., Flores, Erika, Baum, Marc M., VanderVelde, David G., and Russell, Michael J. 2019. Redox and pH Gradients Drive Amino Acid Synthesis in Iron Oxyhydroxide Mineral Systems. *Proc. Natl. Acad. Sci.*, **116**(11), 4828–4833.

Barge, Laura M., Rodriguez, Laura E., Weber, Jessica M., and Theiling, Bethany P. 2022. Determining the 'Biosignature Threshold' for Life Detection on Biotic, Abiotic, or Prebiotic Worlds. *Astrobiology*, **22**(4), 481–493.

Barkana, Rennan. 2016. The Rise of the First Stars: Supersonic Streaming, Radiative Feedback, and 21-cm Cosmology. *Phys. Rep.*, **645**, 1–59.

Barnes, Jason W., Turtle, Elizabeth P., Trainer, Melissa G. et al. 2021. Science Goals and Objectives for the Dragonfly Titan Rotorcraft Relocatable Lander. *Planet. Sci. J.*, **2**(4), 130.

Barnes, Rory. 2017. Tidal Locking of Habitable Exoplanets. *Celest. Mech. Dyn. Astron.*, **129**(4), 509–536.

Baross, J. A., Benner, S. A., Cody, G. D. et al. 2007. *The Limits of Organic Life in Planetary Systems*. Washington, DC: National Academies Press.

Baross, John A., and Hoffman, Sarah E. 1985. Submarine Hydrothermal Vents and Associated Gradient Environments as Sites for the Origin and Evolution of Life. *Orig. Life*, **15**(4), 327–345.

Bartlett, Stuart, Li, Jiazheng,, Gu, Lixiang et al. 2022. Assessing Planetary Complexity and Potential Agnostic Biosignatures Using Epsilon Machines. *Nat. Astron.*, **6**, 387–392.

Batygin, Konstantin, Adams, Fred C., Brown, Michael E., and Becker, Juliette C. 2019. The Planet Nine Hypothesis. *Phys. Rep.*, **805**, 1–53.

Beatty, Thomas G. 2022. The Detectability of Nightside City Lights on Exoplanets. *Mon. Not. R. Astron. Soc.*, **513**(2), 2652–2662.

Becker, Sidney, Feldmann, Jonas, Wiedemann, Stefan et al. 2019. Unified Prebiotically Plausible Synthesis of Pyrimidine and Purine RNA Ribonucleotides. *Science*, **366**(6461), 76–82.

Bell, Elizabeth A., Boehnke, Patrick, Harrison, T. Mark, and Mao, Wendy L. 2015. Potentially

Biogenic Carbon Preserved in a 4.1 billion-Year-Old Zircon. *Proc. Natl. Acad. Sci.*, **112**(47), 14518–14521.

Benner, Steven A. 2014. Paradoxes in the Origin of Life. *Orig. Life Evol. Biosph.*, **44**(4), 339–343.

Benner, Steven A. 2017. Detecting Darwinism from Molecules in the Enceladus Plumes, Jupiter's Moons, and Other Planetary Water Lagoons. *Astrobiology*, **17**(9), 840–851.

Benner, Steven A., Ricardo, Alonso, and Carrigan, Matthew A. 2004. Is There a Common Chemical Model for Life in the Universe? *Curr. Opin. Chem. Biol.*, **8**(6), 672–689.

Benner, Steven A., Kim, Hyo-Joong, and Biondi, Elisa. 2019. Prebiotic Chemistry that Could Not *Not* Have Happened. *Life*, **9**(4), 84.

Benner, Steven A., Bell, Elizabeth A., Biondi, Elisa et al. 2020. When Did Life Likely Emerge on Earth in an RNA-First Process? *ChemSystemsChem*, **2**(2), e1900035.

Berdyugina, S. V., and Kuhn, J. R. 2019. Surface Imaging of Proxima b and Other Exoplanets: Albedo Maps, Biosignatures, and Technosignatures. *Astrophys. J.*, **158**(6), 246.

Bergner, Jennifer B., Rajappan, Mahesh, and Öberg, Karin I. 2022. HCN Snow Lines in Protoplanetary Disks: Constraints from Ice Desorption Experiments. *Astrophys. J.*, **933**(2), 206.

Bergström, Anders, Stringer, Chris, Hajdinjak, Mateja, Scerri, Eleanor M. L., and Skoglund, Pontus. 2021. Origins of Modern Human Ancestry. *Nature*, **590**(7845), 229–237.

Berliner, Aaron J., Mochizuki, Tomohiro, and Stedman, Kenneth M. 2018. Astrovirology: Viruses at Large in the Universe. *Astrobiology*, **18**(2), 207–223.

Bernal, J. D. 1951. *The Physical Basis of Life*. London: Routledge and Keegan Paul.

Betts, Holly C., Puttick, Mark N., Clark, James W., et al. 2018. Integrated Genomic and Fossil Evidence Illuminates Life's Early Evolution and Eukaryote Origin. *Nat. Ecol. Evol.*, **2**(10), 1556–1562.

Biczysko, Malgorzata, Bloino, Julien, and Puzzarini, Cristina. 2018. Computational Challenges in Astrochemistry. *Wiley Interdiscip. Rev. Comput. Mol. Sci.*, **8**(3), e1349.

Birnstiel, T., Fang, M., and Johansen, A. 2016. Dust Evolution and the Formation of Planetesimals. *Space Sci. Rev.*, **205**(1–4), 41–75.

Blackmond, Donna G. 2020. Autocatalytic Models for the Origin of Biological Homochirality. *Chem. Rev.*, **120**(11), 4831–4847.

Blankenship, R. E. 2014. *Molecular Mechanisms of Photosynthesis*. 2nd ed. Chichester: Wiley-Blackwell.

Blount, Zachary D., Lenski, Richard E., and Losos, Jonathan B. 2018. Contingency and Determinism in Evolution: Replaying Life's Tape. *Science*, **362**(6415), eaam5979.

Blum, J., and Wurm G. 2008. The Growth Mechanisms of Macroscopic Bodies in Protoplanetary Disks. *Annu. Rev. Astron. Astrophys.*, **46** 21–56.

Bobe, René, and Wood, Bernard. 2022. Estimating Origination Times from the Early Hominin Fossil Record. *Evol. Anthropol.*, **31**(2), 92–102.

Boe, Benjamin, Jedicke, Robert, Meech, Karen J. et al. 2019. The Orbit and Size-Frequency Distribution of Long Period Comets Observed by Pan-STARRS1. *Icarus*, **333**, 252–272.

Bond, A., and Martin, A. R. 1984. World Ships - An Assessment of the Engineering Feasibility. *J. Br. Interplanet. Soc.*, **37** 254.

Bond, A., and Martin, A. R. 1986. Project Daedalus Reviewed. *J. Br. Interplanet. Soc.*, **39**(9), 385–390.

Bond, David P. G., and Grasby, Stephen E. 2017. On the Causes of Mass Extinctions. *Palaeogeogr. Palaeoclimatol. Palaeoecol.*, **478**, 3–29.

Borgue, Olivia, and Hein, Andreas M. 2021. Near-Term Self-Replicating Probes: A Concept Design. *Acta Astronaut.*, **187**, 546–556.

Borlina, Cauê S., Weiss, Benjamin P., Lima, Eduardo A., et al. 2020. Reevaluating the Evidence for a Hadean–Eoarchean Dynamo. *Sci. Adv.*, **6**(15), eaav9634.

Boston, Penelope J., Ivanov, Mikhail V., and McKay, Christopher P. 1992. On the Possibility of Chemosynthetic Ecosystems in Subsurface Habitats on Mars. *Icarus*, **95**(2), 300–308.

Bostrom, Nick. 2002. Existential Risks: Analyzing Human Extinction Scenarios and Related Hazards. *J. Evol. Technol.*, **9**, 1.

Bostrom, Nick, and Ćirković, Milan M. 2008. *Global Catastrophic Risks*. Oxford: Oxford University Press.

Bottke, William F., and Norman, Marc D. 2017. The Late Heavy Bombardment. *Annu. Rev. Earth Planet. Sci.*, **45**(1), 619–647.

Boulting, William. 1914. *Giordano Bruno: His Life, Thought, and Martyrdom*. London: Kegan Paul, Trench, Trübner.

Bowman, John L. 2022. The Origin of a Land Flora. *Nat. Plants*, **8**(12), 1352–1369.

Boyajian, T. S., LaCourse, D. M., Rappaport, S. A., et al. 2016. Planet Hunters IX. KIC 8462852 - Where's the Flux? *Mon. Not. R. Astron. Soc.*, **457**(4), 3988–4004.

Bracewell, R. N. 1960. Communications from Superior Galactic Communities. *Nature*, **186**(4726), 670–671.

Brain, David A., and Jakosky, Bruce M. 1998. Atmospheric Loss since the Onset of the Martian Geologic Record: Combined Role of Impact Erosion and Sputtering. *J. Geophys. Res.*, **103**(E10), 22689–22694.

Branch, David, and Wheeler, J. Craig. 2017. *Supernova Explosions*. Astronomy and Astrophysics Library. Berlin: Springer-Verlag.

Branscomb, E., Biancalani, T., Goldenfeld, N., and Russell, M. 2017. Escapement Mechanisms and the Conversion of Disequilibria; *the Engines of Creation. Phys. Rep.*, **677**, 1–60.

Briot, D., Schneider, J., and Arnold, L. 2004. G. A. Tikhov, and the Beginnings of Astrobiology. Pages 219–220 of: Beaulieu, J., Lecavelier Des Etangs, A., and Terquem, C. (eds.), *Extrasolar Planets: Today and Tomorrow*. Astronomical Society of the Pacific Conference Series, vol. 321. San Francisco: Astronomical Society of the Pacific.

Briot, Danielle, and Schneider, Jean. 2018. Prehistory of Transit Searches. Pages 35–49 of: Deeg, Hans J., and Belmonte, Juan Antonio (eds.), *Handbook of Exoplanets*. Cham: Springer.

Broadley, Michael W., Bekaert, David V., Piani, Laurette, Füri, Evelyn, and Marty, Bernard. 2022. Origin of Life-Forming Volatile Elements in the Inner Solar System. *Nature*, **611**(7935), 245–255.

Brocks, Jochen J., Jarrett, Amber J. M., Sirantoine, Eva, et al. 2017. The Rise of Algae in Cryogenian Oceans and the Emergence of Animals. *Nature*, **548**(7669), 578–581.

Brown, M. E., and Hand, K. P. 2013. Salts and Radiation Products on the Surface of Europa. *Astron. J.*, **145**(4), 110.

Brown, Michael, Johnson, Tim, and Gardiner, Nicholas J. 2020. Plate Tectonics and the Archean Earth. *Annu. Rev. Earth Planet. Sci.*, **48**(May), 291–320.

Brown, Michael E. 2012. The Compositions of Kuiper Belt Objects. *Annu. Rev. Earth Planet. Sci.*, **40**(1), 467–494.

Brügger, N., Burn, R., Coleman, G. A. L., Alibert, Y., and Benz, W. 2020. Pebbles versus Planetesimals. The Outcomes of Population Synthesis Models. *Astron. Astrophys.*, **640**, A21.

Brunet, Thibaut, and King, Nicole. 2017. The Origin of Animal Multicellularity and Cell Differentiation. *Dev. Cell*, **43**(2), 124–140.

Bryant, Donald A., and Frigaard, Niels-Ulrik. 2006. Prokaryotic Photosynthesis and Illuminated. *Trends Microbiol.*, **14**(11), 488–496.

Budisa, Nediljko, and Schulze-Makuch, Dirk. 2014. Supercritical Carbon Dioxide and Its Potential as a Life-Sustaining Solvent in a Planetary Environment. *Life*, **4**(3), 331–340.

Burcar, Bradley T., Barge, Laura M., Trail, Dustin, et al. 2015. RNA Oligomerization in Laboratory Analogues of Alkaline Hydrothermal Vent Systems. *Astrobiology*, **15**(7), 509–522.

Burkhardt, Christoph, Spitzer, Fridolin, Morbidelli, Alessandro, et al. 2021. Terrestrial Planet Formation from Lost Inner Solar System Material. *Sci. Adv.*, **7**(52), eabj7601.

Butterfield, Nicholas J. 2011. Animals and the Invention of the Phanerozoic Earth System. *Trends Ecol. Evol.*, **26**(2), 81–87.

Cable, Morgan L., Hörst, Sarah M., Hodyss, Robert, et al. 2012. Titan Tholins: Simulating Titan Organic Chemistry in the Cassini-Huygens Era. *Chem. Rev.*, **112**(3), 1882–1909.

Cable, Morgan L., Porco, Carolyn, Glein, Christopher R., et al. 2021. The Science Case for a Return to Enceladus. *Planet. Sci. J.*, **2**(4), 132.

Cabrol, Nathalie A., and Grin, Edmond A. 1999. Distribution, Classification, and Ages of Martian Impact Crater Lakes. *Icarus*, **142**(1), 160–172.

Calcott, Brett, and Sterelny, Kim (eds). 2011. *The Major Transitions in Evolution Revisited.* Cambridge, MA: The MIT Press.

Campbell, Bruce, Walker, G. A. H., and Yang, S. 1988. A Search for Substellar Companions to Solar-Type Stars. *Astrophys. J.*, **331** 902.

Canup, Robin M., and Asphaug, Erik. 2001. Origin of the Moon in a Giant Impact near the End of the Earth's Formation. *Nature*, **412**(6848), 708–712.

Carlson, R. W., Calvin, W. M., Dalton, J. B., et al. 2009. Europa's Surface Composition. Pages 283–328 of: Pappalardo, Robert T., McKinnon, William B., and Khurana, Krishan K. (eds.), *Europa*. Tucson: University of Arizona Press.

Carnahan, Evan, Vance, Steven D., Cox, Rónadh, and Hesse, Marc A. 2022. Surface-To-Ocean Exchange by the Sinking of Impact Generated Melt Chambers on Europa. *Geophys. Res. Lett.*, **49**(24), e2022GL100287.

Carr, M. H. 2012. The Fluvial History of Mars. *Phil. Trans. R. Soc. A*, **370**(1966), 2193–2215.

Carr, M. H., and Head, J. W. 2015. Martian Surface/Near-Surface Water Inventory: Sources, Sinks, and Changes with Time. *Geophys. Res. Lett.*, **42**(3), 726–732.

Carré, Lorenzo, Zaccai, Giuseppe, Delfosse, Xavier, Girard, Eric, and Franzetti, Bruno. 2022. Relevance of Earth-Bound Extremophiles in the Search for Extraterrestrial Life. *Astrobiology*, **22**(3), 322–367.

Carrera, Daniel, Davies, Melvyn B., and Johansen, Anders. 2016. Survival of Habitable Planets in Unstable Planetary Systems. *Mon. Not. R. Astron. Soc.*, **463**(3), 3226–3238.

Carrier, B. L., Beaty, D. W., Meyer, M. A., et al. 2020. Mars Extant Life: What's Next? Conference Report. *Astrobiology*, **20**(6), 785–814.

Carrigan, Richard A., Jr. 2009. IRAS-Based Whole-Sky Upper Limit on Dyson Spheres. *Astrophys. J.*, **698**(2), 2075–2086.

Carroll-Nellenback, Jonathan, Frank, Adam, Wright, Jason, and Scharf, Caleb. 2019. The Fermi Paradox and the Aurora Effect: Exo-civilization Settlement, Expansion, and Steady States. *Astron. J.*, **158**(3), 117.

Cartwright, J. A., Hodges, K. V., and Wadhwa, M. 2022. Evidence against a Late Heavy Bombardment Event on Vesta. *Earth Planet. Sci. Lett.*, **590**, 117576.

Castellano, Marco, Fontana, Adriano, Treu, Tommaso, et al. 2022. Early Results from GLASS-JWST. III. Galaxy Candidates at z 9-15. *Astrophys. J. Lett.*, **938**(2), L15.

Castillo-Rogez, Julie, Weiss, Benjamin, Beddingfield, Chloe, et al. 2023. Compositions and Interior Structures of the Large Moons of Uranus and Implications for Future Spacecraft Observations. *J. Geophys. Res. Planets*, **128**(1), e2022JE007432.

Catling, David C., and Kasting, James F. 2017. *Atmospheric Evolution on Inhabited and Lifeless Worlds*. Cambridge: Cambridge University Press.

Catling, David C., and Zahnle, Kevin J. 2020. The Archean Atmosphere. *Sci. Adv.*, **6**(9), eaax1420.

Catling, David C., Glein, Christopher R., Zahnle, Kevin J., and McKay, Christopher P. 2005. Why O_2 Is Required by Complex Life on Habitable Planets and the Concept of Planetary 'Oxygenation Time'. *Astrobiology*, **5**(3), 415–438.

Catling, David C., Krissansen-Totton, Joshua, Kiang, Nancy Y., et al. 2018. Exoplanet Biosignatures: A Framework for Their Assessment. *Astrobiology*, **18**(6), 709–738.

Cattaneo, Fausto, and Hughes, David W. 2022. How Was the Earth-Moon System Formed? New Insights from the Geodynamo. *Proc. Natl. Acad. Sci.*, **119**(44), e2120682119.

Cavalazzi, Barbara, and Westall, Frances (eds). 2019. *Biosignatures for Astrobiology*. Cham: Springer.

Chaigne, Agathe, and Brunet, Thibaut. 2022. Incomplete Abscission and Cytoplasmic Bridges in the Evolution of Eukaryotic Multicellularity. *Curr. Biol.*, **32**(8), R385–R397.

Chaisson, Eric J. 2001. *Cosmic Evolution: The Rise of Complexity in Nature*. Cambridge, MA: Harvard University Press.

Chamberlain, J. W., and Hunten, D. M. 1987. *Theory of Planetary Atmospheres: An Introduction to Their Physics and Chemistry*. 2nd ed. San Diego: Academic Press, Inc.

Chan, Marjorie A., Hinman, Nancy W., Potter-McIntyre, Sally L., et al. 2019. Deciphering Biosignatures in Planetary Contexts. *Astrobiology*, **19**(9), 1075–1102.

Changela, Hitesh G., Chatzitheodoridis, Elias, Antunes, Andre, et al. 2021. Mars: New Insights and Unresolved Questions. *International Journal of Astrobiology*, **20**(6), 394–426.

Chapman, Clark R., and Morrison, David. 1994. Impacts on the Earth by Asteroids and Comets: Assessing the Hazard. *Nature*, **367**(6458), 33–40.

Charnay, Benjamin, Wolf, Eric T., Marty, Bernard, and Forget, François. 2020. Is the Faint Young Sun Problem for Earth Solved? *Space Sci. Rev.*, **216**(5), 90.

Chen, Guoxiong, Kusky, Timothy, Luo, Lei, Li, Quanke, and Cheng, Qiuming. 2023. Hadean Tectonics: Insights from Machine Learning. *Geology*, **51**(8), 718–722.

Chen, Howard, Zhan, Zhuchang, Youngblood, Allison, et al. 2021a. Persistence of Flare-Driven Atmospheric Chemistry on Rocky Habitable Zone Worlds. *Nat. Astron.*, **5**, 298–310.

Chen, Muhao, Goyal, Raman, Majji, Manoranjan, and Skelton, Robert E. 2021b. Review of Space Habitat Designs for Long Term Space Explorations. *Prog. Aerosp. Sci.*, **122**, 100692.

Chen, Zhong-Qiang, and Benton, Michael J. 2012. The Timing and Pattern of Biotic Recovery Following the End-Permian Mass Extinction. *Nat. Geosci.*, **5**(6), 375–383.

Chevrier, Vincent F., and Rivera-Valentin, Edgard G. 2012. Formation of Recurring Slope Lineae by Liquid Brines on Present-Day Mars. *Geophys. Res. Lett.*, **39**(21), L21202.

Chevrier, Vincent F., Fitting, Alec B., and Rivera-Valentín, Edgard G. 2022. Limited Stability of Multicomponent Brines on the Surface of Mars. *Planet. Sci. J.*, **3**(5), 125.

Chou, Luoth, Mahaffy, Paul, Trainer, Melissa, et al. 2021. Planetary Mass Spectrometry for Agnostic Life Detection in the Solar System. *Front. Astron. Space Sci.*, **8**, 173.

Christensen, U. R. 2010. Dynamo Scaling Laws and Applications to the Planets. *Space Sci. Rev.*, **152**(1-4), 565–590.

Chyba, Christopher, and Sagan, Carl. 1992. Endogenous Production, Exogenous Delivery and Impact-Shock Synthesis of Organic Molecules: An Inventory for the Origins of Life. *Nature*, **355**(6356), 125–132.

Chyba, Christopher F., and Hand, Kevin P. 2001. Life without Photosynthesis. *Science*, **292**(5524), 2026–2027.

Cirkovic, Milan M. 2004. The Temporal Aspect of the Drake Equation and SETI. *Astrobiology*, **4**(2), 225–231.

Cirkovic, Milan M. 2018. *The Great Silence: Science and Philosophy of Fermi's Paradox*. Oxford: Oxford University Press.

Ćirković, Milan M., and Balbi, Amedeo. 2020. Copernicanism and the Typicality in Time. *International Journal of Astrobiology*, **19**(2), 101–109.

Clark, Benton C., Kolb, Vera M., Steele, Andrew, et al. 2021. Origin of Life on Mars: Suitability and Opportunities. *Life*, **11**(6), 539.

Clarke, Andrew. 2014. The Thermal Limits to Life on Earth. *Int. J. Astrobiol.*, **13**(2), 141–154.

Cleaves, H. James. 2008. The Prebiotic Geochemistry of Formaldehyde. *Precambrian Res.*, **164**(3-4), 111–118.

Cleaves, H. James, Chalmers, John H., Lazcano, Antonio, Miller, Stanley L., and Bada, Jeffrey L. 2008. A Reassessment of Prebiotic Organic Synthesis in Neutral Planetary Atmospheres. *Orig. Life Evol. Biosph.*, **38**(2), 105–115.

Cleaves, H. James, Scott, Andrea Michalkova, Hill, Frances C., et al. 2012. Mineral Organic Interfacial Processes: Potential Roles in the Origins of Life. *Chem. Soc. Rev.*, **41**(16), 5502–5525.

Cleaves, H. James, Hystad, Grethe, Prabhu, Anirudh, et al. 2023. A Robust, Agnostic Molecular Biosignature Based on Machine Learning. *Proc. Natl. Acad. Sci.*, **120**(41), e2307149120.

Cleaves, Henderson James, Butch, Christopher, Burger, Pieter Buys, Goodwin, Jay, and Meringer, Markus. 2019. One among Millions: The Chemical Space of Nucleic Acid-Like Molecules. *J. Chem. Inf. Model.*, **59**(10), 4266–4277.

Cleland, Carol E. 2019. *The Quest for a Universal Theory of Life: Searching for Life As We Don't Know It*. Cambridge Astrobiology, vol. 11. Cambridge: Cambridge University Press.

Cloutier, Ryan, and Menou, Kristen. 2020. Evolution of the Radius Valley around Low-mass

Stars from Kepler and K2. *Astron. J.*, **159**(5), 211.

Cocconi, Giuseppe, and Morrison, Philip. 1959. Searching for Interstellar Communications. *Nature*, **184**(4690), 844–846.

Cockell, C. S., Bush, T., Bryce, C., et al. 2016. Habitability: A Review. *Astrobiology*, **16**(1), 89–117.

Cockell, Charles, and Blaustein, Andrew R. (eds). 2001. *Ecosystems, Evolution, and Ultraviolet Radiation*. New York: Springer-Verlag.

Cockell, Charles S. 2014. Trajectories of Martian Habitability. *Astrobiology*, **14**(2), 182–203.

Cockell, Charles S. 2020. *Astrobiology: Understanding Life in the Universe*. 2nd ed. Hoboken: John Wiley.

Cockell, Charles S., Catling, David C., Davis, Wanda L., et al. 2000. The Ultraviolet Environment of Mars: Biological Implications Past, Present, and Future. *Icarus*, **146**(2), 343–359.

Cockell, Charles S., Stevens, Adam H., and Prescott, R. 2019. Habitability Is a Binary Property. *Nat. Astron.*, **3**, 956–957.

Cole, Devon B., Mills, Daniel B., Erwin, Douglas H., et al. 2020. On the Co-evolution of Surface Oxygen Levels and Animals. *Geobiology*, **18**(3), 260–281.

Colose, Christopher M., Del Genio, Anthony D., and Way, M. J. 2019. Enhanced Habitability on High Obliquity Bodies near the Outer Edge of the Habitable Zone of Sun-like Stars. *Astrophys. J.*, **884**(2), 138.

Connelly, James N., Bizzarro, Martin, Krot, Alexander N., et al. 2012. The Absolute Chronology and Thermal Processing of Solids in the Solar Protoplanetary Disk. *Science*, **338**(6107), 651.

Conrad, Pamela G., and Nealson, Kenneth H. 2001. A Non-Earthcentric Approach to Life Detection. *Astrobiology*, **1**(1), 15–24.

Conway, Susan J., de Haas, Tjalling, and Harrison, Tanya N. 2019. Martian Gullies: a Comprehensive Review of Observations, Mechanisms and Insights from Earth Analogues. *Geol. Soc. Spec. Publ.*, **467**(1), 7–66.

Cooper, John F., Johnson, Robert E., Mauk, Barry H., Garrett, Henry B., and Gehrels, Neil. 2001. Energetic Ion and Electron Irradiation of the Icy Galilean Satellites. *Icarus*, **149**(1), 133–159.

Cordiner, M. A., Villanueva, G. L., Wiesemeyer, H., et al. 2022. Phosphine in the Venusian Atmosphere: A Strict Upper Limit From SOFIA GREAT Observations. *Geophys. Res. Lett.*, **49**(22), e2022GL101055.

Covone, Giovanni, Ienco, Riccardo M., Cacciapuoti, Luca, and Inno, Laura. 2021. Efficiency of the Oxygenic Photosynthesis on Earth-Like Planets in the Habitable Zone. *Mon. Not. R. Astron. Soc.*, **505**(3), 3329–3335.

Cowie, Robert H., Bouchet, Philippe, and Fontaine, Benoît. 2022. The Sixth Mass Extinction: Fact, Fiction or Speculation? *Biol. Rev.*, **97**(2), 640–663.

Crawford, Ian A. 2018. Direct Exoplanet Investigation Using Interstellar Space Probes. Pages 3413–3431 of: Deeg, Hans J., and Belmonte, Juan Antonio (eds.), *Handbook of Exoplanets*. Cham: Springer International.

Crick, Francis. 1970. Central Dogma of Molecular Biology. *Nature*, **227**(5258), 561–563.

Crill, Brendan P., and Siegler, Nicholas. 2017 (Sept.). Space Technology for Directly Imaging and Characterizing Exo-Earths. Page 103980H of: *Society of Photo-Optical Instrumentation Engineers (SPIE) Conference Series*. Society of Photo-Optical Instrumentation Engineers (SPIE) Conference Series, vol. 10398. San Diego, CA: Society of Photo-Optical Instrumentation Engineers.

Crowe, Michael J. 1986. *The Extraterrestrial Life Debate 1750–1900: The Idea of a Plurality of Worlds from Kant to Lowell*. Cambridge: Cambridge University Press.

Crowe, Michael J. 2008. *The Extraterrestrial Life Debate, Antiquity to 1915: A Source Book*. Notre Dame: University of Notre Dame Press.

Crutzen, P. J., Isaksen, I. S. A., and Reid, G. C. 1975. Solar Proton Events: Stratospheric Sources of Nitric Oxide. *Science*, **189**(4201), 457–459.

Dal Corso, Jacopo, Song, Haijun, Callegaro, Sara, et al. 2022. Environmental Crises at the Permian-Triassic Mass Extinction. *Nat. Rev. Earth Environ.*, **3**(3), 197–214.

Damer, Bruce, and Deamer, David. 2020. The Hot Spring Hypothesis for an Origin of Life. *Astrobiology*, **20**(4), 429–452.

D'Angelo, M., Cazaux, S., Kamp, I., Thi, W. F., and Woitke, P. 2019. Water Delivery in the Inner Solar Nebula. Monte Carlo Simulations of Forsterite Hydration. *Astron. Astrophys.*, **622**, A208.

Daval, Damien, Choblet, Gaël., Sotin, Christophe, and Guyot, François. 2022. Theoretical Considerations on the Characteristic Timescales of Hydrogen Generation by Serpentinization Reactions on Enceladus. *J. Geophys. Res. Planets*, **127**(2), e06995.

Davies, P. C. W., and Wagner, R. V. 2013. Searching for Alien Artifacts on the Moon. *Acta Astronaut.*, **89**, 261–265.

Davies, Paul C. W., Benner, Steven A., Cleland, Carol E., et al. 2009. Signatures of a Shadow Biosphere. *Astrobiology*, **9**(2), 241–249.

Davila, Alfonso F., and McKay, Christopher P. 2014. Chance and Necessity in Biochemistry: Implications for the Search for Extraterrestrial Biomarkers in Earth-Like Environments. *Astrobiology*, **14**(6), 534–540.

Davis, E. W. 2013. Faster-Than-Light Space Warps, Status and Next Steps. *J. Br. Interplanet. Soc.*, **66**, 68–84.

Dawkins, Richard, and Wong, Yan. 2016. *The Ancestor's Tale: A Pilgrimage to the Dawn of Evolution.* 2nd ed Boston: Mariner Books.

de Duve, Christian. 2005. *Singularities: Landmarks on the Pathways of Life.* Cambridge: Cambridge University Press.

De Marco, Paolo. 2004. Methylotrophy Versus Heterotrophy: A Misconception. *Microbiology*, **150**(6), 1606–1607.

de Pater, Imke, and Lissauer, Jack J. 2015. *Planetary Sciences.* Updated 2nd ed. Cambridge: Cambridge University Press.

Deamer, David, Dworkin, Jason P., Sandford, Scott A., Bernstein, Max P., and Allamandola, Louis J. 2002. The First Cell Membranes. *Astrobiology*, **2**(4), 371–381.

Deamer, David, Cary, Francesca, and Damer, Bruce. 2022. Urability: A Property of Planetary Bodies That Can Support an Origin of Life. *Astrobiology*, **22**(7), 889–900.

Des Marais, David J., Nuth, Joseph A., Allamandola, Louis J., et al. 2008. The NASA Astrobiology Roadmap. *Astrobiology*, **8**(4), 715–730.

Dick, Steven J. 1982. *Plurality of Worlds : The Origins of the Extraterrestrial Life Debate from Democritus to Kant.* Cambridge: Cambridge University Press.

Dodd, Matthew S., Papineau, Dominic, Grenne, Tor, et al. 2017. Evidence for Early Life in Earth's Oldest Hydrothermal Vent Precipitates. *Nature*, **543**(7643), 60–64.

Dohnanyi, J. S. 1969. Collisional Model of Asteroids and Their Debris. *J. Geophys. Res.*, **74**, 2531–2554.

Domagal-Goldman, Shawn D., Wright, Katherine E., Adamala, Katarzyna, et al. 2016. The Astrobiology Primer v2.0. *Astrobiology*, **16**(8), 561–653.

Dong, Chuanfei, Lee, Yuni, Ma, Yingjuan, et al. 2018. Modeling Martian Atmospheric Losses over Time: Implications for Exoplanetary Climate Evolution and Habitability. *Astrophys. J. Lett.*, **859**(1), L14.

Donoghue, Philip C. J., Kay, Chris, Spang, Anja, et al. 2023. Defining Eukaryotes to Dissect Eukaryogenesis. *Curr. Biol.*, **33**(17), R919–R929.

Dormand, John R., and Woolfson, Michael M. 1989. *The Origin of the Solar System: The Capture Theory.* New York:Halsted Press.

Dorn, Evan D., Nealson, Kenneth H., and Adami, Christoph. 2011. Monomer Abundance Distribution Patterns as a Universal Biosignature: Examples from Terrestrial and Digital Life. *J. Mol. Evol.*, **72**(3), 283–295.

Douce, Alberto Patiño. 2011. *Thermodynamics of the Earth and Planets.* Cambridge: Cambridge University Press.

Drake, F. D. 1961. Project Ozma. *Phys. Today*, **14**(4), 40.

Drake, Frank D. 1965. The Radio Search for Intelligent Extraterrestrial Life. Pages 323–345 of: Mamikunian, G., and Briggs, M. H. (eds.), *Current Aspects of Exobiology.* Oxford: Pergamon Press.

Drazkowska, J., Bitsch, B., Lambrechts, M., et al. 2023. Planet Formation Theory in the Era of ALMA and Kepler: From Pebbles to Exoplanets. Page 717 of: Inutsuka, S., Aikawa, Y., Muto, T., Tomida, K., and Tamura, M. (eds.), *Astronomical Society of the Pacific Conference Series.* Astronomical Society of the Pacific Conference

Series, vol. 534. San Francisco: Astronomical Society of the Pacific.

Droser, Mary L., Tarhan, Lidya G., and Gehling, James G. 2017. The Rise of Animals in a Changing Environment: Global Ecological Innovation in the Late Ediacaran. *Annu. Rev. Earth Planet. Sci.*, **45**(1), 593–617.

Dundas, Colin M., McEwen, Alfred S., Chojnacki, Matthew, et al. 2017. Granular Flows at Recurring Slope Lineae on Mars Indicate a Limited Role for Liquid Water. *Nat. Geosci.*, **10**(12), 903–907.

Durante, Marco, and Cucinotta, Francis A. 2011. Physical Basis of Radiation Protection in Space Travel. *Rev. Mod. Phys.*, **83**(4), 1245–1281.

Dworkin, Jason P., Deamer, David W., Sandford, Scott A., and Allamandola, Louis J. 2001. Self-assembling Amphiphilic Molecules: Synthesis in Simulated Interstellar/Precometary Ices. *Proc. Natl. Acad. Sci.*, **98**(3), 815–819.

Dyson, Freeman J. 1960. Search for Artificial Stellar Sources of Infrared Radiation. *Science*, **131**(3414), 1667–1668.

Dyson, Freeman J. 1968. Interstellar Transport. *Phys. Today*, **21**(10), 41–45.

Dyson, Freeman J. 1982. A Model for the Origin of Life. *J. Mol. Evol.*, **18**(5), 344–350.

Ehlmann, B. L., Anderson, F. S., Andrews-Hanna, J., et al. 2016. The Sustainability of Habitability on Terrestrial Planets: Insights, Questions, and Needed Measurements from Mars for Understanding the Evolution of Earth-Like Worlds. *Journal of Geophysical Research (Planets)*, **121**(10), 1927–1961.

Ehlmann, Bethany L., and Edwards, Christopher S. 2014. Mineralogy of the Martian Surface. *Annu. Rev. Earth Planet. Sci.*, **42**(1), 291–315.

Ehlmann, Bethany L., Mustard, John F., Murchie, Scott L., et al. 2011. Subsurface Water and Clay Mineral Formation during the Early History of Mars. *Nature*, **479**(7371), 53–60.

Elkins-Tanton, Linda T. 2012. Magma Oceans in the Inner Solar System. *Annu. Rev. Earth Planet. Sci.*, **40**(1), 113–139.

Eme, Laura, Spang, Anja, Lombard, Jonathan, Stairs, Courtney W., and Ettema, Thijs J. G. 2017. Archaea and the Origin of Eukaryotes. *Nat. Rev. Microbiol.*, **15**(12), 711–723.

Emsenhuber, Alexandre, Mordasini, Christoph, and Burn, Remo. 2023. Planetary Population Synthesis and the Emergence of four Classes of Planetary System Architectures. *Eur. Phys. J. Plus*, **138**(2), 181.

Erastova, Valentina, Degiacomi, Matteo T., Fraser, Donald G., and Greenwell, H. Chris. 2017. Mineral Surface Chemistry Control for Origin of Prebiotic Peptides. *Nat. Commun.*, **8**, 2033.

Estrela, Raissa, and Valio, Adriana. 2018. Superflare Ultraviolet Impact on Kepler-96 System: A Glimpse of Habitability When the Ozone Layer First Formed on Earth. *Astrobiology*, **18**(11), 1414–1424.

Evans, E., Waugh, R., and Melnik, L. 1976. Elastic Area Compressibility Modulus of Red Cell Membrane. *Biophys. J.*, **16**(6), 585–595.

Falkowski, Paul G., and Raven, John A. 2007. *Aquatic Photosynthesis*. 2nd ed. Princeton University Press, Princeton.

Fassett, Caleb I., and Head, James W. 2008. The Timing of Martian Valley Network Activity: Constraints from Buffered Crater Counting. *Icarus*, **195**(1), 61–89.

Feinberg, G., and Shapiro, R. 1980. *Life beyond Earth: The Intelligent Earthling's Guide to Life in the Universe*. New York: William Morrow.

Feinstein, Adina D., Montet, Benjamin T., Ansdell, Megan, et al. 2020. Flare Statistics for Young Stars from a Convolutional Neural Network Analysis of TESS Data. *Astron. J.*, **160**(5), 219.

Feldman, W. C., Prettyman, T. H., Maurice, S., et al. 2004. Global Distribution of near-Surface Hydrogen on Mars. *J. Geophys. Res. Planets*, **109**(E9), E09006.

Firsoff, Valdemar Axel. 1963. *Life beyond the Earth; a Study in Exobiology*. New York: Basic Books.

Fischer, Woodward W., Hemp, James, and Johnson, Jena E. 2016. Evolution of Oxygenic Photosynthesis. *Annu. Rev. Earth Planet. Sci.*, **44**, 647–683.

Fixsen, D. J. 2009. The Temperature of the Cosmic Microwave Background. *Astrophys. J.*, **707**(2), 916–920.

Fogg, M. J. 1987. Temporal Aspects of the Interaction among the First Galactic Civilizations: The "Interdict Hypothesis". *Icarus*, **69**(2), 370–384.

Fogg, M. J. 1998. Terraforming Mars: A Review of Current Research. *Adv. Space Res.*, **22**(3), 415–420.

Foley, Bradford J., and Smye, Andrew J. 2018. Carbon Cycling and Habitability of Earth-Sized Stagnant Lid Planets. *Astrobiology*, **18**(7), 873–896.

Forgan, Duncan, Dayal, Pratika, Cockell, Charles, and Libeskind, Noam. 2017. Evaluating Galactic Habitability Using High-Resolution Cosmological Simulations of Galaxy Formation. *Int. J. Astrobiol.*, **16**(1), 60–73.

Fortney, Jonathan J., Dawson, Rebekah I., and Komacek, Thaddeus D. 2021. Hot Jupiters: Origins, Structure, Atmospheres. *J. Geophys. Res. Planets*, **126**(3), e06629.

Forward, R. L. 1982. Antimatter Propulsion. *J. Br. Interplanet. Soc.*, **35**, 391–395.

Forward, R. L. 1984. Roundtrip Interstellar Travel Using Laser-Pushed Lightsails. *J. Spacecr. Rockets*, **21**(2), 187–195.

Fox-Powell, Mark G., Hallsworth, John E., Cousins, Claire R., and Cockell, Charles S. 2016. Ionic Strength Is a Barrier to the Habitability of Mars. *Astrobiology*, **16**(6), 427–442.

Fraústo Da Silva, J. J. R., and Williams, R. J. P. 2001. *The Biological Chemistry of the Elements: The Inorganic Chemistry of Life*. 2nd ed. Oxford: Oxford University Press.

Freer, M., and Fynbo, H. O. U. 2014. The Hoyle State in ^{12}C. *Prog. Part. Nucl. Phys.*, **78**(Sept.), 1–23.

Freitas, R. A., Jr., and Valdes, F. 1985. The Search for Extraterrestrial Artifacts (SETA). *Acta Astronaut.*, **12**(12), 1027–1034.

Frenkel-Pinter, Moran, Samanta, Mousumi, Ashkenasy, Gonen, and Leman, Luke J. 2020. Prebiotic Peptides: Molecular Hubs in the Origin of Life. *Chem. Rev.*, **120**(11), 4707–4765.

Frenkel-Pinter, Moran, Rajaei, Vahab, Glass, Jennifer B., Hud, Nicholas V., and Williams, Loren Dean. 2021. Water and Life: The Medium Is the Message. *J. Mol. Evol.*, **89**(1–2), 2–11.

Fried, Stephen D., Fujishima, Kosuke, Makarov, Mikhail, Cherepashuk, Ivan, and Hlouchova, Klara. 2022. Peptides before and during the Nucleotide World: An Origins Story Emphasizing Cooperation between Proteins and Nucleic Acids. *J. R. Soc. Interface*, **19**(187), 20210641.

Fry, Iris. 2000. *The Emergence of Life on Earth: A Historical and Scientific Overview*. New Brunswick: Rutgers University Press.

Fry, Iris. 2011. The Role of Natural Selection in the Origin of Life. *Orig. Life Evol. Biosph.*, **41**(1), 3–16.

Fujii, Yuka, Angerhausen, Daniel, Deitrick, Russell, et al. 2018. Exoplanet Biosignatures: Observational Prospects. *Astrobiology*, **18**(6), 739–778.

Furukawa, Yoshihiro, Chikaraishi, Yoshito, Ohkouchi, Naohiko, et al. 2019. Extraterrestrial Ribose and Other Sugars in Primitive Meteorites. *Proc. Natl. Acad. Sci.*, **116**(49), 24440–24445.

Gaillard, Fabrice, Michalski, Joseph, Berger, Gilles, McLennan, Scott M., and Scaillet, Bruno. 2013. Geochemical Reservoirs and Timing of Sulfur Cycling on Mars. *Space Sci. Rev.*, **174**(1-4), 251–300.

Gajjar, Vishal, Perez, Karen I., Siemion, Andrew P. V., et al. 2021. The Breakthrough Listen Search For Intelligent Life near the Galactic Center. I. *Astron. J.*, **162**(1), 33.

Gánti, Tibor. 2003. *The Principles of Life*. Oxford: Oxford University Press.

Gehrels, Neil, Laird, Claude M., Jackman, Charles H., et al. 2003. Ozone Depletion from Nearby Supernovae. *Astrophys. J.*, **585**(2), 1169–1176.

Georgiou, Christos D., and Deamer, David W. 2014. Lipids as Universal Biomarkers of Extraterrestrial Life. *Astrobiology*, **14**(6), 541–549.

Gerya, T. V., Stern, R. J., Baes, M., Sobolev, S. V., and Whattam, S. A. 2015. Plate Tectonics on the Earth Triggered by Plume-Induced Subduction Initiation. *Nature*, **527**(7577), 221–225.

Gibard, Clémentine, Bhowmik, Subhendu, Karki, Megha, Kim, Eun-Kyong, and Krishnamurthy, Ramanarayanan. 2018. Phosphorylation, Oligomerization and Self-assembly in Water under Potential Prebiotic Conditions. *Nat. Chem.*, **10**(2), 212–217.

Gilbert, Walter. 1986. Origin of Life: The RNA World. *Nature*, **319**(6055), 618.

Gillen, Catherine, Jeancolas, Cyrille, McMahon, Sean, and Vickers, Peter. 2023. The Call for a

New Definition of Biosignature. *Astrobiology*, **23**(11), 1228–1237.

Gillmann, Cedric, Way, M. J., Avice, Guillaume, et al. 2022. The Long-Term Evolution of the Atmosphere of Venus: Processes and Feedback Mechanisms. *Space Sci. Rev.*, **218**(7), 56.

Gillon, Michaël, Triaud, Amaury H. M. J., Demory, Brice-Olivier, et al. 2017. Seven Temperate Terrestrial Planets around the Nearby Ultracool Dwarf Star TRAPPIST-1. *Nature*, **542**(7642), 456–460.

Girichidis, Philipp, Offner, Stella S. R., Kritsuk, Alexei G., et al. 2020. Physical Processes in Star Formation. *Space Sci. Rev.*, **216**(4), 68.

Glavin, Daniel P., Burton, Aaron S., Elsila, Jamie E., Aponte, José C., and Dworkin, Jason P. 2019. The Search for Chiral Asymmetry as a Potential Biosignature in Our Solar System. *Chem. Rev.*, **120**(11), 4660–4689.

Godolt, Mareike, Tosi, Nicola, Stracke, Barbara, et al. 2019. The Habitability of Stagnantlid Earths around Dwarf Stars. *Astron. Astrophys.*, **625**, A12.

Goldreich, Peter, and Ward, William R. 1973. The Formation of Planetesimals. *Astrophys. J.*, **183**, 1051–1062.

Gomes, R., Levison, H. F., Tsiganis, K., and Morbidelli, A. 2005. Origin of the Cataclysmic Late Heavy Bombardment Period of the Terrestrial Planets. *Nature*, **435**(7041), 466–469.

Gómez, Felipe, Cavalazzi, Barbara, Rodríguez, Nuria, et al. 2019. Ultra-small Microorganisms in the Polyextreme Conditions of the Dallol Volcano, Northern Afar, Ethiopia. *Sci. Rep.*, **9**, 7907.

Gomez Barrientos, Jonathan, MacDonald, Ryan J., Lewis, Nikole K., and Kaltenegger, Lisa. 2023. In Search of the Edge: A Bayesian Exploration of the Detectability of Red Edges in Exoplanet Reflection Spectra. *Astrophys. J.*, **946**(2), 96.

Gong, Shengping, and Macdonald, Malcolm. 2019. Review on Solar Sail Technology. *Astrodynamics*, **3**(2), 93–125.

Gonzalez, Guillermo, Brownlee, Donald, and Ward, Peter. 2001. The Galactic Habitable Zone: Galactic Chemical Evolution. *Icarus*, **152**(1), 185–200.

Goodwin, Arthur, Garwood, Russell J., and Tartèse, Romain. 2022. A Review of the 'Black Beauty' Martian Regolith Breccia and Its Martian Habitability Record. *Astrobiology*, **22**(6), 755–767.

Gordon, Kenneth E., Karalidi, Theodora, Bott, Kimberly M., et al. 2023. Polarized Signatures of a Habitable World: Comparing Models of an Exoplanet Earth with Visible and Near-infrared Earthshine Spectra. *Astrophys. J.*, **945**(2), 166.

Goudge, Timothy A., Head, James W., Mustard, John F., and Fassett, Caleb I. 2012. An Analysis of Open-Basin Lake Deposits on Mars: Evidence for the Nature of Associated Lacustrine Deposits and Post-Lacustrine Modification Processes. *Icarus*, **219**(1), 211–229.

Gough, D. O. 1981. Solar Interior Structure and Luminosity Variations. *Sol. Phys.*, **74**(1), 21–34.

Gould, Stephen J. 1989. *Wonderful Life: The Burgess Shale and the Nature of History*. New York: W. W. Norton.

Gowanlock, M. G., Patton, D. R., and McConnell, S. M. 2011. A Model of Habitability within the Milky Way Galaxy. *Astrobiology*, **11**(9), 855–873.

Gowanlock, Michael G. 2016. Astrobiological Effects of Gamma-ray Bursts in the Milky Way Galaxy. *Astrophys. J.*, **832**(1), 38.

Gözen, Irep, Köksal, Elif Senem, Põldsalu, Inga, et al. 2022. Protocells: Milestones and Recent Advances. *Small*, **18**(18), 2106624.

Granick, S. 1957. Speculations on the Origins and Evolution of Photosynthesis. *Ann. N. Y. Acad. Sci.*, **69**(2), 292–308.

Grasby, Stephen E., Them, Theodore R., Chen, Zhuoheng, Yin, Runsheng, and Ardakani, Omid H. 2019. Mercury as a Proxy for Volcanic Emissions in the Geologic Record. *Earth Sci. Rev.*, **196**, 102880.

Gray, Robert H. 2020. The Extended Kardashev Scale. *Astron. J.*, **159**(5), 228.

Greaves, Jane S., Richards, Anita M. S., Bains, William, et al. 2021. Phosphine Gas in the Cloud Decks of Venus. *Nat. Astron.*, **5**, 655–664.

Greaves, Jane S., Rimmer, Paul B., Richards, Anita M. S., et al. 2022. Low Levels of sulphur dioxide Contamination of Venusian Phosphine Spectra. *Mon. Not. R. Astron. Soc.*, **514**(2), 2994–3001.

Green, James, Hoehler, Tori, Neveu, Marc, et al. 2021. Call for a Framework for Reporting

Evidence for Life beyond Earth. *Nature*, **598**(7882), 575–579.

Greenberg, Richard. 2010. Transport Rates of Radiolytic Substances into Europa's Ocean: Implications for the Potential Origin and Maintenance of Life. *Astrobiology*, **10**(3), 275–283.

Greene, Thomas P., Bell, Taylor J., Ducrot, Elsa, et al. 2023. Thermal Emission from the Earth-Sized Exoplanet TRAPPIST-1 b Using JWST. *Nature*, **618**(7963), 39–42.

Grimaldi, Claudio. 2017. Signal Coverage Approach to the Detection Probability of Hypothetical Extraterrestrial Emitters in the Milky Way. *Sci. Rep.*, **7**, 46273.

Grimaldi, Claudio. 2023. Inferring the Rate of Technosignatures from 60 yr of Nondetection. *Astron. J.*, **165**(5), 199.

Grosberg, Richard K., and Strathmann, Richard R. 2007. The Evolution of Multicellularity: A Minor Major Transition? *Annu. Rev. Ecol. Evol. Syst.*, **38**, 621–654.

Guillochon, James, and Loeb, Abraham. 2015. SETI via Leakage from Light Sails in Exoplanetary Systems. *Astrophys. J. Lett.*, **811**(2), L20.

Gumsley, Ashley P., Chamberlain, Kevin R., Bleeker, Wouter, et al. 2017. Timing and tempo of the Great Oxidation Event. *Proc. Natl. Acad. Sci.*, **114**(8), 1811–1816.

Gunell, Herbert, Maggiolo, Romain, Nilsson, Hans, et al. 2018. Why An Intrinsic Magnetic Field Does Not Protect a Planet against Atmospheric Escape. *Astron. Astrophys.*, **614** L3.

Guttenberg, Nicholas, Chen, Huan, Mochizuki, Tomohiro, and Cleaves, H. James. 2021. Classification of the Biogenicity of Complex Organic Mixtures for the Detection of Extraterrestrial Life. *Life*, **11**(3), 234.

Haberle, Robert M. 2013. Estimating the Power of Mars' Greenhouse Effect. *Icarus*, **223**(1), 619–620.

Halevy, Itay, and Head, James W., III. 2014. Episodic Warming of Early Mars by Punctuated Volcanism. *Nat. Geosci.*, **7**(12), 865–868.

Hallam, A., and Wignall, P. B. 1997. *Mass Extinctions and Their Aftermath*. Oxford: Oxford University Press.

Halliday, Alex N., and Canup, Robin M. 2023. The Accretion of Planet Earth. *Nat. Rev. Earth Environ.*, **4**(1), 19–35.

Hallsworth, John E. 2021. Mars' Surface is Not Universally Biocidal. *Environ. Microbiol.*, **23**(7), 3345–3350.

Hallsworth, John E., Koop, Thomas, Dallas, Tiffany D., et al. 2021. Water Activity in Venus's Uninhabitable Clouds and Other Planetary Atmospheres. *Nat. Astron.*, **5**, 665–675.

Hamilton, Trinity L. 2019. The Trouble with Oxygen: The Ecophysiology of Extant Phototrophs and Implications for the Evolution of Oxygenic Photosynthesis. *Free Radic. Biol. Med.*, **140**, 233–249.

Hamlin, Trevor A., Poater, Jordi, Fonseca Guerra, Célia, and Bickelhaupt, F. Matthias. 2017. B-DNA Model Systems in Non-terran Bio-solvents: Implications for Structure, Stability and Replication. *Phys. Chem. Chem. Phys.*, **19**(26), 16969–16978.

Hand, K. P., Chyba, C. F., Priscu, J. C., Carlson, R. W., and Nealson, K. H. 2009. Astrobiology and the Potential for Life on Europa. Pages 589–629 of: Pappalardo, Robert T., McKinnon, William B., and Khurana, Krishan K. (eds.), *Europa*. Tucson: University of Arizona Press.

Hand, K. P., Sotin, C., Hayes, A., and Coustenis, A. 2020. On the Habitability and Future Exploration of Ocean Worlds. *Space Sci. Rev.*, **216**(5), 95.

Hand, Kevin P., and Chyba, Christopher F. 2007. Empirical Constraints on the Salinity of the Europan Ocean and Implications for a Thin Ice Shell. *Icarus*, **189**(2), 424–438.

Hansen, C. J., Bourke, M., Bridges, N. T., et al. 2011. Seasonal Erosion and Restoration of Mars' Northern Polar Dunes. *Science*, **331**(6017), 575.

Hao, Jihua, Knoll, Andrew H., Huang, Fang, et al. 2020. Cycling Phosphorus on the Archean Earth: Part II. Phosphorus Limitation on Primary Production in Archean Ecosystems. *Geochim. Cosmochim. Acta*, **280**, 360–377.

Hao, Jihua, Glein, Christopher R., Huang, Fang, et al. 2022. Abundant Phosphorus Expected for Possible Life in Enceladus's Ocean. *Proc. Natl. Acad. Sci.*, **119**(39), e2201388119.

Haqq-Misra, Jacob. 2019. Does the Evolution of Complex Life Depend on the Stellar Spectral

Energy Distribution? *Astrobiology*, **19**(10), 1292–1299.

Haqq-Misra, Jacob, and Kopparapu, Ravi Kumar. 2012. On the Likelihood of Nonterrestrial Artifacts in the Solar System. *Acta Astronaut.*, **72**, 15–20.

Haqq-Misra, Jacob, Kopparapu, Ravi, Fauchez, Thomas J., et al. 2022a. Detectability of Chlorofluorocarbons in the Atmospheres of Habitable M-dwarf Planets. *Planet. Sci. J.*, **3**(3), 60.

Haqq-Misra, Jacob, Fauchez, Thomas J., Schwieterman, Edward W., and Kopparapu, Ravi. 2022b. Disruption of a Planetary Nitrogen Cycle as Evidence of Extraterrestrial Agriculture. *Astrophys. J. Lett.*, **929**(2), L28.

Harris, Alan W., and D'Abramo, Germano. 2015. The Population of near-Earth Asteroids. *Icarus*, **257**, 302–312.

Harrison, Stuart A., Rammu, Hanadi, Liu, Feixue, et al. 2023. Life as a Guide to Its Own Origins. *Annu. Rev. Ecol. Evol. Syst.*, **54**, 327–350.

Harrison, T. Mark. 2020. *Hadean Earth*. Cham: Springer International.

Hart, Michael H. 1975. Explanation for the Absence of Extraterrestrials on Earth. *Q. J. R. Astron. Soc.*, **16**, 128.

Hartman, Hyman. 1975. Speculations on the Origin and Evolution of Metabolism. *J. Mol. Evol.*, **4**(4), 359–370.

Hartmann, William K. 2019. History of the Terminal Cataclysm Paradigm: Epistemology of a Planetary Bombardment That Never (?) Happened. *Geosciences*, **9**(7), 285.

Hartogh, Paul, Lis, Dariusz C., Bockelée-Morvan, Dominique, et al. 2011. Ocean-Like Water in the Jupiter-Family Comet 103P/Hartley 2. *Nature*, **478**(7368), 218–220.

Harwit, Martin. 2006. *Astrophysical Concepts*. 4th ed. New York: Springer-Verlag.

Hassler, Donald M., Zeitlin, Cary, Wimmer-Schweingruber, Robert F., et al. 2014. Mars' Surface Radiation Environment Measured with the Mars Science Laboratory's Curiosity Rover. *Science*, **343**(6169), 1244797.

Hatzes, A. P. 2020. *The Doppler Method for the Detection of Exoplanets*. Bristol: IOP.

Hawkesworth, Chris J., Cawood, Peter A., and Dhuime, Bruno. 2020. The Evolution of the Continental Crust and the Onset of Plate Tectonics. *Front. Earth Sci.*, **8**, 326.

Hayashi, C. 1981. Structure of the Solar Nebula, Growth and Decay of Magnetic Fields and Effects of Magnetic and Turbulent Viscosities on the Nebula. *Prog. Theor. Phys. Suppl.*, **70**, 35–53.

Hayes, Alexander G. 2016. The Lakes and Seas of Titan. *Annu. Rev. Earth Planet. Sci.*, **44**, 57–83.

Hayes, Alexander G., Lorenz, Ralph D., and Lunine, Jonathan I. 2018. A Post-Cassini View of Titan's Methane-Based Hydrologic Cycle. *Nat. Geosci.*, **11**(5), 306–313.

Hays, Lindsay E., Graham, Heather V., Des Marais, David J., et al. 2017. Biosignature Preservation and Detection in Mars Analog Environments. *Astrobiology*, **17**(4), 363–400.

Hazen, Robert M., and Morrison, Shaunna M. 2022. On the Paragenetic Modes of Minerals: A Mineral Evolution Perspective. *Am. Mineral.*, **107**(7), 1262–1287.

Head, J. W., III, Hiesinger, H., Ivanov, M. A., et al. 1999. Possible Ancient Oceans on Mars: Evidence from Mars Orbiter Laser Altimeter Data. *Science*, **286**(5447), 2134–2137.

Hein, A. M., Pak, M., Putz, D., Buhler, C., and Reiss, P. 2012. World Ships - Architectures & Feasibility Revisited. *J. Br. Interplanet. Soc.*, **65**, 119–133.

Heller, René. 2020. Habitability Is a Continuous Property of Nature. *Nat. Astro.*, **4**(Apr.), 294–295.

Hendrix, Amanda R., Hurford, Terry, Patterson, Gerald W., et al. 2021. Potential Ocean Worlds. *Bull. Am. Astron. Soc.*, **53**(May), 212.

Henning, Thomas, and Semenov, Dmitry. 2013. Chemistry in Protoplanetary Disks. *Chem. Rev.*, **113**(12), 9016–9042.

Herbst, Eric. 2017. The Synthesis of Large Interstellar Molecules. *Int. Rev. Phys. Chem.*, **36**(2), 287–331.

Herbst, Eric, and van Dishoeck, Ewine F. 2009. Complex Organic Interstellar Molecules. *Annu. Rev. Astron. Astrophys.*, **47**(1), 427–480.

Hesse, Marc A., Jordan, Jacob S., Vance, Steven D., and Oza, Apurva V. 2022. Downward Oxidant Transport through Europa's Ice Shell by Density-Driven Brine Percolation. *Geophys. Res. Lett.*, **49**(5), e2021GL095416.

Higgs, Paul G., and Lehman, Niles. 2015. The RNA World: Molecular Cooperation at the Origins of Life. *Nat. Rev. Genet.*, **16**(1), 7–17.

Hippke, Michael. 2018. Benchmarking Information Carriers. *Acta Astronaut.*, **151**, 53–62.

Hoehler, Tori M., Mankel, Dylan J., Girguis, Peter R., et al. 2023. The Metabolic Rate of the Biosphere and Its Components. *Proc. Natl. Acad. Sci.*, **120**(25), e2303764120.

Hoffman, Paul F., Kaufman, Alan J., Halverson, Galen P., and Schrag, Daniel P. 1998. A Neoproterozoic Snowball Earth. *Science*, **281**, 1342.

Hoffman, Paul F., Abbot, Dorian S., Ashkenazy, Yosef, et al. 2017. Snowball Earth Climate Dynamics and Cryogenian Geologygeobiology. *Sci. Adv.*, **3**(11), e1600983.

Holden, Dylan T., Morato, Nicolás M., and Cooks, R. Graham. 2022. Aqueous Microdroplets Enable Abiotic Synthesis and Chain Extension of Unique Peptide Isomers from Free Amino Acids. *Proceedings of the National Academy of Science*, **119**(42), e2212642119.

Holtom, Philip D., Bennett, Chris J., Osamura, Yoshihiro, Mason, Nigel J., and Kaiser, Ralf I. 2005. A Combined Experimental and Theoretical Study on the Formation of the Amino Acid Glycine (NH_2CH_2COOH) and Its Isomer ($CH_3NHCOOH$) in Extraterrestrial Ices. *Astrophys. J.*, **626**(2), 940–952.

Horneck, Gerda, Klaus, David M., and Mancinelli, Rocco L. 2010. Space Microbiology. *Microbiol. Mol. Biol. Rev.*, **74**(Jan.), 121–156.

Hörst, S. M. 2017. Titan's Atmosphere and Climate. *J. Geophys. Res. Planets*, **122**(3), 432–482.

Hosokawa, Takashi, Hirano, Shingo, Kuiper, Rolf, et al. 2016. Formation of Massive Primordial Stars: Intermittent UV Feedback with Episodic Mass Accretion. *Astrophys. J.*, **824**(2), 119.

Howard, Andrew, Horowitz, Paul, Mead, Curtis, et al. 2007. Initial Results from Harvard All-Sky Optical SETI. *Acta Astronaut.*, **61**(1–6), 78–87.

Howard, Andrew W., Horowitz, Paul, Wilkinson, David T., et al. 2004. Search for Nanosecond Optical Pulses from Nearby Solar-Type Stars. *Astrophys. J.*, **613**(2), 1270–1284.

Howard, Ward S., Tilley, Matt A., Corbett, Hank, et al. 2018. The First Naked-Eye Superflare Detected from Proxima Centauri. *Astrophys. J. Lett.*, **860**(2), L30.

Howell, Samuel M. 2021. The Likely Thickness of Europa's Icy Shell. *Planet. Sci. J.*, **2**(4), 129.

Howell, Samuel M., and Pappalardo, Robert T. 2020. NASA's Europa Clipper – a Mission to a Potentially Habitable Ocean World. *Nat. Commun.*, **11**, 1311.

Hsu, Hsiang-Wen, Postberg, Frank, Sekine, Yasuhito, et al. 2015. Ongoing Hydrothermal Activities within Enceladus. *Nature*, **519**(7542), 207–210.

Huber, Claudia, and Wächtershäuser, Günter. 2006. α-Hydroxy and α-Amino Acids under Possible Hadean, Volcanic Origin-of-Life Conditions. *Science*, **314**(5799), 630–632.

Hughes, A. Meredith, Duchêne, Gaspard, and Matthews, Brenda C. 2018. Debris Disks: Structure, Composition, and Variability. *Annu. Rev. Astron. Astrophys.*, **56**, 541–591.

Hull, Pincelli M., Bornemann, André, Penman, Donald E., et al. 2020. On Impact and Volcanism across the Cretaceous-Paleogene Boundary. *Science*, **367**(6475), 266–272.

Huwe, Björn, Fiedler, Annelie, Moritz, Sophie, et al. 2019. Mosses in Low Earth Orbit: Implications for the Limits of Life and the Habitability of Mars. *Astrobiology*, **19**(2), 221–232.

Ianeselli, Alan, Salditt, Annalena, Mast, Christof, et al. 2023. Physical Non-equilibria for Prebiotic Nucleic acid Chemistry. *Nat. Rev. Phys.*, **5**, 185–195.

Ikoma, Masahiro, Nakazawa, Kiyoshi, and Emori, Hiroyuki. 2000. Formation of Giant Planets: Dependences on Core Accretion Rate and Grain Opacity. *Astrophys. J.*, **537**(2), 1013–1025.

Imai, Ei-Ichi, Honda, Hajime, Hatori, Kuniyuki, Brack, Andre, and Matsuno, Koichiro. 1999. Elongation of Oligopeptides in a Simulated Submarine Hydrothermal System. *Science*, **283**(Feb.), 831.

Ioppolo, S., Fedoseev, G., Chuang, K. J., et al. 2021. A Non-energetic Mechanism for Glycine Formation in the Interstellar Medium. *Nat. Astron.*, **5**, 197–205.

Isaacson, Howard, Siemion, Andrew P. V., Marcy, Geoffrey W., et al. 2017. The Breakthrough Listen Search for Intelligent Life: Target Selection of Nearby Stars and Galaxies. *Publ. Astron. Soc. Pac.*, **129**(975), 054501.

Ishigaki, Miho N., Tominaga, Nozomu, Kobayashi, Chiaki, and Nomoto, Ken'ichi. 2018. The Initial Mass Function of the First Stars Inferred from Extremely Metal-poor Stars. *Astrophys. J.*, **857**(1), 46.

Izidoro, Andre, and Piani, Laurette. 2022. Origin of Water in the Terrestrial Planets: Insights from Meteorite Data and Planet Formation Models. *Elements*, **18**(3), 181–186.

Izidoro, Andre, Dasgupta, Rajdeep, Raymond, Sean N., et al. 2022. Planetesimal Rings as the Cause of the Solar System's Planetary Architecture. *Nat. Astron.*, **6**, 357–366.

Jacobsen, Stein B. 2005. The Hf-W Isotopic System and the Origin of the Earth and Moon. *Annu. Rev. Earth Planet. Sci.*, **33**, 531–570.

Jakosky, B. M., Slipski, M., Benna, M., et al. 2017. Mars' Atmospheric History Derived from Upperatmosphere Measurements of $^{38}Ar/^{36}Ar$. *Science*, **355**(6332), 1408–1410.

Jakosky, Bruce M. 2019. The CO_2 Inventory on Mars. *Planet. Space Sci.*, **175**, 52–59.

Jakosky, Bruce M., and Edwards, Christopher S. 2018. Inventory of CO_2 Available for Terraforming Mars. *Nat. Astron.*, **2**(July), 634–639.

Jakosky, Bruce M., and Phillips, Roger J. 2001. Mars' Volatile and Climate History. *Nature*, **412**(6843), 237–244.

Jakosky, Bruce M., and Treiman, Allan H. 2023. Mars Volatile Inventory and Outgassing History. *Icarus*, **402**, 115627.

Janhunen, Pekka. 2004. Electric Sail for Spacecraft Propulsion. *J. Propuls. Power*, **20**(4), 763–764.

Javaux, Emmanuelle J. 2019. Challenges in Evidencing the Earliest Traces of Life. *Nature*, **572**(7770), 451–460.

Jeans, J. H. 1916. On the Theory of Star-Streaming and the Structure of the Universe. (Second Paper.). *Mon. Not. R. Astron. Soc.*, **76**(7), 552–572.

Jelen, Benjamin I., Giovannelli, Donato, and Falkowski, Paul G. 2016. The Role of Microbial Electron Transfer in the Coevolution of the Biosphere and Geosphere. *Annu. Rev. Microbiol.*, **70**, 45–62.

Jensen, Bent Borg, and Cox, Raymond P. 1983. Direct Measurements of Steady-State Kinetics of Cyanobacterial N 2 Uptake by Membrane-Leak Mass Spectrometry and Comparisons between Nitrogen Fixation and Acetylene Reduction. *Appl. Environ. Microbiol.*, **45**(4), 1331–1337.

Jia, Xianzhe, Kivelson, Margaret G., Khurana, Krishan K., and Kurth, William S. 2018. Evidence of a Plume on Europa from Galileo Magnetic and Plasma Wave Signatures. *Nat. Astron.*, **2**, 459–464.

Johansen, Anders, Ronnet, Thomas, Bizzarro, Martin, et al. 2021. A Pebble Accretion Model for the Formation of the Terrestrial Planets in the Solar System. *Sci. Adv.*, **7**(8), eabc0444.

Johnson, Benjamin W., and Wing, Boswell A. 2020. Limited Archaean Continental Emergence Reflected in an Early Archaean ^{18}O-enriched Ocean. *Nat. Geosci.*, **13**(3), 243–248.

Johnson, Jarrett L., and Li, Hui. 2012. The First Planets: The Critical Metallicity for Planet Formation. *Astrophys. J.*, **751**(2), 81.

Johnson, Jennifer A. 2019. Populating the Periodic Table: Nucleosynthesis of the Elements. *Science*, **363**(6426), 474–478.

Johnson, Sarah S., Anslyn, Eric V., Graham, Heather V., Mahaffy, Paul R., and Ellington, Andrew D. 2018. Fingerprinting Non-terran Biosignatures. *Astrobiology*, **18**(7), 915–922.

Johnson, Tim E., Kirkland, Christopher L., Lu, Yongjun, et al. 2022. Giant Impacts and the Origin and Evolution of Continents. *Nature*, **608**(7922), 330–335.

Jones, Clive G., Lawton, John H., and Shachak, Moshe. 1994. Organisms as Ecosystem Engineers. *Oikos*, 373–386.

Jones, Eric M. 1985. Where is Everybody? *Phys. Today*, **38**(8), 11.

Jordan, Sean F., Rammu, Hanadi, Zheludev, Ivan N., et al. 2019. Promotion of Protocell Selfassembly from Mixed Amphiphiles at the Origin of Life. *Nat. Ecol. Evol.*, **3**(12), 1705–1714.

Jørgensen, Jes K., Belloche, Arnaud, and Garrod, Robin T. 2020. Astrochemistry during the Formation of Stars. *Annu. Rev. Astron. Astrophys.*, **58**, 727–778.

Journaux, Baptiste, Kalousová, Klára, Sotin, Christophe, et al. 2020. Large Ocean Worlds with High-Pressure Ices. *Space Sci. Rev.*, **216**(1), 7.

Joyce, Gerald F. 1994. Foreword. In: Deamer, David W., and Fleischaker, Gail R. (eds.), *Origins of Life: The Central Concepts*. Boston: Jones & Bartlett Pub, x–xvi.

Judson, Olivia P. 2017. The Energy Expansions of Evolution. *Nat. Ecol. Evol.*, **1**(6), 0138, x–xvi.

Kadoya, Shintaro, Krissansen-Totton, Joshua, and Catling, David C. 2020. Probable Cold and Alkaline Surface Environment of the Hadean Earth Caused by Impact Ejecta Weathering. *Geochemistry, Geophys. Geosyst.*, **21**(1), e2019GC008734.

Kaltenegger, Lisa, and Lin, Zifan. 2021. Finding Signs of Life in Transits: High-Resolution Transmission Spectra of Earth-Line Planets around FGKM Host Stars. *Astrophys. J. Lett.*, **909**(1), L2.

Kamionkowski, Marc, and Riess, Adam G. 2023. The Hubble Tension and Early Dark Energy. *Annu. Rev. Nucl. Part. Sci.*, **73**(Sept.), 153–180.

Kanas, N., and Manzey, D. 2008. *Space Psychology and Psychiatry*. 2nd ed. Space Technology Library, vol. 22. Dordrecht: Springer.

Kang, Wanying, Mittal, Tushar, Bire, Suyash, Campin, Jean-Michel, and Marshall, John. 2022. How Does Salinity Shape Ocean Circulation and Ice Geometry on Enceladus and Other Icy Satellites? *Sci. Adv.*, **8**(29), eabm4665.

Kardashev, N. S. 1964. Transmission of Information by Extraterrestrial Civilizations. *Soviet Astron.*, **8**, 217–221.

Kasting, James F. 1991. CO_2 Condensation and the Climate of Early Mars. *Icarus*, **94**(1), 1–13.

Kasting, James F., Whitmire, Daniel P., and Reynolds, Ray T. 1993. Habitable Zones around Main Sequence Stars. *Icarus*, **101**(1), 108–128.

Kattenhorn, Simon A., and Prockter, Louise M. 2014. Evidence for Subduction in the Ice Shell of Europa. *Nat. Geosci.*, **7**(10), 762–767.

Kauffman, Stuart. 1995. *At Home in the Universe: The Search for the Laws of Self-Organization and Complexity*. Oxford: Oxford University Press.

Kauffman, Stuart A. 1986. Autocatalytic Sets of Proteins. *J. Theor. Biol.*, **119**(1), 1–24.

Kauffman, Stuart A. 2011. Approaches to the Origin of Life on Earth. *Life*, **1**(1), 34–48.

Kennedy, A. 2006. Interstellar Travel: The Wait Calculation and the Incentive Trap of Progress. *J. Br. Interplanet. Soc.*, **59**, 239–246.

Kereszturi, A., Möhlmann, D., Berczi, Sz., et al. 2011. Possible Role of Brines in the Darkening and Flow-Like Features on the Martian Polar Dunes Based on HiRISE Images. *Planet. Space Sci.*, **59**(13), 1413–1427.

Khawaja, N., Postberg, F., Hillier, J., et al. 2019. Low-Mass Nitrogen-, Oxygen-Bearing, and Aromatic Compounds in Enceladean Ice Grains. *Mon. Not. R. Astron. Soc.*, **489**(4), 5231–5243.

Kiang, Nancy Y., Siefert, Janet, Govindjee, and Blankenship, Robert E. 2007a. Spectral Signatures of Photosynthesis. I. Review of Earth Organisms. *Astrobiology*, **7**(1), 222–251.

Kiang, Nancy Y., Segura, Antígona, Tinetti, Giovanna, et al. 2007b. Spectral Signatures of Photosynthesis. II. Coevolution with Other Stars and with Atmosphere on Extrasolar Worlds. *Astrobiology*, **7**(1), 252–274.

Kimura, Jun, and Kitadai, Norio. 2015. Polymerization of Building Blocks of Life on Europa and Other Icy Moons. *Astrobiology*, **15**(6), 430–441.

Kimura, Tadahiro, and Ikoma, Masahiro. 2022. Predicted Diversity in Water Content of Terrestrial Exoplanets Orbiting M Dwarfs. *Nat. Astron.*, **6**, 1296–1307.

Kipfer, K. A., Ligterink, N. F. W., Bouwman, J., et al. 2022. Toward Detecting Polycyclic Aromatic Hydrocarbons on Planetary Objects with ORIGIN. *Planet. Sci. J.*, **3**(2), 43.

Kirschvink, Joseph L., Gaidos, Eric J., Bertani, L. Elizabeth, et al. 2000. Paleoproterozoic Snowball Earth: Extreme Climatic and Geochemical Global Change and Its Biological Consequences. *Proc. Natl. Acad. Sci.*, **97**(4), 1400–1405.

Kitadai, Norio, and Maruyama, Shigenori. 2018. Origins of Building Blocks of Life: A Review. *Geosci. Front.*, **9**(4), 1117–1153.

Kite, Edwin S. 2019. Geologic Constraints on Early Mars Climate. *Space Sci. Rev.*, **215**(1), 10.

Kite, Edwin S., Mayer, David P., Wilson, Sharon A., et al. 2019. Persistence of Intense, Climate-Driven Runoff Late in Mars History. *Sci. Adv.*, **5**(3), eaav7710.

Kite, Edwin S., Steele, Liam J., Mischna, Michael A., and Richardson, Mark I. 2021. Warm Early Mars Surface Enabled by High-Altitude Water Ice Clouds. *Proc. Natl. Acad. Sci.*, **118**(18), e2101959118.

Klein, H. P. 1979. The Viking Mission and the Search for Life on Mars. *Rev. Geophys. Space Phys.*, **17**, 1655–1662.

Knoll, Andrew, Osborn, Mary Jane, Baross, John, et al. 1999. *Size Limits of Very Small Microorganisms: Proceedings of a Workshop*. Washington, DC: National Academy Press.

Knoll, Andrew H. 2011. The Multiple Origins of Complex Multicellularity. *Annu. Rev. Earth Planet. Sci.*, **39**, 217–239.

Knoll, Andrew H. 2021. *A Brief History of Earth: Four Billion Years in Eight Chapters*. New York: Custom House.

Knoll, Andrew H., and Bambach, Richard K. 2000. Directionality in the History of Life: Diffusion from the Left Wall or Repeated Scaling of the Right? *Paleobiology*, **26**(S4), 1–14.

Knoll, Andrew H., and Carroll, Sean B. 1999. Early Animal Evolution: Emerging Views from Comparative Biology and Geology. *Science*, **284**(5423), 2129–2137.

Knoll, Andrew H., and Nowak, Martin A. 2017. The Timetable of Evolution. *Sci. Adv.*, **3**(5), e1603076.

Kobayashi, Kensei, Ise, Jun-ichi, Aoki, Ryohei, et al. 2023. Formation of Amino Acids and Carboxylic Acids in Weakly Reducing Planetary Atmospheres by Solar Energetic Particles from the Young Sun. *Life*, **13**(5), 1103.

Kodama, T., Genda, H., O'ishi, R., Abe-Ouchi, A., and Abe, Y. 2019. Inner Edge of Habitable Zones for Earth-Sized Planets with Various Surface Water Distributions. *J. Geophys. Res. Planets*, **124**(8), 2306–2324.

Koll, Daniel D. B., Malik, Matej, Mansfield, Megan, et al. 2019. Identifying Candidate Atmospheres on Rocky M Dwarf Planets via Eclipse Photometry. *Astrophys. J.*, **886**(2), 140.

Kondepudi, Dilip, and Prigogine, Ilya. 2015. *Modern Thermodynamics: From Heat Engines to Dissipative Structures*. 2nd ed. Chichester: John Wiley.

Koonin, Eugene V., and Wolf, Yuri I. 2009. Is Evolution Darwinian or/and Lamarckian? *Biol. Direct*, **4**, 42.

Kopp, Robert E., Kirschvink, Joseph L., Hilburn, Isaac A., and Nash, Cody Z. 2005. The Paleoproterozoic Snowball Earth: A Climate Disaster Triggered by the Evolution of Oxygenic Photosynthesis. *Proc. Natl. Acad. Sci.*, **102**(32), 11131–11136.

Kopparapu, Ravi, Arney, Giada, Haqq-Misra, Jacob, Lustig-Yaeger, Jacob, and Villanueva, Geronimo. 2021. Nitrogen Dioxide Pollution as a Signature of Extraterrestrial Technology. *Astrophys. J.*, **908**(2), 164.

Kopparapu, Ravi Kumar, Ramirez, Ramses, Kasting, James F., et al. 2013. Habitable Zones around Main-Sequence Stars: New Estimates. *Astrophys. J.*, **765**(2), 131.

Kopparapu, Ravi Kumar, Ramirez, Ramses M., SchottelKotte, James, et al. 2014. Habitable Zones around Main-Sequence Stars: Dependence on Planetary Mass. *Astrophys. J. Lett.*, **787**(2), L29.

Korenaga, Jun. 2013. Initiation and Evolution of Plate Tectonics on Earth: Theories and Observations. *Annu. Rev. Earth Planet. Sci.*, **41**, 117–151.

Korenaga, Jun. 2021. Was There Land on the Early Earth? *Life*, **11**(11), 1142.

Koshland Jr., Daniel E. 2002. The Seven Pillars of Life. *Science*, **295**(5563), 2215–2216.

Krasnokutski, S. A., Chuang, K. J., Jäger, C., Ueberschaar, N., and Henning, Th. 2022. A Pathway to Peptides in Space through the Condensation of Atomic Carbon. *Nat. Astron.*, **6**, 381–386.

Krause, Alexander J., Mills, Benjamin J. W., Merdith, Andrew S., Lenton, Timothy M., and Poulton, Simon W. 2022. Extreme Variability in Atmospheric Oxygen Levels in the Late Precambrian. *Sci. Adv.*, **8**(41), eabm8191.

Kreidberg, Laura, Koll, Daniel D. B., Morley, Caroline, et al. 2019. Absence of a Thick Atmosphere on the Terrestrial Exoplanet LHS 3844b. *Nature*, **573**(7772), 87–90.

Krijt, S., Kama, M., McClure, M., et al. 2023. Chemical Habitability: Supply and Retention of

Life's Essential Elements during Planet Formation. Page 1031 of: Inutsuka, S., Aikawa, Y., Muto, T., Tomida, K., and Tamura, M. (eds.), *Astronomical Society of the Pacific Conference Series*. Astronomical Society of the Pacific Conference Series, vol. 534. San Francisco: Astronomical Society of the Pacific.

Krissansen-Totton, Joshua, Schwieterman, Edward W., Charnay, Benjamin, et al. 2016. Is the Pale Blue Dot Unique? Optimized Photometric Bands for Identifying Earth-Like Exoplanets. *Astrophys. J.*, **817**(1), 31.

Krissansen-Totton, Joshua, Olson, Stephanie, and Catling, David C. 2018. Disequilibrium Biosignatures over Earth History and Implications for Detecting Exoplanet Life. *Sci. Adv.*, **4**(1), eaao5747.

Krissansen-Totton, Joshua, Thompson, Maggie, Galloway, Max L., and Fortney, Jonathan J. 2022. Understanding Planetary Context to Enable Life Detection on Exoplanets and Test the Copernican Principle. *Nat. Astron.*, **6**(Feb.), 189–198.

Kuhn, Jeff R., and Berdyugina, Svetlana V. 2015. Global Warming as a Detectable Thermodynamic Marker of Earth-Like Extrasolar Civilizations: The Case for a Telescope like Colossus. *Int. J. Astrobiol.*, **14**(3), 401–410.

Kumar, Pawan, and Zhang, Bing. 2015. The Physics of Gamma-Ray Bursts & Relativistic Jets. *Phys. Rep.*, **561**, 1–109.

Kurokawa, Hiroyuki, Kurosawa, Kosuke, and Usui, Tomohiro. 2018. A Lower Limit of Atmospheric Pressure on Early Mars Inferred from Nitrogen and Argon Isotopic Compositions. *Icarus*, **299**, 443–459.

Laakso, T. A., and Schrag, D. P. 2017. A Theory of Atmospheric Oxygen. *Geobiology*, **15**(3), 366–384.

Lainey, V., Rambaux, N., Tobie, G., Cooper, N., Zhang, Q., Noyelles, B., and Baillie, K. (2024). A recently formed ocean inside Saturn's moon Mimas. *Nature*, **626**(7998), 280–282.

Lamers, Henny J. G. L. M., and Levesque, Emily M. 2017. *Understanding Stellar Evolution*. Bristol: IOP.

Lammer, Helmut, Zerkle, Aubrey L., Gebauer, Stefanie, et al. 2018. Origin and Evolution of the Atmospheres of Early Venus, Earth and Mars. *Astron. Astrophys. Rev.*, **26**(1), 2.

Lammer, Helmut, Brasser, Ramon, Johansen, Anders, Scherf, Manuel, and Leitzinger, Martin. 2021. Formation of Venus, Earth and Mars: Constrained by Isotopes. *Space Sci. Rev.*, **217**(1), 7.

Lancet, Doron, Segrè, Daniel, and Kahana, Amit. 2019. Twenty Years of 'Lipid World': A Fertile Partnership with David Deamer. *Life*, **9**(4), 77.

Landis, G. A. 1998. The Fermi Paradox: An Approach Based on Percolation Theory. *J. Br. Interplanet, Soc.*, **51**(5), 163–166.

Lane, Nick. 2002. *Oxygen: The Molecule that Made the World*. Oxford: Oxford University Press.

Lane, Nick. 2009. *Life Ascending: The Ten Great Inventions of Evolution*. London: Profile Books.

Lane, Nick, and Martin, William. 2010. The Energetics of Genome Complexity. *Nature*, **467**(7318), 929–934.

Lapôtre, Mathieu G. A., Bishop, Janice L., Ielpi, Alessandro, et al. 2022. Mars as a Time Machine to Precambrian Earth. *J. Geol. Soc.*, **179**(5), jgs2022.

Laskar, J., Joutel, F., and Robutel, P. 1993. Stabilization of the Earth's Obliquity by the Moon. *Nature*, **361**(6413), 615–617.

Latham, David W., Mazeh, Tsevi, Stefanik, Robert P., Mayor, Michel, and Burki, Gilbert. 1989. The Unseen Companion of HD114762: A Probable Brown Dwarf. *Nature*, **339**(6219), 38–40.

Laughlin, Gregory, and Adams, Fred C. 2000. The Frozen Earth: Binary Scattering Events and the Fate of the Solar System. *Icarus*, **145**(2), 614–627.

Lawton, John H., and May, Robert M. (eds). 1995. *Extinction Rates*. Oxford: Oxford University Press.

Lazcano, Antonio. 2010. Historical Development of Origins Research. *Cold Spring Harb. Perspect. Biol.*, **2**(11), a002089.

Lazcano, Antonio, and Bada, Jeffrey L. 2003. The 1953 Stanley L. Miller Experiment: Fifty Years of Prebiotic Organic Chemistry. *Orig. Life Evol. Biosph.*, **33**(3), 235–242.

Le Maistre, Sébastien, Rivoldini, Attilio, Caldiero, Alfonso, et al. 2023. Spin State and Deep

Interior Structure of Mars from InSight Radio Tracking. *Nature*, **619**, 733–737.

Lecar, M., Podolak, M., Sasselov, D., and Chiang, E. 2006. On the Location of the Snow Line in a Protoplanetary Disk. *Astrophys. J.*, **640**(2), 1115–1118.

Lecavelier des Etangs, A., and Lissauer, Jack J. 2022. The IAU Working Definition of an Exoplanet. *New Astron. Rev.*, **94**, 101641.

Lecavelier des Etangs, Alain, Cros, Lucie, Hébrard, Guillaume, et al. 2022. Exocomets Size Distribution in the β ? Pictoris Planetary System. *Sci. Rep.*, **12**(Apr.), 5855.

Lederberg, J. 1960. Exobiology: Approaches to Life beyond the Earth. *Science*, **132**(3424), 393–400.

Lehmer, Owen R., Catling, David C., Parenteau, Mary N., and Hoehler, Tori M. 2018. The Productivity of Oxygenic Photosynthesis around Cool, M Dwarf Stars. *Astrophys. J.*, **859**(2), 171.

Leman, Luke, Orgel, Leslie, and Ghadiri, M. Reza. 2004. Carbonyl Sulfide-Mediated Prebiotic Formation of Peptides. *Science*, **306**(5694), 283–286.

Lenton, Timothy M., Crouch, Michael, Johnson, Martin, Pires, Nuno, and Dolan, Liam. 2012. First Plants Cooled the Ordovician *Nat. Geosci.*, **5**(2), 86–89.

Lenton, Timothy M., Boyle, Richard A., Poulton, Simon W., Shields-Zhou, Graham A., and Butterfield, Nicholas J. 2014. Co-evolution of Eukaryotes and Ocean Oxygenation in the Neoproterozoic Era. *Nature Geoscience*, **7**(4), 257–265.

Lenton, Timothy M., Daines, Stuart J., Dyke, James G., et al. 2018. Selection for Gaia across Multiple Scales. *Trends Ecol. Evol.*, **33**(8), 633–645.

Levin, Gilbert V., and Straat, Patricia Ann. 2016. The Case for Extant Life on Mars and Its Possible Detection by the Viking Labeled Release Experiment. *Astrobiology*, **16**(10), 798–810.

Levy, Matthew, and Miller, Stanley L. 1998. The Stability of the RNA Bases: Implications for the Origin of Life. *Proc. Natl. Acad. Sci.*, **95**(14), 7933–7938.

Lewis, Geraint F., and Barnes, Luke A. 2016. *A Fortunate Universe: Life in a Finely Tuned Cosmos*. Cambridge: Cambridge University Press.

Lewis, John S. 2004. *Physics and Chemistry of the Solar System*. 2nd ed. Burlington: Academic Press.

Lewis, Simon L., and Maslin, Mark A. 2018. *The Human Planet: How We Created the Anthropocene*. New Haven: Yale University Press.

Li, Aigen. 2020. Spitzer's Perspective of Polycyclic Aromatic Hydrocarbons in Galaxies. *Nat. Astron.*, **4**, 339–351.

Lichtenberg, Tim, Schaefer, Laura K., Nakajima, Miki, and Fischer, Rebecca A. 2023. Geophysical Evolution during Rocky Planet Formation. Page 907 of: Inutsuka, S., Aikawa, Y., Muto, T., Tomida, K., and Tamura, M. (eds.), *Astronomical Society of the Pacific Conference Series*. Astronomical Society of the Pacific Conference Series, vol. 534. San Francisco: Astronomical Society of the Pacific.

Lichtenegger, H. I. M., Dyadechkin, S., Scherf, M., et al. 2022. Non-thermal Escape of the Martian CO_2 Atmosphere over Time: Constrained by Ar Isotopes. *Icarus*, **382**, 115009.

Limaye, Sanjay S., Mogul, Rakesh, Smith, David J., et al. 2018. Venus' Spectral Signatures and the Potential for Life in the Clouds. *Astrobiology*, **18**(9), 1181–1198.

Limaye, Sanjay S., Mogul, Rakesh, Baines, Kevin H., et al. 2021. Venus, an Astrobiology Target. *Astrobiology*, **21**(10), 1163–1185.

Lin, Henry W., Gonzalez Abad, Gonzalo, and Loeb, Abraham. 2014. Detecting Industrial Pollution in the Atmospheres of Earth-Like Exoplanets. *Astrophys. J. Lett.*, **792**(1), L7.

Lincowski, Andrew P., Meadows, Victoria S., Crisp, David, et al. 2018. Evolved Climates and Observational Discriminants for the TRAPPIST-1 Planetary System. *Astrophys. J.*, **867**(1), 76.

Lincowski, Andrew P., Meadows, Victoria S., Crisp, David, et al. 2021. Claimed Detection of PH_3 in the Clouds of Venus Is Consistent with Mesospheric SO_2. *Astrophys. J. Lett.*, **908**(2), L44.

Lineweaver, Charles H., and Egan, Chas A. 2008. Life, Gravity and the Second Law of Thermodynamics. *Phys. Life Rev.*, **5**(4), 225–242.

Lineweaver, Charles H., Fenner, Yeshe, and Gibson, Brad K. 2004. The Galactic Habitable Zone and the Age Distribution of Complex Life in the Milky Way. *Science*, **303**(5654), 59–62.

Lingam, Manasvi. 2021a. A Brief History of the Term 'Habitable Zone' in the 19th Century. *Int. J. Astrobiol.*, **20**(5), 332–336.

Lingam, Manasvi. 2021b. Theoretical Constraints Imposed by Gradient Detection and Dispersal on Microbial Size in Astrobiological Environments. *Astrobiology*, **21**(7), 813–830.

Lingam, Manasvi, and Loeb, Abraham. 2017a. Natural and Artificial Spectral Edges in Exoplanets. *Mon. Not. R. Astron. Soc.*, **470**(1), L82–L86.

Lingam, Manasvi, and Loeb, Abraham. 2017b. Risks for Life on Habitable Planets from Superflares of Their Host Stars. *Astrophys. J.*, **848**(1), 41.

Lingam, Manasvi, and Loeb, Abraham. 2018a. Implications of Tides for Life on Exoplanets. *Astrobiology*, **18**(7), 967–982.

Lingam, Manasvi, and Loeb, Abraham. 2018b. Is Extraterrestrial Life Suppressed on Subsurface Ocean Worlds due to the Paucity of Bioessential Elements? *Astron. J.*, **156**(4), 151.

Lingam, Manasvi, and Loeb, Abraham. 2019a. Physical Constraints for the Evolution of Life on Exoplanets. *Rev. Mod. Phys.*, **91**(2), 021002.

Lingam, Manasvi, and Loeb, Abraham. 2019b. Photosynthesis on Habitable Planets around Low-Mass Stars. *Mon. Not. R. Astron. Soc.*, **485**(4), 5924–5928.

Lingam, Manasvi, and Loeb, Abraham. 2019c. Subsurface Exolife. *Int. J. Astrobiol.*, **18**(2), 112–141.

Lingam, Manasvi, and Loeb, Abraham. 2020a. Potential for Liquid Water Biochemistry Deep under the Surfaces of the Moon, Mars, and beyond. *Astrophys. J. Lett.*, **901**(1), L11.

Lingam, Manasvi, and Loeb, Abraham. 2020b. What's in a Name: The Etymology of Astrobiology. *Int. J. Astrobiol.*, **19**(5), 379–385.

Lingam, Manasvi, and Loeb, Abraham. 2021. *Life in the Cosmos: From Biosignatures to Technosignatures*. Cambridge, MA: Harvard University Press.

Lingam, Manasvi, Dong, Chuanfei, Fang, Xiaohua, Jakosky, Bruce M., and Loeb, Abraham. 2018. The Propitious Role of Solar Energetic Particles in the Origin of Life. *Astrophys. J.*, **853**(1), 10.

Lingam, Manasvi, Ginsburg, Idan, and Bialy, Shmuel. 2019. Active Galactic Nuclei: Boon or Bane for Biota? *Astrophys. J.*, **877**(1), 62.

Lingam, Manasvi, Balbi, Amedeo, and Mahajan, Swadesh M. 2021. Excitation Properties of Photopigments and Their Possible Dependence on the Host Star. *Astrophys. J. Lett.*, **921**(2), L41.

Lingam, Manasvi, Balbi, Amedeo, and Mahajan, Swadesh M. 2023a. A Bayesian Analysis of Technological Intelligence in Land and Oceans. *Astrophys. J.*, **945**(1), 23.

Lingam, Manasvi, Hein, Andreas M., and Eubanks, T. Marshall. 2023b. Chasing Nomadic Worlds: A New Class of Deep Space Missions. *Acta Astronaut.*, **212**, 517–533.

Lingam, Manasvi, Haqq-Misra, Jacob, Wright, Jason T., et al. 2023c. Technosignatures: Frameworks for Their Assessment. *Astrophys. J.*, **943**(1), 27.

Linnartz, Harold, Ioppolo, Sergio, and Fedoseev, Gleb. 2015. Atom Addition Reactions in Interstellar Ice Analogues. *Int. Rev. Phys. Chem.*, **34**(2), 205–237.

Linsky, Jeffrey. 2019. *Host Stars and Their Effects on Exoplanet Atmospheres*. Lecture Notes in Physics, vol. 955. Cham: Springer Nature.

Lis, Dariusz C., Bockelée-Morvan, Dominique, Güsten, Rolf, et al. 2019. Terrestrial Deuterium-to-Hydrogen Ratio in Water in Hyperactive Comets. *Astron. Astrophys.*, **625**, L5.

Lissauer, Jack J., Barnes, Jason W., and Chambers, John E. 2012. Obliquity Variations of a Moonless Earth. *Icarus*, **217**(1), 77–87.

Little, John D. C. 1961. A Proof for the Queuing Formula: $L = \lambda W$. *Oper. Res.*, **9**(3), 383–387.

Liu, Yang, Makarova, Kira S., Huang, Wen-Cong, et al. 2021. Expanded Diversity of Asgard Archaea and Their Relationships with Eukaryotes. *Nature*, **593**(7860), 553–557.

Loeb, Abraham, and Turner, Edwin L. 2012. Detection Technique for Artificially Illuminated Objects in the Outer Solar System and beyond. *Astrobiology*, **12**(4), 290–294.

Loeb, Avi. 2022. On the Possibility of an Artificial Origin for 'Oumuamua'. *Astrobiology*, **22**(12), 1392–1399.

Long, K. F., Obousy, R. K., and Hein, A. 2011. Project Icarus: Optimisation of Nuclear Fusion Propulsion for Interstellar Missions. *Acta Astronaut.*, **68**, 1820–1829.

Lovelock, J. E. 1965. A Physical Basis for Life Detection Experiments. *Nature*, **207**(4997), 568–570.

Lovelock, James E., and Margulis, Lynn. 1974. Atmospheric Homeostasis By and for the Biosphere: the Gaia Hypothesis. *Tellus*, **26**(1–2), 2–10.

Lubin, P. 2016. A Roadmap to Interstellar Flight. *J. Br. Interplanet. Soc.*, **69**, 40–72.

Luisi, Pier Luigi. 2016. *The Emergence of Life: From Chemical Origins to Synthetic Biology.* 2nd ed. Cambridge: Cambridge University Press.

Lv, Kong-Peng, Norman, Lucy, and Li, Yi-Liang. 2017. Oxygen-Free Biochemistry: The Putative CHN Foundation for Exotic Life in a Hydrocarbon World? *Astrobiology*, **17**(11), 1173–1181.

Lynden-Bell, D., and Pringle, J. E. 1974. The Evolution of Viscous Discs and the Origin of the Nebular Variables. *Mon. Not. R. Astron. Soc.*, **168**, 603–637.

Lyons, Timothy W., Reinhard, Christopher T., and Planavsky, Noah J. 2014. The Rise of Oxygen in Earth's Early Ocean and Atmosphere. *Nature*, **506**(7488), 307–315.

Lyons, Timothy W., Diamond, Charles W., Planavsky, Noah J., Reinhard, Christopher T., and Li, Chao. 2021. Oxygenation, Life, and the Planetary System during Earth's Middle History: An Overview. *Astrobiology*, **21**(8), 906–923.

MacIver, Malcolm A., and Finlay, Barbara L. 2022. The Neuroecology of the Water-to-Land Transition and the Evolution of the Vertebrate Brain. *Phil. Trans. R. Soc. B*, **377**(1844), 20200523.

MacKenzie, Shannon M., Neveu, Marc, Davila, Alfonso F., et al. 2021a. The Enceladus Orbilander Mission Concept: Balancing Return and Resources in the Search for Life. *Planet. Sci. J.*, **2**(2), 77.

MacKenzie, Shannon M., Birch, Samuel P. D., Hörst, Sarah, et al. 2021b. Titan: Earth-like on the outside, Ocean World on the inside. *Planet. Sci. J.*, **2**(3), 112.

Madhusudhan, Nikku, Piette, Anjali A. A., and Constantinou, Savvas. 2021. Habitability and Biosignatures of Hycean Worlds. *Astrophys. J.*, **918**(1), 1.

Madhusudhan, Nikku, Sarkar, Subhajit, Constantinou, Savvas, et al. 2023. Carbon-Bearing Molecules in a Possible Hycean Atmosphere. *Astrophys. J. Lett.*, **956**(1), L13.

Magnabosco, C., Lin, L. H., Dong, H., et al. 2018. The Biomass and Biodiversity of the Continental Subsurface. *Nat. Geosci.*, **11**(10), 707–717.

Maire, Jérôme, Wright, Shelley A., Barrett, Colin T., et al. 2019. Search for Nanosecond Near-infrared Transients around 1280 Celestial Objects. *Astron. J.*, **158**(5), 203.

Maire, Jérôme, Wright, Shelley A., Werthimer, Dan, et al. 2020. Panoramic SETI: On-Sky Results from Prototype Telescopes and Instrumental Design. Page 114543C of: Holland, Andrew D., and Beletic, James (eds.), *X-Ray, Optical, and Infrared Detectors for Astronomy IX*. Society of Photo-Optical Instrumentation Engineers (SPIE) Conference Series, vol. 11454. San Diego, CA: Society of Photo-Optical Instrumentation Engineers.

Maity, Surajit, Kaiser, Ralf I., and Jones, Brant M. 2015. Formation of Complex Organic Molecules in Methanol and Methanol-Carbon Monoxide Ices Exposed to Ionizing Radiation - a Combined FTIR and Reectron Time-of-Flight Mass Spectrometry Study. *Phys. Chem. Chem. Phys.*, **17**(5), 3081–3114.

Malin, Michael C., and Edgett, Kenneth S. 1999. Oceans or Seas in the Martian Northern Lowlands: High Resolution Imaging Tests of Proposed Coastlines. *Geophys. Res. Lett.*, **26**(19), 3049–3052.

Malin, Michael C., and Edgett, Kenneth S. 2000. Evidence for Recent Groundwater Seepage and Surface Runoff on Mars. *Science*, **288**(5475), 2330–2335.

Mallama, Anthony, Krobusek, Bruce, and Pavlov, Hristo. 2017. Comprehensive Wide-Band Magnitudes and Albedos for the Planets, with Applications to Exoplanets and Planet Nine. *Icarus*, **282**(Jan.), 19–33.

Manabe, Syukuro, and Wetherald, Richard T. 1967. Thermal Equilibrium of the Atmosphere with a

Given Distribution of Relative Humidity. *J. Atmos. Sci.*, **24**(3), 241–259.

Mancinelli, Rocco L., and Klovstad, Melisa. 2000. Martian Soil and UV Radiation: Microbial Viability Assessment on Spacecraft Surfaces. *Planet. Space Sci.*, **48**(11), 1093–1097.

Mand, Thomas D., and Metcalf, William W. 2019. Energy Conservation and Hydrogenase Function in Methanogenic Archaea, in Particular the Genus *Methanosarcina*. *Microbiol. Mol. Biol. Rev.*, **83**(4), e00020–19.

Marchi, S., Bottke, W. F., Elkins-Tanton, L. T., et al. 2014. Widespread Mixing and Burial of Earth's Hadean Crust by Asteroid Impacts. *Nature*, **511**(7511), 578–582.

Marconi, A., Risaliti, G., Gilli, R., et al. 2004. Local Supermassive Black Holes, Relics of Active Galactic Nuclei and the X-ray Background. *Mon. Not. R. Astron. Soc.*, **351**(1), 169–185.

Marks, Joshua H., Wang, Jia, Fortenberry, Ryan C., and Kaiser, Ralf I. 2022. Preparation of Methanediamine ($CH_2(NH_2)_2$) – A Precursor to Nucleobases in the Interstellar Medium. *Proc. Natl. Acad. Sci.*, **119**(51), e2217329119.

Marshall, Charles R. 2006. Explaining the Cambrian 'Explosion' of Animals. *Annu. Rev. Earth Planet. Sci.*, **34**, 355–384.

Marshall, Stuart M., Mathis, Cole, Carrick, Emma, et al. 2021. Identifying Molecules as Biosignatures with Assembly Theory and Mass Spectrometry. *Nat. Commun.*, **12**, 3033.

Martel, Jan, Young, David, Peng, Hsin-Hsin, Wu, Cheng-Yeu, and Young, John D. 2012. Biomimetic Properties of Minerals and the Search for Life in the Martian Meteorite ALH84001. *Annu. Rev. Earth Planet. Sci.*, **40**(1), 167–193.

Martin, William, Baross, John, Kelley, Deborah, and Russell, Michael J. 2008. Hydrothermal Vents and the Origin of Life. *Nat. Rev. Microbiol.*, **6**, 805–814.

Martin, William F., Garg, Sriram, and Zimorski, Verena. 2015. Endosymbiotic Theories for Eukaryote Origin. *Phil. Trans. R. Soc. B*, **370**(1678), 20140330.

Martínez, G. M., Newman, C. N., De Vicente-Retortillo, A., et al. 2017. The Modern Near-Surface Martian Climate: A Review of In Situ Meteorological Data from Viking to Curiosity. *Space Sci. Rev.*, **212**(1–2), 295–338.

Marty, Bernard, Avice, Guillaume, Sano, Yuji, et al. 2016. Origins of Volatile Elements (H, C, N, Noble Gases) on Earth and Mars in Light of Recent Results from the ROSETTA Cometary Mission. *Earth and Planetary Science Letters*, **441**, 91–102.

Matsumura, Soko, Ida, Shigeru, and Nagasawa, Makiko. 2013. Effects of Dynamical Evolution of Giant Planets on Survival of Terrestrial Planets. *Astrophys. J.*, **767**(2), 129.

Maynard-Casely, Helen E., Cable, Morgan L., Malaska, Michael J., et al. 2018. Prospects for Mineralogy on Titan. *Am. Mineral.*, **103**(3), 343–349.

Maynard Smith, John, and Szathmáry, Eörs. 1995. *The Major Transitions in Evolution*. Oxford: Oxford University Press.

Mayor, Michel, and Queloz, Didier. 1995. A Jupiter-Mass Companion to a Solar-Type Star. *Nature*, **378**(6555), 355–359.

Mayr, Ernst. 2004. *What Makes Biology Unique?: Considerations on the Autonomy of a Scientific Discipline*. Cambridge: Cambridge University Press.

McCollom, Thomas M. 2013. Miller-Urey and beyond: What Have We Learned about Prebiotic Organic Synthesis Reactions in the Past 60 Years? *Annu. Rev. Earth Planet. Sci.*, **41**, 207–229.

McCollom, Thomas M., and Seewald, Jeffrey S. 2007. Abiotic Synthesis of Organic Compounds in Deep-Sea Hydrothermal Environments. *Chem. Rev.*, **107**(2), 382–401.

McCollom, Thomas M., Klein, Frieder, and Ramba, Mitchell. 2022. Hydrogen Generation from Serpentinization of Iron-Rich Olivine on Mars, Icy Moons, and Other Planetary Bodies. *Icarus*, **372**, 114754.

McCord, Thomas B., Combe, Jean-Philippe, Castillo-Rogez, Julie C., McSween, Harry Y., and Prettyman, Thomas H. 2022. Ceres, a Wet Planet: The View after Dawn. *Geochemistry*, **82**(2), 125745.

McGuire, Brett A. 2022. 2021 Census of Interstellar, Circumstellar, Extragalactic, Protoplanetary Disk, and Exoplanetary Molecules. *Astrophys. J., Suppl. Ser.*, **259**(2), 30.

McKay, C. P., and Smith, H. D. 2005. Possibilities for Methanogenic Life in Liquid Methane on the Surface of Titan. *Icarus*, **178**(1), 274–276.

McKay, Chris P. 2004. What Is Life – and How Do We Search for It in Other Worlds? *PLoS Biol.*, **2**(9), e302.

McKay, Christopher P. 2014. Requirements and Limits for Life in the Context of Exoplanets. *Proc. Natl. Acad. Sci.*, **111**(35), 12628–12633.

McKay, Christopher P. 2016. Titan as the Abode of Life. *Life*, **6**(1), 8.

McKay, Christopher P. 2018. The Search on Mars for a Second Genesis of Life in the Solar System and the Need for Biologically Reversible Exploration. *Biol. Theory*, **13**(2), 103–110.

McKay, Christopher P., and Marinova, Margarita M. 2001. The Physics, Biology, and Environmental Ethics of Making Mars Habitable. *Astrobiology*, **1**(1), 89–109.

McKay, Christopher P., Toon, Owen B., and Kasting, James F. 1991. Making Mars Habitable. *Nature*, **352**(Aug.), 489–496.

McKay, David S., Gibson, Everett K., Jr., Thomas-Keprta, Kathie L., et al. 1996. Search for Past Life on Mars: Possible Relic Biogenic Activity in Martian Meteorite ALH84001. *Science*, **273**(5277), 924–930.

McKee, Christopher F., and Ostriker, Eve C. 2007. Theory of Star Formation. *Annu. Rev. Astron. Astrophys.*, **45**(1), 565–687.

McLendon, Christopher, Opalko, F. Jeffrey, Illangkoon, Heshan I., and Benner, Steven A. 2015. Solubility of Polyethers in Hydrocarbons at Low Temperatures. A Model for Potential Genetic Backbones on Warm Titans. *Astrobiology*, **15**(3), 200–206.

McLennan, Scott M., Grotzinger, John P., Hurowitz, Joel A., and Tosca, Nicholas J. 2019. The Sedimentary Cycle on Early Mars. *Annu. Rev. Earth Planet. Sci.*, **47**(May), 91–118.

McMahon, Sean, and Cosmidis, Julie. 2022. False Biosignatures on Mars: Anticipating Ambiguity. *J. Geol. Soc.*, **179**(2), jgs2021–050.

Meadows, Victoria S., Reinhard, Christopher T., Arney, Giada N., et al. 2018. Exoplanet Biosignatures: Understanding Oxygen as a Biosignature in the Context of Its Environment. *Astrobiology*, **18**(6), 630–662.

Meadows, Victoria S., Arney, Giada N., Schmidt, Britney E., and Des Marais, David J. (eds). 2020. *Planetary Astrobiology*. Space Science Series. Tucson: University of Arizona Press.

Meech, Karen J., Weryk, Robert, Micheli, Marco, et al. 2017. A Brief Visit from a Red and Extremely Elongated Interstellar Asteroid. *Nature*, **552**(7685), 378–381.

Melosh, H. J., and Vickery, A. M. 1989. Impact Erosion of the Primordial Atmosphere of Mars. *Nature*, **338**(6215), 487–489.

Melosh, H. Jay. 2011. *Planetary Surface Processes*. Cambridge Planetary Science. Cambridge: Cambridge University Press.

Melott, A. L., Lieberman, B. S., Laird, C. M., et al. 2004. Did a Gamma-Ray Burst Initiate the Late Ordovician Mass Extinction? *Int. J. Astrobiol.*, **3**(1), 55–61.

Melott, Adrian L., and Thomas, Brian C. 2011. Astrophysical Ionizing Radiation and Earth: A Brief Review and Census of Intermittent Intense Sources. *Astrobiology*, **11**(4), 343–361.

Méndez, Abel, and Rivera-Valentín, Edgard G. 2017. The Equilibrium Temperature of Planets in Elliptical Orbits. *Astrophys. J. Lett.*, **837**(1), L1.

Merino, Nancy, Aronson, Heidi S., Bojanova, Diana P., et al. 2019. Living at the Extremes: Extremophiles and the Limits of Life in a Planetary Context. *Front. Microbiol.*, **10**, 780.

Mettler, Jean-Noël, Quanz, Sascha P., Helled, Ravit, Olson, Stephanie L., and Schwieterman, Edward W. 2023. Earth as an Exoplanet. II. Earth's Time-Variable Thermal Emission and Its Atmospheric Seasonality of Bioindicators. *Astrophys. J.*, **946**(2), 82.

Meyer-Vernet, Nicole, and Rospars, Jean-Pierre. 2016. Maximum Relative Speeds of Living Organisms: Why Do Bacteria Perform as Fast as Ostriches? *Phys. Biol.*, **13**(6), 066006.

Michael, Greg, Basilevsky, Alexander, and Neukum, Gerhard. 2018. On the History of the Early Meteoritic Bombardment of the Moon: Was There a Terminal Lunar Cataclysm? *Icarus*, **302**, 80–103.

Michalski, Joseph R., Onstott, Tullis C., Mojzsis, Stephen J., et al. 2018. The Martian Subsurface as a Potential Window Into the Origin of Life. *Nat. Geosci.*, **11**(1), 21–26.

Michaud, Eric, Siemion, Andrew P. V., Drew, Jamie, and Worden, S. Pete. 2021. Lunar Opportunities for SETI. *Bull. Am. Astron. Soc.*, **53**(May), 369.

Mieli, E., Valli, A. M. F., and Maccone, C. 2023. Astrobiology: Resolution of the Statistical Drake Equation by Maccone's Lognormal Method in 50 Steps. *Int. J. Astrobiol.*, **22**(4), 428–537.

Mileikowsky, Curt, Cucinotta, Francis A., Wilson, John W., et al. 2000. Natural Transfer of Viable Microbes in Space. 1. From Mars to Earth and Earth to Mars. *Icarus*, **145**(2), 391–427.

Millar, T. J. 2015. Astrochemistry. *Plasma Sources Sci. Technol.*, **24**(4), 043001.

Miller, Stanley L. 1953. A Production of Amino Acids under Possible Primitive Earth Conditions. *Science*, **117**(3046), 528–529.

Miller, Stanley L., and Urey, Harold C. 1959. Organic Compound Synthesis on the Primitive Earth. *Science*, **130**(3370), 245–251.

Mills, Benjamin J. W., Krause, Alexander J., Jarvis, Ian, and Cramer, Bradley D. 2023. Evolution of Atmospheric O_2 through the Phanerozoic, Revisited. *Annu. Rev. Earth Planet. Sci.*, **51**, 253–276.

Mills, Daniel B., Ward, Lewis M., Jones, CarriAyne, et al. 2014. Oxygen Requirements of the Earliest Animals. *Proc. Natl. Acad. Sci.*, **111**(11), 4168–4172.

Mills, Daniel B., Boyle, Richard A., Daines, Stuart J., et al. 2022. Eukaryogenesis and Oxygen in Earth History. *Nat. Ecol. Evol.*, **6**(5), 520–532.

Milo, Ron. 2013. What Is the Total Number of Protein Molecules Per Cell Volume? A Call to Rethink Some Published Values. *Bioessays*, **35**(12), 1050–1055.

Milo, Ron, and Phillips, Rob. 2016. *Cell Biology by the Numbers*. New York: Garland Science.

Mitchell, Peter. 1961. Coupling of Phosphorylation to Electron and Hydrogen Transfer by a Chemi-Osmotic Type of Mechanism. *Nature*, **191**(4784), 144–148.

Mittelholz, Anna, and Johnson, Catherine L. 2022. The Martian Crustal Magnetic Field. *Front. Astron. Space Sci.*, **9**(May), 895362.

Miyakawa, Shin, Cleaves, H. James, and Miller, Stanley L. 2002. The Cold Origin of Life: B. Implications Based on Pyrimidines and Purines Produced from Frozen Ammonium Cyanide Solutions. *Orig. Life Evol. Biosph.*, **32**(3), 209–218.

Miyazaki, Yoshinori, and Korenaga, Jun. 2022. A Wet Heterogeneous Mantle Creates a Habitable World in the Hadean. *Nature*, **603**(7899), 86–90.

Mogul, Rakesh, Limaye, Sanjay S., Lee, Yeon Joo, and Pasillas, Michael. 2021. Potential for Phototrophy in Venus' Clouds. *Astrobiology*, **21**(10), 1237–1249.

Mojzsis, S. J., Arrhenius, G., McKeegan, K. D., et al. 1996. Evidence for Life on Earth before 3,800 Million Years Ago. *Nature*, **384**(6604), 55–59.

Mojzsis, Stephen J. 2021. Habitable Potentials. *Nat. Astron.*, **5**, 1083–1085.

Monnard, Pierre-Alain, Kanavarioti, Anastassia, and Deamer, David W. 2003. Eutectic Phase Polymerization of Activated Ribonucleotide Mixtures Yields Quasi-Equimolar Incorporation of Purine and Pyrimidine Nucleobases. *J. Am. Chem. Soc.*, **125**(45), 13734–13740.

Moran, Uri, Phillips, Rob, and Milo, Ron. 2010. SnapShot: Key Numbers in Biology. *Cell*, **141**(7), 1262–1262.e1.

Morbidelli, A., Nesvorny, D., Laurenz, V., et al. 2018. The Timeline of the Lunar Bombardment: Revisited. *Icarus*, **305**(May), 262–276.

Mordasini, Christoph. 2018. Planetary Population Synthesis. Pages 2425–2474 of: Deeg, Hans J., and Belmonte, Juan Antonio (eds.), *Handbook of Exoplanets*. Cham: Springer.

Morgan, Joanna V., Bralower, Timothy J., Brugger, Julia, and Wünnemann, Kai. 2022. The Chicxulub Impact and Its Environmental Consequences. *Nat. Rev. Earth Environ.*, **3**(5), 338–354.

Morowitz, H., and Sagan, C. 1967. Life in the Clouds of Venus? *Nature*, **215**(5107), 1259–1260.

Morowitz, Harold J. 1967. Biological Self-replicating Systems. *Prog. Theoret. Biol.*, **1**, 35–58.

Morrison, Ian S., and Gowanlock, Michael G. 2015. Extending Galactic Habitable Zone Modeling to Include the Emergence of Intelligent Life. *Astrobiology*, **15**(8), 683–696.

Muñoz Caro, G. M., Meierhenrich, U. J., Schutte, W. A., et al. 2002. Amino Acids from Ultraviolet Irradiation of Interstellar Ice Analogues. *Nature*, **416**(6879), 403–406.

Muchowska, Kamila B., Varma, Sreejith J., and Moran, Joseph. 2020. Nonenzymatic Metabolic Reactions and Life's Origins. *Chem. Rev.*, **120**(15), 7708–7744.

Mulders, Gijs D., Ciesla, Fred J., Min, Michiel, and Pascucci, Ilaria. 2015. The Snow Line in Viscous Disks around Low-Mass Stars: Implications for Water Delivery to Terrestrial Planets in the Habitable Zone. *Astrophys. J.*, **807**(1), 9.

Mulkidjanian, Armen Y., Bychkov, Andrew Yu., Dibrova, Daria V., Galperin, Michael Y., and Koonin, Eugene V. 2012. Origin of First Cells at Terrestrial, Anoxic Geothermal Fields. *Proc. Natl. Acad. Sci.*, **109**(14), E821–E830.

Müller, Felix, Escobar, Luis, Xu, Felix, et al. 2022. A Prebiotically Plausible Scenario of an RNA-Peptide World. *Nature*, **605**(7909), 279–284.

Müller, Tobias W. A., and Haghighipour, Nader. 2014. Calculating the Habitable Zones of Multiple Star Systems with a New Interactive Web Site. *Astrophys. J.*, **782**(1), 26.

Müller, Volker, and Hess, Verena. 2017. The Minimum Biological Energy Quantum. *Front. Microbiol.*, **8**, 2019.

Musk, Elon. 2017. Making Humans a Multi-Planetary Species. *New Space*, **5**(2), 46–61.

Mutschler, Hannes, Wochner, Aniela, and Holliger, Philipp. 2015. Freeze-Thaw Cycles as Drivers of Complex Ribozyme Assembly. *Nat. Chem.*, **7**(6), 502–508.

Nadeau, Jay, Lindensmith, Chris, Deming, Jody W., Fernandez, Vicente I., and Stocker, Roman. 2016. Microbial Morphology and Motility as Biosignatures for Outer Planet Missions. *Astrobiology*, **16**(10), 755–774.

Nam, Inho, Nam, Hong Gil, and Zare, Richard N. 2018. Abiotic Synthesis of Purine and Pyrimidine Ribonucleosides in Aqueous Microdroplets. *Proc. Natl. Acad. Sci.*, **115**(1), 36–40.

Nazé, Yaël. 2009. *L'astronomie des Anciens*. Paris: Belin.

Neish, Catherine D., Somogyi, Árpád, and Smith, Mark A. 2010. Titan's Primordial Soup: Formation of Amino Acids via Low-Temperature Hydrolysis of Tholins. *Astrobiology*, **10**(3), 337–347.

Nelson, David L., and Cox, Michael M. 2004. *Lehninger Principles of Biochemistry*. 4th ed. New York: W. H. Freeman.

Nesvorný, David. 2018. Dynamical Evolution of the Early Solar System. *Annu. Rev. Astron. Astrophys.*, **56**, 137–174.

Neveu, Marc, Hays, Lindsay E., Voytek, Mary A., New, Michael H., and Schulte, Mitchell D. 2018. The Ladder of Life Detection. *Astrobiology*, **18**(11), 1375–1402.

Neveu, Marc, Anbar, Ariel D., Davila, Alfonso F., et al. 2020. Returning Samples from Enceladus for Life Detection. *Front. Astron. Space Sci.*, **7**, 26.

Nicholls, David G., and Ferguson, Stuart J. 2002. *Bioenergetics 3*. 3rd ed. London: Academic Press.

Nixon, C. A., Lorenz, R. D., Achterberg, R. K., et al. 2018. Titan's Cold Case Files - Outstanding Questions after Cassini–Huygens. *Planet. Space Sci.*, **155**(June), 50–72.

Norman, Lucy H. 2011. Is There Life on … Titan? *Astron. Geophys.*, **52**(1), 1.39–1.42.

Nuevo, Michel, Cooper, George, and Sandford, Scott A. 2018. Deoxyribose and Deoxysugar Derivatives from Photoprocessed Astrophysical Ice Analogues and Comparison to Meteorites. *Nat. Commun.*, **9**(Dec.), 5276.

Nursall, J. R. 1959. Oxygen as a Prerequisite to the Origin of the Metazoa. *Nature*, **183**(4669), 1170–1172.

Nutman, Allen P., Bennett, Vickie C., Friend, Clark R. L., van Kranendonk, Martin J., and Chivas, Allan R. 2016. Rapid Emergence of Life Shown by Discovery of 3,700-Million-Year-Old Microbial Structures. *Nature*, **537**(7621), 535–538.

Oba, Yasuhiro, Takano, Yoshinori, Naraoka, Hiroshi, Watanabe, Naoki, and Kouchi, Akira. 2019. Nucleobase Synthesis in Interstellar Ices. *Nat. Commun.*, **10**(Sept.), 4413.

Öberg, Karin I. 2016. Photochemistry and Astrochemistry: Photochemical Pathways to Interstellar Complex Organic Molecules. *Chem. Rev.*, **116**(17), 9631–9663.

Öberg, Karin I., and Bergin, Edwin A. 2021. Astrochemistry and Compositions of Planetary Systems. *Phys. Rep.*, **893**, 1–48.

Öberg, Karin I., Boogert, A. C. Adwin, Pontoppidan, et al. 2011. The Spitzer Ice

Legacy: Ice Evolution from Cores to Protostars. *Astrophys. J.*, **740**(2), 109.

O'Brien, David P., Izidoro, Andre, Jacobson, Seth A., Raymond, Sean N., and Rubie, David C. 2018. The Delivery of Water during Terrestrial Planet Formation. *Space Sci. Rev.*, **214**(1), 47.

Och, Lawrence M., and Shields-Zhou, Graham A. 2012. The Neoproterozoic Oxygenation Event: Environmental Perturbations and Biogeochemical Cycling. *Earth Sci. Rev.*, **110**(1), 26–57.

Odling-Smee, John, Erwin, Douglas H., Palkovacs, Eric P., Feldman, Marcus W., and Laland, Kevin N. 2013. Niche Construction Theory: A Practical Guide for Ecologists. *Q. Rev. Biol.*, **88**(1), 3–28.

O'Malley-James, Jack T., and Kaltenegger, Lisa. 2019a. Biofluorescent Worlds - II. Biological Fluorescence Induced by Stellar UV Flares, a New Temporal Biosignature. *Mon. Not. R. Astron. Soc.*, **488**(4), 4530–4545.

O'Malley-James, Jack T., and Kaltenegger, Lisa. 2019b. Expanding the Timeline for Earth's Photosynthetic Red Edge Biosignature. *Astrophys. J. Lett.*, **879**(2), L20.

O'Malley-James, Jack T., Greaves, Jane S., Raven, John A., and Cockell, Charles S. 2013. Swansong Biospheres: Refuges for Life and Novel Microbial Biospheres on Terrestrial Planets near the End of Their Habitable Lifetimes. *Int. J. Astrobiol.*, **12**(2), 99–112.

Omran, Arthur, Menor-Salvan, Cesar, Springsteen, Greg, and Pasek, Matthew. 2020. The Messy Alkaline Formose Reaction and Its Link to Metabolism. *Life*, **10**(8), 125.

O'Neill, C., Marchi, S., Zhang, S., and Bottke, W. 2017. Impact-Driven Subduction on the Hadean Earth. *Nat. Geosci.*, **10**(10), 793–797.

O'Neill, G. K. 1977. *The High Frontier: Human Colonies in Space*. New York: William Morrow.

Onofri, Silvano, Selbmann, Laura, Pacelli, Claudia, et al. 2018. Integrity of the DNA and Cellular Ultrastructure of Cryptoendolithic Fungi in Space or Mars Conditions: A 1.5-Year Study at the International Space Station. *Life*, **8**(2), 23.

Ord, Toby. 2020. *The Precipice: Existential Risk and the Future of Humanity*. New York: Hachette Books.

Orgel, Leslie E. 2004. Prebiotic Chemistry and the Origin of the RNA World. *Crit. Rev. Biochem. Mol. Biol.*, **39**(2), 99–123.

Ormel, Chris W. 2017. The Emerging Paradigm of Pebble Accretion. Pages 197–228 of: Pessah, Martin, and Gressel, Oliver (eds.), *Formation, Evolution, and Dynamics of Young Solar Systems*. Astrophysics and Space Science Library, vol. 445. Cham: Springer.

Oró, J. 1960. Synthesis of Adenine from Ammonium Cyanide. *Biochem. Biophys. Res. Commun.*, **2**(6), 407–412.

Orosei, R., Lauro, S. E., Pettinelli, E., et al. 2018. Radar Evidence of Subglacial Liquid Water on Mars. *Science*, **361**(6401), 490–493.

Osborn, Ares, and Bayliss, Daniel. 2020. Investigating the Planet-Metallicity Correlation for Hot Jupiters. *Mon. Not. R. Astron. Soc.*, **491**(3), 4481–4487.

Osinski, G. R., Cockell, C. S., Pontefract, A., and Sapers, H. M. 2020. The Role of Meteorite Impacts in the Origin of Life. *Astrobiology*, **20**(9), 1121–1149.

'Oumuamua ISSI Team. 2019. The Natural History of 'Oumuamua. *Nat. Astron.*, **3**(July), 594–602.

Ozaki, Kazumi, and Reinhard, Christopher T. 2021. The Future Lifespan of Earth's Oxygenated Atmosphere. *Nat. Geosci.*, **14**(3), 138–142.

Pace, Norman R. 2001. The Universal Nature of Biochemistry. *Proc. Natl. Acad. Sci.*, **98**(3), 805–808.

Pacetti, E., Balbi, A., Lingam, M., Tombesi, F., and Perlman, E. 2020. The Impact of Tidal Disruption Events on Galactic Habitability. *Mon. Not. R. Astron. Soc.*, **498**(3), 3153–3157.

Paganini, L., Villanueva, G. L., Roth, L., et al. 2020. A Measurement of Water Vapour amid a Largely Quiescent Environment on Europa. *Nat. Astron.*, **4**, 266–272.

Pahlevan, Kaveh, Schaefer, Laura, and Hirschmann, Marc M. 2019. Hydrogen Isotopic Evidence for Early Oxidation of Silicate Earth. *Earth Planet. Sci. Lett.*, **526**(Nov.), 115770.

Palin, Richard M., Santosh, M., Cao, Wentao, et al. 2020. Secular Change and the onset of Plate Tectonics on Earth. *Earth Sci. Rev.*, **207**, 103172.

Papagiannis, Michael D. 1978. Are We All Alone, or Could They Be in the Asteroid Belt ? *Q. J. R. Astron. Soc.*, **19**, 277.

Paprotny, Z. 1977. Nonradio Methods of SETI. *Postepy Astronautyki*, **10**(3), 39–67.

Paradise, Adiv, Obertas, Alysa, O'Grady, Anna, and Young, Matthew. 2019. The Long Night: Modeling the Climate of Westeros. *arXiv e-prints*, Mar., arXiv:1903.12195.

Patel, Bhavesh H., Percivalle, Claudia, Ritson, Dougal J., Duffy, Colm D., and Sutherland, John D. 2015. Common Origins of RNA, Protein and Lipid Precursors in a Cyanosulfidic Protometabolism. *Nat. Chem.*, **7**(4), 301–307.

Patel, Zarana S., Brunstetter, Tyson J., Tarver, William J., et al. 2020. Red Risks for a Journey to the Red Planet: The Highest Priority Human Health Risks for a Mission to Mars. *npj Microgravity*, **6**, 33.

Pearce, Ben K. D., Molaverdikhani, Karan, Pudritz, Ralph E., Henning, Thomas, and Cerrillo, Kaitlin E. 2022. Toward RNA Life on Early Earth: From Atmospheric HCN to Biomolecule Production in Warm Little Ponds. *Astrophys. J.*, **932**(1), 9.

Penteado, E. M., Walsh, C., and Cuppen, H. M. 2017. Sensitivity Analysis of Grain Surface Chemistry to Binding Energies of Ice Species. *Astrophys. J.*, **844**(1), 71.

Perryman, Michael. 2018. *The Exoplanet Handbook*. 2nd ed. Cambridge: Cambridge University Press.

Petigura, Erik A., Marcy, Geoffrey W., Winn, Joshua N., et al. 2018. The California-Kepler Survey. IV. Metal-rich Stars Host a Greater Diversity of Planets. *Astron. J.*, **155**(2), 89.

Petkowski, Janusz Jurand, Bains, William, and Seager, Sara. 2020. On the Potential of Silicon as a Building Block for Life. *Life*, **10**(6), 84.

Piani, Laurette, Marrocchi, Yves, Rigaudier, Thomas, et al. 2020. Earth's Water May Have Been Inherited from Material Similar to Enstatite Chondrite Meteorites. *Science*, **369**(6507), 1110–1113.

Pierrehumbert, Raymond T., and Hammond, Mark. 2019. Atmospheric Circulation of Tide-Locked Exoplanets. *Annu. Rev. Fluid Mech.*, **51**(1), 275–303.

Piran, Tsvi, and Jimenez, Raul. 2014. Possible Role of Gamma Ray Bursts on Life Extinction in the Universe. *Phys. Rev. Lett.*, **113**(23), 231102.

Planck Collaboration. 2020. Planck 2018 Results. VI. Cosmological Parameters. *Astron. Astrophys.*, **641**(Sept.), A6.

Plankensteiner, Kristof, Reiner, Hannes, and Rode, Bernd M. 2005. Prebiotic Chemistry: The Amino Acid and Peptide World. *Curr. Org. Chem.*, **9**(12), 1107–1114.

Pohorille, A., and Pratt, L. R. 2012. Is Water the Universal Solvent for Life? *Orig. Life Evol. Biosph.*, **42**(5), 405–409.

Porter, Susannah M., and Riedman, Leigh Anne. 2023. Frameworks for Interpreting the Early Fossil Record of Eukaryotes. *Annu. Rev. Microbiol.*, **77**, 173–191.

Portree, David S. F. 2001. *Humans to Mars: Fifty Years of Mission Planning, 1950-2000*. Monographs in Aerospace History Series, no. 20. Washington, D.C: National Aeronautics and Space Administration.

Postberg, Frank, Sekine, Yasuhito, Klenner, Fabian, et al. 2023. Detection of Phosphates Originating from Enceladus's Ocean. *Nature*, **618**(7965), 489–493.

Poulet, F., Bibring, J. P., Mustard, J. F., et al. 2005. Phyllosilicates on Mars and Implications for Early Martian Climate. *Nature*, **438**(7068), 623–627.

Poulton, Simon W., Bekker, Andrey, Cumming, Vivien M., et al. 2021. A 200-Million-Year Delay in Permanent Atmospheric Oxygenation. *Nature*, **592**(7853), 232–236.

Powell, Russell. 2020. *Contingency and Convergence: Toward a Cosmic Biology of Body and Mind*. The Vienna Series in Theoretical Biology. Cambridge, MA: The MIT Press.

Powner, Matthew W., Gerland, Béatrice, and Sutherland, John D. 2009. Synthesis of Activated Pyrimidine Ribonucleotides in Prebiotically Plausible Conditions. *Nature*, **459**(7244), 239–242.

Prantzos, Nikos. 2008. On the 'Galactic Habitable Zone'. *Space Sci. Rev.*, **135**(1-4), 313–322.

Preiner, Martina, Asche, Silke, Becker, Sidney, et al. 2020. The Future of Origin of Life Research: Bridging Decades-Old Divisions. *Life*, **10**(3), 20.

Preston, Louisa J., and Dartnell, Lewis R. 2014. Planetary Habitability: Lessons Learned from

Terrestrial Analogues. *Int. J. Astrobiol.*, **13**(1), 81–98.

Price, P. Buford, and Sowers, Todd. 2004. Temperature Dependence of Metabolic Rates for Microbial Growth, Maintenance, and Survival. *Proc. Natl. Acad. Sci.*, **101**(13), 4631–4636.

Prigogine, Ilya. 1978. Time, Structure, and Fluctuations. *Science*, **201**(4358), 777–785.

Pross, Addy. 2004. Causation and the Origin of Life. Metabolism or Replication First? *Orig. Life Evol. Biosph.*, **34**(3), 307–321.

Pu, Judy P., Macdonald, Francis A., Schmitz, Mark D., et al. 2022. Emplacement of the Franklin Large Igneous Province and Initiation of the Sturtian Snowball Earth. *Sci. Adv.*, **8**(47), eadc9430.

Ramirez, Ramses M. 2018. A More Comprehensive Habitable Zone for Finding Life on Other Planets. *Geosciences*, **8**(8), 280.

Ramirez, Ramses M., and Craddock, Robert A. 2018. The Geological and Climatological Case for a Warmer and Wetter Early Mars. *Nat. Geosci.*, **11**(4), 230–237.

Rampelotto, Pabulo Henrique. 2013. Extremophiles and Extreme Environments. *Life*, **3**, 482–485.

Rampino, Michael R. 2020. Relationship between Impact-Crater Size and Severity of Related extinction Episodes. *Earth Sci. Rev.*, **201**(Feb.), 102990.

Rampino, Michael R., Caldeira, Ken, and Prokoph, Andreas. 2019. What Causes Mass Extinctions? Large Asteroid/Comet Impacts, Flood-Basalt Volcanism, and Ocean anoxia – Correlations and Cycles. Pages 271–302 of: Koeberl, Christian, and Bice, David M. (eds.), *250 Million Years of Earth History in Central Italy: Celebrating 25 Years of the Geological Observatory of Coldigioco*, vol. 542. Boulder: The Geological Society of America.

Ranjan, Sukrit, Wordsworth, Robin, and Sasselov, Dimitar D. 2017. The Surface UV Environment on Planets Orbiting M Dwarfs: Implications for Prebiotic Chemistry and the Need for Experimental Follow-Up. *Astrophys. J.*, **843**(2), 110.

Ranjan, Sukrit, Seager, Sara, Zhan, Zhuchang, et al. 2022. Photochemical Runaway in Exoplanet Atmospheres: Implications for Biosignatures. *Astrophys. J.*, **930**(2), 131.

Raulin, François, Brassé, Coralie, Poch, Olivier, and Coll, Patrice. 2012. Prebiotic-like Chemistry on Titan. *Chem. Soc. Rev.*, **41**(16), 5380–5393.

Rauscher, T., and Patkós, A. 2011. Origin of the Chemical Elements. Pages 611–665 of: Vértes, Attila, Nagy, Sándor, Klencsár, Zoltán, Lovas, Rezső G., and Rösch, Frank (eds.), *Handbook of Nuclear Chemistry*. New York: Springer.

Raval, Parth K., Garg, Sriram G., and Gould, Sven B. 2022. Endosymbiotic Selective Pressure at the Origin of Eukaryotic Cell Biology. *eLife*, **11**, e81033.

Raven, J. A., Kübler, J. E., and Beardall, J. 2000. Put Out the Light, and Then Put Out the Light. *J. Mar. Biol. Assoc. UK*, **80**(1), 1–25.

Ray, Christine, Glein, Christopher R., Waite, J. Hunter, et al. 2021. Oxidation Processes Diversify the Metabolic Menu on Enceladus. *Icarus*, **364**, 114248.

Ray, Pratapa Chandra. 1891. *The Mahabharata of Krishna-Dwaipayana Vyasa*. Vol. 12. Calcutta: Bharat Press.

Ray, Tom. 2012. Losing Spin: The Angular Momentum Problem. *Astron. Geophys.*, **53**(5), 5.19–5.22.

Raymond, Sean N., and Izidoro, Andre. 2017. Origin of Water in the Inner Solar System: Planetesimals Scattered Inward during Jupiter and Saturn's Rapid Gas Accretion. *Icarus*, **297**, 134–148.

Raymond, Sean N., and Morbidelli, Alessandro. 2022. Planet Formation: Key Mechanisms and Global Models. Pages 3–82 of: Biazzo, Katia, Bozza, Valerio, Mancini, Luigi, and Sozzetti, Alessandro (eds.), *Demographics of Exoplanetary Systems, Lecture Notes of the 3rd Advanced School on Exoplanetary Science.* Astrophysics and Space Science Library, vol. 466. Cham: Springer International.

Raymond, Sean N., Quinn, Thomas, and Lunine, Jonathan I. 2004. Making Other Earths: Dynamical Simulations of Terrestrial Planet Formation and Water Delivery. *Icarus*, **168**(1), 1–17.

Rees, Martin. 2018. *On the Future: Prospects for Humanity*. Princeton: Princeton University Press.

Reinhard, Christopher T., Planavsky, Noah J., Gill, Benjamin C., et al. 2017. Evolution of the Global Phosphorus Cycle. *Nature*, **541**(7637), 386–389.

Renaud, Joe P., Lopez, Eric, Brande, Jonathan, et al. 2022. The Exoplanet Modeling and Analysis Center at NASA Goddard. *Res. Notes AAS*, **6**(9), 185.

Rich, P. R. 2003. The Molecular Machinery of Keilin's Respiratory Chain. *Biochem. Soc. Trans.*, **31**(Pt 6), 1095–1105.

Riess, Adam G., Yuan, Wenlong, Macri, Lucas M., et al. 2022. A Comprehensive Measurement of the Local Value of the Hubble Constant with 1 km s^{-1} Mpc^{-1} Uncertainty from the Hubble Space Telescope and the SH0ES Team. *Astrophys. J. Lett.*, **934**(1), L7.

Rimmer, Paul B., Xu, Jianfeng, Thompson, Samantha J., et al. 2018. The Origin of RNA Precursors on Exoplanets. *Sci. Adv.*, **4**(8), eaar3302.

Rimmer, Paul B., Jordan, Sean, Constantinou, Tereza, et al. 2021. Hydroxide Salts in the Clouds of Venus: Their Effect on the Sulfur Cycle and Cloud Droplet pH. *Planet. Sci. J.*, **2**(4), 133.

Rivera-Valentín, Edgard G., Chevrier, Vincent F., Soto, Alejandro, and Martínez, Germán. 2020. Distribution and Habitability of (Meta)Stable Brines on Present-Day Mars. *Nat. Astron.*, **4**(May), 756–761.

Robertson, Michael P., and Joyce, Gerald F. 2012. The Origins of the RNA World. *Cold Spring Harb. Perspect. Biol.*, **4**(5), a003608.

Rode, Bernd Michael. 1999. Peptides and the Origin of Life. *Peptides*, **20**(6), 773–786.

Rosing, Minik T. 1999. ^{13}C-Depleted Carbon Microparticles in >3700-Ma Sea-Floor Sedimentary Rocks from West Greenland. *Science*, **283**, 674.

Ross, David, and Deamer, David. 2023. Template-Directed Replication and Chiral Resolution during Wet–Dry Cycling in Hydrothermal Pools. *Life*, **13**(8), 1749.

Rothschild, Lynn J., and Mancinelli, Rocco L. 2001. Life in Extreme Environments. *Nature*, **409**(6823), 1092–1101.

Rouillard, Joti, van Zuilen, Mark, Pisapia, Céline, and Garcia-Ruiz, Juan-Manuel. 2021. An Alternative Approach for Assessing Biogenicity. *Astrobiology*, **21**(2), 151–164.

Rozenberg, Andrey, Inoue, Keiichi, Kandori, Hideki, and Béjà, Oded. 2021. Microbial Rhodopsins: The Last Two Decades. *Annu. Rev. Microbiol.*, **75**, 427–447.

Rucker, Holly R., Ely, Tucker D., LaRowe, Douglas E., Giovannelli, Donato, and Price, Roy E. 2023. Quantifying the Bioavailable Energy in an Ancient Hydrothermal Vent on Mars and a Modern Earth-Based Analog. *Astrobiology*, **23**(4), 431–445.

Rugheimer, S., Segura, A., Kaltenegger, L., and Sasselov, D. 2015. UV Surface Environment of Earth-Like Planets Orbiting FGKM Stars through Geological Evolution. *Astrophys. J.*, **806**(1), 137.

Rugheimer, Sarah, Kaltenegger, Lisa, Zsom, Andras, Segura, Antígona, and Sasselov, Dimitar. 2013. Spectral Fingerprints of Earth-Like Planets Around FGK Stars. *Astrobiology*, **13**(3), 251–269.

Ruiz-Bermejo, Marta, de la Fuente, José Luis, Pérez-Fernández, Cristina, and Mateo-Martí, Eva. 2021. A Comprehensive Review of HCN-Derived Polymers. *Processes*, **9**(4), 597.

Ruiz-Mirazo, Kepa, Briones, Carlos, and de la Escosura, Andres. 2014. Prebiotic Systems Chemistry: New Perspectives for the Origins of Life. *Chem. Rev.*, **114**(1), 285–366.

Rummel, J. D., Race, M. S., Horneck, G., and Princeton Workshop Participants. 2012. Ethical Considerations for Planetary Protection in Space Exploration: A Workshop. *Astrobiology*, **12**(11), 1017–1023.

Rummel, John D., Beaty, David W., Jones, Melissa A., et al. 2014. A New Analysis of Mars 'Special Regions': Findings of the Second MEPAG Special Regions Science Analysis Group (SR-SAG2). *Astrobiology*, **14**(11), 887–968.

Rushby, Andrew J., Claire, Mark W., Osborn, Hugh, and Watson, Andrew J. 2013. Habitable Zone Lifetimes of Exoplanets around Main Sequence Stars. *Astrobiology*, **13**(9), 833–849.

Russell, Michael J. 2021. The 'Water Problem' (sic), the Illusory Pond and Life's Submarine Emergence: A Review. *Life*, **11**(5), 429.

Russell, Michael J. 2023. A Self-Sustaining Serpentinization Mega-engine Feeds the Fougerite Nanoengines Implicated in the Emergence of Guided Metabolism. *Front. Microbiol.*, **14**, 1145915.

Russell, Michael J., Nitschke, Wolfgang, and Branscomb, Elbert. 2013. The Inevitable Journey to Being. *Phil. Trans. R. Soc. B*, **368**(1622), 20120254.

Russell, Michael J., Murray, Alison E., and Hand, Kevin P. 2017. The Possible Emergence of Life and Differentiation of a Shallow Biosphere on Irradiated Icy Worlds: The Example of Europa. *Astrobiology*, **17**(12), 1265–1273.

Saal, Alberto E., Hauri, Erik H., Van Orman, James A., and Rutherford, Malcolm J. 2013. Hydrogen Isotopes in Lunar Volcanic Glasses and Melt Inclusions Reveal a Carbonaceous Chondrite Heritage. *Science*, **340**(6138), 1317–1320.

Sagan, C., and Newman, W. I. 1983. The Solipsist Approach to Extraterrestrial Intelligence. *Q. J. R. Astron. Soc.*, **24**(June), 113.

Sagan, Carl. 1973a. Planetary Engineering on Mars. *Icarus*, **20**(4), 513–514.

Sagan, Carl. 1973b. *The Cosmic Connection: An Extraterrestrial Perspective*. Garden City: Anchor Press.

Sagan, Carl, and Khare, Bishun N. 1971. Long-Wavelength Ultraviolet Photoproduction of Amino Acids on the Primitive Earth. *Science*, **173**(3995), 417–420.

Sagan, Carl, and Mullen, George. 1972. Earth and Mars: Evolution of Atmospheres and Surface Temperatures. *Science*, **177**(4043), 52–56.

Sagan, Carl, Thompson, W. Reid, Carlson, Robert, Gurnett, Donald, and Hord, Charles. 1993. A Search for Life on Earth from the Galileo Spacecraft. *Nature*, **365**(6448), 715–721.

Sagan, Lynn. 1967. On the Origin of Mitosing Cells. *J. Theor. Biol.*, **14**(3), 225,IN1–274,IN6.

Saha, Arpita, Yi, Ruiqin, Fahrenbach, Albert C., Wang, Anna, and Jia, Tony Z. 2022. A Physicochemical Consideration of Prebiotic Microenvironments for Self-Assembly and Prebiotic Chemistry. *Life*, **12**(10), 1595.

Saladino, Raffaele, Di Mauro, Ernesto, and García-Ruiz, Juan Manuel. 2019. A Universal Geochemical Scenario for Formamide Condensation and Prebiotic Chemistry. *Chem. Eur. J.*, **25**(13), 3181–3189.

Salyk, Colette, and Lewis, Kevin W. 2020. *Introductory Notes on Planetary Science: The Solar System, Exoplanets and Planet Formation*. Bristol: IOP.

Sánchez-Baracaldo, Patricia, Bianchini, Giorgio, Wilson, Jamie D., and Knoll, Andrew H. 2022. Cyanobacteria and Biogeochemical Cycles through Earth History. *Trends Microbiol.*, **30**(2), 143–157.

Sandford, Scott A., Nuevo, Michel, Bera, Partha P., and Lee, Timothy J. 2020. Prebiotic Astrochemistry and the Formation of Molecules of Astrobiological Interest in Interstellar Clouds and Protostellar Disks. *Chem. Rev.*, **120**(11), 4616–4659.

Sandström, H., and Rahm, M. 2020. Can Polarity-Inverted Membranes Self-assemble on Titan? *Sci. Adv.*, **6**(4), eaax0272.

Santosh, M., Arai, T., and Maruyama, S. 2017. Hadean Earth and Primordial Continents: The Cradle of Prebiotic Life. *Geosci. Front.*, **8**(2), 309–327.

Sasselov, Dimitar D., Grotzinger, John P., and Sutherland, John D. 2020. The Origin of Life as a Planetary Phenomenon. *Sci. Adv.*, **6**(6), eaax3419.

Sato, Takao, Okuzumi, Satoshi, and Ida, Shigeru. 2016. On the Water Delivery to Terrestrial Embryos by Ice Pebble Accretion. *Astron. Astrophys.*, **589**, A15.

Scheller, E. L., Ehlmann, B. L., Hu, Renyu, Adams, D. J., and Yung, Y. L. 2021. Long-Term Drying of Mars by Sequestration of Ocean-Scale Volumes of Water in the Crust. *Science*, **372**(6537), 56–62.

Scherf, M., and Lammer, H. 2021. Did Mars Possess a Dense Atmosphere during the First ~400 Million Years? *Space Sci. Rev.*, **217**(1), 2.

Schmidt, Gavin A., and Frank, Adam. 2019. The Silurian Hypothesis: Would It be Possible to Detect an Industrial Civilization in the Geological Record? *Int. J. Astrobiol.*, **18**(2), 142–150.

Schneider, J., Dedieu, C., Le Sidaner, P., Savalle, R., and Zolotukhin, I. 2011. Defning and Cataloging Exoplanets: The Exoplanet.eu Database. *Astron. Astrophys.*, **532**, A79.

Schneider, Jean, Silk, Joseph, and Vakili, Farrokh. 2021. OWL-Moon in 2050 and beyond. *Phil. Trans. R. Soc. A.*, **379**(2188), 20200187.

Schon, Samuel C., Head, James W., and Fassett, Caleb I. 2012. An Overfilled Lacustrine System and Progradational Delta in Jezero Crater, Mars: Implications for Noachian Climate. *Planet. Space Sci.*, **67**(1), 28–45.

Schröder, K. P., and Smith, Robert Connon. 2008. Distant Future of the Sun and Earth Revisited. *Mon. Not. R. Astron. Soc.*, **386**(1), 155–163.

Schrödinger, Erwin. 1944. *What Is Life? The Physical Aspect of the Living Cell*. Cambridge: Cambridge University Press.

Schroeder, Daniel V. 2020. *An Introduction to Thermal Physics*. Oxford: Oxford University Press.

Schubert, Gerald, Turcotte, Donald L., and Olson, Peter. 2001. *Mantle Convection in the Earth and Planets*. Cambridge: Cambridge University Press.

Schuchmann, Kai, and Müller, Volker. 2016. Energetics and Application of Heterotrophy in Acetogenic Bacteria. *Appl. Environ. Microbiol.*, **82**(14), 4056–4069.

Schuerger, Andrew C., and Clark, Benton C. 2008. Viking Biology Experiments: Lessons Learned and the Role of Ecology in Future Mars Life-Detection Experiments. *Space Sci. Rev.*, **135**(1–4), 233–243.

Schulte, Patricia M., Healy, Timothy M., and Fangue, Nann A. 2011. Thermal Performance Curves, Phenotypic Plasticity, and the Time Scales of Temperature Exposure. *Integr. Comp. Biol.*, **51**(5), 691–702.

Schulze-Makuch, D., and Irwin, Louis N. 2018. *Life in the Universe: Expectations and Constraints*. 3rd ed. Cham: Springer.

Schulze-Makuch, Dirk. 2021. The Case (or Not) for Life in the Venusian Clouds. *Life*, **11**(3), 255.

Schulze-Makuch, Dirk, Airo, Alessandro, and Schirmack, Janosch. 2017. The Adaptability of Life on Earth and the Diversity of Planetary Habitats. *Front. Microbiol.*, **8**, 2011.

Schwander, Loraine, Brabender, Max, Mrnjavac, Natalia, et al. 2023. Serpentinization as the Source of Energy, Electrons, Organics, Catalysts, Nutrients and pH Gradients for the Origin of LUCA and Life. *Front. Microbiol.*, **14**.

Schwartz, R. N., and Townes, C. H. 1961. Interstellar and Interplanetary Communication by Optical Masers. *Nature*, **190**(4772), 205–208.

Schwieterman, Edward W., Kiang, Nancy Y., Parenteau, Mary N., et al. 2018. Exoplanet Biosignatures: A Review of Remotely Detectable Signs of Life. *Astrobiology*, **18**(6), 663–708.

Schwieterman, Edward W., Reinhard, Christopher T., Olson, Stephanie L., Harman, Chester E., and Lyons, Timothy W. 2019. A Limited Habitable Zone for Complex Life. *Astrophys. J.*, **878**(1), 19.

Seager, S. (ed). 2010. *Exoplanets*. Space Science Series. Tucson: University of Arizona Press.

Seager, S., Bains, W., and Hu, R. 2013. A Biomass-based Model to Estimate the Plausibility of Exoplanet Biosignature Gases. *Astrophys. J.*, **775**(2), 104.

Seager, S., Bains, W., and Petkowski, J. J. 2016. Toward a List of Molecules as Potential Biosignature Gases for the Search for Life on Exoplanets and Applications to Terrestrial Biochemistry. *Astrobiology*, **16**(6), 465–485.

Scager, Sara, and Deming, Drake. 2010. Exoplanet Atmospheres. *Annu. Rev. Astron. Astrophys.*, **48**, 631–672.

Seager, Sara, Petkowski, Janusz J., Gao, Peter, et al. 2021. The Venusian Lower Atmosphere Haze as a Depot for Desiccated Microbial Life: A Proposed Life Cycle for Persistence of the Venusian Aerial Biosphere. *Astrobiology*, **21**(10), 1206–1223.

Seager, Sara, Petkowski, Janusz J., Huang, Jingcheng, et al. 2023a. Fully Fluorinated Non-carbon Compounds NF_3 and SF_6 as Ideal Technosignature Gases. *Sci. Rep.*, **13**, 13576.

Seager, Sara, Petkowski, Janusz J., Seager, Maxwell D., et al. 2023b. Stability of Nucleic Acid Bases in Concentrated Sulfuric Acid: Implications for the Habitability of Venus' Clouds. *Proc. Natl. Acad. Sci.*, **120**(25), e2220007120.

Seaton, Kenneth Marshall, Cable, Morgan Leigh, and Stockton, Amanda Michelle. 2021. Analytical Chemistry in Astrobiology. *Anal. Chem.*, **93**(15), 5981–5997.

Seckbach, Joseph, Oren, Aharon, and Stan-Lotter, Helga (eds). 2013. *Polyextremophiles: Life under*

Multiple Forms of Stress. Cellular Origin, Life in Extreme Habitats and Astrobiology, vol. 27. Dordrecht: Springer.

Segré, Daniel, Ben-Eli, Dafna, Deamer, David W., and Lancet, Doron. 2001. The Lipid World. *Orig. Life Evol. Biosph.*, **31**, 119–145.

Segura, Antígona, Walkowicz, Lucianne M., Meadows, Victoria, Kasting, James, and Hawley, Suzanne. 2010. The Effect of a Strong Stellar Flare on the Atmospheric Chemistry of an Earth-Like Planet Orbiting an M Dwarf. *Astrobiology*, **10**(7), 751–771.

Selin, Helaine, and Sun, Xiaochun. 2000. *Astronomy across Cultures : The History of Non-western Astronomy*. Dordrecht: Kluwer Academic.

Semyonov, Oleg G. 2014. Relativistic Rocket: Dream and Reality. *Acta Astronaut.*, **99**, 52–70.

Shahar, Anat, Driscoll, Peter, Weinberger, Alycia, and Cody, George. 2019. What Makes a Planet Habitable? *Science*, **364**(6439), 434–435.

Shakura, N. I., and Sunyaev, R. A. 1973. Black Holes in Binary Systems. Observational Appearance. *Astron. Astrophys.*, **24**, 337–355.

Shapiro, Anna V., Brühl, Christoph, Klingmüller, Klaus, et al. 2023. Metal-Rich Stars are Less Suitable for the Evolution of Life on Their Planets. *Nat. Commun.*, **14**, 1893.

Shapiro, Robert. 2007. A Simpler Origin for Life. *Sci. Am.*, **296**(6), 46–53.

Sharma, Abhishek, Czégel, Dániel, Lachmann, Michael, et al. 2023. Assembly Theory Explains and Quantifies Selection and Evolution. *Nature*, **622**(7982), 321–328.

Sheikh, Sofia Z. 2020. The Nine Axes of Merit for Technosignature Searches. *Int. J. Astrobiol.*, **19**(3), 237–243.

Shen, Jianxun, Zerkle, Aubrey L., Stueeken, Eva, and Claire, Mark W. 2019. Nitrates as a Potential N Supply for Microbial Ecosystems in a Hyperarid Mars Analog System. *Life*, **9**(4), 79.

Shih, P. M., Hemp, J., Ward, L. M., Matzke, N. J., and Fischer, W. W. 2017. Crown Group Oxyphotobacteria Postdate the Rise of Oxygen. *Geobiology*, **15**(1), 19–29.

Shikibu, Murasaki. (c. 1010) 2003. *The Tale of Genji*. New York: Penguin Books. Translated by Tyler, Royall.

Shimajiri, Y., André, Ph., Ntormousi, E., et al. 2019. Probing Fragmentation and Velocity Sub-structure in the Massive NGC 6334 Filament with ALMA. *Astron. Astrophys.*, **632**(Dec.), A83.

Shklovskii, I. S., and Sagan, Carl. 1966. *Intelligent Life in the Universe*. Holden-Day: San Francisco.

Siemion, A., Benford, J., Cheng-Jin, J., et al. 2015. Searching for Extraterrestrial Intelligence with the Square Kilometre Array. Page 116 of: *Advancing Astrophysics with the Square Kilometre Array (AASKA14)* Giardini Naxos: SISSA.

Sinha, Navita, Nepal, Sudip, Kral, Timothy, and Kumar, Pradeep. 2017. Survivability and Growth Kinetics of Methanogenic Archaea at Various pHs and Pressures: Implications for Deep Subsurface Life on Mars. *Planet. Space Sci.*, **136**, 15–24.

Sleep, N. H., Meibom, A., Fridriksson, Th., Coleman, R. G., and Bird, D. K. 2004. H_2-rich Fluids from Serpentinization: Geochemical and Biotic Implications. *Proc. Natl. Acad. Sci.*, **101**(35), 12818–12823.

Sleep, N. H., Zahnle, K. J., and Lupu, R. E. 2014. Terrestrial aftermath of the Moon-Forming Impact. *Phil. Trans. R. Soc. A*, **372**(2024), 20130172–20130172.

Sloan, David, Alves Batista, Rafael, and Loeb, Abraham. 2017. The Resilience of Life to Astrophysical Events. *Sci. Rep.*, **7**, 5419.

Smith, Eric, and Morowitz, Harold J. 2016. *The Origin and Nature of Life on Earth*. Cambridge: Cambridge University Press.

Smith, Kelly C., and Mariscal, Carlos. 2020. *Social and Conceptual Issues in Astrobiology*. Oxford: Oxford University Press.

Smoluchowski, M. V. 1916. Drei Vortrage uber Diffusion, Brownsche Bewegung und Koagulation von Kolloidteilchen. *Z. Phys.*, **17**, 557–585.

Sobral, David, Smail, Ian, Best, Philip N., et al. 2013. A Large Hα Survey at z = 2.23, 1.47, 0.84 and 0.40: The 11 Gyr Evolution of Star-Forming Galaxies from HiZELS⋆. *Mon. Not. R. Astron. Soc.*, **428**(2), 1128–1146.

Socas-Navarro, Hector. 2018. Possible Photometric Signatures of Moderately Advanced

Civilizations: The Clarke Exobelt. *Astrophys. J.*, **855**(2), 110.

Socas-Navarro, Hector, Haqq-Misra, Jacob, Wright, Jason T., et al. 2021. Concepts for Future Missions to Search for Technosignatures. *Acta Astronaut.*, **182**, 446–453.

Soderlund, Krista M., Kalousová, Klára, Buffo, Jacob J., et al. 2020. Ice-Ocean Exchange Processes in the Jovian and Saturnian Satellites. *Space Sci. Rev.*, **216**(5), 80.

Sojo, Victor, Herschy, Barry, Whicher, Alexandra, Camprubí, Eloi, and Lane, Nick. 2016. The Origin of Life in Alkaline Hydrothermal Vents. *Astrobiology*, **16**(2), 181–197.

Sotin, C., Grasset, O., and Mocquet, A. 2007. Mass Radius Curve for Extrasolar Earth-Like Planets and Ocean Planets. *Icarus*, **191**(1), 337–351.

Sotin, Christophe, Kalousová, Klára, and Tobie, Gabriel. 2021. Titan's Interior Structure and Dynamics after the Cassini–Huygens Mission. *Annu. Rev. Earth Planet. Sci.*, **49**, 579–607.

Sousa-Silva, Clara, Seager, Sara, Ranjan, Sukrit, et al. 2020. Phosphine as a Biosignature Gas in Exoplanet Atmospheres. *Astrobiology*, **20**(2), 235–268.

Space Studies Board. 2018. *Review and Assessment of Planetary Protection Policy Development Processes*. Washington, DC: National Academies Press.

Space Studies Board. 2019. *An Astrobiology Strategy for the Search for Life in the Universe*. Washington, DC: National Academies Press.

Spacek, Jan, Rimmer, Paul, Owens, Gage E., et al. 2024. Production and Reactions of Organic Molecules in Clouds of Venus. *ACS Earth Space Chem.* **8**(1) 89–98.

Spang, Anja, Saw, Jimmy H., Jørgensen, Steffen L., et al. 2015. Complex Archaea That Bridge the Gap between Prokaryotes and Eukaryotes. *Nature*, **521**(7551), 173–179.

Sparks, William B., Parenteau, Mary Niki, Blankenship, Robert E., et al. 2021. Spectropolarimetry of Primitive Phototrophs as Global Surface Biosignatures. *Astrobiology*, **21**(2), 219–234.

Spitoni, E., Matteucci, F., and Sozzetti, A. 2014. The Galactic Habitable Zone of the Milky Way and M31 from Chemical Evolution Models with

Gas Radial Ows. *Mon. Not. R. Astron. Soc.*, **440**(3), 2588–2598.

Stanley, Steven M. 2005. *Earth System History*. 2nd ed. New York: W. H. Freeman.

Steakley, Kathryn E., Kahre, Melinda A., Haberle, Robert M., and Zahnle, Kevin J. 2023. Impact Induced H_2-Rich Climates on Early Mars Explored with a Global Climate Model. *Icarus*, **394**, 115401.

Steigman, Gary. 2007. Primordial Nucleosynthesis in the Precision Cosmology Era. *Annu. Rev. Nucl. Part. Sci.*, **57**(1), 463–491.

Stern, Jennifer C., Sutter, Brad, Freissinet, Caroline, et al. 2015. Evidence for Indigenous Nitrogen in Sedimentary and Aeolian Deposits from the Curiosity Rover Investigations at Gale Crater, Mars. *Proc. Natl. Acad. Sci.*, **112**(14), 4245–4250.

Stern, Jennifer C., Malespin, Charles A., Eigenbrode, Jennifer L., et al. 2022. Organic Carbon Concentrations in 3.5-Billion-Year-Old Lacustrine Mudstones of Mars. *Proc. Natl. Acad. Sci.*, **119**(27), e2201139119.

Stern, Robert J. 2018. The Evolution of Plate Tectonics. *Phil. Trans. R. Soc. A*, **376**(2132), 20170406.

Sterzik, Michael F., Bagnulo, Stefano, Stam, Daphne M., Emde, Claudia, and Manev, Mihail. 2019. Spectral and Temporal Variability of Earth Observed in Polarization. *Astron. Astrophys.*, **622**(Feb.), A41.

Stevenson, Andrew, Hamill, Philip G., O'Kane, Callum J., et al. 2017. Aspergillus Penicillioides Differentiation and Cell Division at 0.585 Water Activity. *Environ. Microbiol.*, **19**(2), 687–697.

Stevenson, James, Lunine, Jonathan, and Clancy, Paulette. 2015. Membrane Alternatives in Worlds without Oxygen: Creation of an Azotosome. *Sci. Adv.*, **1**(1), 1400067.

Strøm, Paul A., Bodewits, Dennis, Knight, Matthew M., et al. 2020. Exocomets from a Solar System Perspective. *Publ. Astron. Soc. Pac.*, **132**(1016), 101001.

Struve, O. 1952. Proposal for a Project of High-Precision Stellar Radial Velocity Work. *The Observatory*, **72**, 199–200.

Stüeken, E. E., Anderson, R. E., Bowman, J. S., et al. 2013. Did Life Originate from a Global Chemical Reactor? *Geobiology*, **11**(2), 101–126.

Stüeken, Eva E., Buick, Roger, Guy, Bradley M., and Koehler, Matthew C. 2015. Isotopic Evidence for Biological Nitrogen Fixation by Molybdenum-Nitrogenase from 3.2 Gyr. *Nature*, **520**(7549), 666–669.

Suazo, Matías, Zackrisson, Erik, Wright, Jason T., Korn, Andreas J., and Huston, Macy. 2022. Project Hephaistos - I. Upper Limits on Partial Dyson Spheres in the Milky Way. *Mon. Not. R. Astron. Soc.*, **512**(2), 2988–3000.

Suazo, Matías, Zackrisson, Erik, Wright, Jason T., Korn, Andreas J., and Huston, Macy. 2024. Project Hephaistos - II. Dyson sphere candidates from Gaia DR3, 2MASS, and WISE. *Mon. Not. R. Astron. Soc.*, **531**(1), 695–707.

Sutherland, John D. 2017. Opinion: Studies on the Origin of Life the End of the Beginning. *Nat. Rev. Chem.*, **1**(2), 0012.

Svanberg, Sune. 2001. *Atomic and Molecular Spectroscopy: Basic Aspects and Practical Applications*. 3rd ed. Atomic, Optical, and Plasma Physics, vol. 6. Berlin: Springer-Verlag.

Szathmáry, Eörs. 2015. Toward Major Evolutionary Transitions Theory 2.0. *Proc. Natl. Acad. Sci.*, **112**(33), 10104–10111.

Szathmáry, Eörs, and Maynard Smith, John. 1995. The Major Evolutionary Transitions. *Nature*, **374**(6519), 227–232.

Szocik, Konrad (ed). 2019. *The Human Factor in a Mission to Mars: An Interdisciplinary Approach*. Space and Society. Cham: Springer.

Szostak, Jack W. 2012. The Eightfold Path to Non-enzymatic RNA Replication. *J. Syst. Chem.*, **3**, 2.

Szostak, Jack W., Bartel, David P., and Luisi, P. Luigi. 2001. Synthesizing Life. *Nature*, **409**(6818), 387–390.

Takahagi, Wataru, Seo, Kaito, Shibuya, Takazo, et al. 2019. Peptide Synthesis under the Alkaline Hydrothermal Conditions on Enceladus. *ACS Earth Space Chem.*, **3**(11), 2559–2568.

Takeuchi, Taku, Miyama, Shoken M., and Lin, D. N. C. 1996. Gap Formation in Protoplanetary Disks. *Astrophys. J.*, **460**(Apr.), 832–847.

Talley, Kemper, and Alexov, Emil. 2010. The Eightfold Path to Non-enzymatic RNA Replication. *Proteins*, **78**(12), 2699–2706.

Tanaka, Hidekazu, Takeuchi, Taku, and Ward, William R. 2002. Three-Dimensional Interaction between a Planet and an Isothermal Gaseous Disk. I. Corotation and Lindblad Torques and Planet Migration. *Astrophys. J.*, **565**(2), 1257–1274.

Tanford, Charles. 1978. The Hydrophobic Effect and the Organization of Living Matter. *Science*, **200**(4345), 1012–1018.

Tang, Fengzai, Taylor, Richard J. M., Einsle, Josh F., et al. 2019. Secondary Magnetite in Ancient Zircon Precludes Analysis of a Hadean Geodynamo. *Proc. Natl. Acad. Sci.*, **116**(2), 407–412.

Tarduno, John A., Cottrell, Rory D., Davis, William J., Nimmo, Francis, and Bono, Richard K. 2015. A Hadean to Paleoarchean Geodynamo Recorded by Single Zircon Crystals. *Science*, **349**(6247), 521–524.

Tarduno, John A., Cottrell, Rory D., Bono, Richard K., et al. 2023. Hadaean to Palaeoarchaean Stagnant-Lid Tectonics Revealed by Zircon Magnetism. *Nature*, **618**(7965), 531–536.

Tarter, Jill. 2001. The Search for Extraterrestrial Intelligence (SETI). *Annu. Rev. Astron. Astrophys.*, **39**, 511–548.

Tarter, Jill C. 2007. The Evolution of Life in the Universe: are We Alone? *Highlights Astron.*, **14**, 14–29.

Taubner, Ruth-Sophie, Pappenreiter, Patricia, Zwicker, Jennifer, et al. 2018. Biological Methane Production under Putative Enceladus-Like Conditions. *Nat. Commun.*, **9**, 748.

Tellis, Nathaniel K., and Marcy, Geoffrey W. 2017. A Search for Laser Emission with Megawatt Thresholds from 5600 FGKM Stars. *Astron. J.*, **153**(6), 251.

Tera, Fouad, Papanastassiou, D. A., and Wasserburg, G. J. 1974. Isotopic Evidence for a Terminal Lunar Cataclysm. *Earth Planet. Sci. Lett.*, **22**(1), 1–21.

Thauer, Rudolf K., Kaster, Anne-Kristin, Seedorf, Henning, Buckel, Wolfgang, and Hedderich, Reiner. 2008. Methanogenic Archaea: Ecologically Relevant Differences in Energy Conservation. *Nat. Rev. Microbiol.*, **6**(8), 579–591.

The Borexino Collaboration. 2020a. Comprehensive Geoneutrino Analysis with Borexino. *Phys. Rev. D*, **101**(1), 012009.

The Borexino Collaboration. 2020b. Experimental Evidence of Neutrinos Produced in the CNO Fusion Cycle in the Sun. *Nature*, **587**(7835), 577–582.

Thomas, Brian C., and Ratterman, Cody L. 2020. Ozone Depletion-Induced Climate Change Following a 50 pc Supernova. *Phys. Rev. Res.*, **2**(4), 043076.

Thomas, Brian C., and Yelland, Alexander M. 2023. Terrestrial Effects of Nearby Supernovae: Updated Modeling. *Astrophys. J.*, **950**(1), 41.

Thomas, Trent B., Hu, Renyu, and Lo, Daniel Y. 2023. Constraints on the Size and Composition of the Ancient Martian Atmosphere from Coupled CO_2-N_2-Ar Isotopic Evolution Models. *Planet. Sci. J.*, **4**(3), 41.

Thompson, Maggie A., Krissansen-Totton, Joshua, Wogan, Nicholas, Telus, Myriam, and Fortney, Jonathan J. 2022. The Case and Context for Atmospheric Methane as an Exoplanet Biosignature. *Proc. Natl. Acad. Sci.*, **119**(14), e2117933119.

Tikoo, Sonia M., and Evans, Alexander J. 2022. Dynamos in the Inner Solar System. *Annu. Rev. Earth Planet. Sci.*, **50**(May), 99–122.

Tilley, Matt A., Segura, Antígona, Meadows, Victoria, Hawley, Suzanne, and Davenport, James. 2019. Modeling Repeated M Dwarf Flaring at an Earth-like Planet in the Habitable Zone: Atmospheric Effects for an Unmagnetized Planet. *Astrobiology*, **19**(1), 64–86.

Tipler, F. J. 1980. Extraterrestrial Intelligent Beings Do Not Exist. *Q. J. R. Astron. Soc.*, **21**, 267–281.

Tobin, John J., van't Hoff, Merel L. R., Leemker, Margot, et al. 2023. Deuterium-Enriched Water Ties Planet-Forming Disks to Comets and Protostars. *Nature*, **615**(7951), 227–230.

Tong, Kai, Bozdag, G. Ozan, and Ratcliff, William C. 2022. Selective Drivers of Simple Multicellularity. *Curr. Opin. Microbiol.*, **67**, 102141.

Toomre, A. 1964. On the Gravitational Stability of a Disk of Stars. *Astrophys. J.*, **139**, 1217–1238.

Tostevin, Rosalie, and Mills, Benjamin J. W. 2020. Reconciling Proxy Records and Models of Earth's Oxygenation during the Neoproterozoic and Palaeozoic. *Interface Focus*, **10**(4), 20190137.

Trail, Dustin, Watson, E. Bruce, and Tailby, Nicholas D. 2011. The Oxidation State of Hadean Magmas and Implications for Early Earth's Atmosphere. *Nature*, **480**(7375), 79–82.

Treiman, Allan H. 2003. Geologic Settings of Martian Gullies: Implications for Their Origins. *J. Geophys. Res. Planets*, **108**(E4), 8031.

Treiman, Allan H. 2021. Uninhabitable and Potentially Habitable Environments on Mars: Evidence from Meteorite ALH 84001. *Astrobiology*, **21**(8), 940–953.

Trumbo, Samantha K., Brown, Michael E., and Hand, Kevin P. 2019. Sodium Chloride on the Surface of Europa. *Sci. Adv.*, **5**(6), aaw7123.

Tsiganis, K., Gomes, R., Morbidelli, A., and Levison, H. F. 2005. Origin of the Orbital Architecture of the Giant Planets of the Solar System. *Nature*, **435**(7041), 459–461.

Turbet, Martin, Forget, Francois, and Schott, Cédric. 2016. The LAPS Project : A Live 1D Radiative-Convective Model to Explore the Possible Climates of Terrestrial Planets and Exoplanets. Page 419.07 of: *AAS/Division for Planetary Sciences Meeting Abstracts #48*. AAS/Division for Planetary Sciences Meeting Abstracts, vol. 48.

Turner, Simon, Wilde, Simon, Wörner, Gerhard, Schaefer, Bruce, and Lai, Yi-Jen. 2020. An Andesitic Source for Jack Hills Zircon Supports onset of Plate Tectonics in the Hadean. *Nat. Commun.*, **11**, 1241.

Tyrrell, Toby. 2013. *On Gaia: A Critical Investigation of the Relationship between Life and Earth*. Princeton: Princeton University Press.

Ueda, Hisahiro, and Shibuya, Takazo. 2021. Composition of the Primordial Ocean Just after Its Formation: Constraints from the Reactions between the Primitive Crust and a Strongly Acidic, CO2-Rich Fluid at Elevated Temperatures and Pressures. *Minerals*, **11**(4), 389.

Vago, Jorge L., Westall, Frances, Pasteur Instrument Team, et al. 2017. Habitability on Early Mars and the Search for Biosignatures with the ExoMars Rover. *Astrobiology*, **17**(6–7), 471–510.

Vakoch, Douglas A., and Dowd, Matthew F. 2015. *The Drake Equation*. Cambridge: Cambridge University Press.

Van Kranendonk, Martin J., Baumgartner, Raphael, Djokic, Tara, et al. 2021. Elements for the Origin of Life on Land: A Deep-Time Perspective from the Pilbara Craton of Western Australia. *Astrobiology*, **21**(1), 39–59.

Vance, S. D., Hand, K. P., and Pappalardo, R. T. 2016. Geophysical Controls of Chemical Disequilibria in Europa. *Geophys. Res. Lett.*, **43**(10), 4871–4879.

Vance, Steven D., Panning, Mark P., Stähler, Simon, et al. 2018. Geophysical Investigations of Habitability in Ice-Covered Ocean Worlds. *J. Geophys. Res. Planets*, **123**(1), 180–205.

Vermeij, Geerat J. 2017. How the Land Became the Locus of Major Evolutionary Innovations. *Curr. Biol.*, **27**(20), 3178–3182.

Verseux, Cyprien. 2020. Bacterial Growth at Low Pressure. *Front. Astron. Space Sci.*, **7**, 30.

Vidotto, Aline A. 2018. Stellar Coronal and Wind Models: Impact on Exoplanets. Page 1857–1876 of: Deeg, Hans J., and Belmonte, Juan Antonio (eds.), *Handbook of Exoplanets*. Cham: Springer.

Villanueva, G. L., Cordiner, M., Irwin, P. G. J., et al. 2021. No Evidence of Phosphine in the Atmosphere of Venus from Independent Analyses. *Nat. Astron.*, **5**, 631–635.

Visser, R., van Dishoeck, E. F., and Black, J. H. 2009. The Photodissociation and Chemistry of CO Isotopologues: Applications to Interstellar Clouds and Circumstellar Disks. *Astron. Astrophys.*, **503**(2), 323–343.

Vukotić, B., Steinhauser, D., Martinez-Aviles, G., Ćirković, M. M., Micic, M., and Schindler, S. 2016. 'Grandeur in This View of Life': N-body Simulation Models of the Galactic Habitable Zone. *Mon. Not. R. Astron. Soc.*, **459**(4), 3512–3524.

Wächtershäuser, Günter. 1988. Before Enzymes and Templates: Theory of Surface Metabolism. *Microbiol. Rev.*, **52**(4), 452–484.

Wächtershäuser, Günter. 2007. On the Chemistry and Evolution of the Pioneer Organism. *Chem. Biodivers.*, **4**(4), 584–602.

Wackett, Lawrence P., Dodge, Anthony G., and Ellis, Lynda B. M. 2004. Microbial Genomics and the Periodic Table. *Appl. Environ. Microbiol.*, **70**(2), 647–655.

Waite, J. Hunter, Glein, Christopher R., Perryman, Rebecca S., et al. 2017. Cassini Finds Molecular Hydrogen in the Enceladus Plume: Evidence for Hydrothermal Processes. *Science*, **356**(6334), 155–159.

Wakelam, Valentine, Bron, Emeric, Cazaux, Stephanie, et al. 2017. H_2 Formation on Interstellar Dust Grains: The Viewpoints of Theory, Experiments, Models and Observations. *Mol. Astrophys.*, **9**, 1–36.

Wald, George. 1957. The Origin of Optical Activity. *Ann. N. Y. Acad. Sci.*, **69**(2), 352–368.

Wald, George. 1974. Fitness in the Universe: Choices and Necessities. *Orig. Life*, **5**(1–2), 7–27.

Walker, J. C. G., Hays, P. B., and Kasting, J. F. 1981. A Negative Feedback Mechanism for the Long-Term Stabilization of the Earth's Surface Temperature. *J. Geophys. Res.*, **86**, 9776–9782.

Walker, Sara I., Bains, William, Cronin, Leroy, et al. 2018. Exoplanet Biosignatures: Future Directions. *Astrobiology*, **18**(6), 779–824.

Walsh, Kevin J., Morbidelli, Alessandro, Raymond, Sean N., O'Brien, David P., and Mandell, Avi M. 2011. A Low Mass for Mars from Jupiter's Early Gas-Driven Migration. *Nature*, **475**(7355), 206–209.

Walters, C., Hoover, R. A., and Kotra, R. K. 1980. Interstellar Colonization: A New Parameter for the Drake Equation? *Icarus*, **41**(2), 193–197.

Wan, Kirsty Y., and Jékely, Gáspár. 2021. Origins of Eukaryotic Excitability. *Phil. Trans. R. Soc. B*, **376**(1820), 20190758.

Ward, Peter, and Brownlee, Donald. 2000. *Rare Earth: Why Complex Life Is Uncommon in the Universe*. New York: Copernicus Books.

Warren-Rhodes, Kimberley, Cabrol, Nathalie A., Phillips, Michael, et al. 2023. Orbit-to-Ground Framework to Decode and Predict Biosignature Patterns in Terrestrial Analogues. *Nat. Astron.*, **7**, 406–422.

Way, M. J., and Del Genio, Anthony D. 2020. Venusian Habitable Climate Scenarios: Modeling Venus through Time and Applications to Slowly Rotating Venus-Like Exoplanets. *J. Geophys. Res. Planets*, **125**(5), e06276.

Weaver, Erik, Isella, Andrea, and Boehler, Yann. 2018. Empirical Temperature Measurement in Protoplanetary Disks. *Astrophys. J.*, **853**(2), 113.

Webb, Stephen. 2015. *If the Universe Is Teeming with Aliens … Where Is EveryBody?* 2nd ed. Cham: Springer International.

Weintraub, David A. 2014. *Religions and Extraterrestrial Life: How Will We Deal with It?* Cham: Springer.

Weiss, Madeline C., Sousa, Filipa L., Mrnjavac, Natalia, et al. 2016. The Physiology and Habitat of the Last Universal Common Ancestor. *Nature Microbiol.*, **1**(9), 16116.

Westall, F., Hickman-Lewis, K., Hinman, N., et al. 2018. A Hydrothermal-Sedimentary Context for the Origin of Life. *Astrobiology*, **18**(3), 259–293.

Westall, Frances, Foucher, Frédéric, Bost, Nicolas, et al. 2015. Biosignatures on Mars: What, Where, and How? Implications for the Search for Martian Life. *Astrobiology*, **15**(11), 998–1029.

Westall, Frances, Brack, André, Fairén, Alberto G., and Schulte, Mitchell D. 2023. Setting the Geological Scene for the Origin of Life and Continuing Open Questions about Its Emergence. *Front. Astron. Space Sci.*, **9**, 1095701.

Westmoreland, Shawn. 2010. A Note on Relativistic Rocketry. *Acta Astronaut.*, **67**(9), 1248–1251.

Whitler, Lily, Endsley, Ryan, Stark, Daniel P., et al. 2023. On the Ages of Bright Galaxies 500 Myr after the Big Bang: Insights into Star Formation Activity at $z \gtrsim 15$ with JWST. *Mon. Not. R. Astron. Soc.*, **519**(1), 157–171.

Wickramasinghe, N. C., Wickramasinghe, Dayal T., Tout, Christopher A., Lattanzio, John C., and Steele, Edward J. 2019. Cosmic Biology in Perspective. *Astrophys. Space Sci.*, **364**(11), 205.

Wiescher, M., Görres, J., Uberseder, E., Imbriani, G., and Pignatari, M. 2010. The Cold and Hot CNO Cycles. *Annu. Rev. Nucl. Part. Sci.*, **60**(Nov.), 381–404.

Wignall, Paul B. 2015. *The Worst of Times: How Life on Earth Survived Eighty Million Years of Extinctions*. Princeton: Princeton University Press.

Wilde, Simon A., Valley, John W., Peck, William H., and Graham, Colin M. 2001. Evidence from Detrital Zircons for the Existence of Continental Crust and Oceans on the Earth 4.4 Gyr Ago. *Nature*, **409**(6817), 175–178.

Williams, Jonathan P., and Cieza, Lucas A. 2011. Protoplanetary Disks and Their Evolution. *Annu. Rev. Astron. Astrophys.*, **49**(1), 67–117.

Williams, R. M. E., Grotzinger, J. P., Dietrich, W. E., et al. 2013. Martian Fluvial Conglomerates at Gale Crater. *Science*, **340**(6136), 1068–1072.

Wittenmyer, Robert A., Wang, Songhu, Horner, Jonathan, et al. 2020. Cool Jupiters Greatly Outnumber Their Toasty Siblings: Occurrence Rates from the Anglo-Australian Planet Search. *Mon. Not. R. Astron. Soc.*, **492**(1), 377–383.

Woese, Carl R. 2004. A New Biology for a New Century. *Microbiol. Mol. Biol. Rev.*, **68**(2), 173–186.

Woese, Carl R., Kandler, Otto, and Wheelis, Mark L. 1990. Towards a Natural System of Organisms: Proposal for the Domains Archaea, Bacteria, and Eucarya. *Proc. Natl. Acad. Sci.*, **87**(12), 4576–4579.

Wolf, E. T., and Toon, O. B. 2015. The Evolution of Habitable Climates under the Brightening Sun. *J. Geophys. Res. Atmos.*, **120**(12), 5775–5794.

Wolstencroft, R. D., and Raven, J. A. 2002. Photosynthesis: Likelihood of Occurrence and Possibility of Detection on Earth-Like Planets. *Icarus*, **157**(2), 535–548.

Wolszczan, A., and Frail, D. A. 1992. A Planetary System around the Millisecond Pulsar PSR1257 + 12. *Nature*, **355**(6356), 145–147.

Wood, Rachel, Liu, Alexander G., Bowyer, Frederick, et al. 2019. Integrated Records of Environmental Change and Evolution Challenge the Cambrian Explosion. *Nat. Ecol. Evol.*, **3**(4), 528–538.

Woolfson, M. M. 1964. A Capture Theory of the Origin of the Solar System. *Proc. R. Soc. Lond. Ser. A*, **282**(1391), 485–507.

Wordsworth, R., Kerber, L., and Cockell, C. 2019. Enabling Martian Habitability with Silica Aerogel via the Solid-State Greenhouse Effect. *Nat. Astron.*, **3**, 898–903.

Wordsworth, Robin, Knoll, Andrew H., Hurowitz, Joel, et al. 2021. A Coupled Model of Episodic Warming, Oxidation and Geochemical Transitions on Early Mars. *Nat. Geosci.*, **14**(3), 127–132.

Wordsworth, Robin D. 2016. The Climate of Early Mars. *Annu. Rev. Earth Planet. Sci.*, **44**, 381–408.

Wray, James J. 2021. Contemporary Liquid Water on Mars? *Annu. Rev. Earth Planet. Sci.*, **49**, 141–171.

Wright, J. T., Griffith, R. L., Sigurdsson, S., Povich, M. S., and Mullan, B. 2014. The $\hat{G}$ Infrared Search for Extraterrestrial Civilizations with Large Energy Supplies. II. Framework, Strategy, and First Result. *Astrophys. J.*, **792**(1), 27.

Wright, Jason T., Cartier, Kimberly M. S., Zhao, Ming, Jontof-Hutter, Daniel, and Ford, Eric B. 2016. The $\hat{G}$ Search for Extraterrestrial Civilizations with Large Energy Supplies. IV. The Signatures and Information Content of Transiting Megastructures. *Astrophys. J.*, **816**(1), 17.

Wright, Jason T., Kanodia, Shubham, and Lubar, Emily. 2018. How Much SETI Has Been Done? Finding Needles in the n-dimensional Cosmic Haystack. *Astron. J.*, **156**(6), 260.

Wright, Jason T., Haqq-Misra, Jacob, Frank, Adam, et al. 2022. The Case for Technosignatures: Why They May Be Abundant, Long-Lived, Highly Detectable, and Unambiguous. *Astrophys. J. Lett.*, **927**(2), L30.

Wurm, Gerhard, and Teiser, Jens. 2021. Understanding Planet Formation Using Microgravity Experiments. *Nat. Rev. Phys.*, **3**(6), 405–421.

Yadav, Mahipal, Kumar, Ravi, and Krishnamurthy, Ramanarayanan. 2020. Chemistry of Abiotic Nucleotide Synthesis. *Chem. Rev.*, **120**(11), 4766–4805.

Youdin, Andrew N., and Goodman, Jeremy. 2005. Streaming Instabilities in Protoplanetary Disks. *Astrophys. J.*, **620**(1), 459–469.

Young, Hugh D., and Freedman, Roger A. 2018. *University Physics with Modern Physics*. 15th ed. London: Pearson.

Yung, Yuk L., Chen, Pin, Nealson, Kenneth, et al. 2018. Methane on Mars and Habitability: Challenges and Responses. *Astrobiology*, **18**(10), 1221–1242.

Zahnle, Kevin J., Lupu, Roxana, Catling, David C., and Wogan, Nick. 2020. Creation and Evolution of Impact-Generated Reduced Atmospheres of Early Earth. *Planet. Sci. J.*, **1**(1), 11.

Zarka, Philippe. 2007. Plasma Interactions of Exoplanets with Their Parent Star and Associated Radio Emissions. *Planet. Space Sci.*, **55**(5), 598–617.

Zeitlin, C., Hassler, D. M., Cucinotta, F. A., et al. 2013. Measurements of Energetic Particle Radiation in Transit to Mars on the Mars Science Laboratory. *Science*, **340**(6136), 1080–1084.

Zellner, Nicolle E. B. 2017. Cataclysm No More: New Views on the Timing and Delivery of Lunar Impactors. *Orig. Life Evol. Biosph.*, **47**(3), 261–280.

Zendejas, J., Segura, A., and Raga, A. C. 2010. Atmospheric Mass Loss by Stellar Wind from Planets around Main Sequence M Stars. *Icarus*, **210**(2), 539–544.

Zhang, Ke, Bergin, Edwin A., Blake, Geoffrey A., Cleeves, L. Ilsedore, and Schwarz, Kamber R. 2017. Mass Inventory of the Giant-Planet Formation Zone in a Solar Nebula Analogue. *Nat. Astron.*, **1**, 0130.

Zhang, Stephanie J., Duzdevich, Daniel, Ding, Dian, and Szostak, Jack W. 2022. Freeze-Thaw Cycles Enable a Prebiotically Plausible and Continuous Pathway from Nucleotide Activation to Nonenzymatic RNA Copying. *Proc. Natl. Acad. Sci.*, **119**(17), e2116429119.

Zhu, Cheng, Turner, Andrew M., Abplanalp, Matthew J., et al. 2020. An Interstellar Synthesis of Glycerol Phosphates. *Astrophys. J. Lett.*, **899**(1), L3.

Zhu, Wei, and Dong, Subo. 2021. Exoplanet Statistics and Theoretical Implications. *Annu. Rev. Astron. Astrophys.*, **59**(Sept.), 291–336.

Zieba, Sebastian, Kreidberg, Laura, Ducrot, Elsa, et al. 2023. No Thick Carbon Dioxide Atmosphere on the Rocky Exoplanet TRAPPIST-1 c. *Nature*, **620**(7975), 746–749.

Zubrin, Robert M., and Andrews, Dana G. 1991. Magnetic Sails and Interplanetary Travel. *J. Spacecraft Rockets*, **28**(2), 197–203.

Zuckerman, B. 2015. Recognition of the First Observational Evidence of an Extrasolar Planetary System. Pages 291–293 of: Dufour, P., Bergeron, P., and Fontaine, G. (eds.), *19th European Workshop on White Dwarfs*. Astronomical Society of the Pacific Conference Series, vol. 493. San Francisco: Astronomical Society of the Pacific.

Author index

Subject index

planetary radar, 329, 335, 338

planetary-scale origin of life, 75, 117, 118

plastids, 127, 134

plate tectonics, 78, 84, 188, 189, 214, 229

plurality of worlds, 8, 9, 12, 266

Pluto, 174, 232

polarisation spectroscopy, 310, 311

polycyclic aromatic hydrocarbons, 29, 32, 47, 218

polyextremophiles, 166, 170

polysaccharides, 94, 95, 110

porpoises, 243

positive climate feedbacks, 139, 176, 178

pre-main-sequence star, 23, 41, 306

pre-Noachian, 205, 212, 216, 218

pre-RNA worlds, 99, 103

predation, 126, 128, 242

present atmospheric level (PAL), 123, 126, 131

primates, 132

primordial soup, 101

principle of mediocrity, 122, 322, 327

prokaryotes: definition, 96

protein functions, 91

proteinogenic amino acid, 91, 92, 106, 109, 254, 300

Proterozoic, 125, 126, 312

protocells, 90, 103, 109, 112, 115, 134

protometabolism, 100, 101, 113, 114, 156

proton motive force (PMF), 113, 152, 158, 161, 170

Proxima b, 71, 186, 267, 271, 284, 285

Proxima c, 71

psychrophiles, 165, 166, 215, 254

punctuated habitability, 216

purine nucleobases, 93

Purple Earth, 310

Purple non-sulfur bacteria, 159

Purple sulfur bacteria, 159

pyrimidine nucleobases, 94, 108

pyrite, 100, 136

pyruvate, 154, 155

Quaternary, 133

racemic, 91

radial velocity semi-amplitude, 271

radiation belts, 228, 236

radiation pressure, 330, 349, 352

radiation-resistant organisms, 169, 186, 195, 210

radiative forcing, 177

radio auroral emission, 190

radiogenic heat, 76, 78, 233

radiolysis, 156, 208, 229, 239

Rayleigh instability criterion, 45, 71

Rayleigh number, 78, 84, 232

Rayleigh–Jeans limit, 289

reactive oxygen species, 138, 169

recurring slope lineae, 207, 208

red algae, 130

red beds, 136

red giant, 25, 26, 29, 341

redox couples, 151

redox potential, 151, 154, 158, 160

redox reactions: definition, 149

redshift, 15, 269, 271

reducing agent: definition, 151

reduction: definition, 151

remote-sensing biosignatures: definition, 294

replication-first origin of life, 99, 107

reptiles, 132

retinal, 161

reverse gyrase, 165

reverse Krebs cycle, 100–102, 159

rhodopsins, 161, 310

ribonucleotides, 108, 109

ribose, 93–95, 107, 108, 110, 117, 152

ribozymes, 99, 121, 237

ribulose 1,5-bisphosphate carboxylase/oxygenase (RuBisCO), 159

Richmond Mine (USA), 168

RNA functions, 93

RNA world, 99, 103, 107, 116, 165

RNA-peptide world, 103

RNA/DNA nucleobase, 93, 103, 108, 117

rocket equation, 348, 351

runaway greenhouse, 178, 181, 187

runaway growth, 54–57

salinity, 116, 166, 167, 209, 226, 228

Salpeter time, 199

salt-in approach, 167

salt-induced peptide formation, 110

salt-out approach, 167

sea spray, 116

second law of thermodynamics, 118, 119

self-replicating probes, 322, 326, 337

semi-arid aquifers, 116

serine, 106, 248

serpentinisation, 137, 156, 157, 208, 226, 238

SETI: definition, 317

sexual reproduction, 128, 134

Sgr A*, 196, 197, 199

shadow biosphere, 90, 263, 265

Shark Bay (Australia), 243

shock waves: prebiotic chemistry, 106, 109, 111

Siberian Traps, 141

siderites, 297

sievert: definition, 346

signal-to-noise ratio (SNR), 309, 329, 330

signalling molecules, 91, 129

silanes, 255, 256

silicon dioxide, 85, 157, 226, 256, 298, 347

silicones, 256

Silurian, 125, 132

simple multicellularity, 126, 129, 142

SNC meteorites, 218

snow line, 46, 67, 223

Snowball Earth, 126, 131, 139

Soanesville Group, 125

social learning, 243

soda lakes, 166

sodium chloride, 110, 167, 228

solar constant, 178

Space Age, 345

space weather, 173, 186, 305

SpaceX, 346

spark discharge, 102, 105, 120

species richness, 119

specific heat capacity, 76, 87, 204, 247, 248

spectral energy distribution, 20, 40

spectral radiance, 170, 198, 281

Sputnik Planitia, 232

sputtering, 212

stagnant-lid tectonics, 84, 189, 214

standard state, 150, 151

starch, 95

starshade, 285

stellar energetic particles (SEPs), 186, 191, 306

stellar flares: definition, 185

stellar flares: general, 173, 195, 310

stellar flares: habitability, 186, 198

stellar proton events (SPEs), 186, 187, 196

stellar winds: definition, 185

Printed in the United States
by Baker & Taylor Publisher Services